Frozen Food Science and Technology

Frozen Food Science and Technology

Edited by

Judith A. Evans
Food Refrigeration and Process Engineering Research Centre (FRPERC)
University of Bristol, UK

Blackwell
Publishing

Blackwell Publishing editorial offices:
Blackwell Publishing Ltd, 9600 Garsington Road, Oxford OX4 2DQ, UK
Tel: +44 (0)1865 776868
Blackwell Publishing Professional, 2121 State Avenue, Ames, Iowa 50014-8300, USA
Tel: +1 515 292 0140
Blackwell Publishing Asia Pty Ltd, 550 Swanston Street, Carlton, Victoria 3053, Australia
Tel: +61 (0)3 8359 1011

First published 2008 by Blackwell Publishing Ltd

ISBN: 978-1-4051-5478-9

Library of Congress Cataloging-in-Publication Data
Frozen food science and technology / edited by Judith A. Evans.
p. cm.
Includes bibliographical references and index.
ISBN-13: 978-1-4051-5478-9 (hardback : acid-free paper)
ISBN-10: 1-4051-5478-0 (hardback : acid-free paper) 1. Frozen foods.
I. Evans, Judith A. (Judith Anne), 1962-

TP372.3.F768 2008
664′.02853–dc22

2007033156

A catalogue record for this title is available from the British Library

The publisher's policy is to use permanent paper from mills that operate a sustainable forestry policy, and which has been manufactured from pulp processed using acid-free and elementary chlorine-free practices. Furthermore, the publisher ensures that the text paper and cover board used have met acceptable environmental accreditation standards.

For further information on Blackwell Publishing, visit our website:
www.blackwellpublishing.com

Contents

Contributors

Mark Berry
Unilever Plc, Sharnbrook
Bedfordshire, United Kingdom

Teresa R.S. Brandão
Escola Superior de Biotecnologia
Universidade Católica Portuguesa
Porto, Portugal

Giovanni Cortella
Department of Energy Technologies
University of Udine
Udine, Italy

Trygve M. Eikevik
Norwegian University of Science
and Technology, Trondheim, Norway

Kostadin Fikiin
Refrigeration Science and Technology
Technical University of Sofia
Bulgaria

John Fletcher
Unilever Plc, Sharnbrook
Bedfordshire, United Kingdom

H. Douglas Goff
Department of Food Science
University of Guelph
Guelph, Ontario, Canada

Elsa M. Gonçalves
Departamento de Tecnologia das
Indústrias Alimentares
Instituto Nacional de Engenharia,
Tecnologia e Inovação
Lisboa, Portugal

Ronan Gormley
Ashtown Food Research Centre
(Teagasc) Ashtown, Dublin
Ireland

Vidar Hardarsson
SINTEF Energy Research
Trondheim, Norway

Anne K.T. Hemmingsen
SINTEF Energy Research
Trondheim, Norway

Steve James
Food Refrigeration and Process
Engineering Research Centre
(FRPERC), Langford
North Somerset, United Kingdom

Onrawee Laguerre
Refrigerating Process Research Unit
Cemagref,
Antony, France

Alain LeBail
ENITIAA (École Nationale
D'Ingénieurs des
Techniques des Industries Agricoles et
Alimentaires), UMR GEPEA,
Nantes, France

Ola M. Magnussen
SINTEF Energy Research
Trondheim, Norway

Peter McClure
Unilever Plc, Sharnbrook
Bedfordshire, United Kingdom

Paul Nesvadba
Rubislaw Consulting Ltd
Angusfield Avenue
Aberdeen, United Kingdom

Tom S. Nordtvedt
SINTEF Energy Research
Trondheim, Norway

Girolamo Panozzo
Construction Technologies Institute – Italian National Research Council (ITC-CNR)
Padova, Italy

Andy Pearson
Star Refrigeration, Glasgow
United Kingdom

Q. Tuan Pham
School of Chemical Sciences and Engineering
University of New South Wales
Sydney, Australia

Cristina L.M. Silva
Escola Superior de Biotecnologia
Universidade Católica Portuguesa
Porto, Portugal

Andy Stapley
Department of Chemical Engineering
Loughborough University
United Kingdom

Joy Wilkinson
Unilever Plc, Sharnbrook
Bedfordshire,
United Kingdom

Noemi E. Zaritzky
CIDCA (Centro de Investigación y Desarrollo en Criotecnología de Alimentos),
Universidad Nacional de La Plata,
La Plata, Argentina

Preface

Freezing is one of the oldest and most commonly used means of food preservation. It has been known to be an extremely effective means of preserving food for extended periods since Paleolithic and Neolithic times, when man used ice and snow to cool food. The cooling effect of salt and ice was first publicly discussed in 1662 by the chemist Robert Boyle, but this technology was certainly known in Spain, Italy and India in the sixteenth century. The manufacture of ice in shallow lakes using radiant 'night cooling' and the preservation of ice and snow in ice houses was a common practice in large country houses in the Victorian times. Ice was a product only for the privileged, and iced desserts were extremely fashionable and a sign of great wealth.

In more temperate climates the preservation of ice and snow was obviously difficult, and it was only with artificial cooling that frozen food became available more widely. In 1755 William Cullen first made ice without any natural form of cooling by vapourising water at low pressure. This was followed by Jacob Perkins in 1834 who made the first ice-making machine operating on ethyl ether. In the following 30 years refrigeration technology developed rapidly, spearheaded by the likes of Joule and Kelvin, and the first patents related to freezing of food were filed. In 1865 the first cold storage warehouse in New York was built which used brine for cooling. In 1868 a ship's cold air machine was used on board the Anchor line's Circassian and Strathlevan ships that transported meat from New York to Glasgow. This was rapidly followed in the 1880s by the transport of meat from Australia and New Zealand to London.

In the late nineteenth century, refrigeration and the freezing of food underwent rapid developments in terms of the freezing processes and the refrigerants used. In 1880 ammonia was first used as a refrigerant and in 1882 the first plate freezer was developed. Although freezing was an extremely important technology, and a vital means of exporting meat for the troops in World War I, it was only after the war that refrigeration machinery underwent massive developments to improve reliability and efficiency.

In 1928 refrigeration was changed forever when Thomas Midgley invented CFCs (Freons). These were hailed as wonder chemicals and were claimed at the time to be efficient and environmentally harmless. Around the same time (1929) Clarence Birdseye began developing frozen meals. His original intention (that another inventor, a Frenchman called Charles Tellier, had in 1869) was to use freezing to dry foods that would have long-term stability and could be reconstituted by the housewife. When this method was found to produce poor quality results, Birdseye reverted to the fast freezing of food. Uniquely, he understood the beneficial impact of fast freezing on the quality of foods that had until that time often been frozen at slow rates.

Developments in freezing and frozen foods technology developed rapidly in the later half of the twentieth century. With changes in consumers' lifestyles the need for convenience food increased and, coupled with the development of low-cost refrigeration technologies, all households could have access to a freezer to store food. At the end of the twentieth century the market for frozen food was increasing at about 10% per year with approximately 25% of refrigerated food being frozen. This growth has since slowed slightly but sales of certain frozen foods such as fish and seafood are growing. Growth of frozen fish in Russia is reported to be 17% per year (*Cold Chain Experts Newsletter*, January, 2006) and the British Frozen Food federation has recently reported that sales by value increased by 3% in 2005/6 (*Refrigeration and Air Conditioning*, November, 2006).

Successful freezing can now preserve food almost in its original form. This makes it possible to preserve and transport food worldwide. As freezing prevents growth of microbes, frozen food can be stored for long periods; there is no need to use preservatives or additives to extend shelf life. Freezing allows flexibility in manufacture and supply and means that food can be preserved at near its optimum quality for distribution and transportation.

This book describes the current technologies to preserve food and the best practices to ensure production of safe, high-quality frozen food. It also points to some new technologies that are already making waves and are likely to cast an even greater impact on the frozen food industry in the future.

One of the largest upheavals in the refrigeration industry in the last 30 years was caused by the realisation that the chemicals invented by Thomas Midgley are harmful to the environment. The phasing out of CFCs (chlorofluorocarbons) and introducing their replacements – HCFCs (hydrofluorocarbons) – as part of the Montreal and Kyoto protocols, have brought about a paradigm shift in the chemicals used as refrigerants. Many older refrigerants with low ODP (ozone depletion potential) and GWP (global warming potential) have been, or are being, re-evaluated so as to raise their refrigeration potential making use of the modern machinery. For example, the refrigeration technology used on board the first ships, that brought meat to the UK from America and Australasia, was based on the use of air as the refrigerant. This technology, although effective, was based on large and inefficient machinery that could not compete once newer equipment came into the market. With modern compact, efficient turbo-machinery these disadvantages were overcome and air could once again be used as a competitive refrigerant.

As well as addressing these refrigeration issues, the book examines many interesting new freezing technologies such as pressure shift freezing. Although not yet a commercial reality for large-scale production, the possibility of a rapidly frozen product with minimal cell disruption is an exciting prospect for the future.

I hope that you will find that this book provides a comprehensive source of information on freezing and frozen storage of food. Our aim is to provide readers with in-depth knowledge of current and emerging refrigeration technologies and how these technologies can be used to optimise the quality of frozen food. An impressive group of authors, each an expert in their particular field, have contributed to this book. I would like to thank each of them for their help in developing a practical and comprehensive guide to freezing and frozen foods.

Judith Evans

1 Thermal Properties and Ice Crystal Development in Frozen Foods

Paul Nesvadba

1.1 INTRODUCTION – WATER IN FOODS

This book deals with freezing of foods, a process in which the temperature of the food is lowered so that some of its water crystallises as ice. This occurs in freeze-drying, freeze concentration of juices, and firming up meat for slicing or grinding ('tempering'). However, the greatest use of freezing of foods is to preserve them, or to extend their storage life. This is the basis of a huge frozen foods sector, widely established and accepted by the food consumers. Low temperatures (−18°C in domestic freezers, −28°C in primary wholesale cold stores or as low as −60°C in some food cold stores) slow down the spoilage processes (enzymic autolysis, oxidation, and bacterial spoilage) that would otherwise occur at room temperature or even at chill temperatures.

1.1.1 Foods commonly preserved by freezing

Water is a facilitator of biochemical deterioration of foods. Dry foods are much more stable than wet foods, because any water remaining in them has low activity, a_w. Freezing removes water from the food matrix by forming ice crystals. Although the ice crystals remain in the food, the remaining water which is in contact with the food matrix becomes concentrated with solutes and its a_w becomes low. Freezing is therefore akin to drying and this is the rationale for preserving food by freezing. Most micro-organisms cease functioning below the water activity of about 0.7.

The commonly frozen foods are those which contain appreciable amounts of water (Table 1.1).

Living cells, biological materials (plant and animal tissues) in the natural state are able to hold typically 80% water by mass on wet basis. Therefore foods derived from them contain similar high proportions of water. This also applies to 'engineered' foods such as ice cream where water/ice mixture is required to impart texture.

1.1.2 Influence of freezing and frozen storage on quality of foods

Food products thawed after cold storage should ideally be indistinguishable from the fresh product (this obviously does not apply to products such as ice cream that are consumed in the frozen state). This requirement is easier to achieve in some foods than in others. Foods with a delicate structure are more likely to suffer cell damage. However, for the main food commodities (bread, meat, fish, vegetables) the quality of the thawed product is indeed

Table 1.1 Water content ranges of commonly frozen foods.

Food commodity	Water content (% wet mass basis)	Reference
Breads	28–46	Holland *et al.* (1991)
Doughs	5–20	Miller and Kaslow (1963)
Fish[a]	50–80	Love (1982)
Ice cream	59–62	Holland *et al.* (1991)
Meats	35–90	Holland *et al.* (1991)
Vegetables	55–90	Holland *et al.* (1991)
Fruit (strawberries, raspberries)	87–90	Holland *et al.* (1991)
Ready meals	50–85	Kim *et al.* (2007)

Note: [a]Water content of fish is approximately (80% – fat content), Love (1982).

comparable with the fresh product (and in some cases, applying certain criteria, for example, vitamin content, enhances the quality of fresh food sold as chilled).

The formation of ice crystals can downgrade the quality of the food by one of the following three mechanisms:

(a) **Mechanical damage to the food structure.** The specific volume of ice is greater than that of water (greater by about 10%) and therefore the expanding ice crystals compress the food matrix. Ice crystal expansion in some fruits such as strawberry damages them severely, because of their delicate structure (the fruit becomes 'soggy' on thawing). On a macroscopic scale, during rapid cryogenic freezing, thermal stresses due to expansion may crack the food.
(b) **Cross-linking of proteins** (in fish and meat). Decrease in the amount of liquid water available to the proteins and increase of electrolyte concentration during freezing lead to aggregation and denaturation of actomyosin (Connell, 1959; Buttkus, 1970).
(c) **Limited re-absorption of water on thawing.** This is connected with mechanism (b). Again, we can take the example of animal tissue in which the muscle proteins, during frozen storage, become 'denuded' of their hydration water and cross-linked. On thawing, the tissue may not re-absorb the melted ice crystals fully to the water content it had before freezing. This leads to undesirable release of exudate – 'drip loss' – and toughness of texture in the thawed muscle, the main attributes determining quality (Mackie, 1993).

Mechanisms (b) and (c) are usually the main causes of deterioration of quality of frozen foods, which means deterioration of quality is caused mainly by processes taking place in frozen storage rather than during the initial freezing. Rapid freezing is possible only for small samples, not commercial ones. The rate of freezing achievable for large commercial 'samples' is so small that the quality of foods would not be greatly affected by the freezing rate (extracellular ice invariably forms for all samples other than those which are small and frozen in a laboratory by special techniques).

Both damage to food and its consequences for consumer-assessed quality depend on the type of food (its biological makeup and structure). For example, meat is less prone to damage from freezing and frozen storage than fish is. This is because meat protein fibres are more 'robust' and, moreover, meat is cooked for longer than fish. Fish, a cold-blooded animal, starts cooking at 35°C – the body temperature of mammals – whereas meat proteins are more stable (there seems to be a correlation between the temperature of the living animal and the stability of proteins, e.g. tropical sea fish as compared with North Sea fish). Adding cryoprotectants to food reduces deterioration in frozen storage. The section 'Glassy State' discusses this further.

The ability to determine the quality of frozen foods rapidly in their frozen state, without having to thaw the food for analysis, is of great significance. Kent *et al.* (2001, 2004, 2005) developed a microwave method for this. If, in a certain situation, this instrumental method cannot be used, a sensory assessment panel is used. The quality attributes of thawed foods are sensory (appearance, odour, flavour, texture – in cooked products). The attributes that are directly connected with water in foods are water-holding capacity and drip loss.

In the UK, frozen–thawed fish cannot legally be presented for sale as fresh for the quality changes freezing causes. This raises the question of enforcement of the law. Apart from the biochemical methods which are slow (Kitamikado *et al.*, 1990; Salfi *et al.*, 1986), it is preferable to use rapid physical and, in particular, electrical methods that have been developed for fish quality measurement but are also useful to check whether the fish had been frozen at all (Jason and Richards, 1975; Rehbein, 1992).

Another legal issue is 'added water'. During freezing of fish fillets, water sprayed on their surface creates a layer of ice that provides some protection against oxidation in frozen storage. On the other hand, the temptation may be to add too much of water because fish is sold by weight. For this problem, rapid methods to detect the amount of water added have been developed (Kent *et al.*, 2001; Daschner and Knöchel, 2003).

Consumers often ask whether thawing and refreezing is detrimental to food quality. The answer is that when done properly (hygienically, thus preventing microbial contamination during thawing), the effect of multiple freezing on quality (e.g. increased drip) is usually not very serious (Oosterhuis, 1981).

1.1.3 Water-binding capacity (or water-holding ability) of foods

Food holds water by several mechanisms. It may be cells holding the water either with cell membranes or between cells and in pores by capillary forces. Such water could be expressed (removed) by pressing. Water binds to hydrophilic components of foods (proteins, carbohydrates, salts and micronutrients) by van der Waals forces including hydrogen bonding.

Interaction of water with fats (lipids) is small because fats are hydrophobic, not readily soluble in water. On the cellular level, exclusion of water from cells is regulated by both the permeability of cell (or micelle) lipid bilayers and osmotic mechanisms. The molecular force in the hydration shell around proteins increases from the outer to the inner hydration layer. The most tightly bound water may not be removed by freezing; this water is called 'unfreezable water'.

The methods to measure water-binding capacity of foods have great commercial and scientific significance. Trout (1988) reviewed the following methods for measuring water-holding capacity of foods: the press, centrifugal, capillary suction, filter paper, small-scale cook yield test and NMR.

1.2 FREEZING OF FOODS

1.2.1 Freezing curves

Freezing of food starts when the food is placed in contact with a cold medium, which can be solid (for example, heat exchanger plates at −30 to −40°C, solid carbon dioxide (dry ice) at

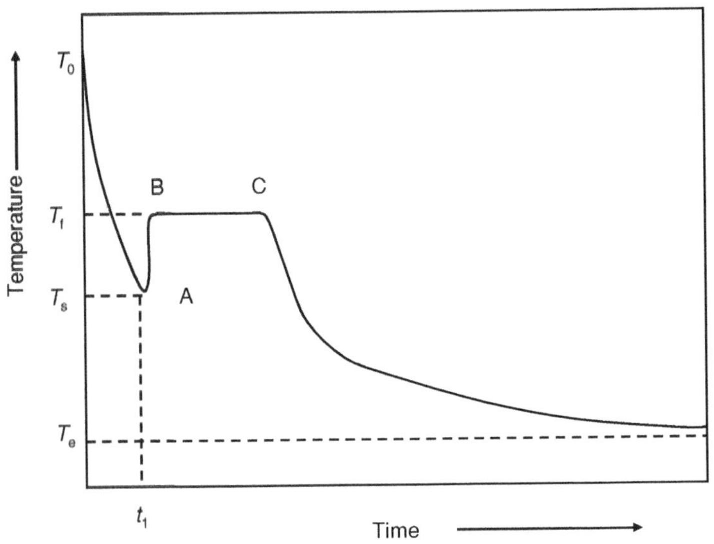

Fig. 1.1 A schematic plot of temperatures in food during freezing, showing the starting temperature, T_0, the initial freezing temperature, T_f, the temperature to which the food may supercool, T_s, the freezing plateau B–C and the equilibrium temperature, T_e.

−78.5°C), liquid (immersion in a cooling mixture or cryogenic fluid such as liquid nitrogen at −196°C) or gas (a stream of air, gaseous nitrogen or CO_2). The surface of the food cools faster than the centre of the food because the heat from the interior of the food has to reach the surface by conduction.

Figure 1.1 shows a typical temperature record during freezing. The temperature at the surface of the food may show supercooling (point A (t_1, T_s)) before increasing momentarily to approximately the initial freezing temperature T_f, and thereafter continuing along the 'thermal arrest' plateau (the B–C part) as transfer of the latent heat of freezing of water (334 kJ/kg for pure free water) from the food begins. The first ice crystals are formed between A and B and further crystals are formed all the way to the final temperature T_e where the temperature of the food equilibrates to the temperature of the cooling medium. No further rapid increase in the amount of ice occurs except for the slow accretion discussed in section 1.2.4.

1.2.2 Supercooling

Below its initial freezing point, a liquid is said to be supercooled. This is a metastable state of the liquid; the liquid can continue to be in this state for a very long time, before nucleation of the first crystal takes place. Following this the crystals grow and spread throughout the volume rapidly. Pure water (free of impurities such as dust particles that would act as nucleation centres) can be supercooled to around −40°C. At lower temperatures water freezes due to homogeneous ice nucleation and growth. In foods the degree of supercooling is much smaller than in pure water because of heterogeneous ice nucleation. Supercooling is important in nature since this is one of the mechanisms by which living plants and animals cope with sub-zero temperatures or minimize the damage of their tissue that ice formation can cause.

1.2.3 Ice nucleation and growth

Ice crystals come to existence as nuclei (seeds) of a critical size that subsequently grow. The critical size is that at which growth of the nucleus results in reduction of surface energy σ as compared with the increase in Gibbs free energy γ due to increase in volume (for a spherical ice crystal of radius r, this happens when $\sigma r^2 < \gamma r^3$).

Nucleation can be homogeneous or heterogeneous. Homogeneous nucleation occurs only in homogeneous particle-free liquids and happens due to random fluctuations of molecules (the random clusters of molecules momentarily assume the configuration of ice and act as seeds). In solid foods the nucleation is heterogeneous, with the cell surfaces acting as nucleation sites. The probability of nucleation at a site is enhanced if the molecular structure of the surface resembles that of ice, i.e. matches the lattice size of the ice crystal and acts as a template. This happens notably with ice nucleation active (INA) proteins found in some bacteria and plants (Govindarajan and Lindow, 1988).

1.2.4 Ice fraction frozen out

Pure water freezes at 0°C (save for the phenomenon of supercooling), but water solutions (in food sodium chloride or other salt solutions) have a lower freezing point, the depression being approximated by Raoult's equation (Miles *et al.*, 1997). During cooling below T_f, the extracellular region forms ice first and then the intracellular region begins to change state. This can be attributed to the fact that the cell (typical diameter 50 μm) membrane prevents growth of external ice into the region inside the cell (called intracellular region) making the intracellular region supercooled (~ -8°C).

Figure 1.2 shows a schematic diagram of an aqueous binary solution. The equilibrium between ice frozen out below T_f and the remaining solution requires the chemical potential of the two to be the same (Pippard, 1961). This leads to a relation between the water activity a_w of the solution and the molecular masses of the components and their fractions. It is possible to show from these thermodynamic considerations (for example, Miles, 1991) that the amount of ice x_i frozen out at each temperature $T < T_f$, is in the first approximation

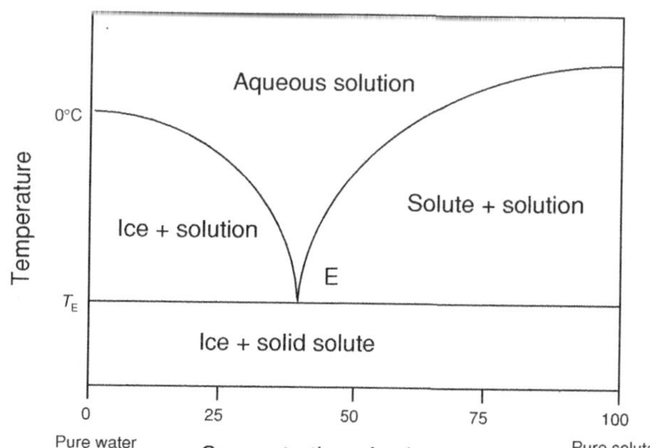

Fig. 1.2 A state diagram, showing schematically the behaviour of an aqueous binary solution with eutectic point E and eutectic temperature T_E.

(assuming an ideal binary solution and small temperature differences $T_f - T$) given by

$$x_i = (x_w - x_u)(1 - T_f/T) \tag{1.1}$$

where T_f and T are in degrees Celsius, x_w is the total water content of the food and x_u is the unfreezable water content. The last one is typically 5% and includes the so-called bound water, so that $x_u > x_b$ where x_b is the content of bound water.

The term 'bound water' is not understood well and not defined clearly. Fennema (1985) defines it in practical terms as

> ... water which exists in the vicinity of solutes and other non-aqueous constituents, exhibits reduced molecular mobility and other significantly altered properties as compared with "bulk water" in the same system, and does not freeze at −40°C.

This definition has two desirable attributes. One, it produces a conceptual picture of bound water, and two, it provides a realistic approach to quantifying the bound water. Water unfreezable at −40°C can be measured with equally satisfying results by either proton NMR or calorimetric procedures.

Figure 1.3 shows the graph of x_i for $T_f = -1°C$ and $x_u = 5\%$. Riedel (1957, 1978) made the first systematic experimental determination of the ice fraction x_i by calorimetric measurements. Other experimental investigations, for example by NMR, confirm that the approximation of x_i by equation (1.1) is acceptable for engineering purposes such as the calculation of thermal properties of frozen food, requiring accuracy of about ±10% (Novikov, 1971).

Equation 1.1 is derived from thermodynamic considerations (see for example Miles (1991)) that do not take into account the fact that even at constant temperature the fraction of ice increases with time, as was observed, for example, by Kent (1975). The time dependence is due to kinetically hindered mobility of the water molecules. Frozen food is not an equilibrium

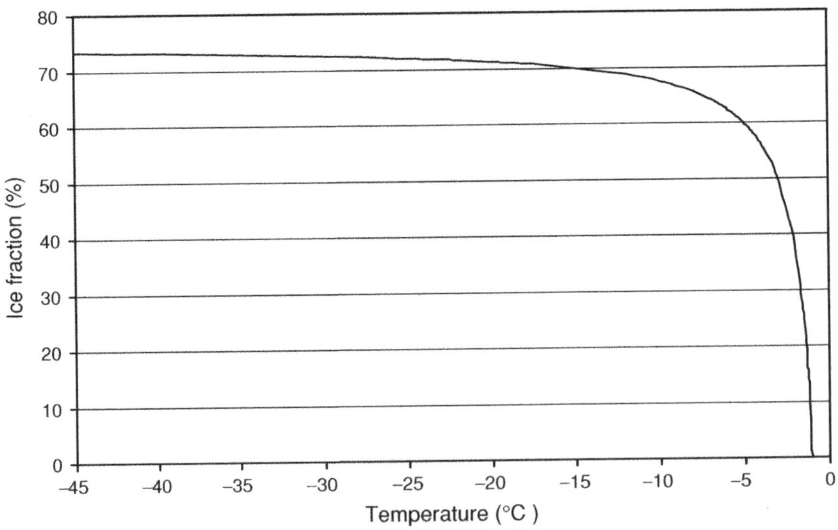

Fig. 1.3 Proportion of water frozen out in food as a function of temperature, calculated for a food with water content x_w of 80% and unfreezable water content x_u of 5%.

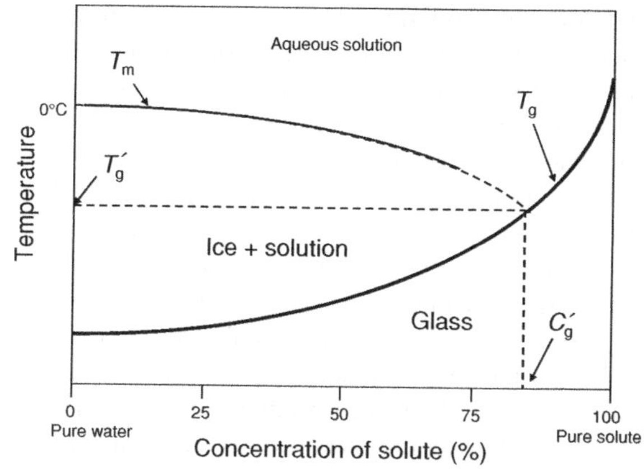

Fig. 1.4 A supplemented phase diagram showing schematically the behaviour of aqueous solution with the melting line T_m, glass transition line, T_g, the concentration of the maximally concentrated solution, C_g' and the corresponding glass transition temperature, T_g'.

system. The water that stays close to the food matrix may be in a glassy state. Then the simple binary diagram in Fig. 1.2 is extended into a 'supplemented' state diagram of foods (Roos, 1992, 1995; Rahman, 2006). This diagram (Fig. 1.4) can incorporate equilibrium melting points, heterogeneous nucleation temperatures, homogeneous nucleation temperatures, glass transition and devitrification temperatures, recrystallisation temperatures and, where appropriate, solute solubilities and eutectic temperatures (MacKenzie *et al.*, 1977). So far only simple binary systems such as water–glucose have been investigated thoroughly enough.

1.2.5 Effect of freezing rate on ice crystal structure

Hayes *et al.* (1984) define the freezing rate in relation to the velocity of movement of the ice-water freezing front. This has also been adopted by the International Institute of Refrigeration in their 'Red book' (Bøgh-Sørensen *et al.*, 2007).

The rates of freezing determine the type, size and distribution of ice formation. These can be extracellular or intracellular ice, dendritic or spherulitic (in rapidly frozen aqueous solutions; Hey *et al.*, 1997), and may be partially constrained by the food matrix. Using very high rates of cooling (up to 10,000°C/min) it is possible to avoid ice formation altogether and instead achieve vitrification leading to glassy state.

Angell (1982), Franks (1982), Garside (1987) and Blanshard and Franks (1987), among others, have reviewed crystallisation in foods. Because of the difficulties in interpreting the results of measurement of ice formation in complex food matrices, most definitive studies have started with simple systems based on aqueous solutions (Bald, 1991). A number of studies of ice formation and its prevention by cryoprotectants or anti-freeze proteins have also been carried out in the context of medical applications, preservation of biological tissue for viability, notably by Mazur (1970, 1984). This clearly shows a considerable 'commonality' between researches in food and medical sciences.

Slow freezing produces fewer larger ice crystals, fast freezing produces a greater number of smaller crystals. Whether large or small crystal size is preferable depends on the purpose of freezing. In ice cream, the ice crystals must be as small as possible so as to make the product as creamy and smooth as possible. However, to concentrate liquid food products, large crystals are easier to separate from the freeze concentrate (Fellows, 2000). In freeze drying (Chapter 12) it is usually desirable to produce a small number of large crystals in order to accelerate the subsequent sublimation process (Fellows, 2000).

When freezing commences, water that is present in the food migrates to join the growing ice crystals. When plant or animal tissues are frozen rapidly (in laboratory conditions, in sufficiently small or thin samples), water does not translocate across the cell membrane and small, uniformly distributed ice crystals are formed within the cell.

In commercial food freezing, the rates of freezing are usually too slow to form intracellular ice. In foods that are frozen slowly, large ice crystals form and ice fills the extracellular space causing dehydration of the cells. The ice crystals force the cells or tissue fibres apart. Although foods that are quick (flash) frozen produce small ice crystals, these ice crystals may grow larger over time through a process known as recrystallisation or Ostwald ripening (Smith and Schwartzberg, 1985). Recrystallisation occurs in frozen foods because larger crystals are thermodynamically more stable (they have a relatively smaller surface energy). Recrystallisation is aided by temperature gradients in the products during freezing or thawing, or temperature fluctuations during extended frozen storage (Chapter 11), distribution (when products are in transit) or domestic storage (home frost-free freezer temperatures may rise to almost 0°C during defrost cycles) (Chapter 15).

1.2.6 Glassy state in frozen foods

When a liquid is cooled rapidly enough to leave insufficient time for crystallisation to occur, and is continued to be cooled this way, the liquid becomes glass by undergoing a second order glass transition, i.e. transition with no release of latent heat (Wunderlich, 1981; Sperling, 1986). This happens in a range of temperatures around T_g, the glass transition temperature. Below T_g the molecules of the liquid (now glass) have much reduced, very low, mobility. The T_g is not a physical constant (such as melting point); it depends on the cooling rate (Hsu *et al.*, 2003). The T_g of pure water is about -140°C.

There are some common misconceptions such as 'glass is a supercooled liquid' or 'glass is a metastable liquid'. Both are wrong because glass is, strictly speaking, a non-equilibrium substance (although it appears to have constant properties when kept at constant temperature for normal observation times). Mobility in glass is extremely low, which makes diffusion of the molecules to a stable (crystalline) configuration extremely limited, so much so that it does not occur for several years, maybe thousands of years.

The concept of glass transitions is well developed in the fields of inorganic glasses and polymer science. Slade *et al.* (1993) were the first proponents of the use of this concept for thermal processing of foods. It explains the behaviour of foods in many food processes (e.g. stickiness of powders produced by spray drying) and the stability of food products in storage. The significance of the glassy state for foods is that they tend to be more stable (less prone to deterioration) if they are kept below T_g of aqueous solution within the food because of the very small mobility of water molecules (hereon we would say 'T_g of food' to mean 'T_g of aqueous solution contained in the food'). The T_g of dry foods is above room temperature and such foods are shelf stable (coffee granules, dry pasta, confectionery). In foods containing large amounts of water (meat, fish, vegetables), and hence in the foods that are preserved by freezing, the T_g is at -28°C or lower.

The concept of T_g is useful when investigating ways of extending the shelf-life of foods in frozen storage. Incorporating ingredients such as cryoprotectants may reduce ice crystal growth and the migration of water molecules from proteins. T_g may be a useful indicator of the effectiveness of the cryoprotectant. Examples of cryoprotectants are monosaccharides, disaccharides, glycerol, sorbitol, phosphate salts, ascorbic acid, carboxymethyl cellulose, gums and trehalose (Anese and Gormley, 1996; Love, 1966; Krivchenia and Fennema, 1988).

Mackie (1993) outlines the possible mechanisms of cryoprotection in proteinaceous foods such as fish:

(a) Preferential exclusion of the cryoprotectant from the protein (Tamiya *et al.*, 1985; Arakawa and Timasheff, 1985; Carpenter and Crowe, 1988). According to this theory the presence of the cryoprotectant increases the chemical potential of both the protein and the cryoprotectant. As a result the protein is stabilised against dissociation and denaturation as these would lead to greater thermodynamically unfavourable contact surface area between the protein and the cryoprotectant.
(b) Preferential hydration of protein molecule via functional –OH or ionic groups, thereby reducing the amount of water removed from the protein on freezing (Matsumoto and Noguchi, 1992).
(c) Decreased molecular mobility in the unfrozen phase surrounding the protein, due to the increased viscosity and formation of a glassy state (Levine and Slade, 1988).

According to the hypothesis of Levine and Slade, adding a cryoprotectant should ideally raise T_g above the storage temperature. This would restrict functioning of the deteriorative processes to a minimum (Goff, 1994). Above T_g the food matrix is usually described as 'rubbery'. Its kinetics follows the William–Landel–Ferry (WLF) equation rather than the Arrhenius law. Even if no cryoprotectant is used, the T_g of the product 'as is' may provide a guide for the economically optimal storage temperature. In Japan −60°C is used for the storage of sensitive high-value products such as tuna species for 'sushi' and 'sashimi' raw fish products. Whether such a general idea applies to all foods has been questioned (Orlien, 2003; Orlien *et al.*, 2003) but nevertheless it provides a useful framework to test the effectiveness of cryoprotectants and stimulates further research in this area.

The T_g hypothesis has been validated so far by many studies: on carbohydrate systems, such as dairy desserts, ice creams and some vegetables (Reid, 1990; Reid *et al.*, 1994, 1995; Roos and Karel, 1991; Roos, 1995) and on systems with globular proteins. It is not yet clear whether the theory applies to the myosin helical protein systems as well, fish muscle for example (Jensen *et al.*, 2003). Herrera and Mackie (2004) and Herrera *et al.* (2000) found that maltodextrins and low molecular weight carbohydrates can inhibit TMAO-demethylase in fish in frozen storage. Rey-Mansilla *et al.* (2001) carried out similar work on fish and Hansen (2004) on pork.

Unlike in medicine (dealing with small samples such as semen, eggs or embryos), the use of cryoprotectants in frozen food technology has been limited due to the difficulties in incorporating cryoprotectants into large samples of food. The process of putting cryoprotectants into food is too slow to rely solely on diffusion, as has been found to happen in strawberry, which necessitates comminution (mincing into small particles), such as the process of making surimi. The other problem (in non-sweet foods) is that the taste of the cryoprotectant can make the food sweet.

Most foods are multi-phase with complex structure and this makes investigation and interpretation of glass transition in them difficult (Roos, 1995). The glass transition is detected from changes in various physical properties associated with changes in molecular mobility and viscosity. These effects are seen in dielectric, mechanical, and thermodynamic properties (enthalpy, free volume, heat capacity and thermal expansion coefficient) (White and Cakebread, 1966; Wunderlich, 1981; Sperling, 1986). Differential scanning calorimetry

(DSC), and especially the new rapid scanning DSC (Saunders *et al.*, 2004), is the most common method used to determine T_g. DSC detects the change in heat capacity c_p occurring over the transition temperature range (Wunderlich, 1981; Kalichevsky *et al.*, 1992; Roos, 1995).

1.3 THAWING OF FROZEN FOODS

Superficially, thawing is the reversal of freezing (energy is supplied to the food in order to melt the ice crystals). However, thawing is more difficult an operation than freezing (and unfortunately mostly left to the consumer at the end of the supply chain). Thawing is difficult and requires care for three reasons:

(1) Thawing creates a region that has a lower thermal conductivity than the still frozen food, thereby impeding the heat flow (Fig. 1.5)
(2) The external medium (or energy source) cannot create as large temperature differences (or gradients) as is possible during freezing without cooking the food during thawing
(3) During thawing there is a higher risk of microbial growth because of temperatures/times allowing bacterial growth.

An emerging method of thawing that does not have the limitation (2) is pressure shift thawing (Cheftel *et al.*, 2002): melting the ice (form III) at temperatures below −15°C under high pressure (200–400 MPa), which serves to bypass the difficulties in conventional thawing such as exposing the surface of the food to temperatures above 0°C.

Thawing carried out on the industrial scale is a step in the processing of semi-finished food materials. However, perhaps most frozen foods are finally thawed at home, shortly before consumption. Thus, ironically, thawing, which is arguably the most difficult operation in the entire chain of operations to produce frozen foods, is ultimately left to the consumer whose handling of the process may negate all the care and strict quality control of the frozen food manufacturing process. Freezing does not kill micro-organisms and therefore the basic rule is to avoid microbial proliferation by thawing foods at chill temperatures, in a domestic refrigerator.

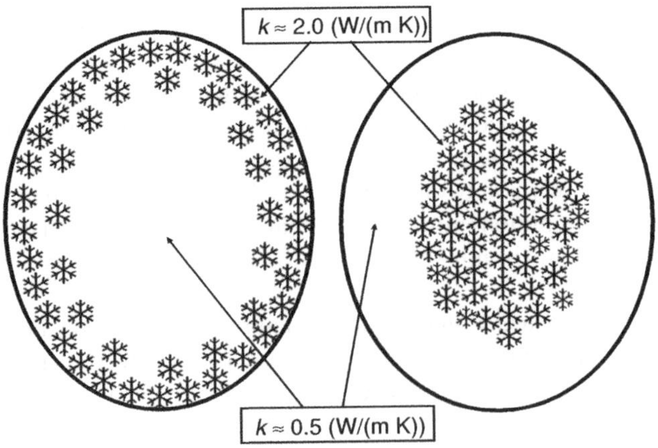

Fig. 1.5 Regions of high and low thermal conductivity during freezing and thawing of foods.

While cooking the food, thawing can sometimes be combined with heating in the oven (either conventional or microwave) if dehydration of the surface is prevented. If it is possible to divide a piece of frozen food into smaller pieces (for example, to separate the slices of bread from a sliced frozen loaf), the rate of heat transfer is quadrupled for each halving of the thickness. This follows from the solution of the heat conduction equation.

Thawing by microwaves has the disadvantage that the electromagnetic waves are preferentially absorbed in the unfrozen (thawed) region of the food. Thawing by ultrasound (domestic thawers have been developed in Japan) is in principle better than thawing by microwaves because ultrasound is absorbed in the compressible frozen region (Miles *et al.*, 1999). A good contact between the food and the ultrasonic source has to be ensured by immersion in water, thus it is suitable only for wet foods of regular shape, which is a disadvantage also of thawing by electric current (ohmic heating).

1.4 THERMOPHYSICAL PROPERTIES DURING FREEZING, THEIR MEASUREMENT AND APPLICATION

Data on thermal properties of foods are essential to design and control the thermal processing of foods and thereby ensure quality and microbiological safety of foods. It is often a difficult task to use the measurement methods correctly and apply the knowledge of thermal processes in industrial applications.

The principal feature of the thermal properties in the frozen range is that they depend strongly on temperature. This is because of the large differences between the properties of ice and liquid water and because of the varying proportion of ice below the initial freezing point, as shown in Fig. 1.3. Figures 1.6, 1.7, 1.8 and 1.9 show graphs of the properties of foods used in heat transfer modelling: density, specific heat capacity, enthalpy and thermal conductivity, respectively.

1.4.1 Specific heat capacity, enthalpy

Water has quite a large specific heat capacity, c_p, (4.18 J/g °C at 20°C) compared with other substances. Ice has a smaller c_p than water, about 2 J/g °C. The latent heat of freezing

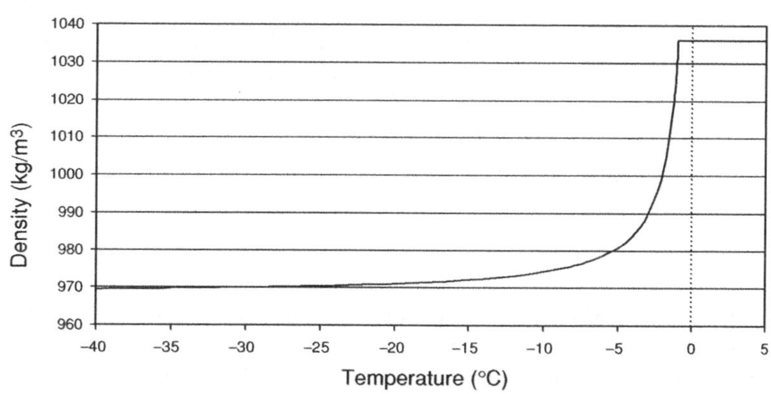

Fig. 1.6 Density of food as a function of temperature calculated with $T_f = -1$°C, $x_w = 0.8$, $x_{protein} = 0.05$, $x_{fat} = 0.075$, $x_{carbohydrate} = 0.075$ and $x_u = 0.05$.

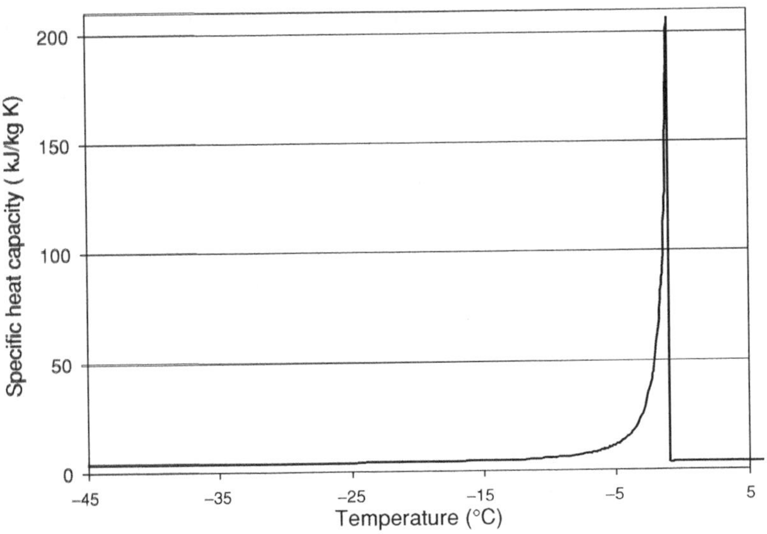

Fig. 1.7 Specific heat capacity of food as a function of temperature, calculated with $T_f = -1°C$, $x_w = 0.8$, $x_{protein} = 0.05$, $x_{fat} = 0.075$, $x_{carbohydrate} = 0.075$ and $x_u = 0.05$.

(or melting) of water (or ice), L, is also large compared with other substances: 334 J/g at 1 bar, 0°C. Because of the large values of c_p and latent heat of water, the energy required for freezing and thawing of foods is large and it increases with increasing water content of food.

Specific heat capacity (and enthalpy), being 'additive' properties, can be calculated by a simple 'mixing' formula:

$$c_p = \sum x_k \cdot c_{pk} \tag{1.2}$$

where c_p is specific heat capacity of food, x_k are the mass fractions of the components (water, ice, protein, carbohydrate, fat, etc.), and c_{pk} are the specific heat capacities of the components at constant pressure. This additive property and independence of structure makes heat capacity much easier to predict than thermal conductivity which depends on the structure of the food.

For frozen foods, x and c_p for water and ice in equation (1.2) vary with temperature; therefore a term has to be added to take into account the specific heat capacity variation due to the changes in proportion of ice: $L(T) \cdot (dx_i/dT)$ (assuming constant pressure). In Fig. 1.7 the steep peak of c_p at the initial freezing point followed by a small, stable value through the remaining part of freezing is due to the latent heat contribution from the gradually frozen-out ice, as shown in Fig. 1.3.

The c_p of foods can be estimated by assuming that the food is a binary solution and using function $x_i(T)$, approximated for example by equation (1.1). The c_p as a function of temperature then has the form

$$c_p(T) = c_s(1 - x_w) + c_w x_w(1 - x_i) + c_i x_i + L x_w(dx_i/dt) \tag{1.3}$$

where the indices s, w and i represent the solid component (dry solid content), water and ice, respectively, c is specific heat capacity and x is mass fraction. Table 1.2 shows the contributions of the sensible and latent heats in equation (1.2), calculated using $x_i(T)$ from equation (1.1).

Table 1.2 Contributions of the sensible and latent heats to the total specific heat capacity of food.

Temperature T(°C)	c_{pw} (kJ/kg°C)	Water $L(T)$[a] (kJ/kg)	Food $L(T)$[b] (kJ/kg)	x_i (kg/kg)	c_p "sensible" (kJ/kg°C)	c_p "latent" (kJ/kg°C)	c_p total (kJ/kg°C)
20	4.182[c]			0.000	3.747[d]	0.000	3.747[d]
15	4.186[c]			0.000	3.742[d]	0.000	3.742[d]
10	4.192[c]			0.000	3.739[d]	0.000	3.739[d]
5	4.202[c]			0.000	3.735[d]	0.000	3.735[d]
0 (water)	4.217[c]			0.000	3.730[d]	0.000	3.730[d]
0 (ice)	2.06[e]	333.6	334	0.000	3.730[d]	0.000	3.730[d]
−1			332	0.000	3.770	199.23	203.00
−2			330	0.375	3.765	49.50	53.26
−5		308.5	323	0.600	3.762	7.77	11.53
−10		284.8	313	0.675	3.761	1.88	5.64
−15		261.6	302	0.700	3.761	0.81	4.57
−20	1.94[e]	241.4	291	0.713	3.760	0.44	4.20
−22[f]		234.8	287	0.716	3.760	0.36	4.12
−40	1.82[e]		245	0.731	3.759	0.09	3.85
−60	1.68[e]		196	0.738	3.761	0.03	3.79
−80	1.54[e]		143	0.741	3.761	0.01	3.77
−100	1.39[e]		87	0.743	3.761	0.01	3.77

Note: Values used in equation (1.1): $x_w = 0.8$, $x_u = 0.05$, $T_f = -1$°C.
Values used in equation (1.3): $c_w(t) = 0.003019T^2 + 0.0586T + 4.285$, mean values for supercooled water, Rasmussen *et al.* (1973).

Sources:

[a] Dorsey (1940), p. 617.
[b] Riedel (1978). $L(T) = 334.1 + 2.05T - 0.00419T^2$.
[c] Kaye and Laby (1986), p. 58.
[d] COSTHERM program with $x_w = 0.8$, $x_{protein} = 0.05$, $x_{carbonydrate} = 0.075$, $x_{fat} = 0\ 075$.
[e] International Critical Tables (1933). $c_i(t) = 0.0067T + 2.073$.
[f] Triple point of water/ice I/ice III.

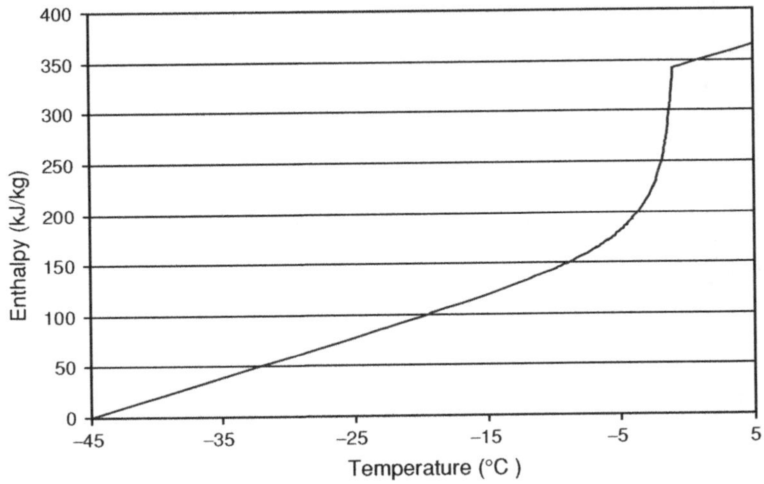

Fig. 1.8 Enthalpy of food as a function of temperature, calculated with $T_f = -1$°C, $x_w = 0.8$, $x_{protein} = 0.05$, $x_{fat} = 0.075$, $x_{carbohydrate} = 0.075$ and $x_u = 0.05$.

1.4.2 Enthalpy

Enthalpy H is the heat content taken with reference to a convenient fixed temperature T_{ref}, usually −40°C (below the range of temperatures usually considered in modelling the behaviour of frozen foods, or 0°C or sometimes the initial freezing point temperature T_f where the change of slope occurs between the frozen and unfrozen ranges, Fig. 1.8). Enthalpy is the integral of the function c_p between T_{ref} and a given temperature:

$$H(T) = \int_{T_{ref}}^{T} c_p(T')dT' \tag{1.4}$$

The function $H(T)$ is more suitable than $c_p(T)$ for use in computer modelling programs because it does not have the sharp peak at T_f. Using the enthalpy method helps us bypass the problem of 'jumping' the peak when advancing the time to the next level in numerical solutions of the heat equation (Albasiny, 1956).

1.4.3 Thermal conductivity

Thermal conductivity of water-containing food is again dominated by the contribution of water and ice, because these have higher thermal conductivities than the food matrix (the dry matter) (Wang and Brennan, 1992). Table 1.3 shows the values of thermal conductivity of water and ice at normal pressure. In comparison, the thermal conductivities of proteins, fats and carbohydrates are significantly smaller, in the range 0.17–0.20 W/(m K) at 0°C (Choi and Okos, 1986).

To estimate the values of thermal conductivity of frozen foods, some assumptions and approximations must be made about the structure of the food and the disposition of the various components dispersed in the food, including any air spaces in porous foods, and the direction (parallel or perpendicular) of heat flow relative to the layers of the components. The simplified models for this are the parallel, perpendicular and dispersed spheres

Table 1.3 Thermal conductivity k of water and ice at normal pressure (1 bar).

Temperature (°C)	k (W/(m K))	Reference
40	0.620	a
20	0.587	a
10	0.620	a
0 (water)	0.554	a
0 (ice)	2.25	b
−10	2.35	c
−15	2.41	c
−20	2.47	c
−40	2.73	c
−50	2.85	b
−100	3.95	b
−150	5.70	b

Sources:
[a]International Critical Tables (1933) $k = 0.587\{1 + 0.00281^* (T - 20)\}$ $0 < T < 80°C$.
[b]Ratcliffe (1962) 'most probable' values from measured data
[c]Ratcliffe (1962) fitted function $k = 780/T_k - 0.615 (T_k > 120$ K is temperature in kelvin).

(Maxwell–Eucken models (Eucken 1932, 1940; Miles *et al.*, 1983; Miles and Morley, 1997)). Figure 1.9 shows the thermal conductivity of food calculated using the parallel model and assuming that the major phase is aqueous binary solution of the same composition and initial freezing point as described in Section 1.4.1. The parallel model has the form

$$k = \sum \frac{\varepsilon_i}{k_i} \tag{1.5}$$

where $\varepsilon_i = \rho(x_i/\rho_i)$ is the volume fraction of the components of the food with overall density ρ (Miles *et al.*, 1983). More complex modelling requires numerical methods for the solution of heat flow equation in dispersed systems (Sakiyama *et al.*, 1990).

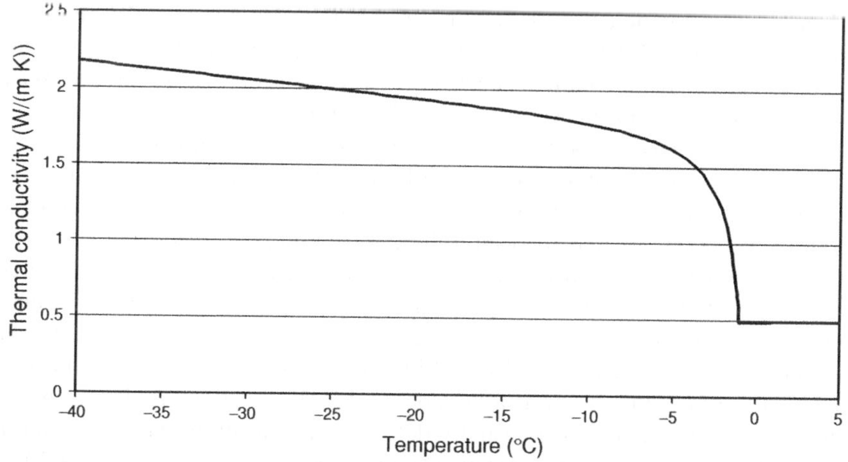

Fig. 1.9 Thermal conductivity of food as a function of temperature, calculated with $T_f = -1°C$, $x_w = 0.8$, $x_{protein} = 0.05$, $x_{fat} = 0.075$, $x_{carbohydrate} = 0.075$ and $x_u = 0.05$.

1.4.4 Density

Water expands by about 10% on freezing and thus with increasing amount of ice formed below the initial freezing point, the food tends to expand (its density decreases, Fig. 1.6) as temperature decreases. This creates thermal stresses and can lead to cracks in the food as the centre freezes last, expands and exerts pressure on the surrounding frozen region.

The basic method of measurement of density is from volume and mass, using a density bottle (pyknometer) for solids and fluids. There is a distinction between bulk and density (assembly of particles, e.g. peas of homogeneous product – density of one pea) and for such assemblies a suitable displacement method determines the volume (Mohsenin, 1970).

1.4.5 Thermal diffusivity

Thermal diffusivity, $a = k/(\rho \cdot c_p)$, is easier to measure than thermal conductivity k and specific heat capacity c_p (Eunson and Nesvadba, 1984). Therefore thermal conductivity can be indirectly determined using thermal diffusivity if the specific heat capacity and density are known, using the equation $k = \rho \cdot a \cdot c_p$ (Nesvadba, 1982).

1.4.6 Surface heat transfer coefficient

For modelling of food freezing (for example, predicting the freezing or thawing times), apart from the thermal properties of the food it is necessary to have an estimate of the surface heat transfer coefficient. The surface heat transfer coefficient h is not an intrinsic property of the food, but reflects the conditions at the boundary between the food and the external heat transfer medium. It is very important in heat transfer calculations (Hallström *et al.*, 1988). Uncertainties in the surface heat transfer coefficient are usually greater than uncertainties in thermal properties of foods and propagate into the calculated temperatures to a greater extent (Meffert, 1983).

In freezing by an external medium (for example, cold air blowing over the food) the rate of heat transfer from the food to the medium depends on the velocity of the moving medium, the shape and surface roughness of the food and other factors. The various factors are very difficult to account for mathematically and instead their overall effect is quantified by the surface heat transfer coefficient h, defined by $q = h(T_s - T_e)$, where q (W/m^2) is the heat transferred per unit area of the food surface per unit time and T_s and T_e are the temperatures of the food surface and the medium, respectively. The dimension of h is therefore $W/(m^2\ K)$. Table 1.4 gives the value of h for various types of freezing.

The surface heat transfer coefficient is correlated with the parameters of the flow of the external heat transfer medium through similarity relationships, involving dimensionless Reynolds (*Re*) and Nusselt (*Nu*) numbers and their correlations (Krokida *et al.*, 2002; Zogzas *et al.*, 2002). Over 400 equations relevant to food processing have been collected in the EU PECO project by van Beek, VeerKamp, and Pol (1997), which make the basis of the predictive software program SURFHEAT (Liu *et al.*, 1997). It was apparent that it is difficult to determine the values of h experimentally because different laboratories obtained different values for nominally the same experimental conditions. Similarly, in industry it is not possible to predict the average value of h without referring to a particular plant as the turbulence and velocity of air can vary. For this reason h is usually determined as a fitting parameter minimising the difference between experimental and measured temperatures (Everington, 1997). Having 'calibrated' the particular plant, the derived value of h can then be used for calculations for a different food product.

Table 1.4 Surface heat transfer coefficient *h* in various methods of food freezing.

Freezing method	*h* (W/(m^2 K))	References
Air – natural convection	6–20	Bird *et al.* (1960)
Air – forced convection	20–90	Becker and Fricke (2004), Hallström *et al.* (1988)
Air impingement	80–160	Anderson and Singh (2006),
	75–250	Soto and Bórquez (2001)
Air – Fluidised bed	90–140	Sheen and Whitney (1990).
Plate heat exchanger	100 (with poor contact, or the effect of the insulating properties of packaging)	Cleland and Earle (1982) Heldman (1980)
	600 (good direct contact)	
Immersion in brine	600	Cleland and Earle (1982),
	900 agitated	Fikiin (2003)
Immersion in liquid nitrogen	170–425	Heldman and Singh (1981)
Immersion in agitated fluid (hydrofluidisation)	160–1500	Verboven *et al.* (2003)
Thawing method		
Air impingement	5–25	Anderson and Singh (2006)
Condensing steam (vacuum thawing)	850–1700	Ling *et al.* (1976)

Influence of packaging that is a barrier to heat (often a required barrier, such as foam polystyrene to provide insulation) is taken into account by the formula based on adding series resistances:

$$Q = A \cdot \Delta T/(1/h_1 + d_1/k_1) \tag{1.6}$$

where Q is the heat flow through area A caused by the temperature difference ΔT, h_1 is the surface heat transfer coefficient at boundary 1, d_1 is the thickness of insulation between the boundary and the food and k_1 is the thermal conductivity of the insulation at boundary 1.

Experimental methods for measuring the surface heat transfer coefficient require much instrumentation (heat flow meters, air velocity meters, multiple thermometers, possibly humidity meters, and other sensors). One simple method of estimating the surface heat transfer coefficient is using a volatile substance (e.g. naphthalene) and periodically weighing the substance to determine its mass loss. The rate of mass loss is in analogy proportional to heat loss (Fellows, 2000).

1.4.7 Sources of data on physical properties of foods

1.4.7.1 *Literature*

There is a vast amount of literature on thermal properties of foods. Unfortunately, the data tend to be scattered and the composition (variety, cultivar), processing conditions and structure of the foods are often not well documented. This detracts attention from the value of the data.

Textbooks giving background information and limited general data are, for example, those of Charm (1978) and Heldman and Singh (1981). Books devoted to physical (including thermal) properties are those of Kostaropoulos (1971), Mohsenin (1980), Rha (1975), Tschubik and Maslow (1973), Houška *et al.* (1994, 1997), Rahman (1996) and Qashou *et al.* (1972). Papers giving comprehensive data sets for groups of food products are those of Lentz (1961), Hill *et al.* (1967), Mellor (1976, 1979), Morley (1966, 1972, 1986) Morley and Miles (1997), Pham and Willix (1989), Sanz *et al.* (1987) and Sweat (1974, 1975, 1985).

1.4.7.2 *Equations and software for predicting thermal properties of foods*

Several computer programs are available for estimating the thermal properties of foods from their proximate chemical composition and density. The most widely distributed is COSTHERM. Hans Pol of the Spelderholt Institute in the Netherlands wrote the first version of this program based on the work of Miles *et al.* (1983) during the EU project COST90. Miles and Morley (1997) updated COSTHERM in the EU project CIPA-CT93-0240. They re-examined the models for thermal conductivity and the initial freezing point. The accuracy of the predictive equations is about ±10%, sufficient for most food engineering calculations. The predictive equations are valid over the temperature range from –40°C to about +90°C.

1.4.7.3 *Offline and online databases of physical properties of foods*

Adam (1969) published the first major bibliography resulting from accumulation of data on physical properties of agro-food materials. The European effort included the EU concerted action projects COST90 and COST90bis (Jowitt *et al.*, 1983). Further work by Houška *et al.* (1994, 1997) has consolidated the methods of measurement and predictive equations relating the physical properties of agro-food materials to the composition and structure of the materials and their processing conditions. Singh (1995) has produced predictive software containing a database, available as a PC program. A larger database has been constructed by Nesvadba *et al.* (2004) and is available at www.nelfood.com.

1.4.7.4 *Measurements of physical properties of foods relevant to freezing*

Measurements (as opposed to modelling) are the primary source of data. Nesvadba (1982) and Ohlsson (1983) reviewed the measurement techniques for thermal properties of foods. Other transient techniques are still being developed, evaluated and standardised (Evitherm, 2003).

1.4.7.4.1 *Measurement of specific heat capacity and enthalpy*

For direct measurement of specific heat capacity or enthalpy some form of calorimetry (Riedel, 1957, 1978) has to be used. DSC is a convenient and rapid technique for measuring the specific heat capacity and phase transitions of small samples (up to 50 mg). Inhomogeneous foods must be homogenized in order to obtain homogeneous samples and valid analytical measurements. Fortunately, homogenisation is usually possible because, unlike for thermal conductivity, altering the structure of the food (pore size, direction of tissue fibres) does not alter the specific heat capacity and latent heat. For small sample sizes (5–50 mg) the DSC gives reliable results rapidly. For larger samples (order of 100 g) adiabatic calorimetry is used with 'home made' (Lindsay and Lovatt, 1994) or commercial (Patrick, 2002) equipment.

1.4.7.4.2 *Measurement of thermal conductivity*

Measurements of thermal conductivity using steady state methods take considerable times to equilibrate and although they have been used (Lentz, 1961; Willix *et al.*, 1998) transient techniques are preferred, of which the heated needle probe (Sweat, 1985) and the heated plane source (Gustafsson, 1991) are the most useful methods for foods.

Figure 1.10 shows the heated probe method. After applying electric current to the heater, the temperature rise of the heated needle, when plotted against logarithm of time, is inversely proportional to the thermal conductivity of the food in which the needle is placed. Care has to be taken that the mathematical assumption of infinitely thin and long probe is satisfied (diameter should be at least 100 times smaller than the length of the probe (Salmon *et al.*, 2003)).

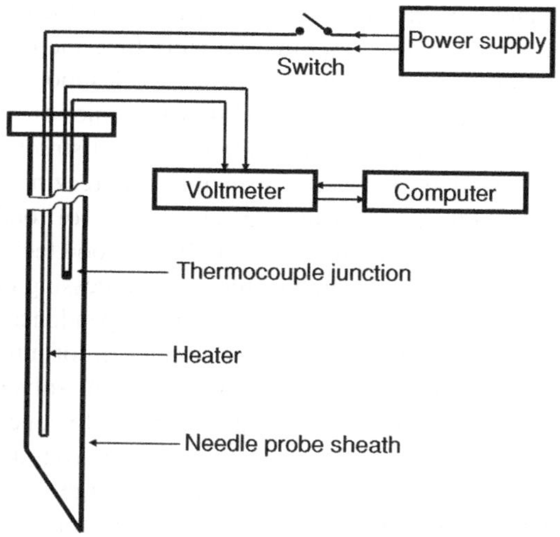

Fig. 1.10 Heated needle probe for thermal conductivity measurements. Thermocouple junction situated at the middle of a long needle probe senses the temperature rise of the needle after switching on the heater.

1.4.7.4.3 *Measurement of other relevant properties*

1.4.7.4.3.1 *Water content.* This is very important in foods and is usually determined gravimetrically, although there exists a range of various other methods, surveyed by Steele and Dang (1983).

1.4.7.4.3.2 *Initial freezing point.* This determines one of the 'anchor points' in the graphs of the thermal properties (Figs. 1.3 and 1.6–1.9). The freezing point is surprisingly difficult to measure. The problem is supercooling when approaching T_f from above and non-equilibrium effects in water re-absorption when approaching T_f from below. Fennema (1973) discusses the methods used to extrapolate T_f from freezing curves. The official AOAC method for products such as milk uses the cryoscope, satisfying EC regulations for milk. DSC is a convenient method of determining the freezing point but it is expensive and also requires extrapolation.

1.5 APPLICATION OF PHYSICAL PRINCIPLES

Understanding of the physical phenomena of heat and mass transfer and ice formation in food freezing often provides immediate pointers to rapid solutions to problems or answers to questions arising in the frozen food industry. The knowledge of numerical values of thermal properties of foods and the surface heat transfer coefficient enables prediction (modelling) of freezing times and refrigeration energy loads.

Figure 1.5 shows an example of the usefulness of physics in aiding understanding of food freezing and thawing. Frozen food is about three times a better conductor of heat than unfrozen food. During freezing a well-conducting external layer is created. During thawing the opposite takes place: a poorly conducting region is created. This is one reason why thawing is much more difficult than freezing. The other reason is that during freezing, temperature differences between the surface of the food and the external (cooling) medium can be higher than in thawing.

Recognition of the water content as the main determinant of the amount of heat required to be removed during freezing and the fact that freezing time increases with the square of product thickness are often good starting points for troubleshooting in the industry. The frozen food industry needs to predict freezing and thawing times of products that are newly formulated, or have new dimensions or composite or layered structure (such as meat pies or ready meals). The knowledge of physical principles of freezing enables estimation of the power or energy required to freeze or thaw given amounts of food in given time and to design freezers, thawers and other equipment used in frozen food technology. The necessary calculations involve some kind of modelling of the food processing operation. These could be simple analytical models, but due to the temperature dependence of thermal properties in the freezing range it is usually necessary to use some form of numerical solution of the heat conduction equation. This can be provided by a range of 'solvers', from simple finite difference algorithms to large commercial packages designed to cope with many situations.

HEATSOLV (Nesvadba, 1997) is an example of a simple program predicting food temperatures during freezing and thawing. An example of a large CFD (Computation Fluid Dynamics) package is COMSOL (2007). There are many other programs available, aimed at both commercial and research markets (see Chapter 3).

In modelling the heat transfer it is necessary to determine the sensitivity of the predicted temperatures to uncertainties in thermal properties and in the surface heat transfer coefficient. Often it is the surface heat transfer coefficient that carries the greatest uncertainty because the conditions on the boundary between the food and the external medium are difficult to quantify (Nicolaï and Baerdemaeker, 1996).

Modelling is of great value even if the surface heat transfer is not known *a priori*, because the model can be 'calibrated' by repeated comparisons between the predicted and experimentally measured temperatures in the food being processed.

REFERENCES

Adam, M. (1969). *Bibliografie fyzikálních vlastností potravin* [Bibliography of physical properties of foodstuffs]. Prague: VÚPP – Food Research Institute, 1969.

Albasiny, E.L. (1956). The solution of non-linear heat conduction problems on the Pilot Ace. Convention on digital computer techniques, The Institution of Electrical Engineers, 9–13 April, 1956. Woking and London: Unwin Brothers Ltd, p. 5.

Anderson, B.A. and Singh, R.P. (2006). Effective heat transfer coefficient measurement during air impingement thawing using an inverse method. *International Journal of Refrigeration* 29(2), 281–293.

Anese, M. and Gormley, R. (1996). Effects of dairy ingredients on some chemical, physico-chemical and functional properties of minced fish during freezing and frozen storage, *Lebensmittel Wissenschaft und Technologie*, 29, 151–157.

Angell, C.A. (1982). Supercooled water. In: F. Franks, ed. *Water – A Comprehensive Treatise*, Vol. 7, *Water in Aqueous Solutions at Subzero Temperatures*. New York: Plenum Press, pp. 1–81.

Arakawa, T. and Timasheff, S.N. (1985). The stabilization of proteins by osmolytes. *Biophysical Journal* 47, 411–414.

Bald, W.B. (1991). Ice crystal formation in idealised freezing systems. In: W.B. Bald, ed. *Food Freezing: Today and Tomorrow*. London: Springer-Verlag, pp. 67–80.

Becker, B.R. and Fricke, B.A. (2004). Heat transfer coefficients for forced-air cooling and freezing of selected foods. *International Journal of Refrigeration* 27(5), 540–551.

Bird, R.B., Stewart, W.E. and Lightfoot, E.N. (1960). In: *Transport Phenomena*. New York and London: John Wiley.

Blanshard, J.M.V. and Franks, F. (1987). Ice crystallization and its control in frozen food systems. In: J.M.V. Blanshard and P. Lillford, eds. *Food Structure and Behaviour* London: Academic Press, pp. 51–56.

Bøgh-Sørensen, L. *et al.*, eds. (International Institute of Refrigeration) (2007). *Recommendations for the Processing and Handling of Frozen Foods* ("*The Red Book*"), 4th edn. Paris: IIR/IIF.

Buttkus, H. (1970). Accelerated denaturation of myosin in frozen solution. *Journal of Food Science* 32, 558–652.

Carpenter, J.F. and Crowe, J.H. (1988). The mechanism of cryoprotection of proteins by solutes. *Cryobiology* 25, 244–255.

Charm, S.E. (1978), *The Fundamentals of Food Engineering*, 3rd edn. Westport, Connecticut, USA: AVI Publishing Company Inc.

Cheftel, J.C., Thiebaud, M. and Dumay, E. (2002). Pressure-assisted freezing and thawing: a review of recent studies. *High Pressure Research* 22, 601–611.

Choi, Y. and Okos, M.R. (1986). Effects of temperature and composition on the thermal properties of foods. In: M. Le Maguer and P. Jelen, eds. *Food Engineering and Process Applications,* Vol. 1, *Transport Phenomena.* New York: Elsevier, pp. 93–101.

Cleland, A.C. and Earle, R.L. (1982). Freezing time prediction for foods – a simplified procedure. *International Journal of Refrigeration* 5(3), 134–140.

COMSOL (2007). Multiphysics Modelling by COMSOL, http://www.comsol.com.

Connell, J.J. (1959). Aggregation of myosin during frozen storage. *Nature* 183, 664–665.

Daschner, F. and Knöchel, R. (2003). Dielectric microwave sensors with multivariate calibration. *Advances in Radio Science* 1, 9–13.

Dorsey, N.E. (1940). *Properties of Ordinary Water Substance.* New York: Reinhold. [Citing on page 617 compilation by R. de Forcrand and L. Gay (1929) and observations by H.C. Dickinson and N.S. Osborne (1915)].

Everington, D.W. (1997). The commercial use of COSTHERM and HEATSOLV in the selection of freezing plant. In: *Modelling of Thermal Properties and Behaviour of Foods During Production, Storage and Distribution.* Proceedings of a conference, Prague, June, 1997, Food Research Institute Prague. pp. 28–33.

Evitherm (2003). *Virtual Institute for Thermal Metrology.* http://www.evitherm.org.

Eucken, A. (1932). Die Wärmeleitfähigkeit keramischer feuerfest Stoffe. Ihre Berechnung aus der Wärmeleitfähigkeit der Betstandteile. (Thermal conductivity of ceramic refractory materials. Its Calculation from the Thermal Conductivity of Constituents), *Forschung auf dem Gebiete des Ingenieurwesens,* 3B, VDI-Forschungsheft No. 353, VDI Verlag, Düsseldorf, March–April 1932, p. 16.

Eucken, A. (1940). Allgemeine Gesetzmassig keiten für das Wärmeleitvermogen veschiedener Stoffarten und Aggregatzustande, *Forschung Gebiete Ingenieurwesens (Ausgabe A),* 11, 6.

Eunson, C. and Nesvadba, P. (1984). Moisture and temperature dependence of thermal diffusivity of cod minces, *Journal of Food Technology,* 19, 585–592.

Fellows, P. (2000). *Food Processing Technology – Principles and Practice,* 2nd edn. Cambridge, UK: Woodhead Publishing.

Fennema, O.R. (1973). Nature of the freezing process. In: O.R. Fennema, W.D. Powrie and E.H. Marth, eds. *Low Temperature Preservation of Food and Living Matter.* New York: Marcel Dekker, pp. 151–227.

Fennema, O.R. (1985). Water and Ice. In: *Food Chemistry*, 2nd edn. Marcel Dekker, pp. 23–67.

Fikiin, K.A. (2003). *Novelties of Food Freezing Research in Europe and Beyond.* Published as the European Commission's Flair-Flow 4 Programme Synthetic Report, July 2003.

Franks, F. (1982). The properties of aqueous solutions at subzero temperatures. In: F. Franks, ed. *Water – a Comprehensive Treatise*, Vol. 7, *Water in Aqueous Solutions at Subzero Temperatures.* New York: Plenum Press, pp. 215–338.

Garside, J. (1987). General principles of crystallization. In: J.M.V. Blanshard and P. Lillford, eds. *Food Structure and Behaviour.* London: Academic Press, pp. 35–99.

Goff, H.D. (1994). Measuring and interpreting the glass transition in frozen foods and model systems. *Food Research International* 27, 187–189.

Govindarajan, A.G. and Lindow, S.E. (1988). The size of bacterial ice-nucleation sites measured *in situ* by radiation inactivation analysis. *Proceedings of the National Academy of Sciences of the USA*, 85(5), 1334–1338.

Gustafsson, S.E. (1991). Transient plane source technique for thermal conductivity and thermal diffusivity measurements of solid materials. *Rev. Sci. Instrum.*, 62, 797–804.

Hallström, B., Skjöldebrand, C. and Trägårdh, C. (1988). *Heat Transfer and Food Products.* New York: Elsevier Applied Science.

Hansen, E. (2004). *Chemical and Physical Changes in Pork during Freezing and Frozen Storage.* PhD thesis, Department of Dairy and Food Science, The Royal Veterinary and Agricultural University, Frederiksberg, Denmark.

Hayes, L.J., Diller, K.R. and Lee, H.S. (1984). On the definition of an average cooling rate during cell freezing. *Cryoletters* 5, 97–110.

Heldman, D.R. (1980). Predicting of Food Freezing Rates. In: *Food Process Engineering,* Vol. 1, London: Applied Science Publishers Ltd.

Heldman, R.R. and Singh R.P. (1981). *Food Process Engineering*, 2nd edn., Westport, CT: AVI Publishing Company.

Herrera, J.R. and Mackie, I.M. (2004). Cryoprotection of frozen-stored actomyosin of farmed rainbow trout (*Oncorhynchus mykiss*) by some sugars and polyols. *Food Chemistry* 84 (1) 91–97.

Herrera, J.R. *et al.* (2000). Inhibition of formaldehyde production in frozen-stored minced blue whiting (*Micromesistius poutassou*) muscle by cryostabilizers: An approach from the glassy state theory, *Journal of Agricultural and Food Chemistry* 48, 5256–5262.

Hey, J.M., Mehl, P.M. and MacFarlane, D.R. (1997). A combined DSC-optical video microscope for crystallization studies. *Journal of Thermal Analysis* 49, 991–998.

Hill, J.E., Leitman J.D. and Sunderland, J.E. (1967). Thermal conductivity of various meats. *Food Technology* 21, 1143–1148.

Holland, B., Welch A.A., Unwin, I.D., Buss, D.M., Paul A.A., and Southgate, D.A.T., eds. (1991). *McCance and Widdowsons' Composition of Foods*, 5th edn. Cambridge: The Royal Society of Chemistry and Ministry of Agriculture, Fisheries and Food.

Houška, M., Adam, A., Celba, J., Havlíček, Z., Jeschke, J., Kubešová, A., Neumannová, J., Pokorný, D., Šesták, J. and Šrámek, P. (1994). *Thermophysical and Rheological Properties of Foods – Milk, Milk Products and Semiproducts*. Food Research Institute Prague, 200 p.

Houška, M., Adam, A., Celba, J., Havlíček, Z., Jeschke, J., Kubešová, A., Neumannová, J., Nesvadba, P., Pokorný, D., Šesták, J. and Šrámek, P. (1997). *Thermophysical and Rheological Properties of Foods – Meat, Meat Products and Semiproducts.* Food Research Institute Prague, 285 p.

Hsu, C.-L., Heldman, D.R., Taylor, T.A. and Kramer, H.L. (2003). Influence of Cooling Rate on Glass Transition Temperature of Sucrose Solutions and Rice Starch Gel. *Journal of Food Science* 68(6), 1970–1975.

Jason, A.C. and Richards, J.C.S. (1975). The development of an electronic fish freshness meter. *J. Phys. E. Sci. Instrum.* 8, 826–830.

Jensen, K.N., Jørgensen, B.M. and Nielsen, J. (2003). Low-temperature transitions in cod and tuna determined by differential scanning calorimetry. *Lebensmittel-Wissenschaft und Technologie* 36(3), 369–374.

Jowitt, R., Escher, F., Hallström, B., Meffert, H.F. Th., Spiess, W.E.L. and Vos, G., eds. (1983). *Physical Properties of Foods.* London and New York: Applied Science Publishers.

Kalichevsky, M.T., Jaroszkiewicz, E.M., Ablett, S., Blanshard, J.M.V. and Lillford, P.J. (1992). The glass transition of amylopectin measured by DSC, DMTA and NMR. *Carbohydrates and Polymers* 18, 77–88.

Kaye, G.W.C. and Laby, T.H. (1986). *Tables of Physical and Chemical Constants*, 15th edn. London: Longman.

Kent, M. (1975). Fish muscle in the frozen state: time dependence of its microwave dielectric properties. *Journal of Food Technology* 10, 91–102.

Kent, M., Knöchel, R., Daschner, F. and Berger, U.-K. (2001). Composition of foods including added water using microwave dielectric spectra. *Food Control* 12(7), 467–482.

Kent, M., Oehlenschlager, J., Mierke-Klemeyer, S., Knöchel, R., Daschner, F. and Schimmer, O. (2004). Estimation of the quality of frozen cod using a new instrumental method. *European Food Research and Technology* 219(5) 540–544.

Kent, M., Knöchel, R., Barr, U.-K., Tejada, M. and Nunes, L. (2005). *A New Method for Measurement of the Quality of Seafood.* Shaker Verlag GmbH, Germany.

Kim, Y., Singh, M. and Kays, S.E. (2007). Near-infrared spectroscopic analysis of macronutrients and energy in homogenized meals. *Food Chemistry* xxx (in press, available online).

Kitamikado, M., Yuan, C.S. and Ueno, R. (1990). An enzymatic method designed to differentiate between fresh and frozen-thawed fish. *Journal of Food Science* 55, 74–76.

Kostaropoulos, A.E. (1971). *Die Wärme- und Temperaturleitfähigkeit gestrockneter Lebensmittel. [Thermal conductivity and diffusivity of dried foodstuffs].* University of Karlsruhe, Faculty of Chemical Engineering, Dissertation, 1971.

Krivchenia, M. and Fennema, O.R. (1988). Effect of cryoprotectants on frozen turbot fillets and comparison with whitefish fillets. *Journal of Food Science* 53, 1004–1008, 1050.

Krokida, M.K., Zogzas, N.P. and Maroulis, Z.B. (2002). Heat transfer coefficient in food processing: Compilation of literature data. *International Journal of Food Properties* 5(2), 435–450.

Lentz, C.P. (1961). Thermal conductivity of meats, fats, gelatin gels and ice. *Food Technology* 15, 243–247.

Levine, H. and Slade, L. (1988). Principle of "cryostabilization" technology from structure/property relationships of carbohydrate/water systems – A review, *Cryo-Letters* 9, 21–63.

Lindsay, D.T. and Lovatt, S.J. (1994). Further enthalpy values of foods measured by an adiabatic calorimeter. *Journal of Food Engineering* 23(4), 609–620.

Ling, A.C., Lund, D.B. and Lightfoot, E.N. (1976). Estimation of the surface heat transfer coefficient with the presence of noncondensable gases. *Journal of Food Science* 41(6), 1345–1348.

Liu, M., Allen, A.R., van Beek, G., Pol, H. and Nesvadba P. (1997). Prediction of the Surface Heat Transfer Coefficients in Food Processing using SURFHEAT, in: Proceedings of Conference/Workshop/Practical Instruction Course *Modelling of Thermal Properties and Behaviour of Foods during Production, Storage and Distribution*, Prague, 23–25 June, 1997. Published by the Food Research Institute Prague, pp. 72–82.

Love, R.M. (1966). The freezing of animal tissue. In: H.T. Meryman, ed. *Cryobiology*, New York: Academic Press, pp. 317–405.

Love, R.M. (1982). Basic Facts about Fish. In: A. Aitken, I.M. Mackie, J.H. Merritt and M.L. Windsor, eds. *Fish Handling and Processing,* 2nd edn. Edinburgh: Ministry of Agriculture Fisheries and Food, Torry Research Station, HMSO, pp. 2–19.

MacKenzie, A.P., Derbyshire, W. and Reid, D.S. (1977). Non-equilibrium freezing behaviour of aqueous systems. *Philosophical Transactions of the Royal Society of London. Series B, Biological Sciences*, 278(959), 167–189. Lecture in: *A Discussion on Water Structure and Transport in Biology,* March 29, 1977, The Royal Society, London.

Matsumoto, J.J. and Noguchi, S.F. (1992). Cryostabilization of protein in surimi. In: Lanier, T.C. and Lee, C.M., Editors, 1992. *Surimi technology*, Marcel Dekker, New York, pp. 357–388.

Mackie, I.M. (1993). The effect of freezing on fish proteins. *Food Review International* 9, 575–610.

Mazur, P. (1970). Cryobiology: The freezing of biological systems. *Science* 168, 939–949.

Mazur, P. (1984). Freezing of living cells: Mechanisms and implications. *American Journal of Physiology* 247, C125–C142.

Meffert, H.F. Th. (1983). History, aims, results and future of thermophysical properties work within COST90. In: Jowitt R. *et al.*, eds. *Physical Properties of Foods*, Chapter 15. London and New York: Applied Science Publishers, pp. 229–267.

Mellor, J.D. (1976). Thermophysical properties of foodstuffs: Introductory Review. *Bulletin of the International Institute of Refrigeration* 56, 51.

Mellor, J.D. (1979). Thermophysical properties of foodstuffs: Part 3 – Measurement. *Bulletin of International Institute of Refrigeration* 59, 551–564.

Miles, C.A., van Beek, G. and Veerkamp, C.H. (1983). Calculation of Thermophysical Properties of Foods. In: R. Jowitt, F. Escher, B. Hallström, H.F.Th. Meffert, W.E.L. Spiess and G. Vos, eds. *Physical Properties of Foods*. New York: Applied Science Publishers. pp. 269–312.

Miles, C.A. (1991). The thermophysical properties of frozen foods. In: W.B. Bald, ed. *Food Freezing: Today and Tomorrow.* London: Springer-Verlag, pp. 45–65.

Miles, C.A. and Morley, M.J. (1997). Estimation of the Thermal Properties of Foods: a Revision of Some of the Equations Used in COSTHERM. In: Proceedings of a Conference/Workshop/Practical Instruction Course: *Modelling of Thermal Properties and Behaviour of Foods during Production, Storage and Distribution*, 23–25 June, 1997, Food Research Institute Prague, pp. 135–143.

Miles, C.A., Mayer, Z., Morley, M.J. and Houška, M. (1997). Estimating the initial freezing point of foods from composition data. *International Journal of Food Science & Technology* 32, 389–400.

Miles, C.A., Morley, M.J. and Randell, M. (1999). High power ultrasonic thawing of frozen foods. *Journal of Food Engineering* 39(2), 151–159.

Miller, B.S. and Kaslow, H.D. (1963). Determination of moisture by NMR and oven methods in wheat, flour, doughs and dried fruit. *Food Technology* 17(5), 142.

Mohsenin, N.N. (1970). *Physical Properties of Plant and Animal Materials,* Vol. 1. London: Gordon and Breach Science Publishers.

Mohsenin, N.N. (1980). *Thermal properties of foods and agricultural materials*. New York: Gordon and Breach Science Publishers.

Morley, M.J. (1966). Thermal conductivity of muscle, fats and bones. *Journal of Food Technology* 1, 303–311.

Morley, M.J. (1972). *Thermal Properties of Meat – Tabulated Data. Special report No. 1,* Agricultural Research Council, Meat Research Institute, Langford, Bristol.

Morley, M.J. (1986). Derivation of physical properties of muscle tissue from adiabatic pressure-induced temperature measurements. *Journal of Food Technology* 21, 269–277.

Morley, M.J. and Miles, C.A. (1997). Modelling the thermal conductivity of starch-water gels. *Journal of Food Engineering* 33, 1–14.

Nesvadba, P. (1982). Methods for the measurement of thermal conductivity and diffusivity of foodstuffs. *Journal of Food Engineering* 1, 93–113.

Nesvadba, P. (1997). Results of the EU PECO project "Modelling of Thermal Properties and Behaviour of Foods during Production, Storage and Distribution", In: P. Nesvadba and M. Houška, eds. Proceeding of a Conference/Workshop/Practical Instruction Course: *Modelling of Thermal Properties and Behaviour of Foods during Production, Storage and Distribution,* Prague, 23–25 June, 1997, Food Research Institute Prague, pp. 155–161.

Nesvadba, P., Houška, M., Wolf, W., Gekas, V., Jarvis, D., Sadd, P.A. and Johns, A.I. (2004). Database of Physical Properties of Agro-Food Materials. *Journal of Food Engineering* 61(4), 497–503.

Nicolaï, B.M. and De Baerdemaeker, J. (1996). Sensitivity Analysis with Respect to the Surface Heat Transfer Coefficient as Applied to Thermal Process Calculations. *Journal of Food Engineering* 28, 21–33.

Novikov, V.M. (1971). Production of frozen fish. In: V.M. Novikov, Trn. from Russian. *Handbook of Fishery Technology,* Vol. 1. [*Spravochnik Tekhnologa Rybnoi Promyshlennosti.*, Moscow, 1971.]. New Delhi: Amerind Publishing, pp. 267–390.

Ohlsson, T. (1983). The Measurement of Thermal Properties. In: R. Jowitt, F. Escher, B. Hallström, H.F.Th. Meffert, W.E.L. Spiess and G. Vos, eds. *Physical Properties of Foods.* London and New York: Applied Science Publishers, pp. 313–328.

Oosterhuis, J.J. (1981). Double freezing of plaice. *Voedingsmiddelentechnologie* 14(2), 11–15.

Orlien, V. (2003). *The Glass Transition Hypothesis – Significance for Chemical Stability in Food Systems.* PhD thesis, Department of Dairy and Food Science, The Royal Veterinary and Agricultural University, Frederiksberg, Denmark.

Orlien, V., Risbo, J., Mogens, L., Andersen, M.L. and Skibsted, L.H. (2003). The Question of High- or Low-Temperature Glass Transition in Frozen Fish. Construction of the Supplemented State Diagram for Tuna Muscle by Differential Scanning Calorimetry. *Journal of Agricultural and Food Chemistry* 51(1), 211–217.

Patrick, A. (2002). Thermal Properties of Food Materials. T^2 *PAC News* (Temperature and Thermophysical Properties Awareness Club of the National Physical Laboratory, UK), Issue 14, Spring 2002, 7–9. Downloadable from http://www.npl.co.uk/tman/newsletters/t2pac_newsletters_archive/downloads/issue14.pdf

Pham, Q.T. and Willix, J. (1989). Thermal conductivity of fresh lamb meat, offals and fat in the range −40 to +30°C: measurements and correlations. *Journal of Food Science* 54(3), 508–515.

Pippard, A.B. (1961). *Elements of classical thermodynamics.* Cambridge: Cambridge University Press.

Qashou, M.S., Vachou, R.I. and Touloukian, Y.S. (1972). Thermal conductivity of foods. *Transactions of ASHRAE* 78, 165.

Rahman, M.S. (1996). *Food Properties Handbook.* Boca Raton, FL: CRC Press.

Rahman, M.S. (2006). State diagram of foods: Its potential use in food processing and product stability. *Trends in Food Science & Technology* 17, 129–141.

Rasmussen, D.H., MacKenzie, A.P., Angell, C.A. and Tucker, J.C. (1973). Anomalous heat capacities of supercooled water and heavy water. *Science* 181, 342–344.

Ratcliffe, E.H. (1962). The Thermal Conductivity of Ice: New Data on the Temperature Coefficient. *Philosophical Magazine* 79, 1197–1203.

Rehbein, H. (1992). Physical and biochemical methods for the differentiation between fresh and frozen-thawed fish or fillets. *Italian Journal of Food Science* 4, 75–86.

Reid, D.S. (1990). Optimizing the quality of frozen foods. *Food Technology* 44(7), 78–82.

Reid, D.S., Kerr, W., and Hsu, J. (1994). The glass transition in the freezing process, *Journal of Food Engineering* 22, 483–494.

Reid, D.S., Hsu, J., and Ju, J. (1995). The importance of the glass transition to the temperature dependence of the stability of frozen foods. In: *The Proceedings of the 19th International Congress on Refrigeration, The Hague, Netherlands*, Vol. 2, Paris: International Institute of Refrigeration, pp. 331–338.

Rey-Mansilla, M.D., Sotelo, C.G., Aubourg, S.P., Rehbein, H., Havemeister, W., Jørgensen, B.M. and Nielsen, M.K. (2001). Localization of formaldehyde production during frozen storage of European hake (*Merluccius merluccius*). *European Food Research and Technology* 213(1), 43–47.

Rha, C.K. (1975). *Theory, determination and control of physical properties of food materials.* Dordrecht, Holland: D. Reidel Publishing Company.

Riedel, L. (1957). Kalorimetrische Untersuchungen über das Gefrieren von Fleisch. *Kältetechnik* 9(2), 38–40.

Riedel, L. (1978). Eine Formel zur berechnung der Enthalpie Fettarmer Lebensmittel in Abhängigkeit von Wassergehalt und Temperatur. *Chemie Mikrobiologie Technologie der Lebensmittel* 5(5), 129–133.

Roos, Y.H. (1992). Phase transitions and transformations in food systems. In: D.R. Heldman and D.B. Lund, eds. *Handbook of Food Engineering.* New York: Marcel Dekker, p. 145.

Roos, Y.H. (1995). *Phase Transitions in Foods.* London: Academic Press.

Roos, Y.H. and Karel, M. (1991). Applying state diagrams to food processing and product development, *Food Technology* 45 (12) 66–71.

Sakiyama, T., Matsushita, Y., Shiinoki, Y. and Yano, T. (1990). Finite-element analysis on the effective thermal conductivity of dispersed systems of spheres of various sizes. *Journal of Food Engineering* 11(4), 317–331.

Salfi, V., Fucetola, F. and Pannunzio, G. (1986). A micromethod for the differentiation of fresh from frozen fish muscle. *Journal of Food Science and Agriculture* 36, 811–814.

Salmon, D., Hammerschmidt, U., van Haneghem, I.A., Kubičár, L., Gustafssson, S. and Tye, R. (2003). *Draft Standard Test Protocol for Measurement of Thermophysical Properties of Materials using Contact Transient Methods based on a Common Principle.* National Physical Laboratory, Teddington, Middlesex, UK.

Sanz, P.D., Alonso, M.D. and Mascheroni, R.H. (1987). Thermophysical properties of meat products: General bibliography and experimental values. *Transactions of ASAE* 30 (1), 283–289, 296.

Saunders, M., Podluii, K., Shergill, S., Buckton, G. and Royall, P. (2004) The potential of high speed DSC (Hyper-DSC) for the detection and quantification of small amounts of amorphous content in predominantly crystalline samples, *International Journal of Pharmaceutics* 274, 35–40.

Sheen, S. and Whitney, L.F. (1990). Modelling heat transfer in fluidized beds of large particles and its applications in the freezing of large food items. *Journal of Food Engineering* 12(4), 249–265.

Singh, R.P. (1995). *Food Properties Database.* Version 2.0 for Windows™. 2nd edn, Boca Raton, FL: CRC Press.

Slade, L. and Levine, H. (1991). Beyond water activity: Recent advances based on an alternative approach to the assessment of food quality and safety. In: F.M. Clydesdale, ed. *Critical Reviews in Food Science and Nutrition*, Vol. 30. Boston: CRC Press, pp. 115–359.

Slade, L., Levine, H., Ievolella, J. and Wang, M. (1993). The glassy state phenomenon in applications for the food industry: Application of the food polymer science approach to structure-function relationships of sucrose in cookie and cracker systems. *J. Sci. Food Agric.* 63, 133–176.

Smith, C.E. and Schwartzberg, H.G. (1985). Ice crystal size changes during ripening in freeze concentration. *Biotechnology Progress* 1, 111–120.

Soto, V. and Bórquez, R. (2001). Impingement jet freezing of biomaterials. *Food Control* 12, 512–522.

Sperling, L.H. (1986). *Introduction to Physical Polymer Science.* New York: John Wiley.

Steele, D.J. and Dang, N.D.P. (1983). *Instrumental methods for the measurement of moisture content in foods: A survey of techniques and application.* Scientific and Technical Survey No. 136, Leatherhead Food Research Association, Leatherhead, UK.

Sweat, V.E. (1974). Experimental values of thermal conductivity of selected fruits and vegetables. *Journal of Food Science* 39, 1080–1083.

Sweat, V.E. (1975). Modelling of thermal conductivity of meats. *Transactions of ASAE* 18, 564–568.

Sweat, V.E. (1985). Thermal properties of low- and intermediate-moisture food. *Transactions of ASHRAE* 91(2A), 369–389.

Tamiya, T., Okahashi, N., Sakuma, R., Aoyama, T., Akahane, T. and Matomoto, J. (1985). Freeze denaturation of enzymes and its prevention with additives. *Cryobiology* 22, 446–456.

Trout, G.R. (1988). Techniques for measuring water-binding capacity in muscle foods – a review of methodology. *Meat Science* 23, 235–232.

Tschubik, I.A. and Maslow, A.M. (1973). *Wärmephysikalische Konstanten von Lebensmitteln und Halbfabrikaten* 1st edn. [Thermophysical Properties of Foodstuffs and Semi-Products, Translation from Russian]. Leipzig: VEB Fachbuchverlag.

van Beek, G., VeerKamp, C.H., and Pol, H. (1997). Database of Surface Heat Transfer Coefficient Equations. Progress report of the EU PECO Project CIPA-CT93-0240 *Modelling of Thermal Properties and Behaviour of Foods during Production, Storage and Distribution,* European Commission, Brussels.

Verboven, P., Scheerlinck, N. and Nicolaï, B.M. (2003). Surface heat transfer coefficients to stationary spherical particles in an experimental unit for hydrofluidisation freezing of individual foods. *International Journal of Refrigeration* 26(3), 328–336.

Wang, N. and Brennan, J.G. (1992). Thermal conductivity of potato as a function of moisture content. *Journal of Food Engineering* 17(2), 153–160.

White, G.W. and Cakebread, S.H. (1966). The glassy state in certain sugar-containing food products. *Journal of Food Technology* 1, 73–82.

Willix, J., Lovatt, S.J. and Amos, N.D. (1998). Additional thermal conductivity values of foods measured by a guarded hot plate. *Journal of Food Engineering* 37, 159–174.

Wunderlich, B. (1981). The basis of thermal analysis. In: E.A. Turi, ed. *Thermal Characterization of Polymeric Materials*, Chapter 2. New York: Academic Press. pp. 91–234.

Zogzas, N.P., Krokida, M.K., Michailidis, P.A. and Maroulis, Z.B. (2002). Literature data of heat transfer coefficients in food processing. *International Journal of Food Properties* 5(2), 391–417.

2 Effects of Freezing on Nutritional and Microbiological Properties of Foods

Mark Berry, John Fletcher, Peter McClure and Joy Wilkinson

2.1 INTRODUCTION

Freezing is a food preservation method that can potentially deliver a high degree of safety, nutritional value, sensory quality and convenience. The original advantages of freezing, compared to other preservation methods, were mainly in providing better quality vegetables, fish and meat at times and places distant from the point of harvest and slaughter. In addition to its value as a preservation method, freezing can supply pleasurable eating experiences. This is exemplified in ice cream, a product whose origins can be traced back to the seventeenth and eighteenth centuries where it was served as a luxury to the aristocracy (Clark, 2004), and which is now a global player in the frozen foods market. Although there is a legal definition for 'ice cream' in many countries, the authors are using a broader (more colloquial) use of the term to include ice cream, sorbets, water ices etc.

In the past, most frozen foods would have been cooked before consumption, providing an important contribution to microbiological safety. More recently, the focus has shifted to convenience and a much wider range of foods are available, many that have been specifically designed to be preserved and distributed in the frozen state. These more 'modern' frozen foods may be pre-cooked meal components and whole meals that simply have to be re-heated before consumption. In addition, many frozen cakes and desserts (including ice cream) are designed to be eaten on thawing or directly from the freezer without any consumer cooking step. The removal of any consumer cooking step has therefore made microbiological safety an essential pre-requisite prior to freezing and determines processes before and during freezing. However, nutritional quality is an increasing concern for consumers and the challenge for the frozen food industry is to maximise the retention of nutrients without compromising microbiological safety.

The essential step in freezing is to lower the temperature of foods with the intention of preventing, or at least minimising, microbial and chemical changes. However, as noted in other chapters of this book, the freezing of natural and fabricated foods results in complex physical and chemical changes. In summary, as the temperature is reduced below 0°C the water in foods begins to be converted into ice. As a result of this, the dissolved solutes become increasingly concentrated in the remaining liquid water, thereby further lowering its freezing point. Depending on their physical structure and chemical composition, natural foods in the frozen state may contain up to 8% water in the liquid phase. This liquid phase contains a complex mixture of cellular metabolites at non-physiologically high concentration. Furthermore, as ice crystals grow in natural food structures they may rupture intercellular and intracellular walls and membranes, resulting in release and mixing of previously compartmentalised substrates and enzymes. Therefore, although maintenance of foods at sub-zero temperatures potentially reduces the rate of reactions with potentially deleterious consequences for safety, quality and nutrition, changes in the concentration of substrate and access to enzymes may act to increase the rate of such reactions. Due to the physical and chemical changes that may continue to occur in the frozen state, a heat treatment before freezing is required for many foods, particularly vegetables (see below) to ensure preservation for an acceptable shelf-life. Heat treatment before freezing, commonly known as 'blanching', is primarily designed to inactivate enzymes that are responsible for deleterious changes in sensory quality. However, blanching also achieves the objective of preserving nutritional value, particularly for nutrients such as ascorbate (vitamin C) that are susceptible to enzymatic oxidation and degradation. The blanching step itself may have significant effects on nutrient retention and therefore the impact of this and other process steps that occur before freezing must be considered in understanding the nutritional value of frozen foods. The early literature describing the factors that influence nutrient retention in frozen foods has been reviewed by Bender (1978, 1993). This review summarises the principles involved and draws attention to more recent studies and to newer products entering the frozen food supply chain.

Micro-organisms of importance in foods are commonly separated into spoilage organisms and those with potential to cause human disease, pathogens. Preservation systems in foods tend to target prevention of growth of spoilage organisms and ensuring absence of harmful levels of pathogens (or their toxins). Since freezing essentially halts the activity of micro-organisms, it can control microbiological spoilage for indefinite periods, provided the temperatures are low enough (e.g. below −10°C). However, many micro-organisms, like many other biological systems, can survive freezing conditions and retain their ability to multiply when conditions become comparatively more favourable. Although there are no 'hard and fast rules' in relation to survival of pathogens under freezing, some groups of organisms differ greatly in their susceptibility or resistance to the effects of freezing. Higher organisms, such as protozoan parasites, are very sensitive to freezing and frozen storage and are destroyed. Gram negative bacteria are more resistant than protozoa but tend to be more susceptible than Gram-positive bacteria. Viruses retain their ability to infect host cells after being frozen and bacterial spores are completely resistant to the effects of freezing. Moulds and yeasts vary in their susceptibility to freezing and frozen storage. For public health, survival of pathogens becomes an important consideration. Generally speaking, if frozen foods have the potential to harbour pathogens at harmful levels, they will require further processing (e.g. cooking) to reduce these pathogens to levels that are not of concern to public health. Where low infectious dose organisms are concerned, the organism has to be eliminated from the frozen food completely.

Other factors that impact on the freezing effects include freezing rate, formulation of the food/ingredient content, packaging material, dimensions of the pack, storage temperatures/time, thawing conditions and physiological state (e.g. growth phase) of the micro-organism during cooling/freezing. Factors that affect resistance or susceptibility to freezing are discussed in detail in a recent review by Archer (2004). The initial stage of freezing, where product is cooled to freezing temperatures may destroy or injure susceptible organisms through cold-shock. Ice crystal formation (intracellular and extracellular) is also known to physically damage cells and further cooling to the final storage temperature may induce further damage, where slow-freezing ice crystals concentrate soluble solids, affecting stability of proteins within microbial cells. Physical damage is often associated with membrane damage. This has been shown in virus particles (*Herpes simplex* virus) where freezing and thawing induced loss of integrity of the membrane glycoprotein structure and viral uncoating (Hansen *et al.*, 2005). Under certain conditions, outflow of water from the cell, related to increase in extracellular osmotic pressure and the membrane-lipid phase transition, can cause cell death (Dumont *et al.*, 2003). It is also thought that oxidative stress may play a role in damage to the cell during freezing and thawing. This has been proposed by Park *et al.* (1998) for *Saccharomyces cerevisiae* and Stead and Park (2000) for *Campylobacter coli*. In *S. cerevisiae*, Tanghe *et al.* (2002) also proposed a role for aquaporin genes (*AQY1* and *AQY2*), suggesting that plasma membrane water transport activity is involved in determining freeze tolerance in yeast. Many organisms have evolved mechanisms that serve to minimise the freeze injury damage. For example, these organisms produce ice nucleation proteins, anti-nucleating proteins and anti-freeze proteins, which in turn minimise freeze injury. The structures and functions of these different proteins have been reviewed by Kawahara (2002). The effects of freezing on pathogens and spoilage micro-organisms concerned with different food types are described in more detail below.

This chapter aims to draw out the consequences of the freezing process on microbiological content and nutritional qualities of foods, looking in particular at the impact of processing before and during freezing. It comprises a review of the more recent literature and a summary of an unpublished study from our own laboratory to illustrate the direction of some future trends in frozen foods.

2.2 MEAT, POULTRY AND FISH

2.2.1 Nutritional aspects

A wide variety of frozen meat, poultry and fish products are now available to the consumer. These range from simply prepared cuts of muscle, whole birds and intact fish to highly processed products containing components derived from several animal tissues. In nutritional terms meat, poultry and fish are particularly important dietary sources of protein and fat; of the minerals iron, zinc, magnesium and selenium (particularly from fish and other sea foods); of the B vitamins, vitamin A (found in liver) and vitamin D (found in oily fish species).

Following slaughter, muscle and other animal tissues undergo complex biochemical changes; these may arise from residual metabolic reactions, or be induced by microbial and/or oxidative spoilage. The objective of freezing is to slow, or to prevent post-mortem changes that may adversely affect microbiological safety, sensory quality and nutritional value. Although the process of freezing by itself does not have significant effects on nutrient levels in meat, poultry and fish foods, it is less certain whether prolonged frozen storage causes significant loss of potentially labile nutrients.

A recent study of frozen storage on protein and fat retention assessed the long-term stability of food samples gathered during clinical feeding trials (Phillips *et al.*, 2001). A mixed food sample was prepared from typical components in an American diet. Storage and thawing of these samples was carried out under 'ideal' conditions – in sealed containers, without temperature fluctuations and without any drip loss after thawing. Although these conditions do not represent normal domestic, or commercial practice, this study confirmed that levels of protein, total fat and individual fatty acids were not significantly reduced by storage at −60°C for up to 50 months.

Studies measuring the effects of frozen storage on animal tissues under more realistic conditions have mainly been directed towards determining effects on sensory quality, rather than on nutrient retention. Early studies on vitamin levels in muscles from pork, beef and poultry indicated that losses were of the order of 10–30% for niacin, pyridoxine, thiamine and riboflavin after storage times of up to a year (Westerman, 1952; Lee *et al.*, 1954; Fennema, 1975; Engler and Bowers, 1976). In a more recent systematic study, levels of thiamine, riboflavin and pyridoxine were measured in pork meat frozen stored for 12 months at –12°C and –24°C (Mikkelsen *et al.*, 1984). Losses compared with samples taken before freezing were 10–20% for all the vitamins measured, but in this, as in earlier studies, a high variability for estimates of loss and lack of consistency between samples and treatments was noted. Many published studies of nutrient loss on frozen storage do not discriminate between loss of nutrients caused by chemical breakdown and physical loss of nutrients in the fluid that exudes from the meat during and after thawing. Following prolonged frozen storage of muscle foods it has been observed that irreversible aggregation of the actin and myosin protein myofibrils may occur. This does not adversely affect the nutritional value of the protein, but may result in a toughening of the meat texture and a reduction in the ability of the muscle structure to hold water. As a consequence, on thawing, frozen intact muscle may lose a significant amount (2–15%) of intracellular and intercellular fluids as so-called 'drip-loss'. If this fluid is not incorporated into the food to be consumed it may represent a significant loss of water-soluble proteins, vitamins and minerals. Many factors may contribute to determining the magnitude of drip loss; these include species, age of the animal, pre-slaughter handling of the animal, freezing rate and thawing rate.

Frozen meat, poultry and fish products stored in contact with air are susceptible to oxidation; this may especially be a problem with fish and poultry that contain nutritionally significant amounts of polyunsaturated fatty acids. Polyunsaturated fatty acids (PUFAs) of the n-3 series, docosahexanoic acid (DHA) and eicosopentanoic acid (EPA) found in fish are particularly prone to oxidation, resulting in formation of volatile products that give rise to the aroma and taste characterised as 'rancid'. Several recent studies have investigated the effects of freezing and frozen storage on n-3 PUFA levels in fish. A significant reduction in the total n-3 PUFA content was reported in Saithe (a lean fish) fillets stored at –20°C for 6 months (Dulavik *et al.*, 1998). Similarly, levels of total n-3 PUFA were reduced in salmon fillets stored at –20°C (Refsgaard *et al.*, 1998) and levels of DHA and EPA were reduced in sardine and mackerel fillets stored for 24 months (Rougerou and Person, 1991). In contrast to these reports of PUFA loss, Polvi *et al.* (1991) found no difference in total n-3 PUFA levels in salmon fillets stored at the relatively high temperature of –12°C for 3 months. Xing *et al.* (1993) also failed to see any losses of DHA and EPA in mackerel and cod fillets stored frozen at –20°C for 28 weeks.

As with many aspects of nutrient stability, the extent of n-3 PUFA loss from frozen fish by oxidation depends on several factors, e.g. access of oxygen to the muscle, handling before freezing and the type of muscle (dark muscle of fish suffers higher rates of iron-catalysed oxidation than white muscle). Although loss of nutritionally important n-3 PUFAs from frozen fish may undoubtedly occur on prolonged frozen storage, in practice this is not likely to be a serious cause for concern. The threshold for sensory detection of rancidity is very low and therefore if frozen fish are consumed within the recommended period of storage, significant proportions of their original content of n-3 PUFAs will not have been lost to oxidation.

In summary, the scientific literature on the systematic effects of freezing and frozen storage on nutrient retention in meat, poultry and fish products is not extensive, it is often decades old and often not of high quality. However, the available data combined with knowledge of the stability of the relevant nutrients in their isolated forms strongly suggest that freezing is an excellent method to preserve the nutritional values of meat, poultry and fish.

2.2.2 Micro-organisms associated with meat, poultry and fish

The micro-organisms associated with raw meat, poultry and fish are varied. They include bacteria, protozoa, viruses and for meats, less well-characterized agents such as prions which cause transmissible spongiform encephalopathies. Many of the contaminating organisms are originally present on the external surface or in the respiratory or intestinal tracts of healthy animals. Contamination occurs and spreads during transport/holding prior to slaughter/killing, during slaughtering and subsequent processing: hide/skin removal, cutting, deboning, evisceration, washing etc. The main microbial hazards of concern in frozen meat and poultry include infectious Gram-negative bacteria such as salmonellae, *Campylobacter* spp., pathogenic *Escherichia coli* and *Yersinia enterocolitica*, Gram-positive infectious bacteria such as *Listeria monocytogenes*, Gram-positive toxicogenic bacteria such as *Staphylococcus aureus*, *Clostridium perfringens* and *C. botulinum*, protozoal parasites such as *Cryptosporidium* spp., *Giardia* spp., *Trichinella spiralis* and *Toxoplasma gondii*. Some of these organisms are found in a wide variety of animal species and cause disease in these animals while the host range of others is more restricted. The hazards that are associated with raw meat and poultry and occur frequently have been discussed in more detail by McClure (2002). Some are present in relatively high numbers (e.g. levels of salmonella up to 10^3/g and *Campylobacter* spp. up to 10^6 colony forming units, CFU, per carcass in poultry) and this is an important consideration for risk assessment and controlling these hazards prior to consumption. As for meat and poultry, microbes of major concern in fish are infectious bacteria and include *Salmonella, Shigella, Listeria monocytogenes, Vibrio cholerae* and other pathogenic vibrios such as *V. parahaemolyticus* and *V. vulnificus*.

Micro-organisms associated with quality or spoilage defects in meat, poultry and fish include pseudomonads, moulds, yeasts, and Gram-positive bacteria such as lactic acid bacteria, *Brochothrix thermosphacta* and micrococci. The density of occurrence of micro-organisms on the skin of meat and poultry can be greater than $10^9/cm^2$. In addition to contamination on the skin, further contamination can result from handling during slaughter, cutting, deboning and packaging, from knives, wash water, hands and clothing of workers and the hides/skins of other animals. Methods for cleaning carcasses are usually ineffective but scalding and fast chilling prior to freezing can reduce numbers of contaminating micro-organisms significantly.

The micro-flora associated with fish and other seafoods typically reflects the flora of the environments in which these have been caught and harvested. As with meat and poultry, if the fish are healthy, then muscle tissue and internal organs are usually sterile although the circulatory system of some shellfish is not 'closed'. For example, the haemolymph of crabs commonly contains marine bacteria, such as vibrios, sometimes at high levels. Other organisms commonly associated with fish and shellfish include pseudomonads, *Micrococcus* spp., members of the *Acinetobacter-Moraxella* genera, corynebacteria, *Geotrichum* spp. and *Rhodotorula* spp. Fish and shellfish associated with coastal waters and estuarine environments tend to harbour a wider variety of organisms, including *Bacillus* spp., members of the Enterobacteriaceae and viruses. Salt-water fish and shellfish harbour halotolerant organisms. When fish are caught, they are often stored in chilled brines or ice, and if caught in salt-water, this dilutes the salt concentration allowing halotolerant micro-organisms, which usually prefer lower concentrations of salt for optimal growth, the opportunity to multiply quickly. The micro-organisms from temperate waters are often psychrotrophic or psychrotolerant while those from tropical waters are not, so speed of chilling and low temperature of storage after capture can have a significant impact on the flora that develops before fish are frozen. In some situations, shellfish, such as shrimp, are cooked soon after capture, so this destroys a large

Table 2.1 Pathogenic organisms and their levels of concern, associated with different raw materials/processes relevant to frozen foods.

Pathogenic micro-organisms or their metabolites	Level causing concern	Relevant foods and raw materials
Bacillus cereus – ingestion of emetic toxin	High (e.g. >10^5 vegetative cells)	Meat, poultry, vegetables (especially rice), dairy (milk, cream)
Bacillus cereus (toxico-infection)	High (e.g. >10^5 vegetative cells or spores)	As above
Campylobacter spp. – infection	Low (e.g. 500 cells)	Meat (beef, pork, lamb), poultry, dairy (milk)
Clostridium botulinum – ingestion of neurotoxins	High (e.g. >10^5 vegetative cells or spores)	Meat, vegetables, fish
Clostridium botulinum (toxico-infection, infant botulism)	Low (e.g. 100 vegetative cells or spores)	Honey, vegetables, meat
Clostridium perfringens (toxico-infection)	High (e.g. >10,000 vegetative cells)	Meat, poultry, dairy (milk)
Escherichia coli (six pathogenic types)	Low–high ((e.g. 10 cells to >10^6), depending on pathotype	Meat, fruit, vegetables, dairy (milk, cheese)
Listeria monocytogenes – infection	Medium (e.g. >1000 vegetative cells), low for immunocompromised individuals	Meat, poultry, fish, ice cream, vegetables, fruit, dairy (milk, cheese)
Salmonella enterica (all sub-species/serotypes) – infection	Low–medium (e.g. 10–10,000 cells)	Meat (beef, pork, lamb), poultry, fish, dairy (milk, cheese), bakery (eggs)
Staphylococcus aureus – ingestion of enterotoxin	High (e.g. >10^5 vegetative cells)	Meat, fish, bakery (process), dairy (milk, cheese)
Yersinia enterocolitica – infection	Not known, probably medium (>10,000 cells)	Meat (particularly pork), poultry, dairy (milk), bakery (eggs), vegetables
Vibrio spp. (*V. cholerae, V. parahaemolyticus, V. vulnificus*) – infection	Low–high (e.g. 500–100,000 cells), depending on strain/species, host susceptibility	Fish, shellfish
Giardia intestinalis (or *lamblia*)	Low (e.g. 25–100 cysts)	Fruit, salad vegetables, handled foods
Cryptosporidium parvum	Low (e.g. 30–132 oocysts)	Fruit, salad vegetables
Cyclospora cayetanensis	Not known, probably low	Fruit, salad vegetables
Trichinella spp. (e.g. *T. spiralis*)	Possibly low	Meat (pork, horse)
Toxoplasma gondii	Probably low	Meat (lamb, horse), poultry
Taenia spp.	Possibly low	Meat (pork)
Anisakis spp.	Possibly low	Fish

proportion of contaminating organisms (vegetative cells) but the handling of these (e.g. during shelling/peeling, sorting) post-cooking often means that contamination levels can return to previous levels and sometimes include other micro-organisms, such as *E. coli* and staphylococci. Levels on shrimp are typically 10^5–10^7 CFU/g at the time of arrival at processing plants, and those from tropical waters may be higher. The key pathogenic micro-organisms associated with foodstuffs, including fish and fish products are listed in Table 2.1. Some micro-organisms such as *Photobacterium phosphoreum*, *Ph. histaminum*,

Carnobacterium piscicola and *Shewanella putrefaciens* cause particular spoilage problems with fish. From a safety perspective, algal toxins (produced in algal blooms) and scombroid poisoning, resulting from bacterial production of biogenic amines (decarboxylation of parent compounds such as amino acids) are additional concerns in seafood, and neither of these is destroyed by consumer cooking. The same is true for other bacterial toxins such as botulinum toxin and staphylococcal enterotoxins, which are not affected by freezing and frozen storage.

2.2.3 Effects of freezing on micro-organisms associated with meat, poultry and fish

The rate of freezing is known to influence the microbiological status of meat, poultry and fish and while it is generally the case that total numbers of micro-organisms decrease slightly during freezing, frozen storage and thawing, growth of some micro-organisms can occur at temperatures below 0°C, with frequent reports of growth down to −5°C and less frequently down to −10°C (Larkin and Stokes, 1968). Frozen poultry meat shows rapid increases in numbers of spoilage micro-organisms when stored at −5°C (Abu Ruwaida *et al.*, 1996) and between −5 and −7°C, where the main organisms of concern are psychrophilic yeasts and moulds, causing whisker-like growth (e.g. *Thamnidium* spp.), black spots (e.g. *Cladosporium* spp.) and white spots (e.g. *Sporotrichum* spp.). Generally speaking, temperatures below −10°C are considered to effectively inhibit growth of micro-organisms associated with foods. Therefore, the rate of cooling is not only relevant to killing or 'cidal' effects on micro-organisms but is also important for preventing growth at sub-zero temperatures, and this is important to prevent spoilage of meat, poultry and fish.

The pattern of survival for different taxonomic groups during freezing and frozen storage of fish is similar to the pattern observed in meat and poultry, with Gram-negative bacteria surviving less well than Gram-positive organisms, and spores showing little or no change in numbers. However, even though freezing inhibits the metabolism of micro-organisms during frozen storage, enzyme activity still continues. This is important in frozen fish where histidine decarboxylase activity results in histamine production, elevated levels of which exceed spoilage limits (ca. 5 ppm) or even safety limits (ca. 50 ppm), particularly if temperature abuse occurs in the supply chain.

The cidal effects of freezing on different types of organisms vary dramatically. Spores of bacteria and moulds remain largely unaffected by freezing although there is some evidence that spores of *C. botulinum* are more susceptible to the effects of ionising radiation following frozen storage for 30 days at −75°C (Lim *et al.*, 2003). Spores are inert and extremely resistant to different stresses and there is little damage afforded by freezing and repeated freeze–thaw cycling. Vegetative bacterial cells vary in their response to freezing. Some organisms (e.g. *E. coli*, *B. subtilis*, *Vibrio cholerae* and some lactic acid bacteria) are capable of responding to lower temperature stress through production of cold-shock proteins (CSPs or cold-induced proteins, CIPs) and a gradual change to colder conditions has been shown to improve the survivability of *E. coli* (Bollman *et al.*, 2001), *L. monocytogenes* (Bayles *et al.*, 2000) and salmonellae (Jeffreys *et al.*, 1998) that are frozen. There is a relatively large body of information on the survival of *L. monocytogenes* subjected to freezing under a range of conditions, showing variable responses and summarised by Archer (2004). *Photobacterium phosphoreum*, associated with spoilage of modified atmosphere-packed cod and salmon, and histamine formation, is very sensitive to freezing and a single freeze–thaw cycle has been shown to dramatically increase shelf-life of these and similar products. However, significant histidine decarboxylase activity of halophilic histamine-forming bacteria can continue after freezing even though cell counts are reduced by more than six $\log_{10}$ cycles (Fuji *et al.*, 1994). There is some evidence that frozen storage of fish for more than 3–6 weeks results in large reductions in levels of *V. cholerae* but there are also several studies reporting recovery of vibrios, including *V. cholerae* and other potentially pathogenic species, from frozen seafood such as frozen raw shrimps and fish. Epidemiological evidence from a large cholera outbreak associated with ice also indicates survival of numbers high enough to cause infection.

Freezing is known to cause sub-lethal damage to vegetative bacteria, such as salmonellae, campylobacters, *L. monocytogenes* and vibrios and that is why care must be taken when investigating these effects, particularly when selective media are used to recover cells in the presence of other micro-organisms. Agents in selective media used to recover specific types of micro-organisms often inhibit damaged cells of the target micro-organisms as well, as shown by Barrell (1988) and more recently by Chang *et al.* (2003). It is widely acknowledged that injured cells in foods may be capable of repair and for this reason, many methods for the detection of pathogens in foods include a recovery step. It is generally concluded that, unless specific effects are shown, such as the effects of repeated freeze–thawing on some bacteria (such as salmonellae), freezing may have little or no effect on the numbers of viable vegetative bacteria.

There is also evidence that prior adaptation (e.g. storage at chill temperatures for only 1.5 h) can have significant effects on survival of *E. coli* (Mihoub *et al.*, 2003) and lactic acid bacteria (Sanders *et al.*, 1999). It is also clear from the former study that some strains are intrinsically more resistant to cold shock injury than other strains, so any additional effect from prior adaptation becomes less important. This is also seen with other stresses such as exposure to low pH. The expression and repression of various proteins and their potential involvement in survival to cold shock is discussed by Mihoub *et al.* (2003).

The absence of CSPs has been proposed as one possible reason for the relatively high susceptibility of *Campylobacter* spp. to the effects of freezing (Park, 2002). A recent study by Georgsson *et al.* (2006) has confirmed results from previous studies, reporting that *Campylobacter* levels in broilers decreased by between 0.7 and 2.9 $\log_{10}$ units after freezing and 31-day storage. Numbers of faecal coliforms, often used as indicator organisms, in the same samples remained largely unchanged. The significant die-off of *Campylobacter* is consistent with previous studies (Humphrey and Cruikshank, 1986; Stead and Park, 2000; Moorhead and Dykes, 2002) which reported approximately 2–3 $\log_{10}$ decreases in numbers after freeze–thaw stress, although, again, there is strain-to-strain variability reported with some strains showing higher sensitivity to freezing (Chan *et al.*, 2001).

Protozoan parasites are considered to be very sensitive to the effects of freezing and this generally results in the elimination of their dormant or quiescent forms. The few data reporting the effects on protozoan parasites known to be transmitted via foods are summarised by Dawson (2005) and Nichols and Smith (2002). Survival for more than 4 weeks under certain freezing conditions, such as slow freezing or freezing in high-osmolyte solutions, has been reported. These data and other information from outbreaks of illness associated with other foods (e.g. ice cream and ice cubes) indicate that survival may occur over short periods, posing a risk in those foods contaminated with oocysts or cysts. Freezing is known to be an effective means to inactivate nematodes such as *Anisakis simplex* and *Trichinella* spp.; freezing is a legally accepted way of inactivating trichinae in pork meat (Murrell *et al.*, 1986). Freezing and frozen storage also destroy fish parasites that are pathogenic to humans.

Transmission of viruses to humans is not generally considered to happen via meat and meat products, although caliciviruses (including Noroviruses and Sapoviruses) infect both humans and other animals. Recent evidence suggests that Norovirus infections often occur in calves and sometimes in pigs and that some animal strains are genetically or antigenically related to human strains (Wang *et al.*, 2006). The significance of this finding is unknown at the present time. More recently, a highly pathogenic (to birds) H5N1 virus was isolated from imported frozen duck meat in Korea (Lu *et al.*, 2003). Although this particular virus appeared to be substantially less pathogenic for mammalian species, it has raised concern about the safety of frozen poultry coming from regions affected by the recent outbreaks of avian influenza around the world, which originated in the Far East. In addition, limited data and information available from studies with poliovirus survival in frozen foods, and epidemiological studies implicating ice with Norovirus-associated gastroenteritis, suggest that at least some viruses retain their infectivity following freezing and frozen storage. This is also true for viruses associated with shellfish, such as oysters, where outbreaks of Norovirus-related gastroenteritis have been linked to frozen products. Prion proteins, because of their innate stability, are likely to remain unchanged during freezing and frozen storage.

2.3 VEGETABLES AND FRUITS

2.3.1 Nutritional aspects

The terms 'vegetable' and 'fruit' in the context of nutrition do not have precise definitions. Plant foods that are commonly known as vegetables and fruits derive from culinary traditions rather than a scientific definition and they consist of a remarkably diverse food group. Under the category of 'vegetables and fruits' humans regularly consume many different plant species and different plant tissues taken from roots, shoots, leaves, fruits, seeds and flowers. Commonly available frozen vegetables and fruits provide a correspondingly large array of nutrients. The most important nutrients for human health and well being found in frozen vegetables and fruits are the vitamins A (as precursor carotenoids), C, K, folate; soluble and insoluble dietary fibres; polyunsaturated fatty acids of the n-6 and n-3 family and the minerals potassium and magnesium. In addition to these nutrients that are by definition essential for life, dietary vegetables and fruits provide a host of other phytochemicals, consumption of which may ensure long-term health and well-being of the consumer (Goldberg, 2003).

Compared to 'fresh' vegetables and fruits, there are generally small differences in the mineral and fibre content of equivalent vegetables supplied as frozen (Polo *et al.*, 1992; Nyman, 1995). However, for some particular fruits, significant changes in fibre profile have been reported, for example, in one variety of mango, cellulose, hemicellulose and lignin contents decreased by 50% during 12 months frozen storage (Cano and Marin, 1995).

There are several factors that can cause significant differences between the vitamin levels in frozen vegetables and fruits, and fresh or otherwise preserved equivalents. The following four sections describe those factors.

2.3.1.1 *Selection of cultivar and time of harvesting*

Particular cultivars and harvest times are chosen to optimise the sensory quality of the produce and these may differ between those selected for freezing and those to be consumed as fresh, canned or dried food. The cultivar and harvest time may affect nutritional value (Shewfelt,

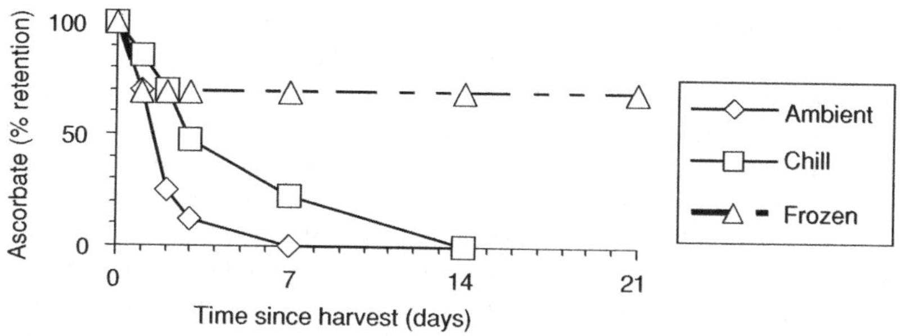

Fig. 2.1 Effects of post-harvest storage and blanching on ascorbate levels in spinach. Adapted from Favell (1998).

1990; Lee and Kader, 2000), for example, peas selected for canning are usually harvested at a more mature stage than those selected for freezing and have approximately a 10% lower ascorbate (vitamin C) concentration (unpublished observations). The type of cultivar may also influence the amount of nutrients lost during processing, reflecting differences between cultivars in morphology and mechanical strength.

2.3.1.2 *Post-harvest treatment*

Plant tissues retain a high metabolic activity after harvest and in addition the processes of harvesting may trigger a series of metabolic responses to physical and physiological stresses. As a consequence vegetables and fruits are relatively unstable after harvest and metabolic reactions may rapidly result in significantly reduced levels of some vitamins.

The magnitude of nutrient losses after harvest and prior to freezing is highly variable, depending on the crop, the method of harvesting, the washing, cutting and peeling steps and the duration and conditions of storage. To preserve the nutritional value of freshly harvested vegetables and fruits it is clearly desirable to minimise the time to blanching and freezing and to cause minimal mechanical damage. The adverse effects of damage are caused by the release of enzymes from intracellular compartments and mixing with potential substrates. For example, vegetables and fruits are important sources of vitamin C, which is vulnerable to oxidation, particularly from the action of the plant enzyme ascorbic acid oxidase (AAO). Pre-freezing processes that cause tissue damage such as bruising, wilting, juicing and pureeing allow AAO, normally sequestered in the extracellular apoplast, to come into contact with ascorbate (concentrated in the cytoplasm) and greatly increase its rate of oxidation (Szeto *et al.*, 2002). Concentrations of ascorbate in spinach may fall to 50% of their initial, pre-harvest level, after 2 days of storage at typical ambient temperature (see Fig. 2.1; Favell, 1998).

2.3.1.3 *Blanching*

Most vegetables and some fruits are customarily blanched before freezing, i.e. they are heated, usually in water or steam, for a variable period in order to inactivate metabolic enzymes. Commercial blanching conditions typically involve heating at 95–100°C for 3–10 minutes, depending on the type and size of material to be blanched. It is mainly to prevent enzyme-mediated oxidation reactions that most vegetables and some fruits are blanched before freezing. Although an additional reason may be to ensure microbiological safety, this objective may be achieved by other means, e.g. use of good agricultural practice and chlorine wash. The advantages of blanching can be illustrated with reference to cauliflower and spinach: if they are frozen without blanching they become unpalatable after only a few months due to the development of 'off' flavours and odours caused primarily by oxidation of membrane lipids. If these vegetables are blanched before freezing they have a storage life of 18–24 months. The conditions of blanching are chosen so as to ensure inactivation of the enzymes responsible for oxidation while minimising loss of sensory quality and nutrients. A great deal of information has been published on losses of labile nutrients during blanching (for review see Clydesdale *et al.*, 1991). Ascorbate is often used as an indicator of potential nutrient loss because of its high solubility, sensitivity to heat and ease of measurement. Typical losses of ascorbate from vegetables during blanching are of the order 5–40% (Favell, 1998; Bender, 1993) and the main mechanism of loss is by leaching into water. In general, nutrient losses are minimised if the raw material is not damaged physically before blanching and if conditions are chosen that keep the temperature, duration of heat exposure and product to water ratio as low as is consistent with denaturing the enzymes responsible for oxidative spoilage.

2.3.1.4 *Frozen storage*

Bender (1993) has summarised the contradictory results of published studies designed to estimate the magnitude of vitamin loss during frozen storage of vegetables and fruits. Even for a particular vegetable, processed and stored under apparently similar conditions, the extent of ascorbate loss has been reported as negligible, or up to 40% after a year of frozen storage (Bender, 1993). As Bender comments, there are many possible sources of experimental variation that may lead to these different conclusions, most notably incomplete denaturation of oxidative enzymes during blanching. Since the review of Bender, no large-scale systematic study addressing this issue has been published. It may be concluded that if vegetables and fruits are adequately blanched and stored at conventional freezer temperatures without undue temperature fluctuations they will still possess nutritionally valuable levels of potentially labile nutrients for a period of at least 12–18 months.

For fruits and other plant foods that are not blanched before freezing there is a potential for ongoing metabolic activity that may influence nutritional value. For example, during frozen storage the growth of ice crystals causes the rupture of cell walls allowing contact of AAO with its substrate ascorbate, leading to oxidation. This process may explain the continuing losses of vitamin C seen during the 4-month frozen storage of orange–carrot juice (Cortés *et al.*, 2005), and 12-month frozen storage of various berry fruits (De Ancos *et al.*, 2000; González *et al.*, 2003; Skrupskis *et al.*, 2003). Pasteurisation (e.g. for 2 min at 85°C), which achieves the same objective of enzyme denaturation as blanching does, inactivates AAO, leading to greater retention of vitamin C. Experiments conducted with mango juice indicated that after a 6-month frozen storage, untreated juice lost 17.4% of its vitamin C, while pasteurised juice lost just 3.7% (Allah and Zaki, 1974). Another example of change in nutritional quality on frozen storage is the change in sugar profile in papaya: sucrose levels decrease while fructose and glucose levels increase, possibly due to continuing activity of invertase in the frozen fruit (Torija *et al.*, 1998).

The choice of whether to blanch fruits and vegetables before freezing depends on a number of factors. Vegetables are generally blanched since they contain high levels of lipoxygenases and other oxidative enzymes (Williams *et al.*, 1986; Barrett and Theerakulkait, 1995), activity of which lead to rapid production of off flavours during frozen storage. Blanching is performed to inactivate these enzymes. Blanching has the advantage of providing a microbiological decontamination step. It also has the advantage of ultimately reducing cooking time for the consumer and thereby increasing convenience. However, blanching is not always a desirable technique as it can destroy the texture and appearance of many kinds of fruits and vegetables. Some fruits, especially the berry fruits, do not ordinarily produce rancid off flavours during frozen storage, which means blanching can be avoided so long as other steps are enough to ensure microbiological safety (e.g. use of good agricultural practice and hygienic handling). However, if blanching is not carried out, enzyme activity will not be destroyed, leading to continuing loss of nutrients such as vitamin C during frozen storage. A way to mitigate this loss, if not stop it fully, is to limit the storage time. This approach will retain maximum possible nutritional value in the absence of blanching. Since properties of fruits and vegetables vary, it is not possible to generalise in terms of the absolute requirements for pre-treatment prior to frozen storage and the subsequent storage stability. Any new vegetable or fruit considered for frozen storage should therefore be tested for its requirements both for blanching and for nutrient stability.

2.3.2 Micro-organisms associated with vegetables and fruits

Vegetables are commonly contaminated with a variety of micro-organisms associated with the soil. Prior to freezing, vegetables are prepared by cleaning and washing, to remove soil and debris, and trimmed and/or cut. Many vegetables are blanched to inactivate enzyme activity during frozen storage and this also impacts on the microbiological quality, inactivating vegetative cells, such as infectious bacteria. Further handling and cooling of the produce may recontaminate vegetables prior to freezing. However, frozen vegetables are not regarded as high-risk foods and are rarely associated with food-borne diseases, since pathogens are unable to proliferate at freezing temperatures and cooking is required prior to consumption. Although contamination with Enterobacteriaceae is commonplace, poor hygienic practice is not the only cause of this since members of this family are commonly associated with plants and plant materials. Moulds are often found on vegetables and can be a good indicator of quality. Lactic acid bacteria are usually present in the highest numbers, which can range between 10^1 and 10^5 CFU/g. Occasionally, high spore loads can be found on crops that have been sprayed with insecticides containing *Bacillus thuringiensis*. This bacterium is closely related to the *B. cereus* subgroup and kills with protein crystals that are toxic to insects. Care should be taken to ensure that insecticide preparations used for crops do not contain enterotoxin-producing spore-formers (Yang *et al.*, 2003) and strains used for this purpose should be screened for this characteristic. Since *B. thuringiensis* cannot easily be distinguished from *B. cereus* and microbiological criteria often include limits for numbers of *B. cereus*, use of such sprays can result in problematic situations, particularly if sprays used contain strains that may produce enterotoxin that is associated with *B. cereus* diarrhoeal syndrome.

Most fruits that are intended to be frozen are picked when ripe, since there is no opportunity for ripening after they have been processed. In addition, frozen fruit is commonly consumed after thawing, without heating. The micro-organisms associated with fruit include moulds, yeasts, protozoa, viruses and bacteria. Fruits, like vegetables, are usually washed (e.g. with chlorinated water) and some are particularly sensitive to handling, and are easily damaged. Some are hand picked and/or sorted and are thereby susceptible to contamination by human pathogens. There are various preparations that are marketed for cleaning/decontaminating fruits and vegetables and these have limited success in removing microbial contaminants and/or inhibiting their growth. Unlike vegetables, blanching of fruits is not commonly applied, except for fruits used in products that undergo further cooking before consumption, such as bakery products. Typical micro-organisms associated with fruits prior to freezing are the same as those present on the raw material at harvesting and also include others introduced through further handling. The pathogens are listed in Table 2.1.

2.3.3 Effects of freezing on micro-organisms associated with vegetables and fruits

The main group of micro-organisms associated with frozen vegetables – lactic acid bacteria – remain largely unaffected by freezing, frozen storage and thawing. Bacterial spores also remain unaffected. The most significant micro-organisms in frozen fruits in terms of number are moulds and yeasts. A key factor influencing the development of flora is pH and in many fruits, this selects for yeasts and moulds that are better able to tolerate the low-pH conditions and presence of organic acids. Bacteria such as infectious pathogens, if present, die off relatively rapidly. Epidemiological studies of food-borne disease outbreaks have shown that viruses survive in frozen fruits. One such study has been recently reported in Denmark, where frozen raspberries contaminated with Noroviruses sourced from Poland were identified as the most likely vehicle. In such situations, contamination is likely to arise from infected handlers and poor hygiene.

2.4 DAIRY

2.4.1 Nutritional aspects

Milk and cheese are important constituents of the diet. Their major contribution in nutritional terms is as a source of calcium and protein.

Milk is a complete food that is able to sustain newborn mammals during their initial growth phase. Milk therefore contains an extensive range of minerals and vitamins, fats, sugars, proteins and functional proteins including immunoglobulins, α-lactalbumin, β-lactoglobulin, lactoferrin, lactoperoxidase, and a variety of growth factors (McIntosh *et al.*, 1998; also see reviews in Shortt and O'Brien, 2004). Though commercially frozen milk is not an important commodity, freezing is used to store human milk to help vulnerable newborns. There have therefore been a number of investigations into the effect of the freezing process and storage on the nutrient content of human milk. The B vitamins biotin, niacin, and pantothenic acid were found to be virtually unchanged up to 3 months of frozen storage, and similarly lysozyme and protease were unaffected (Friend *et al.*, 1983). However, lactoperoxidase activity was found to decrease significantly while lipase activity increased slightly during 3 months of storage. Friend *et al.* (1983) found that the increase in lipase activity correlated with an increase in the level of unesterified fatty acids, indicating that lipolysis was occurring. This effect was most pronounced between 2 and 3 months of frozen storage and may also be responsible for the decrease in the fat-soluble vitamins retinol and α-tocopherol observed in bovine milk frozen for longer time periods (Vidal-Valverde *et al.*, 1992, 1993). Length of time in frozen storage appears to be critical for nutrient retention in human milk: Buss *et al.* (2001) observed that approximately a third of vitamin C content was lost following 1 month of storage, and two-thirds lost following storage for 2 months. They conclude that vitamin C losses were due partly to lactoperoxidase activity. Immunological components, including cells and proteins, in human milk are very important for maintaining the health of newborns. Though macrophages and lymphocytes reduce during freezing, immunologically important proteins were found to be unaffected, though there may be denaturation effects on thawing (Lawrence, 1999). The effect of pasteurisation followed by freezing on nutrients in human milk indicated that though calcium, phosphorous, iron, copper and vitamin A were not affected, there was a potential decrease in the bioavailability of zinc due to a redistribution with a decrease in the whey fraction and an increase in the fat fraction (Góes *et al.*, 2002). The main cause of the zinc redistribution was thought to be the pasteurisation process rather than the freezing process.

Cheese is not generally supplied in a frozen state and freezing in the cheese-making industry is used mainly as a method to maintain starter cultures (for example, see Gilliland, 1981). Freezing of sheep milk cheese has been studied as a way to avoid the seasonal nature of the product but these studies have focussed on changes in sensory properties rather than on nutritional quality. Nine-month frozen storage of the Los Pedroches' brand cheese and of a Manchego-type cheese showed that proteolysis continued slowly during this time, resulting in higher amino acid contents (Tejada *et al.*, 2002; Prados *et al.*, 2006). Similar effects on proteolysis have also been noted for Port Salut Argentino cheese (Verdini *et al.*, 2005). Frozen storage of goats' cheese has also been found to lead to changes in organic acids, with increases in butyric, lactic, propionic, tartaric, and uric acids noted (Park *et al.*, 2006).

2.4.2 Micro-organisms associated with dairy products

The main micro-organisms of concern in dairy produce such as milk and cheese are the spoilage bacteria and the pathogenic bacteria. These arise from contamination of raw materials during milking, from the animals themselves (e.g. flora on the udder or in faecal material), from contaminated equipment/handlers and from the milking environment (e.g. spore-forming bacteria). Milk from cows suffering mastitis or other diseases may contain bacteria not normally found in milk from healthy animals, including *S. aureus*, pyrogenic streptococci and actinomycetes, *Brucella* spp., mycobacteria, salmonellae, *L. monocytogenes*, *Coxiella burnetii*, and some viruses, including foot-and-mouth disease virus (FMDV). Most milk is pasteurised following transport/short-term storage at temperatures/times (e.g. 72°C for 15 s) that inactivate vegetative cells, including infectious bacterial pathogens normally associated with milk, such as salmonellae, *L. monocytogenes*, pathogenic *E. coli*, campylobacters and *Yersinia enterocolitica*. The same type of inactivation is effected by holding at lower temperatures for longer times (e.g. 68°C for 30 min) whereas ultra-high-temperature (UHT) treatment typically uses temperatures of around 149°C for 2 seconds and this will inactivate mesophilic spore-forming bacteria, enabling such milk to be stored at ambient temperature for long periods. A lower temperature/time combination, such as 138°C for 10 seconds, is used for extended shelf-life (referred to as ESL) milk, stored at chill for periods in excess of 4 weeks. In these products, psychrotrophic spore-formers are inactivated but products are not commercially sterile. Some milk undergoes micro-filtration as an alternative to pasteurisation, to remove infectious agents. Some raw milk may contain mycotoxins and is thought to result from animals ingesting contaminated feed.

The spoilage micro-organisms associated with milk include psychrotrophic bacteria such as pseudomonads, *Flavobacterium* spp., *Chromobacterium* spp., *Alcaligenes* spp. and thermophilic spore-forming bacteria that survive pasteurisation. Other ingredients used in frozen desserts include butter, sugar chocolate, fruits, nuts and eggs. These are often added after milk has been pasteurised and assurance of their safety is commonly provided by the suppliers of these materials, some of which (e.g. eggs, chocolate) have been known to be associated with pathogens such as *Salmonella* or their toxins, such as *S. aureus* enterotoxin. Some microbiological contaminants, such as *Listeria monocytogenes*, are associated with wet equipment and manufacturing environments and this poses a risk in the manufacture of ice cream. Environmental monitoring and effective cleaning and disinfection are critical for the safety of these products, particularly when considering post-process handling, e.g. freezing in blast freezers and hardening tunnels.

2.4.3 Effects of freezing on micro-organisms associated with dairy products

The fate of different micro-organisms associated with dairy products generally follows the same trends as already described for other foods. Dairy products that are frozen include ice cream, yoghurts, ice milk, sherbet and other desserts. Ice cream may contain vegetable or non-dairy animal fats and in that sense, are not 'dairy' products, but are considered in this section for convenience. The thermal processes required for pasteurisation of ice cream are higher than those specified for milk since they contain components, such as sugar, that are known to increase the thermal tolerance of micro-organisms. Provided cooling and freezing are relatively rapid, no growth of survivors can take place and there would be no potential for spoilage of product, the major concern considering the most likely surviving micro-organisms. The normal temperature of storage of ice cream is ca. −25°C in factory cold-stores and ca. −18°C in distribution and retail, providing a typical shelf-life of 2 years. There have been occasional outbreaks of diseases linked to frozen dessert products such as ice cream, but these are due to post-process contamination or inadequate pasteurisation, e.g. in home-made ice cream. As for other frozen products, infectious agents may survive longer at freezing temperatures.

There are groups of dairy products that are first fermented, then frozen or supplemented with live micro-organisms and then frozen. Fermented milks have existed for centuries; reports on the health benefits of these products were published as early as the beginning of the twentieth century. Yoghurts are typically fermented with defined species of lactic acid bacteria and more recently there has been a plethora of new product innovations launched in the market that include live bacteria, such as lactic acid bacteria. Frozen products of interest include ice cream (fermented or simply containing added bacteria), frozen yoghurt, yoghurt ice cream and soy desserts. The various health benefits claimed include establishment of healthy gut flora, inhibition of pathogens, improvement of resorption of food components, such as lactose, decreasing prevalence of allergy in susceptible individuals and enhancing the immune system. These products are commonly considered to be beneficial for health and have been recently reviewed by Parvez *et al.* (2006). The bacterial strains used in probiotic (Greek 'for life') products are generally considered as safe and there is no evidence of them causing infections in humans but this possibility remains and therefore requires careful assessment of new strains prior to use and continuous monitoring for any adverse effects.

Survival of probiotic bacteria in frozen foods is extremely important, if they are to retain their claimed functionality, although some beneficial effects of dead cells (e.g. enhancement of immune response) have also been reported. For that reason, a number of recent studies have investigated survival of micro-organisms used as probiotics in an effort to identify strains and/or conditions that lead to better survival during frozen storage. These include organisms such as *Streptococcus thermophilus* (Hong and Marshall, 2001; Fonseca *et al.*, 2001), *L. delbruekii* (Fonseca *et al.*, 2001), *Bifidobacterium lactis* (Heenan *et al.*, 2004; Haynes and Playne, 2002), *L. paracasei* (Heenan *et al.*, 2004; Haynes and Playne, 2002), *L. acidophilus*

(Haynes and Playne, 2002; Godward *et al.*, 2000; Heenan *et al.*, 2004; Wang *et al.*, 2005a; Salem *et al.*, 2006) and *L. rhamnosus* (Heenan *et al.*, 2004; Salem *et al.*, 2006).

2.5 BAKERY

In terms of bakery products, the most important nutrient for consideration is the fortification of flour with folic acid. Folic acid is linked to the prevention of neural tube defects (NTD) in the developing foetus and women of child-bearing age are advised to take supplements to increase intake to a recommended level of 0.4 mg per day. However, since many pregnancies are unplanned, voluntary advice in countries such as the UK has not led to a significant decrease in NTD. Mandatory fortification of flour in the US and Canada since 1998 has led to a reported decrease in the incidence of NTD of 27–50% and the UK is currently considering making fortification of flour mandatory (reviewed in Eichholzer *et al.*, 2006). Folic acid, folates, and folacin are all terms used for a family of compounds with related biological activity; they contain one or more linked molecules of glutamic acid. Commercially available folic acid contains a single glutamic group and has been found to be moderately stable to heat and atmospheric oxygen (Ottaway, 2002). Folate added to flour has been shown to be stable for up to a year at temperatures including freezing (Ranhotra and Keagy, 1995). Frozen storage has also been found to maintain levels of naturally occurring folate found in foods such as strawberries (Vahteristo *et al.*, 1998). However, subsequent heating of folate-containing foods can lead to losses of around 20% (Hoppner *et al.*, 1973; Morgan, 1996).

2.5.1 Micro-organisms associated with bakery products

The raw materials used in bakery products include flour, water, milk, butter, salt, eggs and sugar and occasionally, processed products such as preserved fruits. The flora of flour is more diverse than the wheat from which the flour is made, resulting from contamination during processing. The micro-organisms associated with flour are primarily psychrotrophs, 'flat sour' bacteria (organisms producing acid souring without significant gas production), thermophilic spore-forming bacteria, 'rope' bacilli, moulds and yeasts. The main safety concern is mycotoxin formation and this is controlled by maintaining the moisture content below 12%. Of less concern is contamination with salmonellae and *Bacillus cereus*. Frozen bakery products include par-baked doughs that are intended to be cooked by the consumer. Their processing usually ensures destruction of vegetative cells but bacterial spores remain unaffected, as do some heat-stable metabolites, such as mycotoxins, staphylococcal enterotoxins and emetic toxin of *B. cereus*.

2.5.2 Effects of freezing on micro-organisms associated with bakery products

The micro-organisms remaining in frozen bakery products, such as bacterial spores and other micro-organisms that result from handling product prior to freezing generally remain unaffected during freezing and frozen storage. Since the majority of these products are cooked by the consumer, they pose little risk in terms of consumer safety and they rarely spoil, because of the way they are prepared, which leaves little opportunity for growth to occur. The main area of concern with these products is handling/storage of raw materials and intermediate products prior to factory cooking, and the speed of freezing.

2.6 FUTURE TRENDS

A preliminary study in our own technology programme (previously unpublished) is summarised below to illustrate the future trend of increasing the content of nutrients and health ingredients in frozen food.

2.6.1 Opportunities for design in 'ice cream' formulations and recipes

Ice cream is often considered to address the consumer needs of fun, indulgence and refreshment (Clarke, 2004) rather than as a worthwhile contributor to nutrition. However, the recent drive towards healthier foods has stimulated research into improving the nutrient content of ice cream and the consumer of the future will view ice cream in a very different way. The fact that ice cream is frozen may be less of a driver for providing a cold (and therefore refreshing) treat; and rather more to do with the convenience of having a ready supply (i.e. stably stored in terms of microbiological safety and nutrient content) of portioned food. Furthermore, in contrast to frozen fish, meat, and vegetables, which are 'harvested', ice cream is 'designed'. In other words, there is a very large range of ingredient profiles that can be used to deliver a particular taste, texture, or eating sensation. For example, the fat content can vary from 15% in super-premium ice cream to virtually zero in sorbets and water ices; radically different ingredients can be added from Belgian chocolate to fruit. Whereas there is a notable nutrient value of all dairy products (as discussed in the previous section), the 'designer' aspect of ice cream recipes and ingredient profiles affords an exciting opportunity for adding many more nutrients than is the norm today (including those not traditionally associated with ice cream). In principle, almost any nutrient could be added (subject to constraints from the regulatory authorities). Finally, since ice cream can also be designed to have a particular portion-size, a physiologically significant quantity of each nutrient can be added per portion.

2.6.2 Stability of nutrients during ice cream processing and storage

This is a new technical area because ice cream has traditionally been seen as an occasional sweet treat, rather than a 'pleasurable food'. Moreover, the frequency of ice cream consumption has been too low to have a significant impact (either positive or negative) on nutrition. For example, in many European countries, the average per capita consumption of ice cream is less than 10 litres per annum (Clarke, 2004). Therefore, little attention has been paid to the nutrient content of ice cream either when freshly prepared, during processing, or after storage. The stability of most nutrients during frozen storage can reasonably be assumed to be good in the frozen state (because the low temperature slows the chemistry of deterioration), but the stability of nutrients during the 'freezing process' (i.e. ice cream processing) is almost completely unknown. The industrial process for producing ice cream comprises a number of unit steps including: homogenisation, pasteurisation, 'ageing', freezing and aeration, and 'hardening'. Each of these processing steps could potentially damage nutrient quality. Particularly relevant to this chapter is the freezing process itself, which is a dynamic and continuous process.

The most commonly used 'freezer' is more correctly described as a scrape surface heat exchanger (SSHE). This simultaneously aerates and freezes the ice cream mix and the resultant ice cream is typically extruded at about −5°C. The introduction of air at this stage could also promote the oxidative deterioration of susceptible nutrients. After extrusion, the ice cream is passed through a 'hardening tunnel' (an enclosed chamber where extremely cold air is blown over the ice cream) before storage (typically around −25°C). It is also important to note that there are several different processes that are used in the ice cream industry. For example, there is a process known as 'quiescent-freezing' where the mix (either aerated or non-aerated) is frozen statically; typically in moulds that are immersed in a brine bath (calcium chloride solution) or a mixture of water and ethylene glycol at about −40°C. Each of these different processes has the potential for impacting on nutrient quality in a different way, depending on the processing temperature and the amount of aeration. For more information on ice cream processing, the reader is referred to Clarke (2004) and to Marshall *et al.* (2003).

2.6.3 Preliminary investigation of nutrient stability during freezing and storage of ice cream

By way of example, we have investigated how vitamin D and anti-oxidants were affected by the freezing process and by storage. We selected vitamin D because it is low in many diets, particularly in the winter (Lamberg-Allardt *et al.*, 2001; Diamond *et al.*, 2005). Moreover, there are very few commonly available foods or food ingredients that are rich in vitamin D (cod liver oil being one notable exception), therefore the option of adding vitamin D to a 'designed' food like ice cream is attractive. We selected anti-oxidants because they are potentially beneficial for reducing the risk of cardiovascular disease (Voutilainen *et al.*, 2006); also a high intake of some anti-oxidants has been associated with healthy skin, particularly in the context of reduced ultraviolet light-induced damage (Stahl *et al.*, 2000; Boelsma *et al.*, 2001). Even though it is possible to get anti-oxidants from fruits and vegetables, quite large amounts have to be consumed to significantly increase the anti-oxidant level in blood plasma. Therefore, it is attractive to add anti-oxidants (in the form of refined ingredients) to a 'designed' product. Moreover, there is a patent in the field (Tijburg and Weststrate, 1997) that describes a food containing a cocktail of anti-oxidants such as vitamin E, lycopene, and β-carotene.

To assess the stability of these nutrients in ice cream, pilot scale samples were prepared with nutritionally significant quantities of either vitamin D or anti-oxidants (lycopene and β-carotene). In brief, a SSHE (Technohoy MF75) was used to prepare aerated ice cream (50% air phase by volume). Aerated ice cream was prepared because, as noted above, many chemical reactions that lead to deterioration of nutrients are oxidative. Therefore, this experimental approach was seen as a stringent test case. Nutrient content was measured at several stages during processing and storage (in the pre-mix, post-homogenisation, post-freezing, and after storage for 1 week and 4 weeks at −20°C). It was found that nutrient stability was very high both during ice cream processing and during subsequent storage (see Figs. 2.2 and 2.3).

Our preliminary study resulted in an interesting finding. It suggests that the processing technology used for industrial ice cream manufacture may well be suitable for designing frozen portioned-foods that are more nutritious than is the norm for ice cream today. We propose that this is (at least partly) because the process readily lends itself to the addition of nutrients that (according to our data) are robust enough to survive both aeration and freezing. This is in contrast to the processing of frozen fruits and vegetables where blanching and/or frozen storage can lead to significant loss of nutrients.

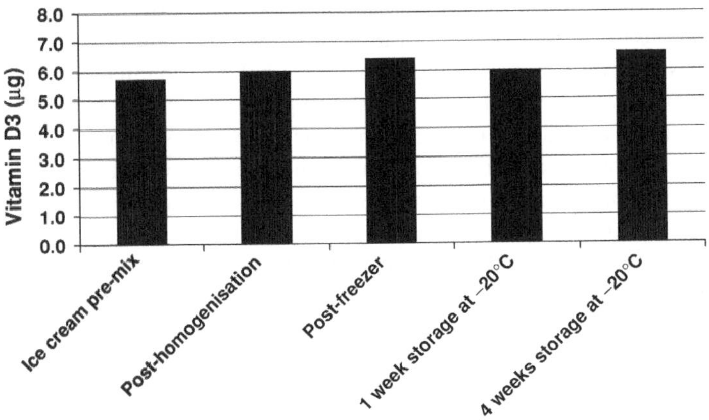

Fig. 2.2 Stability of vitamin D3 during ice cream processing and storage. Amounts are given in micrograms (µg) per 60 g portion of ice cream.

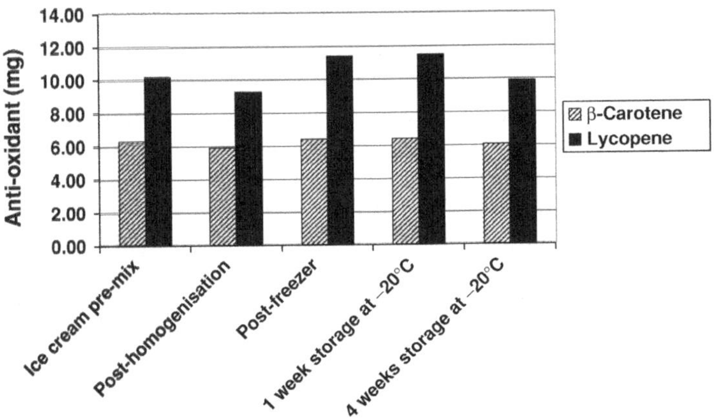

Fig. 2.3 Stability of anti-oxidants during ice cream processing and storage. Amounts of anti-oxidant are given in milligrams (mg) per 60 g portion of ice cream. Values for β-carotene are given in hatched bars, values for lycopene are given in solid bars.

2.7 CONCLUSIONS

2.7.1 Microbiological aspects

It is clear that freezing and frozen storage cannot be relied upon to completely destroy all micro-organisms in foods. However, some micro-organisms, such as protozoa, multi-cellular parasites and less robust bacteria, such as *Campylobacter* are more easily damaged and significant reductions in numbers can be seen in these organisms, many of which are capable of causing foodborne disease if present even in low numbers. These cidal effects are utilised in HACCP plans for some foods, such as beef and pork products, to eliminate the larvae of helminths and adult forms of multi-cellular parasites. In some foods, such as fish intended to be eaten raw, the use of freezing is mandated by regulatory authorities (Schantz, 1989).

According to Stern *et al.* (2003) one of the major factors responsible for the reduction in cases of poultry-borne campylobacteriosis in Iceland was the introduction of a programme for freezing carcasses from *Campylobacter*-positive flocks. Since then Iceland has mandated a policy for the continued use of this procedure. Other countries in Europe, such as Denmark and Norway, have also introduced controls for *Campylobacter* spp. that include freezing of poultry carcasses. To be able to use freezing as a control for unwanted micro-organisms in foods requires good data, including validation 'in-product' using sensitive methods of recovery/detection. A number of areas for future research have been identified by Archer (2004). While these were considered in relation to improved control of food-borne pathogens, many would also provide better understanding of factors that may be used to enhance survival of beneficial micro-organisms in frozen products, such as probiotics. Enhancement of cidal effects against bacteria has been demonstrated by a combination of rapid freezing and slow thawing (Bogh-Sorenson, 2000). Other recent developments that have the potential to impact on microbiological activity in products during freezing/frozen storage include:

- Pre-freeze treatment with preservatives applied to the food (Saurel, 2002), such as potassium lactate and sodium diacetate, that has been shown to extend the lag time of *L. monocytogenes* (Yoon *et al.*, 2004);
- Infusion of carbohydrates, which can increase the glass transition temperature above storage temperature, increasing shelf-life;
- Application of high-pressure (e.g. 40 MPa) freezing, inhibiting enzyme activity in frozen fish;
- Application of high-pressure thawing, allowing enhanced kill or growth inhibition of micro-organisms (Haack and Heinz, 2000).

Combinations of factors that have been shown to increase the survival of beneficial bacteria include addition of inulin to probiotic-fermented ice cream (Akin, 2005). Some studies have also provided a better understanding of the factors that influence survival, such as production of cold-induced proteins (Wang *et al.*, 2005b).

2.7.2 Nutritional aspects

It is increasingly recognised that consumption of vegetables and fruits significantly reduces the risk of some cancers, of cardiovascular disease and many age-related degenerative diseases (Goldberg, 2003). This awareness has given rise to public health campaigns in many countries to increase the amount of vegetables and fruits consumed. Awareness of the importance of vegetables and fruit consumption naturally leads to the question whether preserved and processed produce can be expected to have the same health benefits as raw produce. Because of the wealth of data available showing the excellent nutrient retention of frozen vegetables and fruits compared to fresh equivalents, the US Food and Drug Administration formally announced in 1998 that 'because frozen fruits and vegetable products are nutritionally comparable to the raw versions they would likely have the same inherent beneficial effects as the raw version'. Although it is by no means certain, it appears likely that the beneficial effects of consuming vegetables and fruits are not just a consequence of their recognised nutrients. A large number of potentially beneficial compounds, the so-called phytonutrients, or non-nutrient phytochemicals, are found in vegetables and fruits. It is not yet clear which particular compounds, or even which group of phytochemicals could be responsible for health benefits, but if and when the protective agents are identified, it will be necessary to ascertain the effects of freezing and associated processes on their retention in frozen vegetables and fruits.

In the past, the hallmark of frozen foods has been microbiological safety. The frozen food industry has focussed on the provision of microbiologically safe, convenient, and good value foods. In the future, nutritional qualities will play an ever-increasing role, ranking alongside microbiological safety in its importance to the industry.

The continuing challenge for the frozen food industry is to obtain a balance between acceptable sensory characteristics, convenience, microbiological safety and nutritional quality.

Acknowledgements

The authors gratefully acknowledge the expertise of their colleagues Ian D. Smith and Beate Nicol who carried out the analytical measurements for vitamin D and anti-oxidants.

REFERENCES

Abu Ruwaida, A.S., Sawaya, W.N., Hussein, A.S., Baroon, Z. and Khalafawi, S. (1996). Incidence of microorganisms affecting shelf-life, quality and safety of locally-produced frozen broiler chickens in Kuwait. *Arab Gulf Journal of Scientific Research* 14(3), 609–628.

Akin, M.S. (2005). Effects of inulin and different sugar levels on viability of probiotic bacteria and the physical and sensory characteristics of probiotic fermented ice-cream. *Milchwissenschaft Milk Science International* 60(3), 297–301.

Allah, M.A.A. and Zaki, M.S.A. (1974). Preservation of mango juice by freezing and canning. *Nahrung* 18(2), 207–216.

Archer, D.L. (2004). Freezing: an underutilized food safety technology? *International Journal of Food Microbiology* 90(2), 127–138.

Barrell, R.A.E. (1988). The survival and recovery of *Salmonella typhimurium* phage type U285 in frozen meats and tryptone soya yeast extract broth. *International Journal of Food Microbiology* 6(4), 309–316.

Barrett, D.M. and Theerakulkait (1995). Quality indicators in blanched, frozen, stored vegetables. *Food Technology* January, 62–65.

Bayles, D.O., Tunick, M.H., Foglia, T.A. and Miller, A.J. (2000). Cold shock and its effect on ribosomes and thermal tolerance in *Listeria monocytogenes*. *Applied and Environmental Microbiology* 66(10), 4351–4355.

Bender, A.E. (1978). *Food Processing and Nutrition*. Academic Press.

Bender, A.E. (1993). *Nutritional aspects of frozen foods in frozen food technology*. In: C.P. Mallett, ed. Blackie Academic and Professional, pp. 123–140.

Boelsma, E., Hendriks, H.F.J. and Roza, L. (2001). Nutritional skin care: health effects of micronutrients and fatty acids. *American Journal of Clinical Nutrition* 73, 853–864.

Bogh-Sorenson, L. (2000). Maintaining safety in the cold chain. In: C. Kennedy, ed. *Managing Frozen Foods*. Cambridge: Woodhead Publishing Ltd.

Bollman, J., Ismond, A. and Blank, G. (2001). Survivalo of *Escherichia coli* O157:H7 in frozen foods: impact of the cold shock response. *International Journal of Food Microbiology* 64(1–2), 127–138.

Buss, I.H., McGill, F., Darlow, B.A. and Winterbourn, C.C. (2001). Vitamin C is reduced in human milk after storage. *Acta Paediatric* 90, 813–815.

Cano, M.P. and Marin, M.A. (1995). Effects of freezing preservation on dietary fibre content of mango (*Mangifera indica* L.) fruit. *European Journal of Clinical Nutrition* 49(Suppl 3), S257–S260.

Chan, K.F., Tran, H.L., Kanenaka, R.Y. and Kathariou, S. (2001). Survival of clinical and poultry-derived isolates of *Campylobacter jejuni* at a low temperature (4°C). *Applied and Environmental Microbiology* 67(9), 4186–4191.

Chang, V.P., Mills, E.W. and Cutter, C.N. (2003). Comparison of recovery methods for freeze-injured *Listeria monocytogenes*, *Salmonella typhimurium*, and *Campylobacter coli*, in cell suspensions and associated with pork surfaces. *Journal of Food Protection* 66(5), 798–803.

Clarke, C. (2004). *The Science of Ice Cream.* Cambridge: The Royal Society of Chemistry.

Clydesdale, F.M., Ho, C.T., Lee, C.Y., Mondy, N.J. and Shewfelt, R.L. (1991). Effects of post-harvest treatment and chemical interaction on the bioavailability of ascorbic acid thiamine, vitamin A carotenoids and minerals. *Critical Reviews in Food Science in Food Science and Nutrition* 30, 599–638.

Dawson, D. (2005). Foodborne protozoan parasites. *International Journal of Food Microbiology* 103(2), 207–227.

Cortés, C., Esteve, M.J., Frígola, A. and Torregrosa, F. (2005). Changes in carotenoids including geometrical isomers and ascorbic acid content in orange-carrot juice during frozen storage. *European Food Research and Technology* 221, 125–131.

De Ancos, B., González, E.M. and Cano, M.P. (2000). Ellagic acid, vitamin C, and total phenolic contents and radical scavenging capacity affected by freezing and frozen storage in raspberry fruit. *Journal of Agriculture and Food Chemistry* 48, 4565–4570.

Derzelle, S., Hallet, B., Ferain, T., Delcour, J. and Hols, P. (2003). Improved adaptation to cold-shock, stationary-phase, and freezing stresses in *Lactobacillus plantarum* overproducing cold-shock proteins. *Applied and Environmental Microbiology* 69(7), 4285–4290.

Diamond, T.H., Eisman, J.A., Mason, R.S., Nowson, C.A., Pasco, J.A., Sambrook, P.N. and Wark, J.D. (2005). Vitamin D and adult bone health in Australia and New Zealand: a position statement. *Medical Journal of Australia* 182(6), 281–285.

Dulavik, B., Sorensen, N.K., Barstad, H., Horvil, O. and Olsen, R.L. (1998). Oxidative stability of frozen light and dark muscles of Saithe (*Pollachius virens*). *Journal of Food Lipids* 5, 233–245.

Dumont, F., Mraechal, P.A. and Gervais, P. (2003). Influence of cooling rate on *Saccharomyces cerevisiae* destruction during freezing: unexpected viability at ultra-rapid cooling rates. *Cryobiology* 46(1), 33–42.

Engler, P.P. and Bowers, J.A. (1976). B-vitamin retention in meat during storage and preparation. *Journal of the American Dietetics Association* 69, 253–257.

Eichholzer, M., Tönz, O. and Zimmermann, R. (2006). Folic acid: a public-health challenge. *The Lancet* 367, 1352–1361.

Favell, D.J. (1998). A comparison of the vitamin C content of fresh and frozen vegetables. *Food Chemistry* 62, 59–64.

Fennema, O. (1975). Effects of freeze preservation on nutrients. In: R.S. Harris and E. Karmas, eds. *Nutritional Evaluation of Food Processing*, 2nd edn. Westport, CT: AVI Publishing.

Fonseca, F., Beal, C. and Corrieu, G. (2001). Operating conditions that affect the resistance of lactic acid bacteria to freezing and frozen storage. *Cryobiology* 43(3), 189–198.

Friend, B.A., Shahani, K.M., Long, C.A. and Vaughn, L.A. (1983). The effect of processing and storage on key enzymes, B vitamins, and lipids of mature human milk I: evaluation of fresh samples and effects of freezing and frozen storage. *Pediatric Research* 17, 61–64.

Fuji, T., Kurihara, K. and Okuzumi, M. (1994). Viability and histidine-decarboxylase activity of halophilic histamine-forming bacteria during frozen storage. *Journal of Food Protection* 57(7), 611–613.

Georgsson, F., Þorkesson, Á.E., Geirsdóttir, M., Reiersen, J., Stern, N. (2006). The influence of freezing and duration of storage on *Campylobacter* and indicator bacteria in broiler carcasses. *Food Microbiology* 23(7), 677–683.

Gilliland, S.E. (1981). Factors affecting cottage cheese starter cultures. *Cultured Dairy Products Journal* 16(1), 15–17.

Godward, G., Sultana, K., Kailasapathy, K., Peiris, P., Arumugaswamy, R. and Reynolds, N. (2000). The importance of strain selection on the variability and survival of probiotic bacteria in dairy foods. *Milchwissenschaft* 55(8), 441–445.

Góes, H.C.A., Torres, A.G., Donangelo, C.M. and Trugo, N.M.F. (2002). Nutrient composition of banked human milk in Brazil and influence of processing on zinc distribution in milk fractions. *Nutrition* 18, 590–594.

Goldberg, G. (2003). *Plants: Diet and Health.* Blackwell Publishing.

González, E.M., de Ancos, B. and Cano, M.P. (2003). Relation between bioactive compounds and free radical-scavenging capacity in berry fruits during frozen storage. *Journal of the Science of Food and Agriculture* 83, 722–726.

Haack, E. and Heinz, V. (2000). Improvement of food safety by high pressure processing. II Studies on use in the meat industry. *Fleischwirtshaft* 81(6), 38–41.

Hansen, R.K., Zhai, S., Skepper, J.N., Johnston, M.D., Alpar, H.O. and Slater, N.K.H. (2005). Mechanisms of inactivation of HSV-2 during storage in frozen and lyophilized forms. *Biotechnology Progress* 21(3), 911–917.

Haynes, I.N. and Playne, M.J. (2002). Survival of probiotic cultures in low-fat ice-cream. *Australian Journal of Dairy Technology* 57(1), 10–14.

Heenan, C.N., Adams, M.C., Hosken, R.W. and Fleet, G.H. (2004). Survival and sensory acceptability of probiotic microorganisms in a nonfermented frozen vegetarian dessert. *Lebensmittel Wissenschaft und Technologie* 37(4), 461–466.

Hong, S.H. and Marshall, R.T. (2001). Natural exopolysaccharides enhance survival of lactic acid bacteria in frozen dairy desserts. *Journal of Dairy Science* 84(6), 1367–1364.

Hoppner, K., Lampi, B. and Perrin, D.E. (1973). Folacin activity of frozen convenience foods. *Journal of the American Dietetic Association* 63(5), 536–539.

Humphrey, T.J. and Cruikshank, J.G. (1986). Injury and recovery in freeze or heat-damaged *Campylobacter jejuni*. *Letters in Applied Microbiology* 3, 81–84.

Jeffreys, A.G., Hark, K.M., Steffan, R.J., Foster, J.W. and Bej, A.K. (1998). Growth, survival and characterisation of CspA in *Salmonella enteritidis* following cold shock. *Current Microbiology* 36(1), 29–35.

Kawahara, H. (2002). The structures and functions of ice crystal-controlling proteins from bacteria. *Journal of Bioscience and Bioengineering* 94(6), 492–496.

Lamberg-Allardt, C.J.E., Outila, T.A., Karkkainen, M.U.M., Rita, H.J. and Valsta, L.M. (2001). Vitamin D deficiency and bone health in healthy adults in Finland: could this be a concern in other parts of Europe?. *Journal of Bone Health and Mineral Research* 16(11), 2066–2073.

Larkin, J. and Stokes, J.L. (1968). Growth of psychrophilic microorganisms at subzero temperatures. *Canadian Journal of Microbiology* 14, 97–101.

Lawrence, R.A. (1999). Storage of human milk and the influence of procedures on immunological components of human milk. *Acta Paediatric* 88, 14–18.

Lee, F.A., Brooks, R.F., Pearson, A.M., Miller, J.I. and Wanderstock, J. (1954). Effects of freezing rate on meat; appearance palatability and vitamin content. *Journal of the American Dietetic Association* 30, 351–354.

Lee, S.K. and Kader, A.A. (2000). Preharvest and postharvest factors influencing vitamin C content of horticultural crops. *Postharvest Biology and Technology* 20, 207–220.

Lim, Y.H., Hamdy, M.K. and Toledo, R.T. (2003). Combined effects of ionizing-irradiation and different environments on *Clostridium botulinum* type E spores. *International Journal of Food Microbiology* 89(2–3), 251–263.

Lu, X., Cho, D., Hall, H., Rowe, T., Sung, H., Kim, W., Kang, C., Mo, I., Cox, N., Klimov, A. and Katz, J. (2003). Pathogenicity and antigenicity of a new influenza A (H5N1) virus isolated from duck meat. *Journal of Medical Virology* 69(4), 553–559.

Marshall, R.T., Goff, H.D. and Hartel, R.W. (2003). *Ice Cream*, 6th edn. New York: Kluwer Academic/Plenum Publishers.

McClure, P.J. (2002). Microbiological hazard identification in the meat industry. In: M. Brown, ed. *HACCP in the Meat Industry.* Cambridge, England: Woodhead Publishing Ltd.

McIntosh, G.H., Royle, P.J., Le Leu, R.K., Regester, G.O., Johnson, M.A., Grinsted, R.L., Kenward, R.S. and Smithers, G.W. (1998). Whey proteins as functional food ingredients. *International Dairy Journal* 8, 425–434.

Mihoub, F., Royle, P.J., Le Leu, R.K., Regester, G.O., Johnson, M.A., Grinsted, R.L., Kenward, R.S. and Smithers, G.W. (2003). Cold adaptation of *Escherichia coli*: microbiological and proteomic approaches. *International Journal of Food Microbiology* 89(2–3), 171–184.

Mikkelsen, K., Rasmussen, E.L. and Zinck, O. (1984). Retention of vitamins B1, B2, and B6 in frozen meats. In: P. Zeuthen, ed. *Thermal Processing and Quality of Foods*. Elsevier Applied Science, pp. 777–781.

Moorhead, S.M. and Dykes, G.A. (2002). Survival of *Campylobacter jejuni* on beef trimmings during freezing and frozen storage. *Letters in Applied Microbiology* 34, 72–76.

Morgan, W. (1996). Effects of processing and preparation of foods on folate content. *Australian Journal of Nutrition and Dietetics* 53, S31–S35.

Mullen, W., Stewart, A.J., Lean, M.E.J., Gardner, P., Duthie, G.G. and Crozier, A. (2002). Effect of freezing and storage on the phenolics, ellagitannins, flavonoids, and antioxidant capacity of red raspberries. *Journal of Agriculture and Food Chemistry* 50, 5197–5201.

Murrell, K.D., Fayer, R. and Dubey, J.P. (1986). Parasitic organisms. In: A.M. Pearson and T.R. Dutson, eds. *Advances in Meat Research*, Vol. 2. Westport,CT: AVI Publishing, pp. 311–377.

Nichols, R. and Smith, H. (2002). Parasites: *Cryptosporidium, Giardia* and *Cyclospora* as foodborne pathogens. In: C. de W. Blackburn and P.J. McClure, eds. *Foodborne Pathogens Hazards, Risk Analysis and Control.* Cambridge: Woodhead Publishing Ltd., pp. 453–478.

Nyman, M. (1995). Effects of processing on dietary fibre content of vegetables. *European Journal of Clinical Nutrition* 49, S215–S218.

Ottaway, P.B. (2002). The stability of vitamins during food processing. In: C.J.K. Henry and C. Chapman, eds. *The Nutrition Handbook for Food Processors*, 1st edn. CRC Press, Woodhead Publishing Ltd., pp. 247–264.

Park, J.I., Grant, C.M., Davies, M.J. and Dawes, I.W. (1998). The cytoplasmic Cu, Zn, superoxide dismutase of *Saccharomyces cerevisiae* is required for resistance to freeze-thaw stress: generation of free radicals during freezing and thawing. *Journal of Biological Chemistry* 273(36), 22921–22928.

Park, S.F. (2002). The physiology of *Campylobacter* species and its relevance to their role as food borne pathogens. *International Journal of Food Microbiology* 74(3), 177–188.

Park, Y.W., Lee, J.H. and Lee, S.J. (2006). Effects of frozen and refrigerated storage on organic acid profiles of goat milk plain soft and Monterey Jack cheeses. *Journal of Dairy Science* 89, 862–871.

Parvez, S., Malik, K.A., Kang, S.A. and Kim, H.-Y. (2006). Probiotics and their fermented food products are beneficial for health. *Journal of Applied Microbiology* 100(6), 1171–1185.

Phillips, K.M., Simpkins, A.H., Amanna, K.R., Wolf, W., Stewart, K.K., Clark, S., Kim, K.P., Beecher, G.R. and Holden, J. (2001). Long-term stability of nutrients in a frozen mixed food control material. *Fresnius Journal of Analytical Chemistry* 370, 297–302.

Polo, M.V., Lagarda, M.J. and Farre, R. (1992). The effect of freezing on mineral content of vegetables. *Journal of food Composition and Analysis* 5, 77–83.

Polvi, S.M., Ackman, R.G., Lall, S.P. and Saunders, R.L. (1991). Stability of lipids and omega-3 fatty acids during frozen storage of Atlantic salmon. *Journal of Food Processing and Preservation* 15, 167–181.

Prados, R., Pino, A., Rincón, F., Vioque, M. and Fernández-Salguero, J. (2006). Influence of the frozen storage on some characteristics of ripened Manchego-type cheese manufactured with a powdered vegetable coagulant and rennet. *Food Chemistry* 95, 677–682.

Ranhotra, G.S. and Keagy, P.M. (1995). Adding folic acid to cereal-grain products. *Cereal Foods World* 40, 73–76.

Refsgaard, H.H.F., Brockhoff, P.B. and Jensen, B. (1998). Sensory and chemical changes in farmed Atlantic salmon (*Salmo salar*) during frozen storage. *Journal of Agricultural and Food Chemistry* 46, 3473–3479.

Rougerou, A. and Person, O. (1991). Influence of preservation method on unsaturated fatty acids of nutritional interest in sardines and mackerels. *Medicine et Nutrition* 27, 353–358.

Salem, M.M.E., Fathi, F.A. and Awad, R.A. (2006). Production of functional ice cream. *Deutsche Lebensmittel Rundschau* 102(7), 326–330.

Sanders, J.W., Venema, G. and Kok, J. (1999). Environmental stress response in *Lactococcus lactis*. *FEMS Microbiological Reviews* 23(4), 483–501.

Saurel, R. (2002). Use of vacuum technology to improve processed fruit and vegetables. In: W. Jongen, ed. *Fruit and Vegetable Processing: Improving Quality*. Cambridge: Woodhead Publishing Ltd., pp. 359–363.

Schantz, P.M. (1989). The dangers of eating fish raw. *New England Journal of Medicine* 320, 1143–1145.

Shewfelt, R.L. (1990). Sources of variation in the nutrient content of agricultural commodities from the farm to the consumer. *Journal of Food Quality* 13, 37–54.

Shortt, C. and O'Brien, J., eds. (2004). *Handbook of Functional Dairy Products*. Boca Raton: CRC Press.

Skrupskis, I., Aboltins, A. and Iljins, U. (2003). Durable storage possibility of fruit and berries. *Acta Horticulturae* 599, 719–724.

Stahl, W., Heinrich, U., Jungmann, H., Sies, H. and Tronnier, H. (2000). Carotenoids and carotenoids plus vitamin E protect against ultraviolet light-induced erythema in humans. *American Journal of Clinical Nutrition* 71, 795–798.

Stead, D. and Park, S.F. (2000). Roles of Fe superoxide dismutase and catalase in resistance of *Campylobacter coli* to freeze-thaw stress. *Applied and Environmental Microbiology* 66(7), 110–112.

Stern, N.J., Hiett, K.L., Alfredsson, G.A., Kristensson, K.G., Reierson, J., Hardardottir, H., Briem, H., Gunnarsson, E., Georgsson, F., Lowman, R., Berndtson, E., Lammerding, A.M., Paoli, G.M. and Musgrove, M.T. (2003). *Campylobacter* spp. in Icelandic poultry operations and human disease. *Epidemiology and Infection* 130, 20–32.

Szeto, Y.T., Tomlinson, B. and Benzie, I.F.F. (2002). Total antioxidant and ascorbic acid content of fresh fruit and vegetables: implications for dietary planning and food preservation. *British Journal of Nutrition* 87, 55–59.

Tanghe, A., Van Dijck, P., Dumortier, F., Teunissen, A., Hohmann, S. and Thevelein, J.M. (2002). Aquaporin expression correlates with freeze tolerance in baker's yeast, and overexpression improves freeze tolerance in industrial strains. *Applied and Environmental Microbiology* 68(12), 5981–5989.

Tejada, L., Sánchez, E., Gómez, R., Vioque, M. and Fernández-Salguero, J. (2002). Effect of freezing and frozen storage on chemical and microbiological characteristics in sheep milk cheese. *Journal of Food Science* 67(1), 126–129.

Tijburg, L.B.M. and Weststrate, J.A. (1997). Antioxidant comprising food products. *European Patent*, Publication number EP 0 845 951 B1.

Torija, E., Díez, C., Matallana, C., Camara, M., Camacho, E. and Mazarío, P. (1998). Influence of freezing process on free sugars content of papaya and banana fruits. *Journal of the Science of Food and Agriculture* 76, 315–319.

Vahteristo, L.T., Lehikoinen, K.E., Ollilainen, V., Koivistoinen, P.E. and Varo, P. (1998). Oven-baking and frozen storage affect folate vitamer retention. *Lebensmittel Wissenschaft und Technologie* 31, 329–333.

Verdini, R.A., Zorrilla, S.E. and Rubiolo, A.C. (2005). Effects of the freezing process on proteolysis during the ripening of Port Salut Argentino cheeses. *International Dairy Journal* 15, 363–370.

Vidal-Valverde, C., Ruiz, R. and Medrano, A. (1992). Stability of retinol in milk during frozen and other storage conditions. *Zeitschrift für Lebensmittel-Untersuchung und-Forschung* 195, 562–565.

Vidal-Valverde, C., Ruiz, R. and Medrano, A. (1993). Effects of frozen and other storage conditions on [alpha]-tocopherol content of cow milk. *Journal of Dairy Science* 76, 1520–1525.

Voutilainen, S., Nurmi, T., Mursu, J. and Rissanen, T.H. (2006). Carotenoids and cardiovascular health. *American Journal of Clinical Nutrition* 83, 1265–1271.

Wang, Y., Corrieu, G. and Beal, C. (2005a). Fermentation pH and temperature influence the cryotolerance of *Lactobacillus acidophilus*. *Journal of Dairy Science* 88(1), 21–29.

Wang, Y., Delette, M., Guillot, G. and Beal, C. (2005b). Influence of cooling temperature and duration on cold-adaptation of *Lactobacillus acidophilus*. *Cryobiology* 50(3), 294–307.

Wang, Q.-H., Souza, M., Funk, J.A., Zhang, W. and Saif, L.J. (2006). Prevalence of noroviruses and sapoviruses in swine of various ages determined by reverse transcription-PCR and microwell hybridization assays. *Journal of Clinical Microbiology* 44(6), 2057–2062.

Westerman, B.D. (1952). B-complex vitamins in meat IV; influence of storage time and temperature on vitamins in pork roasts. *Journal of the American Dietetics Association* 28, 331–335.

Williams, D.C., Lim, M.H., Chen, A.O., Pangborn, R.M. and Whitaker, J.R. (1986). Blanching of vegetables for freezing – which indicator enzyme to choose. *Food Technology* June, 130–139.

Xing, Y., Yoo, Y., Kelleher, S.D., Nawar, W.W. and Hultin, H.O. (1993). Lack of changes in fatty acid composition of mackerel and cod during iced and frozen storage. *Journal of Food Lipids* 1, 1–14.

Yang, C.-Y., Pang, J.-C., Kao, S.-S. and Tsen, H.-Y. (2003). Enterotoxigenicity and cytotoxicity of *Bacillus thuringiensis* strains and development of a process for Cry1Ac production. *Journal of Agricultural and Food Chemistry* 51(1), 100–105.

Yoon, K.S., Burnette, C.N., Abou-Zeid, K.A. and Whiting, R.C. (2004). Control of growth and survival of *Listeria monocytogenes* on smoked salmon by combined potassium lactate and sodium diacetate and freezing stress during refrigeration and frozen storage. *Journal of Food Protection* 67(11), 2465–2471.

3 Modelling of Freezing Processes

Q. Tuan Pham

3.1 INTRODUCTION

The mathematical modelling of food freezing processes poses special challenges. Thermophysical properties such as specific heat and thermal conductivity change suddenly around the freezing point, leading to highly non-linear partial differential equations that are difficult to solve. In complex-shaped objects, the progress of the freezing front can be highly unpredictable. Freezing is also associated with large and sudden volume changes, mass transfer, stress and cracking, cellular dehydration, supercooling, and ice nucleation and propagation, phenomena on which detailed studies are only beginning now. The main concern about frozen storage is moisture migration both on the macro (weight loss) and micro scales (recrystallisation).

Methods that have been applied to model the freezing processes can be classified into analytical, empirical and numerical techniques. The choice of technique depends on the objectives of the modellers and the technical means at their disposal. Analytical techniques produce exact results provided their underlying assumptions are fulfilled, which is rarely the case. Their main usefulness is in providing benchmark results for the verification of other methods. Empirical formulas are derived with the objective of providing quick answers, using no more than a hand calculator or spreadsheet, with accuracy (usually 10%) enough for most industrial users. They can be used only in situations similar to those used to derive and validate the formulas. Numerical methods can in principle provide exact or near-exact predictions for almost any situation, although in practice their accuracy is limited by inadequate knowledge of the problem's parameters (product properties, geometry, flow characteristics). For complex problems such as freezing with simultaneous heat and mass transfer, nucleation-controlled freezing or the solidification of liquid products, numerical methods are always the first choice. There are now powerful numerical modelling software packages in the market, allowing non-expert users to model complicated situations. To use them successfully, however, a minimum basic knowledge of heat, mass and momentum transfer and of the principles of numerical modelling is still required of the user.

This chapter starts by reviewing the various types of methods used in modelling freezing processes that involve only heat transfer, the simplest type of freezing. It then addresses more complicated situations involving mass transfer on a macroscale (between food and environment) or microscale (between cells and extracellular phase), supercooling and nucleation, high pressure, or fluid flow. As most readers will use commercial modelling software rather than write their own programs, the modelling of some freezing processes using common commercial software will be described. While trying to be as comprehensive as possible, it is assumed that the reader, whether college student or industrial user, wants practical guidance on how to solve problems, therefore one or two methods of solutions for each sub-problem will be described in detail while most alternatives will be reviewed only briefly.

3.2 ANALYTICAL SOLUTIONS

Let us consider the basic problem of a water-rich solid object being frozen in a cold medium. In the general case, the slab will start at a temperature higher than the freezing point, so there will be a period of precooling, with the surface cooling the fastest. Usually, the surface will supercool by a few degrees before nucleation, when its temperature jumps up back towards the freezing point. The surface temperature then continues to fall towards the external medium's temperature. A frozen layer forms at the surface and advances towards the centre. As the outside layers freeze, the centre loses most of its 'sensible heat,' then continues to be just above or at the freezing point without changing phase: this is called the freezing plateau. After some time the freezing front reaches the centre, signalling the end of the freezing plateau, and the centre temperature falls towards the medium temperature. The whole process is shown in Fig. 3.1. We are ignoring for the moment the fact that the phase change in foods usually takes place over a temperature range rather than at a sharp freezing point.

During this process the Fourier heat conduction equation (Carslaw and Jaeger, 1959, p. 10) applies throughout the product:

$$\rho c \frac{\partial T}{\partial t} = \nabla \cdot (k \nabla T) + q \tag{3.1}$$

When all the phase change and latent heat release takes place at one temperature, it is known as the Stefan problem. Analytical solutions are available for some simple cases. These solutions are generally obtained by solving two simultaneous instances of equation (3.1), one for the frozen zone and one for the unfrozen zone, with the shared boundary condition $T = T_f$ at the freezing front. Having obtained the temperature profile, the speed of the freezing front is then calculated from the energy balance, which states that the heat leaving the front through the frozen phase minus that arriving through the unfrozen phase equals the rate of latent heat

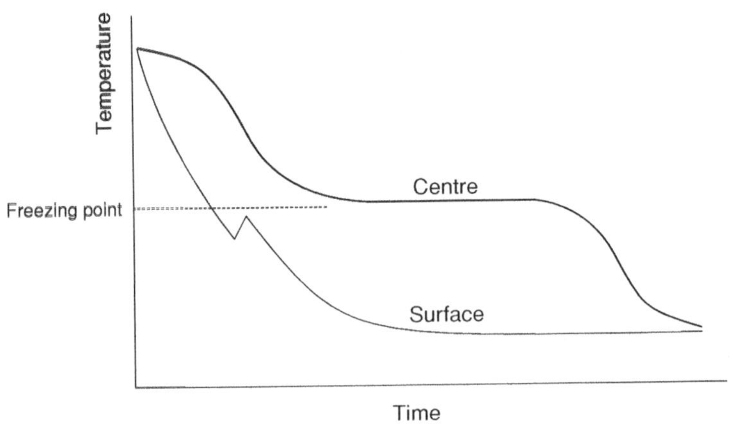

Fig. 3.1 Typical temperature history during freezing.

release:

$$\left(k\frac{dT}{dn}\right)_{\text{frozen}} - \left(k\frac{dT}{dn}\right)_{\text{unfrozen}} = \rho L_f v_f \tag{3.2}$$

where n is the distance normal to the front and v_f the velocity of the front (Fig. 3.2).

Although their practical usefulness is limited, analytical solutions provide useful results for checking the validity of numerical calculations. For the freezing of a semi-infinite body (half space) subjected to a step change in surface temperature, the analytical solution is given in Carslaw and Jaeger (1959, p. 283). It can be used to calculate the approximate temperature change, heat load and freezing rate in the initial period of freezing, when the frozen layer is thin compared to the dimension of the product. The freezing of a corner region subjected to similar conditions has been solved analytically by Budhia and Kreith (1973). Plank's (1913a, 1913b) pseudo-steady state solution, which would apply to the hypothetical case of zero specific heats, can be written in the following form (Pham, 1986a):

$$t_{\text{Plank}} = \frac{\rho L_f R}{Eh(T_f - T_a)}\left(1 + \frac{Bi}{2}\right) \tag{3.3}$$

where the shape factor E is 1 for slabs, 2 for infinite cylinders and 3 for spheres. For multi-dimensional regular shapes (infinite rectangular rods, finite cylinders, rectangular bricks), analytical expressions for E in the form of infinite series (McNabb *et al.*, 1990a, 1990b) and plots (Hossain *et al.*, 1992a) are also available.

An approximate solution has been derived for a sphere with non-zero specific heat, initially at the freezing temperature, using perturbation methods (Riley *et al.*, 1974):

$$t = \frac{3\eta^2 - 2\eta^3}{6} + \frac{\eta^2 Ste}{6} - \frac{\eta^2 Ste^2}{45(1-\eta)} + O(Ste^3) \tag{3.4}$$

where $\eta = \dfrac{R - r_f}{R}$ is the dimensionless frozen thickness and $Ste = c_f\,(T_f - T_a)/L_f$ is the Stefan number. The equation is applicable for up to $\eta \approx 0.8$, or 99% frozen fraction.

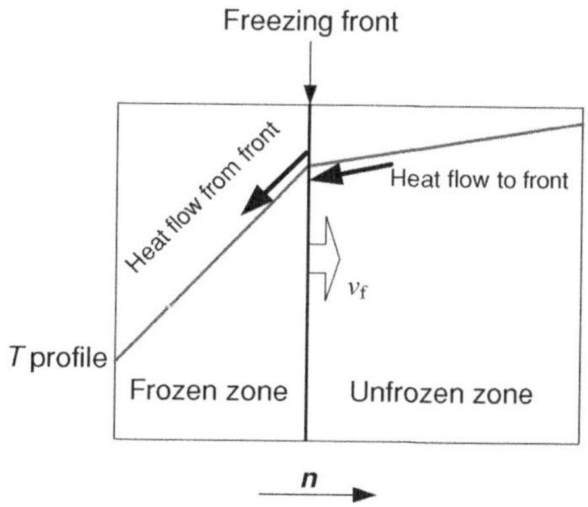

Fig. 3.2 Energy balance at the freezing front.

3.3 EMPIRICAL FORMULAE

3.3.1 Freezing time

3.3.1.1 *Simple shapes*

Cleland and Earle (1977, 1979a, 1979b, 1982, 1984) were the first to take a systematic approach to developing an empirical freezing time prediction method that takes into account the sensible heat effects, backed up by carefully collected data covering most of the range of interest to industry. Cleland and Earle's final equation (1984) can be written as follows for the three basic shapes (infinite slabs, infinite cylinders and spheres):

$$t_f = \frac{\rho \Delta H_{10}}{E(T_f - T_a)}\left(\frac{2P_1 R}{h} + \frac{4P_2 R^2}{k_f}\right)\left[1 - \frac{1.65 Ste}{k_f}\ln\left(\frac{T_c - T_a}{T_{ref} - T_a}\right)\right] \tag{3.5}$$

where

$P_1 = 0.5[1.026 + 0.5808Pk + Ste(0.2296Pk + 0.1050)]$

$P_2 = 0.125[1.202 + Ste(3.41Pk + 0.7336)]$

$Pk = \dfrac{c_u(T_i - T_f)}{L_f}$

$Ste = \dfrac{c_f(T_f - T_a)}{L_f}$

E is the shape factor (1 for slabs, 2 for infinite cylinders, 3 for spheres). Pk and Ste are dimensionless numbers which express the magnitude of precooling and post-cooling effects, respectively. Many other empirical freezing times methods have since been developed. A survey by Cleland (1990) lists 62 such methods, and many more have been proposed since then. Some of the methods can be considered semi-analytical, in that they are based on approximate analytical reasoning and try to correlate experimental data with a minimum of empirical parameters; among these are methods based on the addition of precooling, phase change and post-cooling times by de Michelis and Calvelo (1982) and Pham (1984, 1986a).

The ultimate criterion for choosing an empirical correlation is that it should yield an acceptably accurate result with as little effort as possible. The most thorough validation of various methods against data was performed by Cleland (1990, p. 114) and Pham and Willix (1990). Cleland compared the predictions of six well-known methods to a composite dataset formed by merging the most reliable experimental data available, as well as to finite differences calculations. Pham and Willix tested four methods against a composite dataset consisting of 280 previous data and 32 new data from their own tests. These systematic comparisons showed that Pham's methods (1984, 1986a) were the most accurate ones of all the methods tested. Hossain *et al.* (1992a, 1992b, 1992c) have also shown that Pham's methods give accurate results for complex shapes when combined with appropriate shape factors.

Pham's (1986a) method can be written as follows:

$$t_f = \frac{R}{Eh}\left(\frac{\Delta H_1}{\Delta T_1} + \frac{\Delta H_2}{\Delta T_2}\right)\left(1 + \frac{Bi}{2}\right) \tag{3.6}$$

where ΔH_1 and ΔT_1 are the specific enthalpy change and temperature difference, respectively, for the precooling period, and ΔH_2 and ΔT_2 those for the combined freezing—post-cooling period, calculated from:

$$\Delta H_1 = H_i - H_{fm} = \rho_u c_u (T_i - T_{fm}) \quad (3.6a)$$

$$\Delta T_1 = \frac{(T_i + T_{fm})}{2 - T_a} \quad (3.6b)$$

$$\Delta H_2 = H_{fm} - H_c = \rho_f [L_f + c_f (T_{fm} - T_c)] \quad (3.6c)$$

$$\Delta T_2 = T_{fm} - T_a \quad (3.6d)$$

T_{fm} is the 'mean freezing temperature' and the following equation is probably valid for most water-rich biological materials:

$$T_{fm} = 1.8 + 0.263 T_c + 0.105 T_a \quad (3.6e)$$

It can be readily seen that equation (3.6) is an extension to Plank's equation, equation (3.3), with the term $\Delta H_1/\Delta T_1$ representing the precooling time, $\Delta H_2/\Delta T_2$ representing the phase change–post-cooling time, and the term $1 + Bi/2$ representing the effect of internal resistance to heat transfer. T_{fm} is a 'mean freezing temperature' which represents some kind of time-averaged product temperature during the combined phase change and post-cooling period. In the expression for T_{fm}, the term involving T_c expresses the effect of the post-cooling period on the mean product temperature, while the term involving T_a can be interpreted as a correction for the temperature profile in the frozen product during the phase change and post-cooling periods. The three constants in T_{fm} are the only empirical parameters in Pham's method.

Equation (3.6) has been extended to the freezing of foodstuffs with variations in environmental conditions (Pham, 1986b) and to the asymmetric freezing of slabs (Pham, 1987a).

3.3.1.2 *Multi-dimensional shapes*

The freezing of an irregular shape can be approximated by that of an equivalent regular shape; for example, Illicali and Holacar (1990) approximated an ellipsoid by an equivalent sphere. The technique has also been frequently used on specific products such as lamb legs. This approach is likely to be useful only when the variation in Biot number is not great, since the effect of shape depends greatly on this parameter, and the deviation from the standard shape not too large.

Based on the observation that, under the same conditions, the freezing time of a slab is twice that of an infinite cylinder with the same thickness/diameter and three times that of a sphere, Cleland and Earle (1982) proposed that a similar ratio can be found for other shapes. They termed this ratio the 'equivalent heat transfer dimensionality', E. Thus:

$$t_f\,(\text{any shape}) = \frac{t_f\,(\text{slab})}{E} \quad (3.7)$$

For regular shapes (infinite rectangular rods, finite cylinders, rectangular bricks), analytical expressions for E were derived by McNabb *et al.* (1990a, 1990b) and reproduced in Hossain *et al.* (1992a) who also plotted them for $Bi = 0.005$–50 and aspect ratios up to 10. Simplified approximate equations for shape factors for common cases have also been proposed and are listed below. It should be kept in mind that these tend to be more accurate at moderate aspect ratios and low to moderate Biot numbers.

Table 3.1 Geometric constants for calculating the freezing shape factors *E* (Cleland *et al.*, 1987a).

Shape	G_1	G_2	G_3
Finite cylinder, height < diameter	1	2	0
Finite cylinder, diameter < height	2	0	1
Rectangular rod	1	1	0
Rectangular brick	1	1	1

3.3.1.2.1 *Infinite rectangular rods, bricks, finite cylinders*
The empirical method by Cleland *et al.* (1987a, 1987b) can be used:

$$E = G_1 + G_2E_1 + G_3E_2 \tag{3.8}$$

where G_1, G_2 and G_3 are constants that depend on the type of shape, as listed in Table 3.1, while E_1 and E_2 are empirical functions of Bi and the aspect ratios:

$$E_1 = X\left(2.32\beta_1^{-1.77}\right)\frac{1}{\beta_1} + \frac{1 - X\left(2.32\beta_1^{-1.77}\right)}{\beta_1^{1.47}}\frac{0.73}{\beta_1^{2.50}} \tag{3.9}$$

$$E_2 = X\left(2.32\beta_2^{-1.77}\right)\frac{1}{\beta_2} + \frac{1 - X\left(2.32\beta_2^{-1.77}\right)}{\beta_2^{1.47}}\frac{0.50}{\beta_2^{3.69}} \tag{3.10}$$

where $X(x)$ is a function defined by

$$X(x) = \frac{x}{(Bi^{1.34} + x)} \tag{3.11}$$

3.3.1.2.2 *Elliptical cylinders*
McNabb *et al.* (1990b) derived the following approximate expression, which is accurate to within about −7% to +3% (Hossain *et al.*, 1992a):

$$E = 1 + \frac{1 + 2/Bi}{\beta^2 + 2\beta_1/Bi} \tag{3.12}$$

3.3.1.2.3 *Three-dimensional ellipsoids*
The three-dimensional version of McNabb *et al.*'s equation can be used and is accurate to within about −7% to +5% (Hossain *et al.*, 1992b):

$$E = 1 + \frac{1 + 2/Bi}{\beta_1^2 + 2\beta_1/Bi} + \frac{1 + 2/Bi}{\beta_2^2 + 2\beta_2/Bi} \tag{3.13}$$

3.3.1.2.4 *Irregular shapes*
Irregular shapes can often be approximated by ellipses or ellipsoids, and the freezing time equations for these can be applied by provided that β_1 and β_2 are replaced by $S_1/\pi R^2$ and $S_2/\pi R^2$, respectively (Cleland *et al.*, 1987a, 1987b; Hossain *et al.*, 1992a, 1992b). For 2D

irregular shapes, S_1 is the cross-sectional area and R is the smallest half-dimension. For 3D irregular shapes, S_1 represents the areas of the (approximately) smallest cross-section and S_2 represents the cross-section through the smallest diameter orthogonal to the first. Some subjective judgement has to be used for highly irregular or non-symmetrical shapes, and thin protrusions may have to be ignored.

3.3.2 Freezing heat load

Knowledge of the dynamic product heat load is important for designing refrigeration systems. Commercial food freezing processes usually start with a precooling period before phase change occurs. The dynamic heat load during the precooling period can be estimated following Pham (2001). For the phase-change period, the method of Lovatt *et al.* (1993a, 1993b) can be used. A slightly simplified version of this method (lumping the freezing and post-cooling heat load together) will be presented here. By considering the movement of the freezing front, the following approximate equation for heat load is obtained:

$$Q = -\Delta H_2 \frac{dr_f}{dt}\frac{dV_f}{dr_f} \tag{3.14}$$

where ΔH_2 is the latent + post-cooling heat as defined earlier, dr_f/dt is the freezing front velocity and dV_f/dr_f the change of frozen volume with frozen depth. The former can be calculated from Plank's (1913a) equation or Pham's (1986a) equation:

$$\frac{dr_f}{dt} = \frac{-(T_a - T_f)}{\Delta H_2 r_f^{E-1}\left[(1/hR^{E-1}) - \left(r_f^{2-E} - R^{2-E}/k_f(2-E)\right)\right]} \tag{3.15}$$

while the latter is a function of geometry and can be calculated from

$$\frac{dV_f}{dr_f} = Z\frac{V}{R}\left(\frac{r_f}{R}\right)^{Z-1} \tag{3.16}$$

The freezing time shape factor E can be calculated as before. For simple shapes (infinite slabs, infinite cylinders and spheres), $Z = E$, while for other shapes

$$Z \approx \frac{AR}{V} \tag{3.17}$$

For very irregular shapes with protrusions AR/V may be larger than 3, which would give unrealistic results, hence the following upper limit should be imposed:

$$Z \leq 3$$

Lovatt *et al.*'s method has been validated against numerical computations and experimental data. The simplification presented in this chapter (lumping together the latent and post-cooling heat loads) is not expected to affect its validity, since the post-cooling load is small (10% or less) compared to the latent load.

3.4 NUMERICAL SOLUTION OF THE CONDUCTION EQUATION

3.4.1 Discretisation of the conduction equation

The numerical solution of equation (3.1) involves two steps: discretising the space domain to obtain a set of ordinary differential equations (ODE) relating the nodal temperatures, then solving this set of ODEs. The ODEs can be written in matrix form as

$$\mathbf{C}\frac{d\mathbf{T}}{dt} + \mathbf{KT} = \mathbf{f} \tag{3.18}$$

where **T** is a vector of nodal temperatures, **C** is the global capacitance matrix containing the specific heat c, **K** the global conductance matrix containing the thermal conductivity k, and **f** the global forcing matrix containing known terms arising from heat generation and boundary conditions. The exact form of equation (3.18) depends on which method is used for discretising space. The most common such methods are finite difference (FDM), finite element (FEM) and finite volume (FVM).

FDM is most convenient and efficient for problems involving simple geometries. For shapes that do not deviate too much from a regular geometry, the use of a boundary fitted grids (Thompson *et al.*, 1982) extends the use of FDM, which is much faster than the unstructured meshes of FEM or FVM. FDM involves superimposing a grid of structured (arrayed) nodes on the calculation domain. For each node, equation (3.1) is written down, with the temperature gradients around each node calculated by central differences. When all such equations are put together, we obtain the matrix equation (3.18).

With FDM there are problems in discretising a Neumann (Newton's cooling law) boundary condition. The surface flux applies at the surface node itself, while the flux on the inner side of the surface node applies at a point $\Delta x/2$ away, causing an asymmetric situation. Another practical problem is that with very high heat transfer coefficients, the temperature of the surface nodes may be subjected to instabilities due to the excessive surface heat flux. The boundary condition is handled better by the control-volume formulation of FDM (actually an FVM method), where the solid is first divided into control volumes and the nodes are placed in the middle of each control volume.

Having obtained a set of ODEs in time relating the nodal temperatures (equation (3.18)), solution will proceed in a series of time steps starting from the known initial conditions:

$$\mathbf{C}\frac{\mathbf{T}^{\text{New}} - \mathbf{T}}{\Delta t} = -\mathbf{K}\overline{\mathbf{T}} + \overline{\mathbf{f}} \tag{3.19}$$

On the right-hand side of equation (3.19), some time-averaged value of the nodal temperatures must be used, such as:

$$\overline{\mathbf{T}} = \alpha\mathbf{T}^{\text{New}}(1-\alpha)\mathbf{T} \tag{3.20}$$

where the superscript 'New' refers to temperatures at the end of the time step. α is a parameter varying between 0 and 1. The most popular stepping method is Crank–Nicolson ($\alpha = 0.5$) in view of its combination of unconditional stability and second-order accuracy (Crank and Nicolson, 1947). The Lees three-level scheme (1976) is also popular in the food freezing literature and is often believed to be particularly suited for handling variable properties.

However, the author has not found any advantage of this scheme over the Crank–Nicolson scheme (Pham, 1987b). In regular shapes with two or three dimensions (finite cylinders, rods or brick shapes), the tridiagonal matrix algorithm can be used to efficiently solve equation (3.19) by applying the alternating direction method (Peaceman and Rachford, 1955).

For shapes that cannot be represented by an orthogonal grid, FEM and FVM are used. In FEM (Segerlind, 1984; Zienkiewicz, 1991), the object is divided into small elements. Adjacent elements share some of their nodes. Within each element, the temperature field at **x** is approximated by interpolation:

$$T(\mathbf{x}, t) = \mathbf{N}^T(\mathbf{x})\mathbf{T}(t) \tag{3.21}$$

where $\mathbf{T}(t)$ are the vector of the temperatures at the nodes and $\mathbf{N}(\mathbf{x})$ is a vector of 'shape functions', or interpolating functions. In the Galerkin FEM, we require that the weighted residuals of the heat conduction equation, equation (3.1), are zero when the shape functions are also used as weighting functions:

$$\int_\Omega \mathbf{N}\left[\rho c\frac{\partial T}{\partial t} = \nabla \cdot (k\nabla T) - q\right] d\Omega = 0 \tag{3.22}$$

Substituting equation (3.21) into equation (3.22) and solving it yield a relationship between the nodal temperatures:

$$\mathbf{C}^{(e)}\frac{d\mathbf{T}^{(e)}}{dt} + \mathbf{K}^{(e)}\mathbf{T}^{(e)} = \mathbf{f}^{(e)} \tag{3.23}$$

where the superscript (e) indicates that this is a relationship between the nodes in one element only. Summing over all elements give the standard matrix equation (3.18).

In FEM, the physical significance of the non-diagonal capacitance terms in equation (3.18) is not intuitively obvious and arises from the fact that heat capacity is distributed over the element. However, a rough physical interpretation is possible and helps us understand how the method works. Within an element, the thermal energy ρcT at each point is attributed to the nodes according to the shape function, i.e. more towards the nearest node and less towards the furthest. When the temperature at a node i is changed, this affects the temperature profile and hence the thermal energy distribution throughout the element, and C_{ij} represents the effect of a change in T_i on the thermal energy attributed to node j.

In the *lumped capacitance* version of FEM, *all* the thermal energy change due to a change in T_i is attributed to node i itself, hence $\mathbf{C}$ becomes a diagonal matrix. In other words, the mass of each element is assumed to be concentrated at the nodes. In effect, this is a form of FVM, where each node holds the thermal mass of the control volume surrounding it. This formulation has some advantages over the Galerkin formulation in terms of simplicity and stability (Banaszek, 1989) and is particularly useful for dealing with the latent heat peak during freezing, as will be seen later.

In FVM, the domain considered is divided into control volumes, each associated with a node at the centre. The heat conservation equation is assumed to hold over each control volume as a whole by setting its residual to 0:

$$\int_\Omega \left[\rho c\frac{\partial T}{\partial t} - \nabla \cdot (k\nabla T) - q\right] d\Omega = 0 \tag{3.24}$$

The first term in the volume integral is calculated by assuming that T is uniform in the control volume and equal to the nodal temperature. Using the divergence theorem, the second term can be transformed into the net heat flow over the volume's boundaries. The heat flow through each surface is assumed to be a linear function of the nodal temperatures in the vicinity of the surface. Thus, equation (3.24) can be replaced by

$$\delta V \rho c \frac{\partial T_i}{\partial t} = \sum_{j=1}^{N} k B_{ij} T_j + q \delta V \tag{3.25}$$

where B_{ij} are coefficients that depend on the nodal arrangement.

Assembling all these nodal again yields equation (3.18). As with FDM or lumped capacitance FEM (but unlike Galerkin FEM), in this case **C** is a diagonal matrix, which will present some advantages.

3.4.2 Dealing with latent heat release

The major difficulty in the numerical solution of the heat transport equation is in dealing with the large latent heat, which evolves over a very small temperature range. Voller (1996) reviews the techniques for dealing with phase change. These methods can be divided into *fixed grid* methods and *moving grid* methods. In the latter, the object is divided into a frozen zone and an unfrozen zone. Some nodes, element boundaries or control volume boundaries are put on the freezing front and allowed to move with it.

Moving grid methods can give precise, non-oscillating solutions for the temperature and ice front position. They have been applied by Califano and Zaritsky (1997) to the freezing of foods with irregular two-dimensional geometry. However, moving grid methods are less flexible than fixed grid methods because most foods do not have a sharp phase change temperature but freeze gradually, hence it is not clear how the freezing front should be defined. With foods of complex shape, the calculation of the front's position and subsequent grid adjustment is a major task. Therefore, this article will concentrate on fixed grid methods. The freezing front can still be located in these methods by carrying out an interpolation to locate the position where the freezing temperature applies (Pham, 1986c; Udaykumar, 2002).

3.4.2.1 *Source term methods*

In one type of fixed grid methods (Voller and Swaminathan, 1991), latent heat is treated as a source term, i.e. is included in the term q in equation (3.1):

$$q = \rho \frac{\Delta g}{\Delta t} \tag{3.26}$$

where Δt is the time step and Δg is the latent heat per unit mass released over Δt. Since Δg depends on the temperature change $T^{\mathrm{New}} - T$, an iterative method such as Newton–Raphson has to be used for the solution: the new nodal temperature are calculated, the source terms re-estimated, and the equations solved again for nodal temperatures. Fundamentally, there is little distinction between the source term methods and the apparent specific heat methods, apart from the details of the iteration procedure, since the discretised equation (equation (3.19)) in both methods is the same. Source term methods are popular in the modelling of metal casting, where a well-defined freezing range exists and the latent heat can be clearly distinguished from the specific heat.

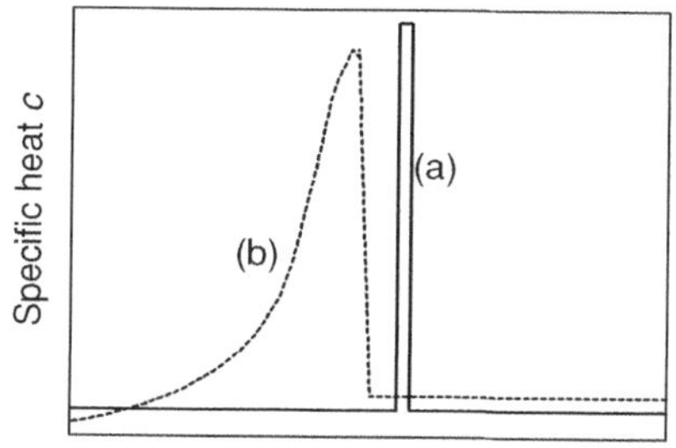

Fig. 3.3 Apparent specific heat for (a) a material with sharp phase change, (b) a material with gradual phase change.

3.4.2.2 *Apparent specific heat method*

In the apparent specific heat methods, latent heat is merged with sensible heat to produce a specific heat curve with a large peak around the freezing point (Fig. 3.3). Because of the large variations in specific heat, iteration must be carried out at every step. It is difficult to obtain convergence with this technique, and there is always a chance that the latent heat is underestimated. This happens when the temperature at a node steps over the peak in the apparent specific heat curve: the mean specific heat between the initial and final temperature is then always less than the peak, and the temperature change will therefore be overestimated. Many approximate methods have been proposed to calculate the effective specific heat near the freezing point (Comini and Del Giudice, 1976; Morgan *et al.*, 1978; Lemmon, 1979; Cleland *et al.*, 1984; Comini *et al.*, 1989) but none is entirely satisfactory (Pham, 1995). For this reason the apparent specific heat method is not recommended. However, in general-purpose commercial software such as COMSOL it is often the only method available. To ensure convergence in reasonable time, the user should 'smear out' the latent heat peak over as wide a temperature range as possible, while ensuring that this will not affect accuracy. For example, when freezing pure water ($T_f = 0°C$) in a medium at −40°C, the temperature driving force over most of the process is of the order 40°C. Therefore, if the latent heat load is assumed to be released over a range $T_{f1} = -0.5°C$ to $T_{f2} = +0.5°C$, say, the predicted freezing time is not expected to be greatly affected. The following apparent specific heat table could therefore be used:

$$T \leq T_{f_1}, c = c_{ice}$$

$$T_{f_1} < T \leq T_{f_2}, c = \frac{c_{ice} + c_{water}}{2} + \frac{L_f}{T_{f_2} - T_{f_1}}$$

$$T > T_{f_2}, c = c_{water}$$

3.4.2.3 *Enthalpy method*

The basic conduction equation can be written as

$$\rho \frac{\partial H}{\partial t} = \nabla \cdot (k \nabla T) + q \tag{3.27}$$

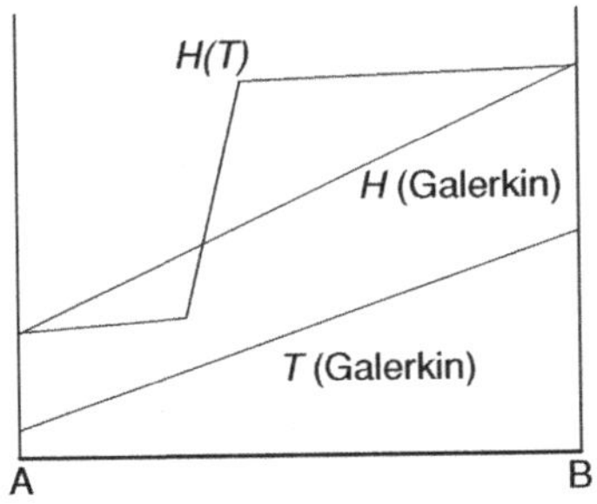

Fig. 3.4 Temperature and enthalpy approximations in a 1D element around the freezing point.

where H is the (specific) enthalpy. After the usual FDM, FEM or FVM manipulations, we obtain the matrix equations

$$\mathbf{M}\frac{d\mathbf{H}}{dt} + \mathbf{KT} = \mathbf{f} \tag{3.28}$$

where **M** is a (constant) mass matrix and **H** is the vector of nodal enthalpies.

In FDM, FVM and lumped capacitance FEM, **M** is a diagonal matrix and the nodal enthalpies H_i can be calculated one by one by using the explicit (Euler) stepping scheme:

$$H_i^{\text{New}} = H_i^{\text{Old}} + \Delta t \frac{f_i^{\text{Old}} - \sum_j K_{ij} T_j^{\text{Old}}}{M_i} \tag{3.29}$$

This approach was first proposed for FDM by Eyres *et al.* (1946).

To obtain an exact solution to any implicit ($\alpha > 0$) solution of equation (3.28), iteration must be carried out at every time step. The enthalpy change vector $\Delta\mathbf{H} \equiv \mathbf{H}^{\text{New}} - \mathbf{H}$ at each time step is iteratively adjusted until the residual of equation (3.28) becomes acceptably small. A successive substitution scheme such as Gauss–Seidel can be used, but convergence is very slow and various over-relaxation schemes have been proposed (Voller *et al.*, 1990). A better approach is to use a Newton–Raphson iteration.

The discretisation of highly non-linear problems such as the phase change problem with Galerkin FEM poses serious conceptual difficulties. The Galerkin approach assumes that temperature is distributed over the element according to the shape function, i.e. $T = \mathbf{N}^{\mathrm{T}}\mathbf{T}$ where **T** is the vector of nodal temperatures. Since H is a non-linear function of T, it cannot be assumed that H is distributed according to $H = \mathbf{N}^{\mathrm{T}}\mathbf{H}$ as well. In fact, this interpolation will be very inaccurate around the freezing point (see Fig. 3.4 where the nodal temperatures in a 1D element are just above and below the freezing point). Yet, in the enthalpy method, this assumption has to be made when transforming equation (3.27) into a Galerkin FEM equation. In the effective specific heat method, the Galerkin FEM user is faced with how to calculate an effective specific heat over both time (Δt) and space (the element's domain): numerical averaging methods (which uses some sampling procedure over the element's domain) fail when $c(T)$ has a very sharp peak (Fig. 3.4). For these reasons, the use of FVM or lumped capacitance FEM is recommended over Galerkin FEM.

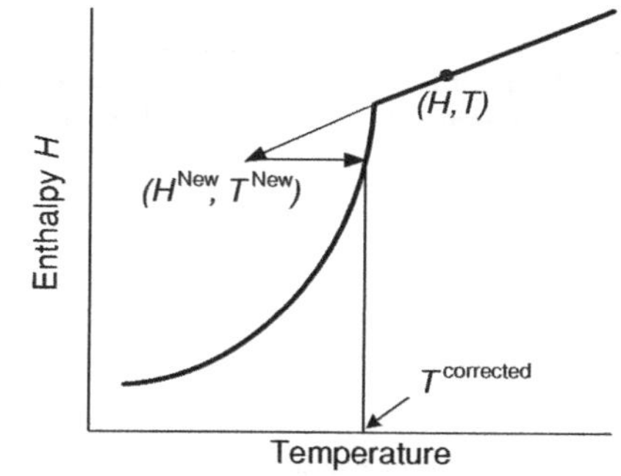

Fig. 3.5 Illustration of Pham's (1985) temperature correction step.

3.4.2.4 *Quasi-enthalpy (temperature correction) method*

Pham (1985) proposed a simple method which is effective in dealing with the latent heat peak but avoids the need for iteration. The method was first applied to FDM but was subsequently extended to lumped capacitance FEM (Pham, 1986) and to Galerkin FEM (Comini *et al.*, 1989), and can be readily implemented in FVM. The approach uses the apparent specific heat method, but each time adds a specific heat estimation step and a temperature correction step. Further investigations (Voller, 1996; Pham, 1995) indicate that temperature correction is the crucial step. After a set of new nodal temperatures T_i^{New} is obtained from equation (3.19), the changes in nodal enthalpies are calculated from $c_i(T_i^{\text{New}} - T_i)$. The new nodal enthalpies H_i^{New} are then calculated from the old nodal enthalpies H_i:

$$H_i^{\text{New}} = H_i + c_i\left(T_i^{\text{New}} - T_i\right) \tag{3.30}$$

and the new nodal temperatures are corrected to

$$T_i^{\text{Corrected}} = T\left(H_i^{\text{New}}\right) \tag{3.31}$$

The temperature correction technique is illustrated in Fig. 3.5. Equation (3.30) ensures that enthalpy is always conserved, hence this method can be termed the quasi-enthalpy method.

Pham (1995) compared ten fixed-grid methods in terms of accuracy, time interval for convergence, heat balance error (a measure of whether the latent heat load peak has been missed), and computing effort. He concluded that the (non-iterative) lumped capacitance FEM with Pham's quasi-enthalpy method performed best in terms of most of the above criteria. Voller (1996) also concluded that this method is an 'excellent scheme' for fixed grids.

To summarise, the methods mentioned here for dealing with latent heat can be classified and compared as shown in Table 3.2.

3.4.3 Dealing with variable thermal conductivity

The rapid change in thermal conductivity around the freezing point contributes to the difficulty in the numerical modelling of phase change. In computing the heat flux between two

Table 3.2 Summary of methods for dealing with latent heat.

Method	Time stepping	Material applicability	Programming difficulty	Accuracy	Speed
Moving grid		Sharp phase change	Very hard	Good	—
Fixed grid					
Source term methods					
	Implicit iterative	All materials	Hard	Good	Slow
Apparent specific heat methods					
	Euler	Gradual phase change	Very easy	Poor	Slow
	Implicit non-iterative	Gradual phase change	Easy	Poor	Slow
	Implicit iterative	Gradual phase change	Easy	Poor	Slow
Enthalpy methods					
	Euler	All materials	Very easy	Good	Slow
	Implicit iterative	All materials	Hard	Good	Slow
Pham's quasi-enthalpy method					
	Implicit non-iterative	All materials	Easy	Good	Fast

nodes at different temperatures, it is unclear what value should be used for k when it varies with temperature. A rigorous formulation is obtained by using the Kirchhoff transformation (Carslaw and Jaeger, 1959 (with Kirchhoff, 1894 quoted on p. 11); Fikiin, 1996):

$$u = \int_{T_{\text{REF}}}^{T} k d\theta \tag{3.32}$$

which, when substituted into the Fourier equation, equation (3.1), gives

$$\frac{\rho c}{k}\frac{\partial u}{\partial t} = \nabla^2 u + q \tag{3.33}$$

The ratio $\rho c/k$ depends on temperature, and therefore on u. This technique groups all the non-linearities into a single factor, after which the equation can be solved by FDM, FEM or FVM using the enthalpy method (Voller and Swaminathan, 1991) or Pham's quasi-enthalpy method (1985, 1986c). Scheerlinck (2000) and Scheerlinck *et al.* (2001) found that the Kirchhoff transformation leads to a significant reduction in computation time when using an iterative method, because the **K** matrix becomes constant and does not have to be recomputed.

With composite materials, the Kirchhoff transformation may cause some problems in the modelling of boundaries between different materials, particularly with FEM. For example, when two adjacent elements made of different materials share the same nodes, the values of u at these nodes will be different depending on whether they are viewed from one element or the other. The elemental equations (equation (3.23)) cannot be assembled into a global matrix equation in the usual manner. Instead, each node that is shared by two materials must be treated as two separate nodes related by a 'phase equilibrium' relationship.

3.5 COUPLED HEAT AND MASS TRANSFER

3.5.1 Governing equations

In food freezing, heat transfer is always accompanied by mass transfer, which may have important implications on weight loss and product quality. We will concentrate on the transfer of moisture only, which is the most common situation, although solute transfer also happens in immersion freezing.

When mass transfer occurs, conduction is not the only mode of heat transfer. Thermal energy is also conveyed by the diffusing substance, necessitating the addition of a second transport term. This can be expressed with the enthalpy form of the heat transport equation:

$$\frac{\partial(\rho H)}{\partial t} = \nabla \cdot (k\nabla T) + \nabla(H_w \dot{m}_w) \tag{3.34}$$

$\dot{m}_w$ is the mass flux and H_w the enthalpy of the diffusing substance. The mass flux is assumed to follow Fick's law:

$$\dot{m}_w = D_w \nabla W \tag{3.35}$$

where W is the mass concentration of the diffusing substance (kg water m^{-3}) and D_w its (effective) diffusivity. The governing equation for mass transfer is therefore

$$\frac{\partial W}{\partial t} = \nabla \cdot (D_w \nabla W) \tag{3.36}$$

Mechanical effects (gravity, pressure gradient) have been ignored, as well as the mass diffusion due to temperature gradient (Soret effect) and heat diffusion due to concentration gradient (Dufour effect). The second term in the heat transport equation can usually be neglected in dense foods due to the very slow moisture diffusion rate, but not in porous foods.

In the freezing of non-porous foods such as meat, water evaporates from the surface and is replenished by deep water diffusing towards the surface, until freezing occurs. Thereafter the water sublimes from the ice front at or near the surface, but there is no significant water movement in the food. Mass transfer occurs in a rather thin layer near the surface only, in contrast to heat transfer which happens throughout. In the freezing of porous food such as bread and dough, moisture movement continues right through the freezing process deep inside the food. Each situation requires a different modelling approach.

3.5.2 Freezing of dense foods

Moisture in foods can simultaneously exist in several phases: vapour, free liquid and various types of bound moisture, each of which has its own diffusion rate. However, it is commonly assumed that moisture movement in foods can be described by a single-phase diffusion equation, equation (3.36), with an effective diffusivity D_w. This equation is of the same form as the heat conduction equation, equation (3.1), and can be solved by the same methods (FDM, FEM or FVM). In dense food, moisture diffusion is very slow and its contribution to heat transport can be neglected, hence the second term in equation (3.34) can be neglected, except at the evaporating surface itself.

The problems to be considered are the changes in boundary conditions and the differences in scale between heat and mass transfer. Prior to surface freezing, water evaporates from the surface at a rate proportional to $(P_s - P_a)$ where P_s is the water partial pressure at the food surface and P_a that in the surroundings, and is replenished by moisture diffusing from the inside. The boundary conditions contain P_s, which is a non-linear function of surface temperature and moisture. If explicit time stepping is used, this poses no particular problem, but for any other stepping method, P_s has to be linearised around the present temperature and moisture value (Davey and Pham, 1997), or an iteration must be carried out at every time step.

Due to the big difference between moisture and heat diffusivity in dense foods, by the time the freezing process is completed, only a very thin layer near the surface has undergone any change in moisture. To model moisture movement accurately, therefore, would require an extremely fine grid and hence very small time intervals. Pham and Karuri (1999) proposed resolving this difficulty by using a two-grid method, where a second, very fine, one-dimensional finite volume grid just under the surface is used to model the mass transfer. At each time step, the heat flow equation is solved first using the main grid which extends over the whole product, then the mass transfer equation is solved using the second grid. The approach was implemented by Trujillo (2004) in modelling the chilling of a beef side, using the FVM-based CFD software FLUENT.

Once the surface has frozen, water becomes immobilised and internal diffusion stops. Moisture then sublimes, at first from the surface, then through a layer of desiccated food that gradually thickens as the ice front recedes, at a rate determined by (Pham and Willix, 1984; Pham and Mawson, 1997)

$$\dot{m} = \frac{P_{\text{sat}}(T_{\text{s}}) - P_{\text{a}}}{\dfrac{1}{k_{\text{g}}} + \dfrac{R_{\text{g}}T}{M_{\text{w}}}\dfrac{\delta}{D_{\delta}}} \tag{3.37}$$

where δ is the desiccated thickness and D_δ the diffusivity of water through it. This equation correlates satisfactorily the weight loss data of lamb in frozen storage, the resistance to mass transfer increasing linearly with total weight loss as expected (Pham and Willix, 1984). The model was solved numerically for 1D geometry using a front-tracking finite difference method (Campañone *et al.*, 2001). The dehydrated zone is modelled by a flexible grid with distance increments increasing proportionately to the depth of the freezing zone. Because the desiccated layer is normally very thin in the freezing of dense foods, modelling it numerically requires a very thin grid. In fact, at the moment freezing starts, δ may be zero, which would cause difficulty. In the author's view, when the rate of sublimation is very slow and the desiccated layer very thin (thinner than, say, half the thickness of a control volume), we can assume pseudo-steady state (i.e. the water vapour profile in the desiccated layer is as if the sublimating front was stationary), and it is sufficiently accurate to use an ODE approach to calculate its thickness:

$$\frac{d\delta}{dt} = \frac{\dot{m}_{\text{w}}}{\rho_{\text{s}} W(\delta)} \tag{3.38}$$

where $W(\delta)$ is the moisture content at depth δ.

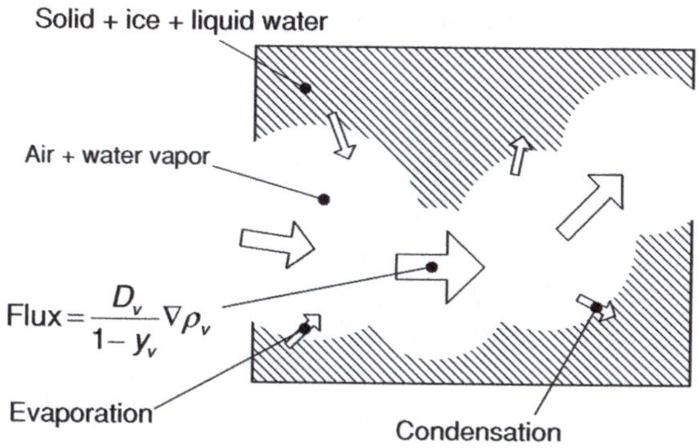

Fig. 3.6 Diffusion and evaporation–condensation in porous material.

3.5.3 Air freezing of porous foods

Mass diffusion is much faster in porous food than in dense foods and may take place by several mechanisms: molecular diffusion of gases in the pores, Darcy flow, and capillary diffusion. During freezing of foods at atmospheric pressure, vapour diffusion in the pores is the main mechanism. Water evaporates from the warm inner parts of the food and diffuses towards the outside. When the freezing point is reached, the vapour condenses into ice. This situation has been modelled by van der Sluis (1993) and Hamdami *et al.* (2004a, 2004b) for bread freezing, in view of the suspected influence of ice formation under the crust on crust detachment.

The evaporation–condensation model (De Vries, 1952, 1958, 1987) is often used to calculate heat transport in porous foods. Evaporating moisture picks up latent heat, and when it recondenses this heat is released (Fig. 3.6). It can be shown that evaporation–condensation gives rise to an additional effective thermal conductivity:

$$k_{\text{eva-con}} = \frac{\varepsilon}{\tau}\frac{L_V M_W D_V a_W}{R_g T}\frac{P_{atm}}{P_{atm} - a_W P_{sat}}\frac{dP_{sat}}{dT} \tag{3.39}$$

where ε is the void fraction, τ the tortuosity of the diffusion path, ρ_v the mass concentration of water vapour in the pores, and D_V the diffusivity of water vapour in air. The evaporation–condensation model is theoretically well supported and has been found to correctly predict the large variations of effective thermal conductivity with temperature in porous foods Hamdami *et al.* (2004a, 2004b). The factor φ must usually be determined empirically by curve fitting.

3.5.4 Immersion freezing of porous foods

In immersion freezing the food is in contact with a liquid below its freezing temperature. The liquid may be a non-aqueous refrigerant or water containing an acceptable solute such as a salt, sugar, alcohol or mixture of these, at a concentration high enough to maintain a liquid phase at a low temperature. The use of an ice slurry (Fikiin and Fikiin, 1998) greatly enhances the process's efficiency by making use of the latent heat of the ice. If the food being frozen is not wrapped in an impervious film, solutes will diffuse into the food and water out of it, at the same time that heat is transferred. The objective may be to minimise this mass diffusion or to aim at some optimal value. For example, by freezing fruit in sugar–ethanol aqueous

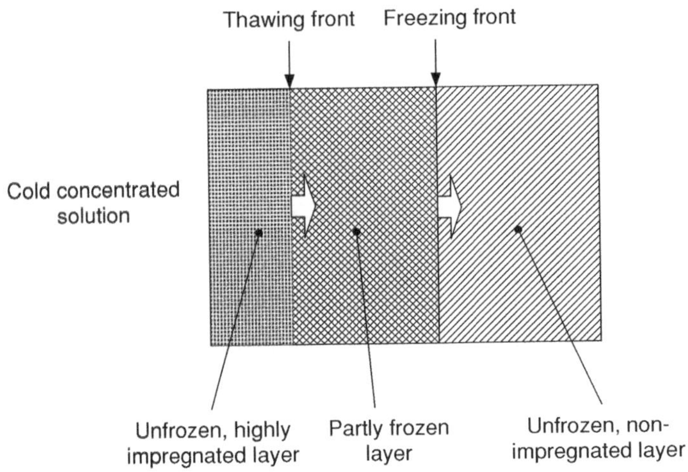

Fig. 3.7 Immersion freezing of porous slab.

solutions or ice slurries, new dessert products can be formulated with beneficial effect on colour, flavour and texture, due to the enzyme-inhibiting action of the sugar (Fikiin *et al.*, 2003). The objective of modelling in these cases is to predict both the temperature and the concentration profiles and histories in the food. As solutes penetrate the tissue, the freezing point will be depressed to various extents and it is important to take this effect into account.

Lucas *et al.* (2001) modelled the immersion freezing in a concentrated solution of a one-dimensional slab of inert porous material impregnated with dilute aqueous solution. The equations to be solved are those for diffusion of solute (equation (3.36)) and heat conduction (equation (3.1)). The porous medium is treated as a homogeneous phase, with average transport properties calculated from the fractions of ice, bead solid, water and solute. In particular, diffusion takes place in the liquid channels only and hence the effective mass diffusivities must take into account the void fraction and tortuosity as in equation (3.39) (Lucas *et al.*, 1999). The transport equations are then solved by finite difference. The model is verified by the brine immersion freezing of a bed of glass beads impregnated with dilute NaCl solution. The model shows that at the start of the process, solute diffusion can prevent freezing at the surface and enhance the thawing of recently frozen layers. This results in a non-frozen surface layer co-existing with an inner frozen layer or frozen core (Fig. 3.7). Continuing solute diffusion in the non-frozen layers causes the adjacent ice crystals to melt. If this process is allowed to continue, the product thaws completely and its solute concentration approximates that of the solution. A similar model was developed by Zorrilla and Rubiolo (2005) who use the enthalpy formulation for the heat transport equation.

3.6 SUPERCOOLING AND NUCLEATION EFFECTS

So far we have assumed that the freezing rate is entirely controlled by heat transfer. However, in many cases, the dynamics of nucleation and mass transfer have observable effects. Some degree of supercooling is observed in most food freezing processes, where the surface dips briefly below freezing point before suddenly coming up to the freezing temperature. Once nucleation has happened, in most industrial freezing processes, the freezing process reverts

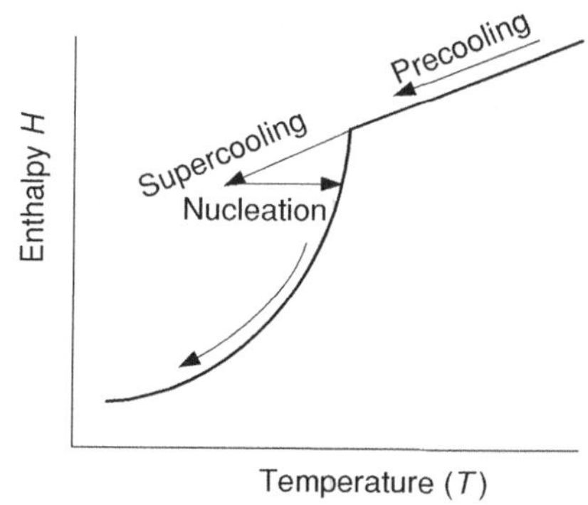

Fig. 3.8 Modelling supercooling and nucleation on the H–T diagram.

to being heat transfer controlled. Pham (1989) has modelled this type of behaviour with a finite difference model using the quasi-enthalpy method and validated the model with data from Menegalli *et al.* (1978). Miyawaki *et al.* (1989) independently used the apparent specific heat technique to solve the same problem. To simulate supercooling, the specific heat above freezing is assumed to continue to apply below the initial freezing point, until the coldest node reaches nucleation temperature. At that point, all the nodes that have an enthalpy value H below freezing are assumed to freeze instantaneously, releasing enough latent heat for the node to warm up to the equilibrium temperature $T(H)$ (Fig. 3.8). Incidentally, Fig. 3.8 resembles Fig. 3.5, which illustrates Pham's temperature correction step.

Pham (1989) found that supercooling normally has negligible effect on freezing time. However, this conclusion may not hold for all types of foods. When water is held as small droplets in an emulsion, such as in ice cream or butter, each ice crystal cannot grow beyond its droplet and each droplet has to crystallise separately. In such cases the freezing process may be very gradual and a freezing plateau may not even be present, as was observed experimentally (Nahid *et al.*, 2004). This may have important implication on the heat load and product temperatures in freezers and cold stores. The product may undergo internal warming due to gradual latent heat release during cold storage, causing higher than normal product temperature and heat load. Nahid *et al.* (2006) use the explicit finite difference method to model the freezing of butter blocks in air at −11°C and −23°C, assuming that nucleation takes place in individual water droplets according to the model of Avrami (1939, 1940, 1941). At constant temperature, Avrami's model predicts that

$$F = 1 - \exp(Kt^n) \tag{3.40}$$

where F is the frozen fraction of water and K describes the combined effect of the rate of nucleation and the rate of crystal growth. The former depends on the degree of supercooling (Michelmore and Franks, 1982), while the latter is proportional to the difference between the equilibrium freezing temperature of the liquid phase and the crystal temperature (Müller *et al.*, 2004). When temperature is not constant, Avrami's equation can be differentiated to give the rate of freezing:

$$\frac{dF}{dt} = nK^{1/n}(1 - F)[-\ln(1 - F)]^{1-1/n} \tag{3.41}$$

Nahid *et al.* obtained good agreement between calculated and measured product temperatures, both exhibiting a cooling period followed by a temperature peak as freezing occurs, rather than the usual temperature plateau.

Nucleation and crystal growth may also affect food quality and drip loss. Maximum drip loss in meat is believed to happen when a large intracellular crystal forms in each cell, which causes distortion and damage to the cell wall (Bevilacqua *et al.*, 1979). This happens at an intermediate freezing rate, since faster freezing causes the formation of multiple small intracellular crystals, while slower freezing leads to extra-cellular freezing. Devireddy *et al.* (2002) developed a finite volume model to predict the formation of intracellular ice in biological tissues in the context of cryosurgery. The material is divided into two phases, extra- and intracellular. Extracellular liquid is assumed to freeze without supercooling. As it does so, the extracellular solute concentration increases, causing an osmotic pressure to develop and water to diffuse across the cell membrane (Mazur, 1963). Due to this diffusion, intracellular solute concentration increases, but at a rate slower than extracellular concentration. However, the temperature in the cell falls just as fast as that of the extracellular medium. There is thus some degree of supercooling inside the cell. The probability of intracellular nucleation is obtained from the amount of supercooling, using a model by Toner (1993).

Because of the mass transfer process between extra- and intracellular spaces and the supercooling of the latter, equation (3.1) cannot be solved directly as in heat transfer–controlled freezing. Instead an iterative procedure has to be carried out at every time step to satisfy the heat balance as well as the intra- and extracellular mass balances:

Guess the rate of ice formation in each control volume;
Iterate:
- Calculate the new nodal temperatures throughout the domain from the heat conduction equation (3.1) and the amount of ice formed over the time step;
- Calculate the extracellular ice from the temperatures, assuming thermodynamic equilibrium in the extracellular space;
- Calculate the probabilities of intracellular nucleation, i.e. the increase in the number of cells with internal ice;
- Calculate the amount of intracellular ice;
- Calculate the total latent heat released by extra- and intra-cellular ice until convergence is reached.

An approximate analytical equation for predicting crystal size from dendritic growth theory was presented by Woinet *et al.* (1998a, 1998b), assuming a Neumann boundary condition, and validated against data from agar gel freezing. Udaykumar *et al.* (2002) present and validate a finite volume technique for computing dendritic growth of crystals from pure melts, assuming diffusion control. An entirely different modelling approach is the use of cellular automata or hybrid automata (Brown, 1998; Jarvis *et al.*, 2000; Raabe, 2002), where the material is modelled as a collection of microscopic elements that change phase stochastically depending on the state of the surrounding elements. So far, little work has been done in the food-freezing field on the modelling of nucleation and crystal growth, although it seems to be a promising area of investigation. The main difficulty in applying these models is the difficulty of getting data on relevant parameters such as cell membrane mass transfer resistance and nucleation factors. Nucleation can be either homogeneous (occurring in pure water) or heterogeneous (occurring due to contact with other substances). Homogeneous nucleation is very rare in food freezing, while heterogeneous nucleation depends on the composition and microstructure of the food and cannot easily be predicted.

3.7 HIGH-PRESSURE FREEZING AND THAWING

High-pressure freezing, and in particular pressure shift freezing, are gaining attention as a freezing method for high-quality or freeze-sensitive foods (Le Bail *et al.*, 2002). In pressure shift freezing, the food is cooled under high pressure to sub-zero temperatures. Because the freezing point decreases with pressure, phase change does not take place. When the product temperature has more or less equilibrated, pressure is released suddenly. The food is now supercooled by several degrees and nucleation takes place spontaneously throughout the supercooled product, causing an instantaneous temperature rise. There may be a short period of equilibration where some water remains supercooled (Otero and Sanz, 2000). The uniform nucleation ensures evenly small crystal size and minimal textural damage. High-pressure thawing has also been investigated as a fast thawing method. Because of the lowering of the freezing point, the difference between product and ambient temperature is increased, hence a larger heat flux is obtained.

In high-pressure thawing and freezing, the effect of pressure on thermal properties (latent heat, freezing point, thermal conductivity) must be taken into account. Freezing point is decreased by pressure according to Clapeyron's equation. Latent heat is also decreased. Thermal conductivity below zero will also be different, since there is no ice.

Chourot *et al.* (1997) modelled high-pressure thawing of an infinite cylinder of pure water, using FDM with Crank–Nicolson stepping and the apparent specific heat approach. The latent heat is assumed to contribute a triangular peak spanning 1 K at the base. Thermal conductivity is assumed to be constant above and below the phase change range, and vary linearly over this range. The total latent heat and mean phase change temperature are given as polynomial functions of pressure. The entire thawing process takes place under pressure.

Denys *et al.* (1997) modelled pressure shift freezing using FDM with explicit stepping and apparent specific heat formulation. At the moment of pressure release, the temperature rise from T_i to T_{new} is calculated by an enthalpy balance. The product is assumed to be at uniform temperature when pressure is released; however, in a subsequent paper (Denys *et al.*, 2000) this restriction is relaxed, and the energy balance is carried out node by node.

In the light of what has been discussed earlier on the handling of latent heat and the shortcomings of the apparent specific heat approach, it can be seen that a simpler, more efficient and flexible program, which can handle any temperature and pressure regime, can be written by using the enthalpy or quasi-enthalpy formulations, i.e. at every time step:

- calculate nodal enthalpies from the enthalpy or quasi-enthalpy method
- calculate nodal temperatures from nodal enthalpies and pressure.

Figure 3.9 shows a pressure shift process on the enthalpy–temperature diagram. The food is cooled at high pressure from A to B, pressure is released along BC causing partial freezing, then freezing is completed along CD. From the programming point of view this presents no extra complication over a 'standard' freezing program, apart from the need to solve for T from H and P instead of from H alone. If only a finite number of pressure levels are used (usually two), the property subroutine needs only contain the H–T relationship at these pressures. Also, if it is not intended that nucleation happens during the high-pressure stage, there is no need to know the frozen section (dotted curve) of the H–T curve for high pressure.

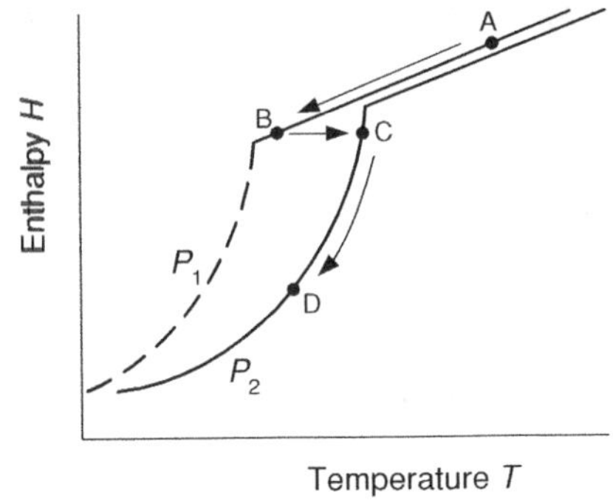

Fig. 3.9 Pressure shift freezing on the enthalpy–temperature diagram.

3.8 THERMOMECHANICAL EFFECTS DURING FREEZING

Water expands by about 9% by volume when turning into ice, causing considerable stresses in foods during freezing. In cryogenic freezing, this expansion is followed by a significant thermal contraction, of the order of 0.5% in linear terms or 1.5% in volumetric terms (Rabin *et al.*, 1998). Frozen food is brittle and these stresses may cause cracking in the food, especially at high cooling rates such as in cryogenic freezing. Rubinsky *et al.* (1980) carried out an approximate analytical analysis of thermal stresses during the freezing of organs but neglected phase-related volume change. Rabin and Steif (1998) calculated thermal stresses in freezing a sphere, taking both phase change expansion and thermal contraction into account, but neglecting the property changes due to freezing. Shi *et al.* (1998, 1999) carried out thermal strain and stress calculations using the commercial software ABAQUS (ABAQUS, Inc., Rhode Island, USA). They used both an elastic model and a viscoelastic model, but neglected thermal contraction. Pham *et al.* gave numerical solutions (Pham *et al.*, 2005) and analytical solutions (Pham *et al.*, 2006) of strains and stresses in a freezing elastic sphere.

At moderate values of strains and stresses, thermophysical properties can be assumed to be unaffected, therefore the analysis can be carried out in two stages: the thermal history is calculated first, using any of the methods listed earlier, followed by stress and strain calculation. The stress analysis assumes that total strain is the sum of thermal strain (due to temperature change) and mechanical strain (due to mechanical stresses):

$$\varepsilon_{ij} = \varepsilon_{ij}^{(\mathrm{m})} + \varepsilon_{ij}^{(\mathrm{T})} \tag{3.42}$$

where the thermal strain can be calculated as a function temperature and includes both the phase change expansion and thermal contraction of ice.

In general it can be said that strain and stress calculations are a specialised field best left to experts and specialised software. However, in the case of spherical foods, the problem is greatly simplified by the disappearance of most of the terms in the stress and strain equations except two: the radial stress and the tangential stress. This leaves two ODEs that can easily be solved. However, there is uncertainty as to whether a solid food behaves as a free-flowing liquid (Rabin and Steif, 1998) or as a rigid solid (Pham *et al.*, 2005, 2006) just before it changes phase. The two assumptions lead to quite different predictions in the resulting strains and stresses. In the former case, the unfrozen food is squeezed out of the advancing ice front, increasing core pressure; in the latter case, the food freezes as a solid and simply tries to expand, causing internal tension in the core.

3.9 FREEZING OF LIQUID FOODS

The freezing of liquid foods such as ice cream freezing or the freeze concentration of fruit juices is a complex problem that has received little attention to date. Due to the need to simultaneously solve the flow equations and the energy transport equation (and perhaps the species transport equation as well), it must be modelled numerically, using computational fluid dynamics (CFD) software. The energy equation, equation (3.1), now incorporates a convection term $\rho c\vec{v} \cdot \nabla T$ and becomes:

$$\rho c \frac{\partial T}{\partial t} + \rho c\vec{v} \cdot \nabla T = \nabla \cdot (k\nabla T) + q \tag{3.43}$$

or in terms of enthalpy (equation 3.27)

$$\rho \frac{\partial H}{\partial t} + \rho\vec{v} \cdot \nabla H = \nabla \cdot (k\nabla T) + q \tag{3.44}$$

while additional transport equations must be solved for continuity and momentum transport (one for each direction of space), and, if the flow is turbulent, for the variables characterising the turbulence field (such as turbulence intensity k and turbulence dissipation ε in the k-ε model). The energy equation can be solved using any of the methods described earlier.

The main problem with the momentum equation is how to deal with the solidification of the fluid, which causes velocity to vanish. The momentum equation can be written as

$$\rho \frac{\partial \vec{v}}{\partial t} + \rho\vec{v} \cdot \nabla\vec{v} = \mu_{\text{eff}}\nabla^2\vec{v} - \nabla p + \vec{S} \tag{3.45}$$

where μ_{eff} is the effective viscosity (including turbulent viscosity) and $\vec{S}$ is the momentum source term (usually the gravitational force). Bennon and Incropera (1987) review the various methods of handling solidification. In one class of methods, the solid and liquid phases are treated as different domains with a clear boundary between them (similarly to the moving grid method for solving the freezing of solid foods). The method is complicated and cannot be applied to materials which do not exhibit a sharp phase change temperature.

A second approach involves changing the viscosity. The viscosity of the fluid is specified as a function of temperature and increases more or less gradually by several order of magnitudes as the temperature falls below the freezing point. A reasonable freezing temperature range must be defined to avoid too sharp a change, which would cause convergence problem. The method is easy to implement (commercial numerical software normally allow viscosity to be entered as a function of temperature) and intuitively attractive, since it reflects the behaviour of most real materials. However, if a realistic viscosity–temperature curve is used, convergence will be very slow or impossible due to the very large and steep change in viscosity over a small temperature range. Therefore a much smaller value of solid viscosity may have to be used, and the solution will only be approximate, with a small but finite residual velocity in the frozen region.

A third method termed the *source method*, or *porosity method*, is employed in commercial CFD software packages such as CFX and Fluent, which are extensively used to model the casting of metals and polymers. From the temperature field, the frozen fraction at each point is calculated. The frozen fraction is visualised as a porous solid phase, which causes a resistance to the flow of the liquid fraction. This resistance is entered into the momentum equation as a negative source term, which is proportional to velocity and increases with the frozen fraction F according to the Kozeny–Carman relationship:

$$\vec{S} = -A\frac{F^2}{(1-F)^3}\vec{v} \tag{3.46}$$

A is a proportionality constant which is related to the viscosity and drag coefficient. A large value of A causes a sharper transition from liquid to solid behaviour, but may cause convergence problems. A small number is usually added to the denominator of equation (3.46) to prevent the source term from going to infinity.

3.10 CONCLUSIONS

The numerical modelling of the classical 'pure thermal' freezing problem can be considered solved in principle. An enthalpy or quasi-enthalpy (temperature correction) method is recommended, in conjunction with FVM or lumped capacitance FEM. Explicit time stepping is recommended for small or one-off problems, Pham's quasi-enthalpy method for those who want speed as well as uncomplicated programming. However, it may be difficult for the user of commercial software to apply the enthalpy or quasi-enthalpy methods, in which case an iterative apparent specific heat technique has to be used. Iterative techniques are also a viable alternative in the solution of coupled equations (freezing with mass transfer, freezing of liquids) since iteration is required in any case.

Freezing is almost always coupled with other physical processes, such as mass transfer, nucleation, crystal growth, mass transfer across cell membrane, vitrification, thermal expansion, mechanical strain and stress and cracking. Even in normal freezing, internal pressure may have some effect on the freezing point that has up till now been neglected. The food engineer is interested not only in freezing times or heat loads, but also in food quality factors: drip, colour, texture, flavour, distortion and cracks and microbial growth. To predict these, detailed modelling is needed on physical processes other than heat transfer.

To model non-thermal phenomena successfully, data on some food properties hitherto neglected by food technologists (moisture diffusivity, absorption isotherm, nucleation parameters, cell size, cell membrane permeability, viscoelastic properties, tortuosity factor in porous foods, etc.) will have to be collected. For the modelling of high-pressure freezing and thawing, there is a need for more data and prediction methods for thermal properties as a function of pressure. More and better data on heat transfer coefficients and thermal properties and better methods for their prediction will also be an ongoing area of research. The use of computational fluid dynamics (CFD) for calculating heat transfer coefficients in food refrigeration is increasingly becoming popular, but the lack of a satisfactory turbulence model for many practical situations (circulating flows, natural or mixed convection) means that CFD results cannot yet be completely trusted. More sophisticated turbulence models are on the way, but will not become practically useful until much faster computers are available.

NOMENCLATURE

a_w water activity
c specific heat ($J\ kg^{-1}\ K^{-1}$)
$\mathbf{B}$ $\nabla^T\mathbf{N}$
Bi Biot number (hR/k)
$\mathbf{C}$ capacitance matrix
d_l limiting crystal size (m)
d_m average crystal size (m)
d_s average size of non-ice inclusions (m)
D_v diffusivity of water vapour in air ($m^2\ s^{-1}$)
D_w diffusivity of water ($m^2\ s^{-1}$)
D_δ effective diffusivity of water vapour in desiccated layer ($m^2\ s^{-1}$)
E Shape factor for freezing
$\mathbf{f}$ forcing vector
h heat transfer coefficient ($W\ m^{-2}\ K^{-1}$)
H specific enthalpy ($J\ kg^{-3}$)
$\mathbf{H}$ vector of nodal enthalpies ($J\ kg^{-3}$)
i, j node number
k thermal conductivity ($W\ m^{-1}\ K^{-1}$)
k_1, k_2 rate constants (s^{-1})
$k_{eva\text{-}con}$ thermal conductivity due to evaporation–condensation ($W\ m^{-1}\ K^{-1}$)
k_g mass transfer coefficient ($kg\ m^{-2}\ s^{-1}\ Pa^{-1}$)
L_f Latent heat of freezing ($J\ kg^{-1}$)
$\mathbf{K}$ conductance matrix
L_v latent heat of vapourisation ($J\ kg^{-1}$)
$\dot{m}$ mass flux ($kg\ m^{-2}\ s^{-1}$)
$\mathbf{M}$ mass matrix (kg)
M_w molecular mass of water ($kg\ kmol^{-1}$)
$\mathbf{n}$ vector normal to the freezing front and pointing into the frozen region (m)
$\mathbf{N}$ vector of shape functions
P partial pressure of water (Pa)
q heat generation ($J\ m^{-3}$)
$\mathbf{r}$ position vector (m)
R half dimension of object (m)
R_g universal gas constant (8314.4 $J\ kmol^{-1}\ K^{-1}$)
t time (s)
Δt time step (s)
T temperature (K)
T_a environment temperature (K)
T_f initial freezing temperature (K)
$\mathbf{T}$ nodal temperature vector (K)
u Kirchhoff transform variable, $\mu = \int_{T_{REF}}^{T} k\,dQ$ ($W\ m^{-1}$)

$\vec{v}$ fluid velocity vector (m s^{-1})
v_f velocity of freezing front (m s^{-1})
δV control volume (m^3)
W total water concentration (kg m^{-3})
$\mathbf{x}$ position vector (m)
α weighting coefficient for time stepping
δ thickness of desiccated layer (m)
ε_{ij} strain
φ factor for diffusion in porous media
ρ density (kg m^{-3})
ρ_s bulk density of dry solid (kg m^{-3})
Ω element or control volume domain

Subscripts
a value in bulk air
eff effective
i, j node number
sat saturation value for pure water
w water

Superscripts
(e) belonging to one element
(m) mechanical
New value at the end of time step
(T) thermal.

REFERENCES

Avrami, M. (1939). Kinetics of phase change I. *Journal of Chemical Physics* 7, 1103–1112.

Avrami, M. (1940). Kinetics of phase change II. *Journal of Chemical Physics* 8, 212–224.

Avrami, M. (1941). Kinetics of phase change III. *Journal of Chemical Physics* 9, 177–184.

Banaszek, J. (1989). Comparison of control volume and Galerkin finite element methods for diffusion-type problems. *Numerical Heat Transfer* 16, 59–78.

Bennon, W.D. and Incropera, F.P. (1987). A continuum model for momentum, heat and species transport in binary solid-liquid phase change systems – I. Model formulation. *International Journal of Heat and Mass Transfer* 30, 2161–2170.

Bevilacqua, A., Zaritsky, N.E. and Calvelo, A. (1979). Histological measurements of ice in frozen beef. *Journal of Food Technology* 1, 237–251.

Brown, S.G.R. (1998). Simulation of diffusional composite growth using the cellular automaton finite difference (CAFD) method. *Journal of Materials Science* 33, 4769–4773.

Budhia, H. and Kreith, F. (1973). Heat transfer with melting or freezing in a wedge. *International Journal of Heat and Mass Transfer* 16, 195–211.

Califano, A.N. and Zaritzky, N.E. (1997). Simulation of freezing or thawing heat conduction in irregular two dimensional domains by a boundary fitted grid method. *Lebensmittel-Wissenschaft und- Technologie* 30, 70–76.

Campañone, L.A., Salvadori, V.O. and Mascheroni, R.H. (2001). Weight loss during freezing and storage of unpackaged foods. *Journal of Food Engineering* 47, 69–79.

Carslaw, H.S. and Jaeger, J.C. (1959). *Conduction of Heat in Solids*, 2nd edn. Oxford: Clarendon Press.

Chourot, J.M., Boillereaux, L., Havet, M. and Le Bail, A. (1997). Numerical modeling of high pressure thawing: application to water thawing. *Journal of Food Engineering* 34, 63–75.

Cleland, A.C. (1990). *Food Refrigeration Processes Analysis, Design and Simulation*. London: Elsevier.
Cleland, D.J., Cleland, A.C. and Earle, R.L. (1987a). Prediction of freezing & thawing times for multidimensional shapes by simple formulae. Part 1: Regular shapes. *International Journal of Refrigeration* 10, 157–164.
Cleland, D.J., Cleland, A.C. and Earle, R.L. (1987b). Prediction of freezing & thawing times for multidimensional shapes by simple formulae. Part 2: Irregular shapes. *International Journal of Refrigeration* 10, 234–240.
Cleland, D.J., Cleland, A.C., Earle, R.W. and Byrne, S.J. (1984). Prediction of rates of freezing, thawing and cooling in solids of arbitrary shape using the finite element method. *International Journal of Refrigeration* 7, 6–13.
Cleland, A.C. and Earle, R.L. (1977). A comparison of analytical and numerical methods for predicting the freezing times of foods. *Journal of Food Science* 42, 1390–1395.
Cleland, A.C. and Earle, R.L. (1979a). A comparison of methods for predicting the freezing times of cylindrical and spherical foodstuffs. *Journal of Food Science* 44, 958–963.
Cleland, A.C. and Earle, R.L. (1979b). Prediction of freezing times for foods in rectangular packages. *Journal of Food Science* 44, 964–970.
Cleland, A.C. and Earle, R.L. (1982). A simplified method for prediction of heating and cooling rate in solids of various shapes. *International Journal of Refrigeration* 5, 98–106.
Cleland, A.C. and Earle, R.L. (1984). Assessment of freezing time prediction methods. *Journal of Food Science* 49, 1034–1042.
Comini, G. and Del Giudice, S. (1976). Thermal aspects of cryosurgery. *Journal of Heat Transfer* 98, 543–549.
Comini, G., Del Giudice, S. and Saro, O. (1989). Conservative equivalent heat capacity methods for non-linear heat conduction. In: R.W. Lewis and K. Morgan, eds. *Numerical Methods in Thermal Problems*, Vol. 6, Part 1. Swansea: Pineridge Press, pp. 5–15.
Crank, J. and Nicolson, P. (1947). A practical method for numerical integration of solutions of partial differential equations of heat conduction type. *Proceedings of Cambridge Philosophical Society* 43, 50–67.
Davey, L.M. and Pham, Q.T. (1997). Predicting the dynamic product heat load and weight loss during beef chilling using a multi-region finite difference approach. *International Journal of Refrigeration* 20, 470–482.
de Michelis, A. and Calvelo, A. (1982). Mathematical models for non-symmetric freezing of beef. *Journal of Food Science* 47, 1211–1217.
De Vries, D.A. (1952). The thermal conductivity of granular materials. *Annex 1952-1 Bulletin of the International Institute of Refrigeration* pp. 115–131.
De Vries, D.A. (1958). Simultaneous transfer of heat and moisture in porous media. *Transactions of the American Geophysical Union* 39, 909–916.
De Vries, D.A. (1987). The theory of heat and moisture transfer in porous media revisited. *International Journal of Heat and Mass Transfer* 30, 1343–1350.
Denys, S., Van Loey, A.M. and Hendrickx, M.E. (1997). Modeling heat transfer during high-pressure freezing and thawing. *Biotechnology Progress* 13, 416–423.
Denys, S., Van Loey, A.M. and Hendrickx, M.E. (2000). Modeling conductive heat transfer during high-pressure thawing processes: determination of latent heat as a function of pressure. *Biotechnology Progress* 16, 447–455.
Devireddy, R.V., Smith, D.J. and Bischof, J.C. (2002). Effect of microscale mass transport and phase change on numerical prediction of freezing in biological tissues. *Journal of Heat Transfer* 124, 365–374.
Eyres, N.R., Hartree, D.R., Ingham, J., Jackson, R., Sarjant, R.J. and Wagstaff, J.B. (1946). The calculation of variable heat flow in solids. *Transactions of the Royal Society* A240, 1.
Fikiin, K.A. (1996). Generalised numerical modelling of unsteady heat transfer during cooling and freezing using an improved enthalpy method and quasi-one-dimensional formulation. *International Journal of Refrigeration* 19, 132–140.
Fikiin, K.A. and Fikiin, A.G. (1998). Individual quick freezing of foods by hydrofluidisation and pumpable ice slurries. In: *Advances in the Refrigeration Systems, Food Technologies and Cold Chain*, Proceedings of IIR Conference, Sofia (Bulgaria), Refrigeration Science and Technology, International Institute of Refrigeration, pp. 319–326. Published also in *AIRAH Journal* (2001), 55(11), 15–18.
Fikiin, K., Tsvetkov, O., Laptev, Yu., Fikiin, A. and Kolodyaznaya, V. (2003). Thermophysical and engineering issues of the immersion freezing of fruits in ice slurries based on sugar-ethanol aqueous solution. *Ecolibrium* August, 10–14.

Hamdami, N., Monteau, J.Y. and Le Bail, A. (2004a). Heat and mass transfer simulation during freezing in bread. In: *Proceedings ICEF9 9th International Congress on Engineering and Foods, Montpellier* (CDROM).

Hamdami, N., Monteau, J.Y. and Le Bail, A. (2004b). Simulation of coupled heat and mass transfer during freezing of a porous humid matrix. *International Journal of Refrigeration* 27, 595–603.

Hossain, Md.M., Cleland, D.J. and Cleland, A.C. (1992a). Prediction of freezing & thawing times for foods of regular multidimensional shape by using an analytically derived geometric factor. *International Journal of Refrigeration* 15, 227–234.

Hossain, Md.M., Cleland, D.J. and Cleland, A.C. (1992b). Prediction of freezing and thawing times for foods of two-dimensional irregular shape by using a semi-analytical geometric factor. *International Journal of Refrigeration* 15, 235–240.

Hossain, Md.M., Cleland, D.J. and Cleland, A.C. (1992c). Prediction of freezing and thawing times for foods of three-dimensional irregular shape by using a semi-analytical geometric factor. *International Journal of Refrigeration* 15, 241–246.

Illicali, C. and Holacar, M. (1990). A simplified approach for predicting the freezing times of foodstuffs of anomalous shape. In: W.E.L. Spiess and H. Schubert, eds. *Engineering and Food*, Vol. 2. London: Elsevier, pp. 418–425.

Jarvis, D.J., Brown, S.G.R. and Spittle, J.A. (2000). Modelling of non-equilibrium solidification in ternary alloys: comparison of 1D, 2D and 3D cellular automaton-finite difference simulation. *Materials Science and Technology* 16, 1420–1424.

Carslaw, H.S. and Jaeger, J.C., eds. (1959). *Conduction of Heat in Solids*, 2nd edn. Oxford: Clarendon Press, p. 11 (Kirchhoff, G. (1894). *Vorlesungen uber die Theorie der Warme*).

Le Bail, A., Chevalier, D., Mussa, D.M. and Ghoul, M. (2002). High pressure freezing and thawing of foods: a review. *International Journal of Refrigeration*. 25, 504–513.

Lees, M. (1976). A linear three-level difference scheme for quasilinear parabolic equations. *Mathematics of Computation* 20, 516–522.

Lemmon, E.C. (1979). Phase change technique for finite element conduction code. In: R.W. Lewis and K. Morgan, eds. *Numerical Methods in Thermal Problems*. Swansea: Pineridge Press, pp. 149–158.

Lovatt, S.J., Pham, Q.T., Loeffen, M.P.F. and Cleland, A.C. (1993a). A new method of predicting the time-variability of product heat load during food cooling, Part 1: Theoretical considerations. *Journal or Food Engineering* 18, 13–36.

Lovatt, S.J., Pham, Q.T., Loeffen, M.P.F. and Cleland, A.C. (1993b). A new method of predicting the time-variability of product heat load during food cooling, Part 2: Experimental testing. *Journal or Food Engineering* 18, 37–62.

Lucas, T., Flick, D. and Raoult-Wack, A.L. (1999). Mass and thermal behaviour of the food surface during immersion freezing. *Journal of Food Engineering* 41, 23–32.

Lucas, T., Chourot, J.M., Bohuon, Ph. and Flick, D. (2001). Freezing of a porous medium in contact with a concentrated aqueous freezant: numerical modelling of coupled heat and mass transport. *International Journal of Heat and Mass Transfer* 44, 2093–2106.

Mazur, P. (1963). Kinetics of water loss from cells at subzero temperature and the likelihood of intracellular freezing. *Journal of General Physiology* 47, 347–69.

McNabb, A., Wake, G.C., Hossain, Md.M. and Lambourne, R.D. (1990a). Transition times between steady states for heat conduction, Part I: General theory and some exact results. *Occasional Pubs in Maths & Statistics* No. 20, Massey University.

McNabb, A., Wake, G.C., Hossain, Md.M. and Lambourne, R.D. (1990b). Transition times between steady states for heat conduction, Part II: Approximate solutions and examples. *Occasional Pubs in Maths & Statistics* No. 21, Massey University.

Menegalli, F.C. and Calvelo, A. (1978). Dendritic growth of ice crystals during freezing of beef. *Meat Science* 3, 179–199.

Michelmore, R.W. and Franks, F. (1982). Nucleation rates of ice in undercooled water and aqueous solutions of polyethylene glycol. *Cryobiology* 19, 163–171.

Miyawaki, O., Abe, T. and Yano, T. (1989). A numerical model to describe freezing of foods when supercooling occurs. *Journal of Food Engineering* 9, 143–151.

Morgan, K., Lewis, R.W. and Zienkiewicz, O.C. (1978). An improved algorithm for heat conduction problems with phase change. *International Journal of Numerical Methods in Engineering* 12, 1191–1195.

Müller, G., Métois, J. and Rudolph, P. (2004). *Crystal Growth from Fundamentals to Technology*. Amsterdam: Elsevier.

Nahid, A., Bronlund, J.E., Cleland, D.J., Oldfield, D.J., Philpott, B. (2004). Prediction of thawing and freezing of bulk palletised butter. In: *Proceedings of the 9th International Congress on Engineering and Foods (ICEF 9)*. Montpellier CDROM; paper print 668 p.

Nahid, A., Bronlund, J.E., Cleland, D.J. and Philpott, B.A. (2006). Modelling the freezing of single cartons of butter. In: *Proceedings 11th Asian Pacific Confederation of Chemical Engineering (APCChe) Congress*. Kuala Lumpur, Malaysia, 27–30/8/2006.

Otero, L. and Sanz, P. (2000). High pressure shift freezing, Part 1: Amount of ice instantaneously formed in the process. *Biotechnology Progress* 16, 1030–1036.

Peaceman, D.W. and Rachford, H.H. (1955). The numerical solution of parabolic and elliptic differential equations. *Journal of the Society for Industrial and Applied Mathematics* 3, 28–41.

Pham, Q.T. (1984). An extension to Plank's equation for predicting freezing times of foodstuffs of simple shapes. *International Journal of Refrigeration* 7, 377–383.

Pham, Q.T. (1985). A fast, unconditionally stable finite-difference method for heat conduction with phase change. *International Journal of Heat and Mass Transfer* 28, 2079–2084.

Pham, Q.T. (1986a). Simplified equation for predicting the freezing time of foodstuffs. *Journal of Food Technology* 21, 209–221.

Pham, Q.T. (1986b). Freezing of foodstuffs with variations in environmental conditions. *International Journal of Refrigeration* 9, 290–295.

Pham, Q.T. (1986c). The use of lumped capacitances in the finite-element solution of heat conduction with phase change. *International Journal of Heat and Mass Transfer* 29, 285–292.

Pham, Q.T. (1987a). A converging-front model for the asymmetric freezing of slab-shaped foodstuffs. *Journal of Food Science* 52, 795–800.

Pham, Q.T. (1987b). A comparison of some finite-difference methods for phase-change problems. *Proceedings 17th International Congress of Refrigeration*. Vienna. Vol. B, pp. 267–271.

Pham, Q.T. (1989). Effect of supercooling on freezing times due to dendritic growth of ice crystals. *International Journal of Refrigeration* 12, 295–300.

Pham, Q.T. (1995). Comparison of general purpose finite element methods for the Stefan problem. *Numerical Heat Transfer Part B – Fundamentals* 27, 417–435.

Pham, Q.T. (2001). Prediction of cooling/freezing/thawing time and heat load. In: D.-W. Sun, ed. *Advances in Food Refrigeration*. Leatherhead: Leatherhead Food RA, pp. 110–152.

Pham, Q.T., Le Bail, A., Hayert, M. and Tremeac, B. (2005). Stresses and cracking in freezing spherical foods: a numerical model. *Journal of Food Engineering* 71, 408–418.

Pham, Q.T., Le Bail, A. and Tremeac, B. (2006). Analysis of stresses during the freezing of spherical foods. *International Journal of Refrigeration* 29, 125–133.

Pham, Q.T. and Karuri, N. (1999). A computationally efficient technique for calculating simultaneous heat and mass transfer during food chilling. *20th International Congress of Refrigeration*. Sydney, 19–24 September, 1999.

Pham, Q.T. and Mawson, R. (1997). Moisture migration and ice recrystallization in frozen foods. In: M.C. Erickson and Y.C. Hung, eds. *Quality in Frozen Foods*. New York: Chapman and Hall, pp. 67–91.

Pham, Q.T. and Willix, J. (1984). A model for food desiccation in frozen storage. *Journal of Food Science* 49, 1275–1281, 1294.

Pham, Q.T. and Willix, J. (1990). Effect of Biot number and freezing rate on the accuracy of some food freezing time prediction methods. *Journal of Food Science* 55, 1429–1434.

Plank, R. (1913a). Beitrage zur Berechnung und Bewertung der Gefriergeschwindigkeit von Lebensmitteln. *Zeitschrift fur die gesamte Kalte Industrie*, Reihe 3 Heft 10, 1–16.

Plank, R. (1913b). Die Gefrierdauer von Eisblocken. *Zeitschrift fur die gesamte Kalte Industrie* 20(6), 109–114.

Raabe, D. (2002). Cellular automata in materials science with particular reference to recrystallization simulation. *Annual Review of Materials Research* 32, 53–76.

Rabin, Y. and Steif, P.S. (1998). Thermal stresses in a freezing sphere and its application in cryobiology *Transactions of the ASME* 65, 328–333.

Rabin, Y., Taylor, M.J. and Wolmark, N. (1998). Thermal expansion measurement of frozen biological tissues at cryogenic temperatures. *ASME Journal of Biomechanical Engineering* 120, 259–266.

Riley, D.S., Smith, F.T. and Poots, G. (1974). The inward solidification of spheres and circular cylinders. *International Journal of Heat and Mass Transfer* 17, 1507–1516.

Rubinsky, B., Cravalho, E.G. and Mikic, B. (1980). Thermal stresses in frozen organs. *Cryobiology* 17, 66–73.

Scheerlinck, N. (2000). *Uncertainty Propagation Analysis of Coupled and Non-Linear Heat and Mass Transfer Models*. PhD thesis, Katholieke Universteit Leuven, pp. 180–181.

Scheerlinck, N., Verboven, P., Fikiin, K.A., de Baerdemaeker, J. and Nicolai, B.M. (2001). Finite-element computation of unsteady phase change heat transfer during freezing or thawing of food using a combined enthalpy and Kirchhoff transform method. *Transactions of the American Society of Agricultural Engineers* 44, 429–438.

Segerlind, L. (1984). *Applied Finite Element Analysis*, 2nd edn. New York: John Wiley.

Shi, X., Datta, A.K. and Mukherjee, Y. (1998). Thermal stresses from large volumetric expansion during freezing of biomaterials. *Transactions of the ASME* 120, 720–726.

Shi, X., Datta, A.K. and Mukherjee, Y. (1999). Thermal fracture in a biomaterial during rapid freezing. *Journal of Thermal Stresses* 22, 275–292.

Thompson, J.F., Warsi, Z.U.A. and Mastin, C.W. (1982). Boundary-fitted coordinate systems for numerical solution of partial differential equations—A review. *Journal of Computational Physics* 47, 1–108.

Toner, M. (1993). Nucleation of ice crystals inside biological cells. In: P. Steponkus, ed. *Advances in Low Temperature Biology*. London: JAI Press, pp. 1–52.

Trujillo, F.J. (2004). *A Computational Fluid Dynamic Model of Heat and Moisture Transfer during Beef Chilling*. PhD thesis, School of Chemical Engineering and Industrial Chemistry, University of New South Wales, Sydney, Australia.

Udaykumar, H.S., Mao, L. and Mittal, R. (2002). A finite-volume sharp interface scheme for dendritic growth simulations: comparison with microscopic solvability theory. *Numerical Heat Transfer, Part B* 42, 389–409.

Van der Sluis, S.M. (1993). Cooling and freezing simulation of bakery products. *Proceedings of International Institute of Refrigeration Meeting Commissions B1, B2, D1, D2/3*. Palmerston North, New Zealand, 1993.

Voller, V.R. (1996). An overview of numerical methods for solving phase change problems. In: W.J. Minkowycz and E.M. Sparrow, eds. *Advances in Numerical Heat Transfer*, Vol. 1. London: Taylor & Francis, pp. 341–375.

Voller, V.R. and Prakash, C. (1987). A fixed-grid numerical modeling methodology for convection-diffusion mushy region phase-change problems. *International Journal of Heat and Mass Transfer* 30, 1709–1720.

Voller, V.R. and Swaminathan, C.R. (1991). Generalized source-based method for solidification phase change. *Numerical Heat Transfer Part B* 19, 175–189.

Voller, V.R., Swaminathan, C.R. and Thomas, B.G. (1990). Fixed grid techniques for phase change problems: a review. *International Journal of Numerical Methods in Engineering* 30, 875–898.

Woinet, B., Andrieu, J. and Laurent, M. (1998a). Experimental and theoretical study of model food freezing, Part 1: Heat transfer modelling. *Journal of Food Engineering* 35, 381–393.

Woinet, B., Andrieu, J., Laurent, M. and Min, S.G. (1998b). Experimental and theoretical study of model food freezing, Part 2: Characterization and modelling of ice crystal size. *Journal of Food Engineering* 35, 394–407.

Zienkiewicz, O.C. and Taylor, R.L. (1991). *The Finite Element Method*. London: McGraw-Hill.

Zorrilla, S.E. and Rubiolo, A.C. (2005). Mathematical modeling for immersion chilling and freezing of foods. Part I: Model development. *Journal of Food Engineering* 66, 329–338.

4 Specifying and Selecting Refrigeration and Freezer Plant

Andy Pearson

4.1 INTRODUCTION

The refrigeration system may be one of the largest users of electricity in a food factory, and certainly will be in a storage and distribution facility where it can account for up to 70% of the site electrical consumption, so it is important to specify requirements accurately and unambiguously in order to ensure efficient operation of the installed plant. Where food is being frozen, whether from raw or after cooking, the way in which it is frozen, including temperatures and freezing rate can have a significant effect on product appearance, quality and shelf life. Slow freezing, perhaps as a result of equipment not meeting the design requirements, can be particularly damaging as it causes the growth of large ice crystals within the product and may damage the cell structure. In some cases the freezing process can significantly reduce weight loss, increasing the value of product sold by weight and in special processes, such as ice cream making, the method of freezing is an integral part of the production process. For all these reasons it follows that great care should be taken when buying a refrigerating plant, not just to avoid expensive mistakes but also to maximise the margin made on the product.

The words 'freezer', 'chiller', 'cold store' and 'chill store' tend to be used loosely in the food production market. For this text a stricter convention, common among refrigeration engineers, will be followed. A 'freezer' is not just a store for cold produce, as the term is used domestically. It denotes a piece of equipment specifically designed to turn fresh products into frozen ones by leading it through a clearly defined freezing process. A 'chiller' is a similar machine, but used to cool product down without taking it through freezing. Note that 'chiller' is also used to denote a unit for cooling liquid, usually water, brine or glycol, but usually the context is sufficiently clear to indicate which type of chiller is being discussed. A 'cold store' is a store-room or warehouse operated below 0°C for holding frozen products and a 'chill room' is similar for chilled products. Cold stores usually run at a temperature somewhere between −10°C and −30°C depending on the product: typically at about −23°C for most foods and down to −29°C for ice cream. Chill store temperatures depend on the product stored, with 0°C usual for fresh meat, 2°C for general chill and dairy produce and 10°C for bakery products. Some fruits and vegetables require particular temperature control, for example, to regulate the ripening of bananas, and often in these cases a strictly controlled ventilation system is required to regulate the quantity of off-gases in the store atmosphere. In some cases the quantity of carbon dioxide in the atmosphere needs to be controlled. In others it is the production of ethylene, which accelerates the ripening process, which is of concern. Significant quantities of fresh air ventilation may be required to control the amounts of these trace gases in the atmosphere. The cooling load required to maintain the correct temperature of ventilation make-up air in summer can often be a significant portion of the total cooling requirement, and in winter heating of the make-up air may be required.

In process areas, where people work for prolonged periods, the temperature control must take account of their comfort requirements. It may be possible to use infrared heating to warm people and not product, and to a certain extent the avoidance of drafts and the stratification of the room air can make the working environment more acceptable, but designers and operators should bear in mind that product quality depends on temperature control, so these requirements must take precedence over personnel comfort. Where possible the temperature control equipment in these areas must be tamper-proof. If the equipment produces a noticeable draught across the work area it is much more likely to generate complaints about temperature. Low air velocity can be achieved by using ducts or fabric sock systems to distribute the chilled air more evenly, or a displacement ventilation system may be used to introduce the chilled air at a low level. However, the work area must be cleanable, so underfloor ducting is not used and fabric socks, designed to be easily removed for regular washing, are preferred to fixed ductwork in most applications. Ducts are less suited to oily or greasy atmospheres, where regular steam cleaning of fixed ductwork may be the only acceptable way to maintain hygiene standards for ducted systems.

The type of equipment chosen by the system designer will have a significant effect on the profitability of the business. Systems that offer good temperature control will minimise product spoilage and can increase the shelf life and the value of the product, but if they use more energy to run, or if they require frequent maintenance and adjustment, they will add unnecessary costs to the production process. Often there is a trade off between energy and maintenance costs in the overall operating cost of the plant. For example, a reciprocating compressor might offer higher efficiency than a screw compressor, but it will also need more frequent and more expensive overhauls to keep it in good condition. If this essential maintenance is neglected then an expensive breakdown might occur and the whole factory's output is put at risk. Another key choice is between evaporative and air-cooled condensers. Air-cooled condensers require very little maintenance, but typically increase the electrical consumption of a plant by up to 30%. Evaporative condensers enable more efficient operation, but they are required by law to be properly maintained with appropriate water treatment to ensure that the system is kept free from legionella bacteria. Even worse, if the water treatment is neglected it can result in expensive damage to the condenser through the use of water that is either too hard or too soft. This chapter explains the basic requirements of a refrigeration system for a food factory and then details some of the key decisions to be taken when the plant is being specified. It goes on to examine issues which may arise during the tendering process, and explains the benefits and disadvantages of various methods of tendering. The effect of some key decisions on the long-term life of the plant is also explained.

4.2 REFRIGERATION CYCLES

4.2.1 Cryogenic fluids

Some freezing processes use cryogenic fluid, usually liquid nitrogen, to achieve very quick low-temperature freezing. This is particularly appropriate for small-scale production in the early stages of product and process development because the capital cost for the nitrogen freezer is very low compared to mechanical refrigeration, and the system is easy to assemble, modify, expand and relocate as the development progresses. Nitrogen freezers may require less floor space than a mechanical refrigeration system, making them easier to set up on a trial basis. However, liquid nitrogen costs are high, as the nitrogen production process is energy intensive. As energy costs rise the use of liquid nitrogen becomes less attractive and the payback period for investment in refrigeration plant becomes shorter. If there is insufficient space in the factory for a mechanical freezer, the user may be locked into a very expensive production process once the product is established. Some consideration should be given to process development when the initial trials are being established. Much of the equipment required for a nitrogen freezer is provided on a lease by the nitrogen supply company, and it may be necessary to commit to a supply contract, either for a minimum quantity of nitrogen or a minimum time period, in order to recoup the supply company's set-up costs. Carbon dioxide is also used for some products, although it forms dry powder when liquid is sprayed into the atmosphere, which limits its application. It is also possible to use synthetic liquefied air (SLA), which has the advantage of maintaining a safe atmosphere with no danger of oxygen deprivation. However, SLA is more expensive than liquid nitrogen, which makes it progressively less attractive as energy prices rise.

4.2.2 Compressor systems

Most industrial refrigeration systems use compressors to pump a cooling fluid to a suitable pressure, usually rejecting heat to ambient. The boiling temperature of liquids dependents on the pressure, with lower boiling points at low pressure, so by compressing the gas evaporated from the cold side of the refrigerating system it is possible to recondense it at a higher temperature. During boiling heat energy is absorbed by the refrigerant at constant temperature, and when the gas recondenses this energy is removed. A mechanical refrigeration cycle therefore has four main elements: an evaporator, a compressor, a condenser and an expander, as shown in Fig. 4.1.

The evaporator cools the product or process by running the refrigerant at a low enough pressure to boil it at a temperature colder than the product. The compressor takes the outlet gas from the evaporator and raises the pressure so that the boiling point of the refrigerant rises. This requires some power input, and the work done in compressing the gas is converted to additional heat in the refrigerant. The condenser brings the high-pressure refrigerant gas in contact with a relatively cool sink for the heat, and as heat is removed from the gas it turns back to liquid (condenses). The heat removed from the gas as it condenses is the sum of the heat which was extracted from the product or process and the heat created during the compression process. The expander is necessary to reduce the liquid pressure back down to the evaporating pressure. Often the expander is just a flow-regulating valve, called an expansion valve, but in more complex systems a work recovery device, acting like a compressor in reverse, can be used to improve the system efficiency. The expansion device is also used to regulate the flow of liquid to the evaporator so that it is not overfilled. In small systems a simple thermostatic valve can be used, where a sensor at the evaporator outlet compares the temperature there with the liquid inlet temperature. This type of system is called 'direct expansion', or 'DX' for short, because the refrigerant flow goes directly from the expansion valve to the evaporator. The flow is controlled to ensure that the outlet is always a few degrees warmer than the inlet, indicating that all the liquid is being evaporated and the outlet is dry, so this arrangement is sometimes also called 'dry expansion'. This evaporator outlet pipe often feeds the compressor directly, so it is important to prevent any liquid flowing into the compressor, where it could cause severe damage by flooding the compression space or by washing lubricant away from

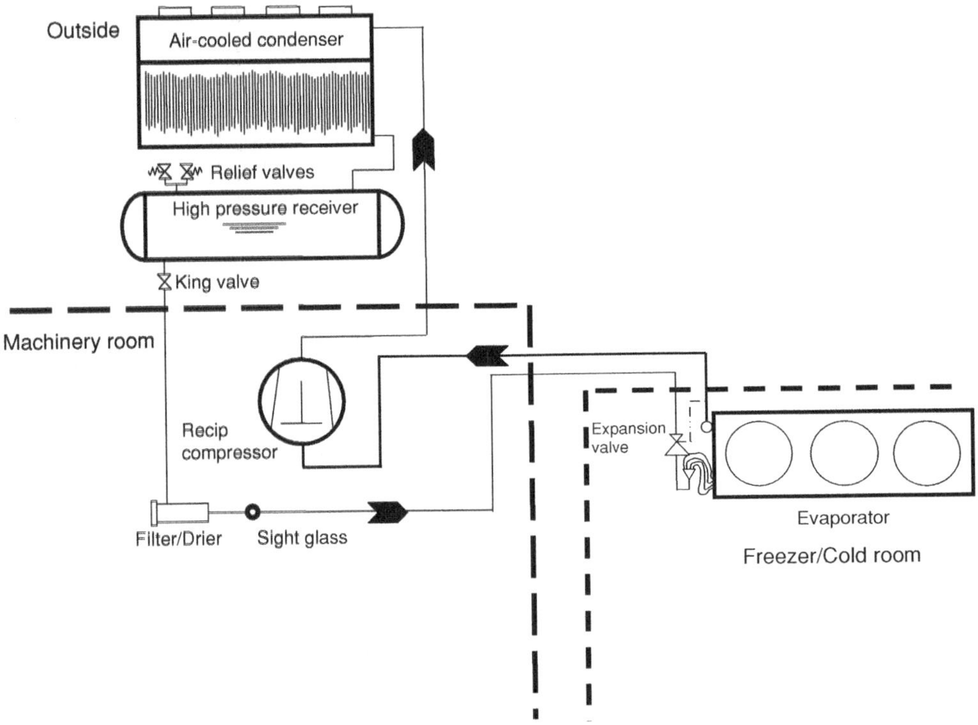

Fig. 4.1 A simplified refrigeration cycle (direct expansion).

the bearing surfaces. Larger systems may use a pumped recirculation system, as shown in Fig. 4.2; in this case the refrigerant flow is fed to a pressure vessel in the low-pressure part of the system and the flow is regulated to ensure that the vessel is not overfilled. When the liquid is pumped the evaporator outlet is usually wet, as the extra liquid is returned to the pump supply pressure vessel, which acts as a liquid separator and ensures that the compressor inlet remains dry. These systems are described in more detail later in the chapter. It should be noted that both of these are called 'direct system' because the refrigerant is in direct contact with the process or product-cooling heat exchanger. This is in contrast to an 'indirect system' where the refrigeration plant cools a heat transfer fluid, typically a glycol or salt solution, and the heat transfer fluid is pumped to the process or product-cooling heat exchanger. Care is required in discriminating between 'direct systems' and 'direct expansion systems'.

4.2.3 Absorption systems

It is also possible to vary the boiling point of the refrigerant by combining it with a second fluid and varying the relative proportions of the two components as they circulate round the system. These are known as absorption systems, and are typically used for large water chillers. The absorption process is also capable of running at lower temperatures for freezer plants, but such systems are very uncommon. The two most common fluid pairs are ammonia/water and water/lithium bromide. In the former pair, ammonia is the refrigerant and the plant is capable of low-temperature operation. Water is the absorber. In the water/lithium bromide

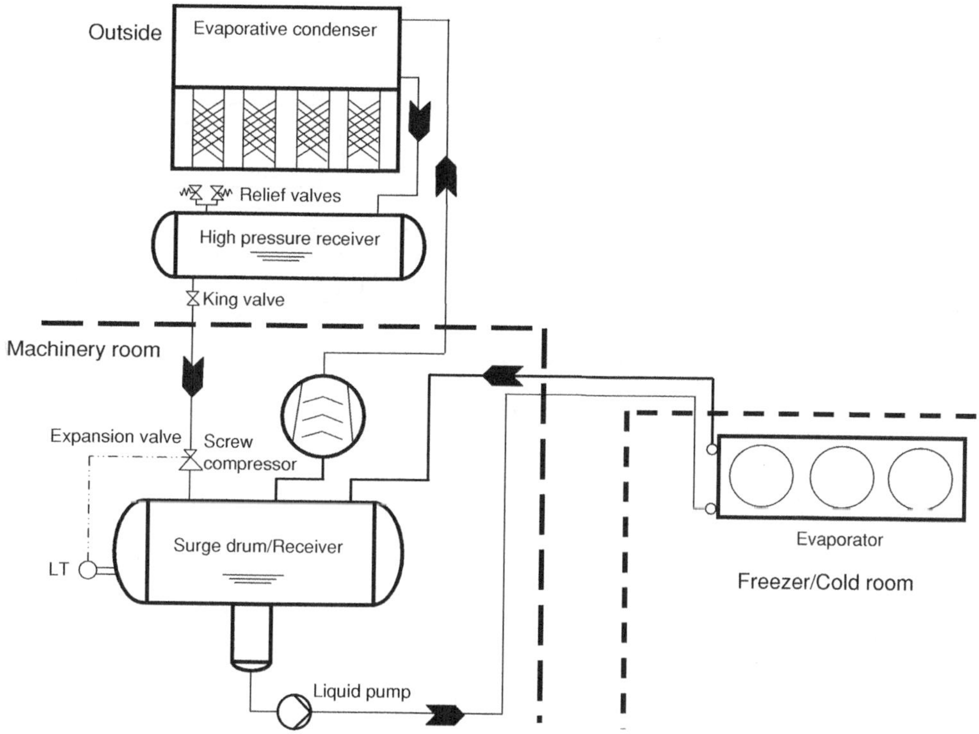

Fig. 4.2 A simplified pumped circulation system.

system, the water is the refrigerant and the operating temperature is limited by the triple point of water, which is 0°C. Lithium bromide plants are therefore only used for water chilling and are not capable of low-temperature operation. The absorption system has a condenser, an expansion device and an evaporator like a vapour compression system, but the compressor is replaced with an absorber and regenerator which vary the relative proportions, and enable heat to be absorbed by low-pressure refrigerant in the evaporator and then rejected from high-pressure refrigerant in the condenser. Gas from the evaporator is absorbed in the carrier fluid and a pump is used to raise the pressure of this weak solution to condensing pressure. At condensing pressure the solution is heated, which drives refrigerant gas off, strengthening, or 'regenerating' the solution. The regeneration process requires quite a lot of heat, so the thermal performance of the absorption system is relatively poor; however, if the heat is freely available as a by-product from another process then the system may be very economical to operate. The condenser system has to reject the heat extracted from the process and the heat used by the regenerator, so will be larger than required for a vapour compression system of the same capacity. In small domestic absorption refrigerators (an old design, not now in common use) the pressure differential between condenser and evaporator is achieved by adding a third fluid, often hydrogen, to the evaporator portion of the circuit. The partial pressure of the refrigerant in the evaporator is therefore lower than in the condenser, providing the necessary temperature differences to make the system work without the use of a pump to transfer solution from absorber to regenerator. A typical industrial absorption system is shown in Fig. 4.3.

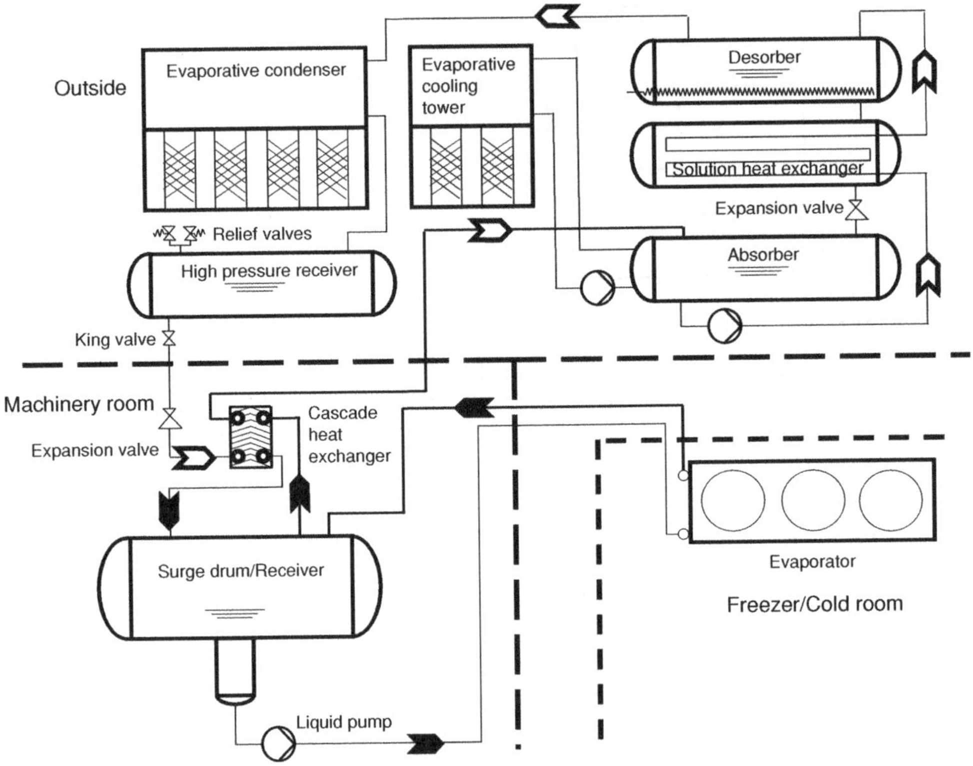

Fig. 4.3 Absorption refrigeration freezer.

4.3 ASSESSING ENERGY EFFICIENCY OF REFRIGERATION CYCLES

System performance is assessed by comparing the ratio of heat removed from the process to the work required to achieve that heat removal. Although they seem very different, the input work and output heat are both forms of energy. The ratio is usually expressed as the rate of heat removal, measured in kW, divided by the power input (rate of work input), also measured in kW. In some applications the total energy input and output, measured in kJ or MJ, is a more appropriate method of calculation. In a batch freezer for example, the operating temperature will be high at the start of the process and will drop as the freezing process progresses to completion. In this case a measure of total energy (MJ/MJ) required to complete the batch is more useful than the instantaneous ratio (kW/kW). The ratio is called the coefficient of performance (CoP). This concept is often extended to include ancillary devices necessary to complete the refrigerating function, such as cooling tower fans and pumps, oil pumps for compressors, evaporator fans and even drain line and defrost heaters. This extended ratio is called the coefficient of system performance (CoSP), and is generally a smaller figure than the CoP, which usually includes only the compressor motor power. In some markets the performance is called 'energy efficiency ratio' (EER) and may be expressed as ton of refrigeration (TR) per motor horse power (bhp). Note that the ratio may also be inverted – typically bhp per TR – such that a larger value represents poorer system performance. In the United States the convention is that heat pump performance (heat/work) is called CoP, but chiller performance (also heat/work) is called EER. Any specified measure of performance must state clearly whether it is work/heat or heat/work, particularly if the SI unit kW is used for both values. For the rest of this chapter the European convention has been followed: CoP is used for refrigeration and heat pump systems, indicating the ratio of heat extracted to work input, and where there is any ambiguity they are denoted CoP_r and CoP_h, respectively.

The main influence on the system efficiency is the extent to which the operating temperature of the refrigerant must be raised from the evaporator to the condenser. In a chill room for example, the evaporator may operate at −5°C and the condenser design condition may be 35°C. In a blast freezer with the same condenser condition the evaporator may operate at −45°C. In the latter case the compressor must provide much more work to raise the operating temperature as the temperature lift is double: 80 K instead of 40 K. As there is a need for some work input for the system to work, it is useful to consider the 'ideal' efficiency, sometimes known as the Carnot efficiency, and compare this with system performance under consideration. The Carnot efficiency indicates the performance that would be achieved by a system using an ideal cycle with no thermodynamic losses, such as friction. It is a simple expression of the evaporating and condensing temperatures. Expressed as a coefficient of performance, it is the ratio of the absolute evaporating temperature to the temperature lift (the difference between the condensing and evaporating temperatures). In the chill room and blast freezer examples above, the Carnot CoPs would be 6.7 and 2.85, respectively. Coefficients achieved in practice are usually much smaller than these, and the difference between the ideal and actual energy consumption is sometimes called 'exergy'.

It follows that the efficiency of a refrigeration system can be improved by ensuring that the temperature lift is minimised. At the design point this is achieved by ensuring that the heat exchangers operate with as small a temperature difference as possible; in practice it is also necessary to ensure that the temperature lift is kept to the minimum practical in the prevailing operating conditions. For example, the condenser temperature can be reduced in cold weather as a means of lowering the average lift. It might also be possible to raise the evaporator pressure, for example when the load on a glycol chilling system is low, in order to improve overall operating efficiency. If heat loads could be reduced, or even eliminated, from the system then the overall efficiency could be further improved.

The efficiency of the basic refrigeration cycle described above can be enhanced in a number of ways, usually by increasing the complexity of the circuit. If the temperature lift is greater than 60 K then it is worth considering a two-stage compressor, or two-stage system (where the two stages are achieved with separate compressors). This is advantageous because compressor efficiency is a function of pressure ratio and pressure difference, so by reducing each of these the compressor performance is improved. The optimum intermediate pressure lies roughly at the geometric mean of the suction and discharge pressures, so that the low-stage and high-stage pressure ratios are the same, but the exact value of the optimum depends on the compressor characteristic and should be calculated for each system. There are significant differences between reciprocating and screw compressor performance characteristics. In isentropic efficiency the screw compressor peak performance is related to the ratio between suction and discharge volumes, expressed as the volume ratio, or V_i. If the compressor has a high volume ratio, it is suitable for operation over a higher pressure ratio, but the peak efficiency would be lower than a machine with a low volume ratio. It is possible in some screw compressors to vary the ratio during operation, by changing the shape of the discharge port but this is of limited value in low-temperature systems because the operating pressure ratio is so high that the volume ratio at these conditions is always far smaller then the optimum value since a typical screw compressor cannot achieve a volume ratio above 5.8. The consequence of operating at a pressure ratio above the value dictated by the compressor geometry is that the pressure in the discharge chamber would be lower than the condensing pressure so when the discharge port opens there would be a flow of gas from the system into the compressor, resulting in some recompression, and hence wasting of some energy. If the volume ratio is too high then the pressure reached in the discharge chamber will be higher than necessary. There will be a sudden rush of gas from the compressor when the port opens, and in this case energy is wasted because the pressure lift within the compressor is higher than necessary. For chill compressors where the discharge pressure is varied to suit the ambient temperature conditions, variable volume ratio can improve the efficiency by several percent in annual energy consumption, but if the discharge pressure does not vary much, for example in a heat recovery system, or if the system only operates in warmer weather, then there will not be much benefit. In a reciprocating compressor, the efficiency is determined by the pressure ratio because the cylinder is not completely evacuated at the end of the compression stroke. When the discharge valve closes and the piston returns to the bottom of its stroke, the residual gas in the clearance volume re-expands. This restricts the amount of gas that can be introduced through the suction valve, reducing the volumetric efficiency, and it also requires to be recompressed, reducing the isentropic efficiency. These compressors are therefore extremely inefficient at high pressure ratios, because the net swept volume is relatively small.

In a two-stage plant it may be appropriate to use a reciprocating compressor on the high-pressure stage, where the swept volume is relatively small, and a screw compressor on the low-pressure stage. In this case the low stage could use a variable volume ratio and high stage could be operated with a fixed pressure ratio to maintain good efficiency across a range of ambient conditions by varying the intermediate pressure. However, using this strategy is often not possible because the plant includes a 'side load', for example a chill store or glycol chilling plant, which requires the intermediate pressure to be fixed in order to provide an appropriate intermediate evaporating temperature. In this case it might be appropriate to provide a stand-alone chiller for the side load, in order not to compromise the energy efficiency strategy for the main plant. Relevant factors to be considered in this case include the relative size of the heat loads, the load-time profile for each part of the system and the efficiency of the possible stand-alone chiller.

4.4 CHOICE OF REFRIGERANT

It is clear that the ideal efficiency of a refrigerating system does not depend on the fluid used as the refrigerant, because it is simply a function of evaporating temperature and condensing temperature. However, some of the factors, which cause the system performance to fall short of the ideal, depend on the fluid properties, and so system operating efficiency is strongly influenced by the choice of refrigerant. Refrigerant choice also affects the capital cost of the equipment required, and may also influence the space required to house the plant. The level of capital expenditure required on heat exchanger surface in order to achieve equivalent efficiency, or on ancillary components such as economisers or oil coolers, may also vary.

Apart from capital costs, space requirements and operating efficiency, the level of maintenance required and the cost of health and safety considerations including appropriate training and the additional personal protective equipment required for some systems may also be influenced by refrigerant choice.

For most industrial systems the choice lies between ammonia and a halocarbon and is determined by consideration of system size, health and safety requirements, capital costs and operational efficiency. In general, ammonia is preferred for large systems because the equipment cost is relatively low, the energy efficiency is high and the cost of the additional safety equipment required is a small proportion of the total equipment cost. In smaller installations the convenience of the non-flammable, non-toxic halocarbons may outweigh the capital and operating cost factors. Ammonia is acutely toxic and a specific hazard analysis should be conducted for the installation, including the effect of a large release of ammonia on the local area. Particular attention should be paid in this analysis to the proximity of offices, shops and houses. However, the risks associated with ammonia releases must be kept in perspective to ensure that the correct system specification is defined. Fatalities and serious injury are extremely rare and without exception they occur in the immediate vicinity of the ammonia release, usually as a consequence of some maintenance intervention or other accident. The European safety standard for refrigeration, EN-378, requires that automatic gas detection is installed in ammonia plant rooms, and that it is linked to a manned post or to automatic electrical isolation of the power supply to the plant room. It is also used to start emergency ventilation of the plant room, usually at relatively high ventilation rates. This is not for toxicity considerations, but because ammonia is flammable within a defined concentration range. The system designer must decide whether the plant specification should go beyond the minimum requirements of the safety standard. For example, it might be appropriate to include ammonia sensors and alarms in a large production area, or to provide power isolation in other parts of the building as well as the machinery room. Alarms at low concentrations are generally not required because ammonia has a highly distinctive odour, and causes discomfort at levels that do not cause injury. This 'self-warning' feature of ammonia provides sufficient warning of a leak to the local population, but it might be appropriate to place sensors in normally unoccupied areas to warn against a dangerous build up of toxic gas. Operating procedures for all maintenance activity should be implemented, including where appropriate evacuation of adjacent areas in case a major release occurs. Design of ammonia plant rooms should also include provision in the drainage system for sufficient capacity to prevent solutions of ammonia reaching the local ground water, because weak ammonia solutions can be highly toxic to fish and other marine life.

The most commonly used halocarbons for industrial refrigeration are R-404A and R-410A. They are not favoured for large systems because it is difficult to keep the plant 'leak tight', and the refrigerant is inherently expensive. They are particularly suited to small systems, and to low temperature freezers and stores, because they operate above atmospheric pressure down to temperatures of about −50°C, whereas ammonia is at atmospheric pressure when boiling at −33°C. If the required cooling capacity can be achieved with semi-hermetic compressors then halocarbons are worth considering. It should be noted that, for halocarbon refrigerants, if the refrigerant charge is greater than the practical limit for any room or chamber served by the refrigeration system, then that room or chamber should be fitted with a suitable gas detector and alarm system. These detectors should trigger an alarm if the refrigerant level in the relevant space rises to the practical limit. This might mean that the detection and alarm system requirements for a large halocarbon system serving a number of small rooms would be significantly more arduous and expensive than the requirements for an ammonia system, which would only require detection and ventilation in the machinery room.

In the event of a large halocarbon release the main hazards encountered are asphyxiation and the effects of toxic products of combustion. The off-gases from burning HCFC and HFC refrigerants are significantly more toxic than ammonia, and extreme caution must be taken when fighting fire in these systems, or when 'hot works' such as welding or brazing are in progress.

There will usually be very little difference in capital cost between medium-sized plants with ammonia or halocarbon – for very small systems halocarbon would probably be cheaper, but for large plants ammonia would be more cost effective.

An additional risk to be considered in the event of a large ammonia release, alongside flammability and personnel toxicity, is the risk of significant product spoilage in the event of contact with ammonia. Reduction or elimination of the ammonia charge should be considered, for example by using a glycol solution as a secondary system. Some recent plants have used carbon dioxide in this application because it has a much higher heat capacity than glycol, so the auxiliary loads on the plant, such as circulating pumps, are greatly reduced. Carbon dioxide operates at high pressures, but this will not present any health and safety risk if the system is designed correctly. There will be a requirement for gas detection and alarms in occupied spaces served by the carbon dioxide system on the same basis as for fluorocarbons.

4.5 SELECTING THE BEST REFRIGERATING SYSTEM FOR A PROCESS

The most appropriate system for a refrigerating process will usually be a compromise between capital and operating costs because in many cases the simplest of systems will not be the most efficient to operate. Some key factors must be considered in laying down the requirements for a plant:

- What range of products is to be handled by the system?
- What quantity of product is to be handled by the system?
- What product condition is required at the system inlet?
- What time constraints will be placed on the process?
- What condition is required at the exit?
- What energy performance is expected of the equipment?
- What space is available for the product cooling equipment?
- What location is available for the plant which feeds the cooling equipment?
- What budget is set for the acquisition of the equipment?
- What budget is set for the operation and maintenance of the equipment?

4.5.1 What range of products is to be handled by the system

In cold and chill stores, where the only requirement of the process is to maintain the product at a prescribed temperature the type of product is not significant, provided it does not add moisture to the storage atmosphere. Wet products such as fish or meat should be wrapped before long-term storage otherwise the loss of moisture will reduce product weight and impair quality, a double loss of value. In freezers and chillers the type of product is much more significant as it determines the heat transfer performance of the product, and hence the freezing capacity of the plant. This is covered in greater detail in subsequent chapters; however, it should be noted here that freezing performance is determined not just by the thermophysical properties of the product, such as density, specific heat capacity, latent heat, thermal conductivity, water content and fat content. The size and shape of the product, the orientation in which it is presented to the airstream and the material in which it is wrapped can all cause significant increases in the freezing time. There may also be limitations to the air velocity that can be applied to the product, delicate confectionary for example, and this may also result in lengthened freezing or chilling times. If possible the designer should be mindful of the full range of possible product to be handled by a chiller or freezer, and should include means of adjustment to the basic system to ensure that the loss of performance is minimal when the product specification changes. This might include variable speed fans to vary the air velocity over the product, adjustable air baffles, variable belt speed or control over the rate of temperature reduction to ensure that thin product is not frozen inefficiently in a machine designed for thicker portions.

4.5.2 What quantity of product is to be handled by the system

In situations where the volume of product is constant but the density is variable the time required to freeze different grades of product can vary greatly. The specification must be unambiguous about the weight of product to be processed, the physical shape it presents and the weight and type of packaging and transport equipment included. The refrigeration load on the system is calculated from the weight of product handled by the system in a given time, not the volume of product.

4.5.3 What product condition is required at the system inlet

There are two classes of chiller and freezer: continuous and batch. Spiral and tunnel freezers are of the continuous type. Product is fed through them on a conveyor, often coming directly from a preparation or cooking process. Blast and plate freezers are of the batch type, where product is loaded into a fixed location where it remains for the duration of the freezing process and is then removed. As a consequence of this mode of operation, time must be allowed for loading and unloading as part of the total batch time. Some plate freezer types can be automated to provide a form of 'semi-continuous' operation, but they are strictly still batch processing because there is a need for defrosting, unloading and refilling in every freeze cycle.

Continuous freezers are generally more suited to smaller products with short freezing times, and to product in its finished form where the presentation of the frozen food is important. The amount of time that the product spends within the freezer is called the residence time, and is adjusted for different products by varying the conveyor speed. Small products with high water content, such as vegetables, will freeze quickly, and it may be difficult to retain the character of the individual items. In this case jets of air may be used to create a fluidised bed to ensure that each item is individually frozen. Larger products and those that contain more air, such as bakery products, will require a longer residence time, and may also require to be handled more gently to avoid damage. As mentioned above, the packaging of food can have a significant effect on the freezing time. For example, the freezing time of cooked meat on trays in a spiral freezer will be affected by the depth and material of the tray. Packaged ready meals are particularly difficult to freeze because there is often an air gap between the top of the meal and the underside of the lid. Where possible the trays should be left open-topped until the freezing process is over, but this may not be possible for hygiene reasons, or simply because it is not possible to seal the package when it is cold. If the product is fed from a cooking system directly to the chiller or freezer then there will be significant advantage in pre-chilling with ambient air before it is placed in the freezer. This might be achieved by including a reasonable length of conveyor from oven to freezer, perhaps with cooling fans blowing ambient air over the product, rather than positioning them right next to each other.

Batch freezers are suitable for larger products, particularly where it has been wrapped or palletised. Plate freezers are often used for boxed beef, and whole pallets are used for loading larger blast freezers, mainly to save time in handling the product. Freezing cycles for batch processes are longer than for continuous freezers because it is economical to group the product into larger bundles (boxes or pallets) to reduce the ratio of handling time to freezing time and maximise the utilisation of the freezer. There is an advantage in achieving the full loading–freezing–unloading–cleaning within a 24-hour period as this matches the typical factory shift pattern. It is often necessary to stack product onto pallets with spacers to enable some airflow through the block in order to reduce freezing times to fit within the required timeframe, but for most products, given the size of the pallet, the thermal conductivity of the product and the insulating effect of the typical packaging material, it is not possible to reduce the batch time to less than 12 hours, so a typical freezer cannot handle more than one batch per day. A typical blast freezer cycle may load the freezer in the late afternoon and start the freeze mid-evening. Assuming an 18-hour freeze has been achieved, the freezer will run overnight and through the next morning until the early afternoon. This enables it to be unloaded and cleaned, ready for the next batch, through the mid-afternoon period. An advantage of this cycle is that it fits well with a typical time-of-day electrical tariff, where the cheapest electricity is available overnight, when the freezer plant is most heavily loaded, and the freezer is off during the most expensive period, typically 16:00 to 19:00 hours. Plate freezer cycle times are typically shorter than blast freezers because the product is presented to the freezer in slabs to suit the shape of the plate, and because direct contact between the plate and the product gives better heat transfer than an air blast freezer. For unwrapped meat products in 75-mm blocks, freezing times of less than 1 hour can be achieved. For boxed product in 150-mm blocks, the freezing process could take up to 12 hours.

4.5.4 What time constraints will be placed on the process

There is a limit to the minimum length of time required to achieve freezing at the core of the product, and it is determined by the size, shape and physical properties of the product as well as the type of wrapping used, if any. Individual peas within a fluidised bed have a diameter of about 4 mm and will be frozen within a few seconds, whereas for slabs of beef packed in corrugated cardboard boxes and stacked on pallets it is much more difficult to transfer the heat from the centre of the block to the surrounding air. In the latter case greatly reducing the temperature of the air is not a viable option; the only thing that will allow the heat to transfer out of the block is time. If a certain product throughput is required in the process, whether it is a batch process such as a blast freezer or a continuous process such as a spiral freezer, the only way that large, slow-freezing products can be accommodated is by providing significant product dwell time within the freezer. This has implications for the size and cost of the freezer, but also has an effect on the size of the refrigeration plant. The plant capacity required to freeze a tonne of peas will be very large because the heat is removed in a matter of seconds. If the product is a tonne of boxed beef packaged on a pallet then the heat removal might take 24 hours or more. However, as approximately the same amount of heat is removed but over a much longer period, the load that the refrigeration plant is required to handle is very much smaller for the case of beef freezing. The refrigeration plant required to remove the same amount of heat in 24 hours in the meat freezer as was removed in 40 seconds in the pea freezer would be only 0.1% of the size of the pea freezer. This in turn means the refrigeration plant would be only 0.1% of the size of the pea freezer. In practice, the ancillary loads such as fans form a significant part of the total, so the difference in plant size is less extreme than this example, but it should be clear that freezing 1 tonne of peas and 1 tonne of boxed beef require refrigeration plants of very different sizes.

4.5.5 What condition is required at the exit

The specification for a freezer or chiller must state clearly the expected output condition of the product. In freezers it is usual to state that the product core temperature must have reached a certain temperature within a defined time period. The surface temperature will drop to within a few kelvins of the air or plate temperature within minutes of starting of the freeze process, and so is a very poor indication of the progress of the freezing process. For many products the core temperature can only be measured by destructive testing with temperature probes, so it is not possible to base the equipment control on this temperature and the freeze cycle is based on a fixed elapsed time, with some destructive testing of a few samples at the end of the cycle to provide verification. A core temperature of −18°C is often used as the acceptance criterion for freezers. This is not an unreasonable value, although it originates from the freezing point of a mixture of salt solution and snow (which defined zero degree on the Fahrenheit scale), and therefore is not directly related to product quality. It is essential that the whole mass of the product has passed through the freezing point phase change, and the −18°C target is often justified on the basis that it allows some leeway for varying freezing rates depending on product position within the freezer and variations in air flow, air temperature and surface heat transfer. Provided there is no adverse effect, the system specifier should consider whether a slightly higher core temperature would be acceptable, as there could be a significant energy saving in accepting the core at −15°C, or even as high as −10°C and either reducing the freeze time or raising the air temperature. The practice of partially freezing the product and then placing it in a cold store for 'finishing' is good neither for product quality nor for efficiency, and is not acceptable. The product must also be sufficiently well frozen so that it can be handled without damage to the product surface or packaging. Packaging can also be damaged by condensation freezing on the surface of the product if it is left for any time in a chill area before being placed in cold storage. If this is unavoidable then consideration should be given to dehumidification of the chill area in the vicinity of the freezer outlet.

4.5.6 What energy performance is expected of the equipment

It is difficult to give specific, accurate requirements for the energy performance of refrigeration equipment because there is a broad band of possible performance depending on factors such as the size and type of evaporators, the way in which the equipment is used and the way in which it is maintained. However, it is important to pay attention to the power consumption figures proposed by system designers and contractors as decisions taken at the purchasing stage are difficult and expensive to modify at a later date. Every tenderer for a project should be required to give details of the sizing of evaporators and condensers, and the energy consumption expected of the compressors and ancillary devices. These should be compared between potential suppliers and appropriate cost options should be included for more efficient alternative designs. The system specification should include a method of measuring the power consumption of the plant. As a minimum this should be a kW h energy meter for each major piece of equipment, and if possible the information should be incorporated into any graphic monitoring system, so that performance analysis is made as easy as possible. These features are not expensive to add to a system at the precontract stage, but they can be very costly to retrofit. Where possible the plant should be configured to record energy consumption in relation to product throughput, and to display the results in tabular or graphical form to facilitate performance comparisons.

4.5.7 What space is available for the product cooling equipment

Freezers and chillers divide into two basic types as described above: batch and continuous. In both cases the total space required for a functional unit is far greater than the physical size of the product being frozen. For blast freezers the aircooler is likely to be large; occupying up to 25% of the footprint of the freezer, and space must also be provided for airflow through the product and back to the aircooler. The access to the cooler must be suitable for the type of materials-handling equipment used: for smaller products trays mounted in racked trolleys are often used, but larger loads typically require pallet trucks or forklifts to move them. The product layout within the freezer must allow room to manoeuvre the truck, but must also ensure even airflow across all the products.

Spiral freezers generally require a smaller footprint but are taller than tunnel freezers. In both cases there must be sufficient space around the freezer for maintenance, including in the case of ammonia equipment the requirement for draining oil from the evaporators. Tunnel freezers are long and narrow, but multi-pass arrangements are possible to reduce the overall length. The airflow is usually across the product path, and if it is possible to divide the tunnel into temperature zones then some of the equipment can run at a higher evaporating temperature. Alternatively, if the airflow is designed to travel along the length of the tunnel then a greater rise in air temperature can be achieved, so the air volume can be reduced, giving a significant reduction in fan power. However, in this case the evaporating condition is determined by the air temperature required at the coldest part of the air circuit.

4.5.8 What location is available for the plant which feeds the cooling equipment

For small freezers, typically of capacity 10 kW or less, the refrigeration equipment can be built into the freezer unit. If the unit is air-cooled then a heat load of roughly twice the cooling capacity will be imposed on the room cooling system. Alternatively, a water-cooled unit connected to the building, chilled water or glycol system can be used. Heat loads up to about 20 kW can be handled by a condensing unit; a package containing compressors and a condenser which can be located outdoors or in a roof void. For larger freezers the refrigeration system will be remote, and may be housed in a dedicated plant room. The capital cost and the energy cost of the installation will be reduced by locating the plant room as close to the freezer as possible, keeping the pipe runs as short as possible. Consideration should always be given when laying out the factory floor at the building design stage to the refrigeration plant location. Large systems are likely to use pumped recirculation, so it must be provided in the pipework design to let liquid drain back to the pump vessel in the plant room. If this is not possible then it may be feasible to locate the pump vessel close to the freezer and run the dry suction pipe back to the plant room where the compressors, condensers and ancillary equipment will be located. The suction pipe location must be co-ordinated with other services to ensure that the refrigeration system functions reliably and efficiently. The other services may also have particular functional requirements, such as traps in steam lines, or cleanability for ductwork, so an integrated approach to services design is necessary to ensure correct operation. This is unlikely to happen if each contractor is allowed to install their equipment where they choose on a 'first come first served' basis.

4.5.9 What budget is set for the acquisition of the equipment

Before going out to tender, the project team should be able to set down their expectation of price of equipment, by breaking their requirement into discrete functional items and estimating the cost to build and install each of them. This estimate, even if it is carefully calculated with expert advice, should always be treated as guidance rather than a fixed sum. Circumstances may change during the tendering process, or additional information may be presented, which offer the possibility of significant system enhancement in plant capacity or efficiency. The project team must also bear in mind that capital costs may escalate if there are specification changes anytime in the life of the project.

The budget should consider the purchase price of the refrigeration equipment, related costs such as electrical infrastructure and materials handling, and the associated structural work, including insulated enclosures if required.

4.5.10 What budget is set for the operation and maintenance of the equipment

As mentioned previously, it is difficult to assess the specific energy consumption of a freezer because it is dependent on product throughput as well as ambient temperature. The maintenance costs are likewise difficult to predict with accuracy; however, this does not mean that valid comparisons between competitive quotations are not possible. Each proposal should be assessed and ranked according to energy cost, maintenance hours required, materials consumption and other operating costs such as water treatment chemicals. Once ranked it is then possible to apply cost factors to each element of the operating budget in order to derive an approximation of the annual operating cost. This approach also gives the dual advantage of enabling comparison between bids and providing some means of assessing the payback for additional expenditure on more efficient or more easily maintained equipment.

4.6 DEVELOPING A SPECIFICATION

The specification must state clearly what is expected of the equipment. The product throughput and range of product types must be given, and where possible the extent to which the plant is expected to cope with variation from these values should be defined. Although this might make the basic design slightly more expensive it is likely to avoid expensive design changes during the construction phase of the project, and it will allow greater flexibility in the use of the plant, once it is in operation. This may avoid the need for expensive upgrading after a short operating life. The specification should define the operating environment, including ambient maximum and minimum temperatures and humidity; however, if these are not known, the designers should be able to make an estimate based on the site location. In these circumstances all tenderers should be asked to quote against the same requirements so that bid comparisons are not distorted. The extent to which the equipment will operate on part load should be defined, as it may necessitate the provision of additional facilities, and these may have a major effect on the efficiency of the system. For example, a system which provides refrigeration for a spiral freezer that runs continually from Monday morning to Friday evening may also feed an adjacent cold store. The system must be capable of providing cooling to the cold store when the freezer is not in operation, or an alternative means of cooling the cold store must be available when the freezer is off-line.

In addition to the initial plant requirement, consideration should be given to future expansion, but in this case unless the future plans are well defined and unlikely to change there is little benefit in spending too much money on provisions. In general it is worth sizing vessels, such as high pressure receivers, for future expansion, but it is probably not worth buying additional compressor, or evaporator capacity in the early stages of a project. Condenser capacity requires additional consideration. If the full condenser capacity is installed in the early stages of a project there will be a short-term benefit in operating cost if it is possible to run the plant at lower condensing pressures. There is also a danger that plant performance is impaired when condensers or receivers are added at a later date. It is possible to ensure correct operation of multiple condensers and receivers, but the interconnecting pipework is complex and there is great scope for inefficiency and instability in operation. There may be grounds for sizing the original steelwork to suit the substitution of the condenser and receiver with larger units at the time of the expansion, but if this is done the system will be out of commission while the change is made.

Space constraints for the installation must be clearly identified in the specification, and the location of the plant room relative to the freezers should be established and fixed as soon as possible. The installation will be most successful if the refrigeration system requirements are fully appreciated by the building and production process design teams, so that a fully integrated design can be achieved.

The specification must also be specific about the legal requirements that are mandated for the project, and the client should be clear about the implications of these. In the United Kingdom most industrial refrigeration installation projects are governed by the Construction Design and Management Regulations (CDM) and if the site installation takes more than 30 days or involves more than 500 person-days then the Health and Safety Executive must be notified. CDM places legal requirements on the process owner, called the 'user' in the regulations, even before the project commences. Industrial installations are probably also covered by the Pressure Systems Safety Regulations (PSSR) and the Pressure Equipment Regulations (PER). PSSR require that a scheme of examination is created for the equipment under the responsibility of the 'user' and that the equipment is examined in accordance with the written scheme before it is put into service and every year thereafter.

4.7 THE TENDERING PROCESS

The traditional way to procure a refrigeration plant was to appoint a team of expert advisors, create a brief for potential suppliers, invite sealed bids for the equipment described in the brief, evaluate the tenders received and select a successful contractor to provide the equipment required. This process is apparently simple, but contains many pitfalls and in some cases is counterproductive. For example, it is now recognised that tendering in this way encourages conflict rather than co-operation between project team members, and can often lead to delays, disputes, or budget over-runs. Each party is prompted by the system to look after their own interests as opposed to those of the client or the project team, and often the only way for a sub-contractor to improve his project's financial performance is cutting corners or skimping in some other way. This tendering process is slow, requiring from 2 to 6 months for preparation of the brief, a further 2 months for bid preparation by the prospective contractors and a further month for tender analysis and contractor selection. During this 9-month process the project will be progressing, and often the design that is required by the time the sub-contract is placed is significantly different to the scheme that has been costed by the sub-contractor. The process also tends to create long chains of command from the client through the professional advisors to the main contractor and on to the sub-contractors. With a specialist package such as refrigeration, the client is often frustrated by his lack of influence over the work being conducted on his behalf by the sub-contract refrigeration contractor, but if he instructs the sub-contractor directly, or even suggests that he might do so, he is open to claims for delay and additional costs which may arise anywhere in the contractual chain. This frustration is partly addressed by the use of a Management Contract rather than a traditional Building Contract. In the Management Contract a single management contractor is appointed, and he places works contract packages with all the trades engaged in the project. This gives the client the opportunity to meet with all the works contractors and encourages a freer exchange of information, making it easier to cope with late changes to the project plan. A good Management Contractor recognises that his role is substantially different from a traditional Main Contractor and will ensure that the full benefits of this style of contract are achieved, but all too often the contract administration is handled in exactly the same way as a building contract, and the potential benefits are not realised. In these cases the client can be excused for wondering why he is paying for an extra link in the chain of command.

At a more general level there are also problems with the traditional tendering route. First, the specification is prepared by a design team who is not ultimately responsible for, nor even involved in, the operation of the system. They are likely to stick with tried and tested design concepts, and will stifle innovation whenever it tends to take them from their comfort zone into the unknown. Their design is then reworked by each of the bidders, potentially introducing multiple options and discrepancies between bids, before one is selected who ultimately re-does the design to suit the requirements now established during the long bidding process. Even worse, for every successful bidder there will be at least two or three unsuccessful ones. Their bidding overhead must be recovered on the projects that they win – so the total cost to industry is pushed up even further. In total each project tendered in this way is designed at least three times and often as many as six, including the efforts of the unsuccessful tenderers.

In recent years much has been written about an alternative approach to construction contracts, generally known as 'partnering'. In a partnering situation the team is appointed before the start of the project, down to the level of specialist sub-contractors. This team signs up to a methodology and then agrees a set of operating parameters and sets a budget for the project. Once this has been done the detailed design is completed by the various specialists exchanging information in conjunction with the client. When the detailed design is complete the budget is reappraised and once it has been accepted the project work commences. From this point on the project can either follow a traditional contract model – usually a Management Contract – or can continue on a less conventional route. In an open book contract there are two stages. For an open book tender the sub-contractor will present a detailed breakdown of costs, including equipment and sub-contract quotes. These are marked up by agreed percentages to give the total price for the project. This becomes the fixed, lump-sum price in the contract. For an open-book contract, the visibility of costs carries through to contract completion, usually with an agreed methodology for handling cost overruns and underspends on individual line items. A common approach is to agree on a 'guaranteed maximum price' where, with the exception of design changes, the project cost is capped. If the result is higher than the cap the contractor pays the difference: if it is less than the cap the result is shared between the client and the sub-contractor. This works well if there is a positive spirit on both sides and provided the original budget is accepted by all as a fair estimate. A further advantage of partnering is that it fosters a long-term relationship, and if the project is the latest in a long line of successful installations then the contractor selection stage is merely a case of validating the previous team. The client should check whether the key members of previous project teams will be involved again, and may ask for references for any new team members, but if there is a case history of previous successful projects between the two parties there is no point wasting energy on investigating what is already well known. One of the major benefits of starting a project on the basis of a partnering agreement is that it enables a high level of trust between all those involved in the process to be developed.

At the opposite end of the trust–suspicion scale there is e-bidding. This is a method of tendering which is designed to get the lowest possible first cost for the client, by requiring all tenderers to post their bids on a website. The tenderers are told what the lowest price is, and where in the ranking their current bid lies. If they choose to do so they can reduce their price and increase their chances of success. In some e-bids the client is obliged to accept the lowest bid, unless there is a problem with the technical submission; in others it is clearly stated that there is no such obligation, and the client may choose whichever bid he likes. The major flaw with e-bidding is that it is founded on two false premises: first, that all suppliers are equally capable and competent to fulfil the client's requirements, and second, that they are all equally hungry for work. The process requires a very detailed specification before the prices are prepared, and all the bidders must confirm that they are tendering in accordance with the specification. There is no means within the e-bid process for one of the bidders to add value to their proposal, for example by offering a larger condenser. Anything that increases the basic price is disadvantageous to them. The bidder's current workload is more subtle a problem: if one contractor is very quiet he is likely to accept a lower margin on the project than a company who are already busy. It is naïve to assume that the contractors are equally competent and equally successful, so it follows that on average the low bidder's business is quiet because he is less successful than his competitor. If a client is asked whether he wants a low price or not he will probably say yes. However, if asked whether he wants to buy from the least successful and therefore most desperate contractor on the list the answer is likely to be different. The answer to this dilemma is to have a thorough specification and vetting procedure before the tender documents are issued; however, this process takes time and costs money, and still leaves the client with a contractor who is looking for every opportunity to save money on his project costs in order to claw back some margin. The obvious conclusion is that e-bidding is a suitable way to buy commodities in a large market, but is not so appropriate for specialist sub-contract construction with a restricted field of suppliers. If the only reason that the client is adopting e-bidding is to ensure his sub-contractor is honest then he might be better advised to try to foster a more positive, trusting relationship.

Whichever tender process is used, the decision must be made before starting the project and cannot be changed midway through the process. Analysis of the tenders received whether as the result of a sealed bid, as the technical submissions supporting an e-bid or as part of the partnering design development phase should always include a review of energy consumption and operating costs, and should incorporate a hazard and operability study to ensure that the concept does not have built-in expensive mistakes. Often the hazop (hazard and operability study) is not conducted until the contract has been placed, by which time design changes become expensive and cause delays. A disadvantage of traditional tendering and e-bidding is that the pre-contract design is often not sufficiently detailed to make a detailed analysis possible. With the hazop integrated into the pre-contract design development, many of the requirements can be incorporated at little or no additional cost.

4.8 HOW TO DESIGN FOR ENERGY EFFICIENCY

The first step in designing an efficient refrigerating system is to eliminate all unnecessary heat loads. These might be as obvious as electrical heaters or other energy inputs, but can also include door openings and other moisture ingress. For example, excessive humidity, which causes condensation or frost on the evaporator represents a substantial heat load. If this can be eliminated by running at a higher temperature, by dehumidifying or by better control of humidifiers then significant savings in system efficiency can be achieved. The next step is to ensure that the temperature lift is as small as possible by correct sizing of the evaporator and condenser heat exchangers. This must take account of part load operation, and with careful design it need not add much to the project capital cost. Each energy consuming element, such as compressors, fans and pumps should be considered in its own right, and selected to be as efficient as possible, including the use, where possible, of efficient motors. The final element in efficient design is the inclusion of provisions necessary to ensure that the original efficiency can be maintained throughout the life of the plant. This will include provision of sufficient monitoring points for temperature, pressure and power to enable the system to be re-commissioned as part of a routine maintenance programme, provision of access for inspection and maintenance and arrangement of equipment to promote efficient operation. Where possible the compressor plant should be located as close to the load as possible to minimise the suction line pressure drop. Care must also be taken with the location of condensers, particularly the air-cooled types, to ensure that recirculation of warm air is minimised.

4.9 HOW TO DESIGN FOR LONG-TERM ENVIRONMENTAL SUSTAINABILITY

In addition to the energy efficiency considerations outlined above there are several other aspects of sustainability that should be considered in the design of refrigeration systems. As the majority of refrigerant loss is reported to be associated with major equipment failures the system should be designed to have as small a refrigerant charge as possible, bearing in mind the need for reliable, efficient operation. It must be possible to decant and recover the refrigerant in a safe and convenient manner. The consequence of accidental release of refrigerant, through drains, leaks or relief valves should be considered, particularly where there is a significant effect on the local environment, and it should be noted that the refrigerant released will also contain oil, and so can cause localised pollution even after it has evaporated. Refrigerant and oil drained from a system should be treated as controlled waste and handled accordingly.

4.10 HOW TO DESIGN FOR FLEXIBILITY IN USAGE

The designer should consider the consequences of changes in the operation of the system when it is first laid out, and where possible should make allowances for future modifications. This will not only consist of the addition of extra heat loads, but might also include changes in the operating temperature level or relocation of factory equipment. Where possible, and where the budget allows, pipe sizes should be kept as large as possible to enable loads to be moved. Insulation thickness on chill pipes should be suitable for low-temperature operation. Consideration should be given to provision of anti-frost heater mats under chill areas if there is any prospect of them being converted to freezers or cold stores in future. The alternative would be that the floor would have to be lifted to enable a heater mat to be laid, which could be time-consuming, disruptive and expensive. Where low-temperature compressors may be used at chill conditions the motor size should be increased accordingly, as the power consumption of a given size of compressor will be much higher when operating at chill conditions. There should also be adequate provision in the electrical supply for operation at chill conditions, if this is likely. These provisions are all expensive, so judgement must be used to determine whether the chance of change justifies the extra cost.

4.11 MAINTAINING AND OPERATING REFRIGERATION EQUIPMENT

Refrigeration equipment ought to be capable of providing years of trouble-free efficient operation, but it takes some careful maintenance as well. The effects of ice, moisture and exposure to weather can have a detrimental effect on plant mechanical integrity, which may lead to unreliable operation. Wear and tear on small components such as solenoid valves and pressure regulators can lead to expensive malfunctions, for example in hot gas defrost valve stations, so such components should be the subject of regular inspection and replacement where necessary. In many cases condition monitoring, for example, of lubricant condition or vibration, can be used to extend the time between expensive overhauls, but care must be taken with some equipment where failure is sudden, without any prior warning and with expensive consequences. For example, reciprocating compressor overhauls are required regularly. If these are not completed following the compressor manufacturer's guidelines then there is a risk that the small springs in the loading gear may break and fall into the compressor cylinder. This can cause serious damage far beyond the value of the overhaul, and there is no condition monitoring that can predict this type of failure. Screw compressors are more suited to condition monitoring, as bearing wear can be monitored and maintenance planned to suit production and load variations.

For fluorocarbon refrigerants (f-gases) there should be a regime of regular leak testing on refrigerating systems with a charge of more than 3 kg. The European f-gas directive, introduced in 2007, requires leak testing annually if the refrigerant content of the system is above 3 kg, every 6 months if it is above 30 kg and every 3 months if it is above 300 kg. Much more frequent leak testing is recommended if expensive losses and inefficient running are to be avoided. The provision of a gas detection system in the machinery room is not a substitute for thorough leak testing, as it is possible to lose substantial quantities of refrigerant without registering anything on the gas detector. For very large distributed f-gas systems it may be appropriate to divide the system into sub-sections and thoroughly test each sub-section by rotation over a month. It is recommended that some sort of leak testing on large systems should form part of the weekly routine.

Evaporators and condensers, which rely on good airflow for their continued operation, should be inspected regularly and cleaned if necessary. Evaporative condenser water treatment for the control of harmful bacteria is required by law, but the effects of scale or corrosion should also be considered in order to ensure long, reliable operation of the equipment. Typically, scaling occurs in hard-water areas and corrosion is more prevalent in soft-water areas, but if water softeners are used there is an increased risk of corrosion even though the mains water is hard.

Functional checks of safety devices are required annually by law under the Pressure Systems Safety Regulations. It is sensible to extend the scope of inspection beyond the letter of the law for these checks and include a system control functional check and a report on plant performance. If necessary, as indicated by poor performance, a re-commissioning exercise should be conducted to bring the plant back to its original efficiency.

Although refrigeration plant presents specific hazards of working with volatile, low-temperature liquids, asphyxiation or poisoning and working with high-pressure fluids there are many other hazards to be considered, including working at height, lifting heavy weights, working with electricity and working in a noisy environment. These hazards must be included in any risk assessment for refrigeration plant operation or maintenance.

5 Emerging and Novel Freezing Processes

Kostadin Fikiin

5.1 NEED FOR PROCESS INNOVATIONS IN THE FOOD FREEZING INDUSTRY TO IMPROVE THE HUMAN WELL-BEING AND QUALITY OF LIFE

As is known, heat and cold are of the same physical nature. In spite of this, they have played different roles in the development of human civilisation. Prometheus, the mythological hero who bestowed the divine fire of Olympus to mankind, is glorified in immortal poetical and musical works. However, the pioneers who created artificial refrigeration and gave it to humanity have not yet been praised in a work of art as a token of gratitude. For millennia, cold has primarily been associated with winter, diseases and people's misery, rather than with its proven capability to preserve biological materials. More recently, in the industrialised world, food refrigeration has become a powerful instrument for improving the quality of life.

Refrigeration does not have a competitive alternative to maintain the nutritional resources of humankind. The worldwide food output amounts nearly 5 billion tonnes per year, some 2 billion of which need refrigerated processing, but only 400 million are effectively refrigerated. Chilling is an indispensable element of almost all post-harvest or post-mortem techniques for handling food commodities of plant or animal origin, while freezing has been established and recognised as the paramount commercial method for long-term preservation of the natural quality attributes of perishable foods, thereby forming a substantial part of the global economy and the well-being of citizens. In terms of money, every year the global investment in refrigerating equipment is over US$ 170 billion, while all refrigerated foodstuffs cost US$ 1200 billion (which exceeds 3.5 times the US military budget). Some 700–1000 million household refrigerators and 300,000,000 m^3 of cold-storage facilities are available over the world. The annual global production of various frozen foods is about 50 million tonnes (plus 20 million tonnes of ice creams and 30 million tonnes of fish), with a remarkable growth of 10% every year. Refrigeration thus accounts for about 15% of the worldwide electricity consumption, thereby determining to a large extent the global economic sustainability (in terms of energy efficiency and environmental friendliness).

Food refrigeration stakeholders and cold chain professionals are part of the FAO/WHO Codex Alimentarius Commissions and played an important role in the historic world leaders' summits in Montreal, Kyoto and Johannesburg (devoted to ozone-depletion, global warming and sustainable development). Refrigeration and the cold chain are among the top priorities of the US Presidential Council for Food Safety. While the economies of industrialised nations waste 25–30% of their perishable food production because of imperfect or lacking cold chain, the disastrous dimension of such agricultural losses in developing countries is a major contributor to malnutrition.

The twenty-first century poses new challenges for the global frozen food sector, which can be summarised as follows:

- Although the promising capabilities of several emerging freezing technologies (detailed in this chapter) attract the attention of researchers and industrialists, many such innovations remain unimplemented in the common industrial practice.
- Both conventional and novel freezing techniques have a substantial potential for further optimisation by involving advanced modelling and experimental tools and enriching theoretical understanding of underlying phenomena (e.g. heat transfer, fluid flow and biochemical processes).
- The manufacture of frozen commodities and related scientific research are still insufficiently attractive for high-skilled experts and young specialists as compared with *hi-tech* branches (e.g. information technologies, electronics, communications, biotechnologies, etc.).
- Developing countries around the world need more refrigeration capacities and inexpensive food freezing equipment to make their economies more competitive on the global markets.

Simultaneously, the high reputation of freezing as one of the safest and most nutritionally valuable preservation techniques should not create a false sense of total security and should not defeat the care and diligence when managing the frozen supply chain (IIR, 1986, 1999; Kennedy, 2000). While freezing drastically reduces detrimental phenomena in foods, a number of physical and biochemical reactions still occur and are accentuated if proper processing and handling conditions are not maintained throughout the entire chain for production, storage, transport, distribution, retail and household handling of various frozen commodities of plant or animal origin by using an integral *farm-to-table* approach.

Contemporary food industry promotes research and innovations, which primarily deal with the sustainable processing, preservation and supply of safe, high-quality, healthier foods and beverages for the consumers. In that context, food-freezing technologies are a crucial and underexploited element of the sustainable food production and preservation and related food chain management. This importance perceptibly increases as a result of the ongoing extension of traditional food chains into new markets and emerging economies around the world. Essential measures should, therefore, be undertaken to raise professional competence and encourage a stronger public commitment to food freezing investigations. It is vital to make public authorities and food policy makers more aware of the topical professional endeavours of refrigeration scientists and industrialists around the world. Relevant funding agencies (such as the European Commission) supported, therefore, a number of successful freezing-related research projects whose deliverables and industrial implementation result in reduced post-harvest losses, extended shelf-life and better quality of frozen foods, lower investments and running costs, higher energy savings, and enhanced environmental friendliness (Fikiin, 2003).

5.2 STATE OF THE ART AND CONVENTIONAL FREEZING MODES

In the early twentieth century, many people were experimenting with mechanical and chemical methods to preserve food. As an industrial process, quick freezing began its history nearly 80 years ago when Clarence Birdseye found a way to flash-freeze foods and deliver them

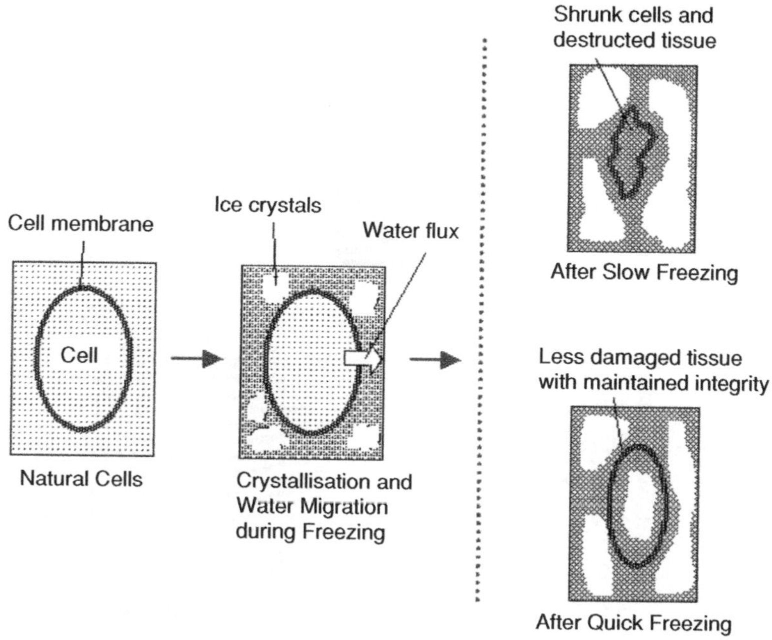

Fig. 5.1 Crucial impact of freezing rate on the end product quality (Fikiin, 2003).

to the public – one of the most important steps forward ever taken in the food industry. During his stay on the Arctic, Birdseye observed that the combination of ice, wind and low temperature almost instantly froze just-caught fish. More importantly, he also found that when such quick-frozen fish were cooked and eaten, they were scarcely different in taste and texture from how they would have been if fresh. After years of work, Birdseye invented a system that packed dressed fish, meat or vegetables into waxed-cardboard cartons, which were flash-frozen under pressure (US Patent No. 1,773,079, 1930). Then, he turned to marketing and a number of ventures have been initiated to manufacture, transport and sell frozen foods (e.g. construction of double plate freezers and grocery display cases; lease of refrigerated boxcars for railway transport; and retail of frozen products in Springfield, Massachusetts, in 1930). These technology achievements constituted the world's first cold chain for frozen foods (Fikiin, 2003).

Thus, quick freezing has further been adopted as a widespread commercial method for long-term preservation of perishable foods, which improved both the health and convenience of virtually everyone in the industrialised countries. Freezing rate affects strongly the quality of frozen foods, in which the predominant water content should quickly be frozen in a fine-grain crystal structure in order to prevent the cellular tissues and to inhibit rapidly the spoiling microbiologic and enzymatic processes (Fig. 5.1).

Basic heat transfer considerations (Fikiin, 2003) clearly suggest that the desired shortening of freezing duration and a resulting high throughput of refrigerating equipment could be achieved by means of: (i) lower refrigerating medium temperature (which generally requires greater investment and running costs for the refrigeration machines to be employed); (ii) enhanced surface heat transfer coefficients (by increased refrigerating medium velocity and boundary layer turbulence, involvement of surface phase-change effects and less packaging);

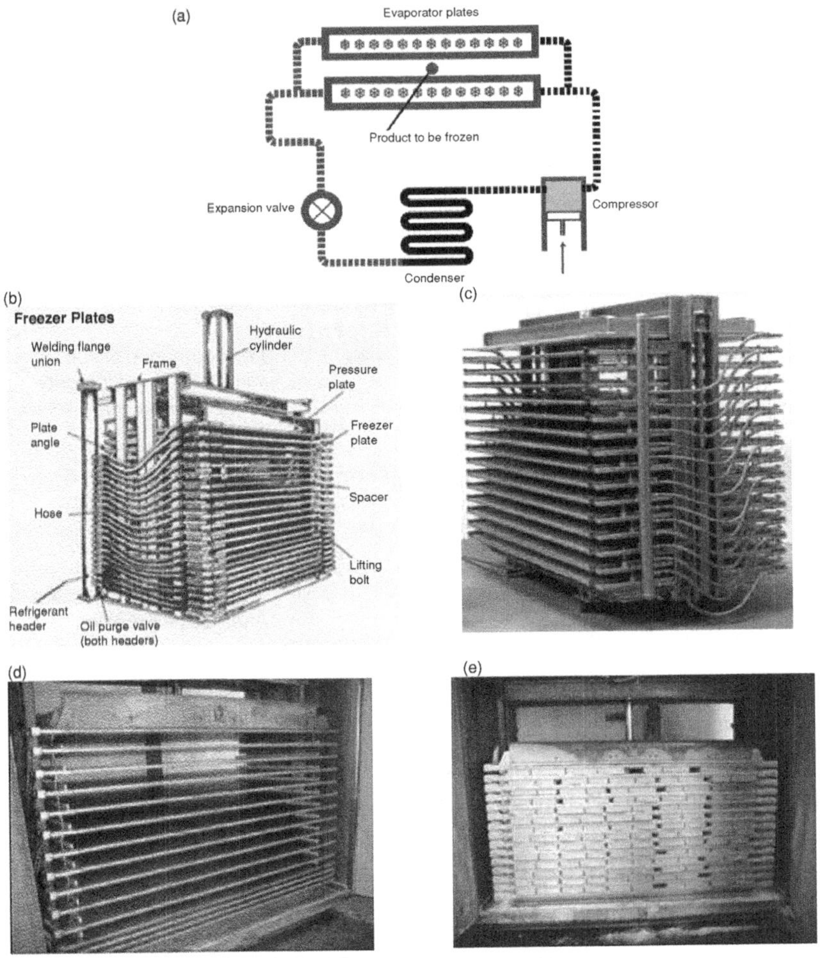

Fig. 5.2 Multi-plate freezing systems.

and (iii) reduced size of the refrigerated objects (by freezing small products individually or appropriately cutting the large ones into minor pieces).

As detailed in Chapters 4 and 6–8, air-blast and multi-plate freezers are most widespread, while air fluidising systems are used for IQF of small products (Figs. 5.2, 5.3 and 5.5). The application of cryogenic IQF (Fig. 5.4) is still very restricted because of the high price of the liquefied gases used.

5.2.1 Fluidised-bed freezing systems

The air fluidisation (Fig. 5.5) was studied extensively and used commercially, with an increasing popularity, during the last 50 years (Fikiin *et al.*, 1965, 1966, 1970; Fikiin, 1969, 1979, 1980). This freezing principle possesses many attractive features, including:

- High freezing rate due to the small sizes and thermal resistance of the IQF products, great overall heat transfer surface of the fluidised foods and high surface heat transfer coefficients.

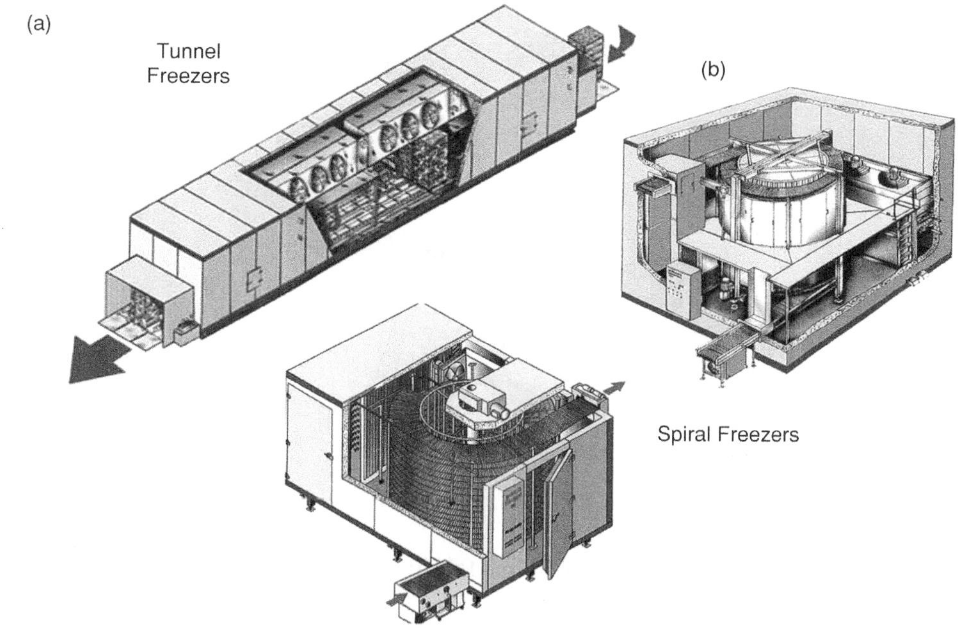

Fig. 5.3 Air-blast freezing systems.

- Good quality of the frozen products that have an attractive appearance and do not stick together.
- Continuity and possibilities for complete automation of the freezing process.

In spite of these advantages the fluidisation freezing by air has some drawbacks, such as:

- Necessity of two-stage refrigerating plants (using large quantities of CFC-, HCFC- or HFC based refrigerants with significant ozone depletion or global warming potentials) to maintain an evaporation temperature of about −45°C, which needs high investment and power costs.

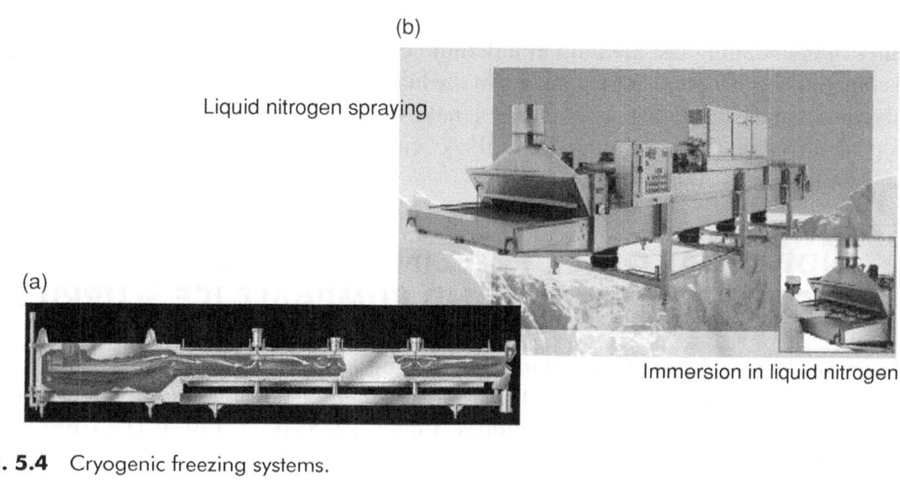

Fig. 5.4 Cryogenic freezing systems.

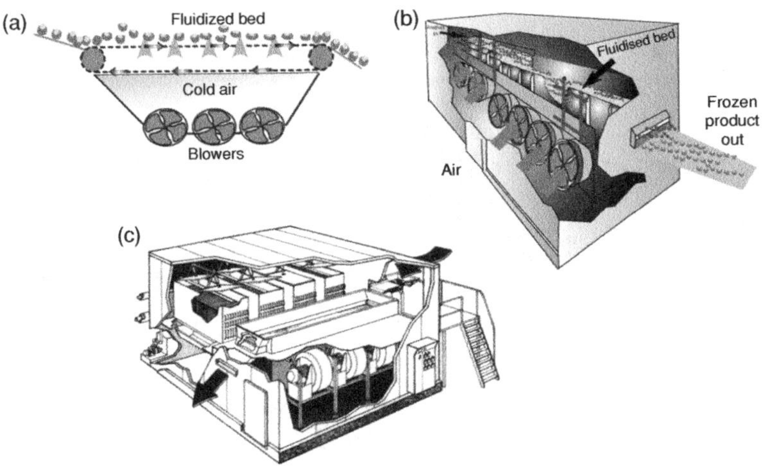

Fig. 5.5 Fluidised-bed freezing systems.

- Lower surface heat transfer coefficients and freezing rates in comparison with the immersion methods.
- Need for a high speed and pressure airflow, which results in a great fan power consumption.
- Some moisture losses from the product surface and a rapid frosting of the air coolers, caused by the great temperature differential between the products and the evaporating refrigerant.
- Excessive sensitivity of the process parameters to the product shape, mass and sizes, which requires careful control, specific for every separate food commodity.

5.2.2 Freezing by immersion

The immersion freezing in non-boiling liquid refrigerating media is a well-known method having several important advantages: high heat transfer rate, fine ice crystal system in foods, great throughput, low investments and operational costs (Tressler, 1968; Fleshland and Magnussen, 1990; Lucas and Raoult-Wack, 1998; Fikiin *et al.*, 2001). Immersion applications have been limited because of the uncontrolled solute uptake by the refrigerated products and operational problems with the immersion liquids (high viscosity at low temperatures, difficulty in maintaining the medium at a definite constant concentration and free from organic contaminants). Recent achievements in the heat and mass transfer, physical chemistry, fluid dynamics and automatic process control make it possible to solve these problems and to develop advanced innovative immersion IQF systems (Fikiin and Fikiin, 1998, 1999, 2002, 2003a, 2003b; Fikiin, 2003).

5.3 INDIVIDUAL QUICK FREEZING OF FOODS BY HYDROFLUIDISATION AND PUMPABLE ICE SLURRIES

The *Hydrofluidisation Method* (HFM) for fast freezing of foods was suggested and patented recently to overcome the drawbacks and to bring together the advantages of both air fluidisation and immersion food freezing techniques (Fikiin, 1985, 1992, 1994). The HFM uses a

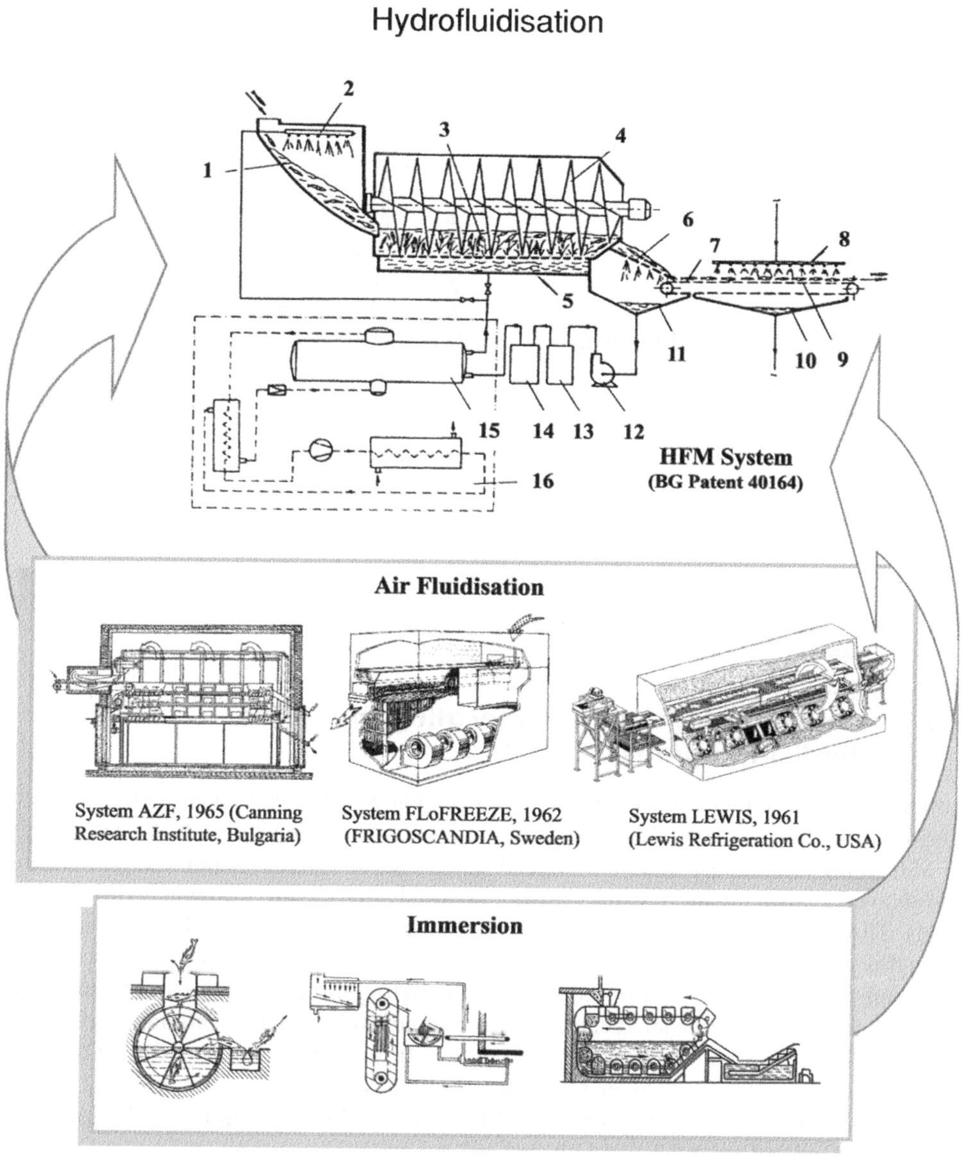

Fig. 5.6 Possible arrangement of a HFM-based freezing system combining the advantages of both air fluidisation and immersion food freezing techniques (Fikiin and Fikiin, 1998, 1999): (1) charging funnel; (2) sprinkling tubular system; (3) refrigerating cylinder; (4) perforated screw; (5) double bottom; (6) perforated grate for draining; (8) sprinkling device for glazing; (7 and 9) netlike conveyor belt; (10 and 11) collector vats; (12) pump; (13 and 14) rough and fine filters; (15) cooler of refrigerating medium; (16) refrigeration plant. Reproduced by permission of the International Institute of Refrigeration: www.iifiir.org.

circulating system that pumps the refrigerating liquid upwards, through orifices or nozzles, in a refrigerating vessel, thereby creating agitating jets. These form a fluidised bed of highly turbulent liquid and moving products, and thus evoke extremely high surface heat transfer coefficients. The principle of operation of a HFM freezing system is illustrated in Fig. 5.6.

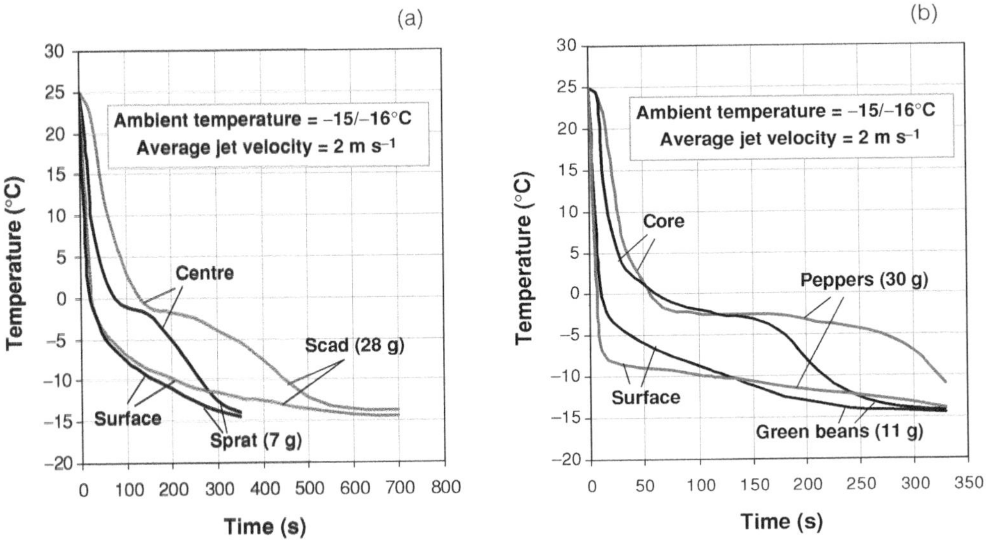

Fig. 5.7 Experimental temperature histories during HFM freezing of fish (a) and vegetables (b) when using sodium chloride solution (without ice slurry) as a fluidising agent (Fikiin, 1992; Fikiin and Fikiin, 1998, 1999). Reproduced by permission of the International Institute of Refrigeration: www.iifiir.org.

5.3.1 Unfreezable liquid refrigerating media as fluidising agents

Although various immersion techniques have been known for long time, until now hydrofluidisation principles have not been used for chilling and freezing of foods. Experiments on HFM freezing of small fish and some vegetables through an aqueous solution of sodium chloride showed a much higher freezing rate as compared to other IQF techniques (Fikiin, 1992, 1994). The maximal surface heat transfer coefficient achieved exceeded 900 W m^{-2} K^{-1}, while this was 378 W m^{-2} K^{-1} when immersing in running liquid, 432 W m^{-2} K^{-1} for sprinkling and 475 W m^{-2} K^{-1} for immersion with bubbling through (Fikiin and Pham, 1985). Even at a slight or moderate jet agitation and a comparatively high refrigerating medium temperature of about −16°C, the scad fish were frozen from 25°C down to −10°C in the centre in 6–7 minutes, sprat fish and green beans in 3–4 minutes and green peas within 1–2 minutes. As an illustration, Fig. 5.7 shows recorded temperature histories during hydrofluidisation freezing of scad and sprat fish, green beans and peppers.

5.3.2 Two-phase ice slurries as fluidising agents

Pumpable ice slurries (known under different trade names, such as *FLO-ICE*, *BINARY ICE*, *Slurry-ICE*, *Liquid ICE*, *Pumpable ICE* or *Fluid ICE*) were proposed recently as environmentally benign secondary coolants circulated to the heat transfer equipment of refrigeration plants, instead of the traditional harmful CFC- or HCFC-based refrigerants (Ure, 1998; Pearson and Brown, 1998). Promising attempts to refrigerate foods by immersion in such slurries have already been carried out. For example, fish chilling in brine-based slurries has potential to replace the traditional use of ice flakes (Fikiin *et al.*, 2001, 2002, 2005). A number of foods immersed in slurries with various ice contents are shown in Fig. 5.8.

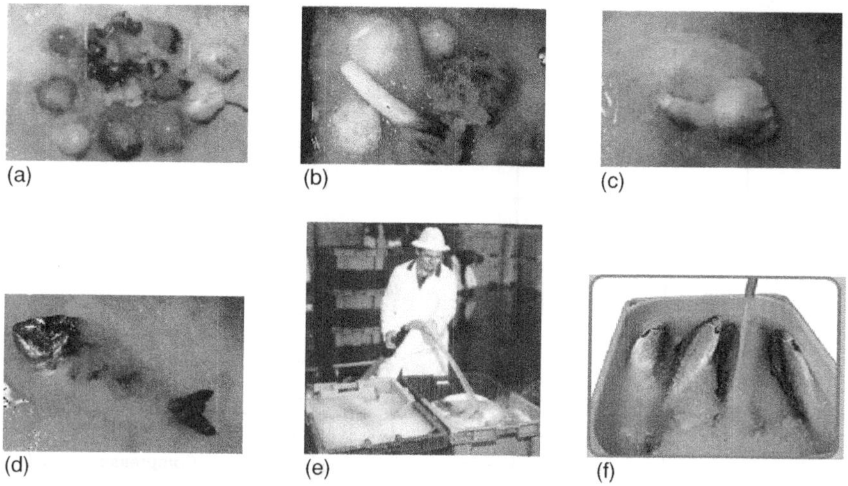

Fig. 5.8 Different foods immersed in slurries with various ice concentrations: (a) fruits; (b) vegetables; (c) chickens; (d), (e) and (f) fish (Fikiin *et al.*, 2002). Reproduced by permission of the International Institute of Refrigeration: www.iifiir.org.

Fikiin and Fikiin (1998, 1999) proposed, therefore, a novel method to enhance the advantages of the hydrofluidisation (described hereafter) by employing two-phase ice suspensions as fluidising media. The ice slurries reveal a great energy potential as HFM refrigerating media whose minute ice particles absorb latent heat when thawing on the product surface. Hence, the goal of the ice slurry involvement is to provide an enormously high surface heat transfer coefficient (of the order of 1000–2000 W m^{-2} K^{-1} or more), excessively short freezing time and uniform temperature distribution in the whole volume of the freezing apparatus. The combination of the HFM with the high heat transfer efficiency of the ice-slurry-based refrigerating media represents a new interdisciplinary research field whose development would advance essentially the refrigerated processing of foods. The HFM freezing with ice slurries can acquire a process rate approaching that of the cryogenic flash freezing modes. For instance, at a refrigerating ice-slurry temperature of −25°C and a heat transfer coefficient of 1000 W m^{-2} K^{-1}, strawberries, apricots and plums can be frozen from 25°C down to an average final temperature of −18°C within 8–9 minutes, raspberries, cherries and morellos within 1.5–3 minutes, and green peas, blueberries and cranberries within approximately 1 minute only. The general layout of an ice-slurry-based system for hydrofluidisation freezing is shown in Fig. 5.9.

5.3.3 Advantages of the hydrofluidisation freezing

As described above, the novelty of the hydrofluidisation method lies in the involvement of unfreezable liquids or pumpable ice slurries as fluidising agents. It is well known that the immersion freezing history began with use of brines to freeze fish, vegetables and meat. Binary or ternary aqueous solutions containing soluble carbohydrates (e.g. sucrose, invert sugar, glucose (dextrose), fructose and other mono- and disaccharides) with additions of ethanol, salts, glycerol, etc., have been studied as possible immersion media. There are practically unlimited possibilities to combine constituents and to formulate appropriate multi-component HFM refrigerating media based on one-phase liquids or two-phase ice slurries, which have

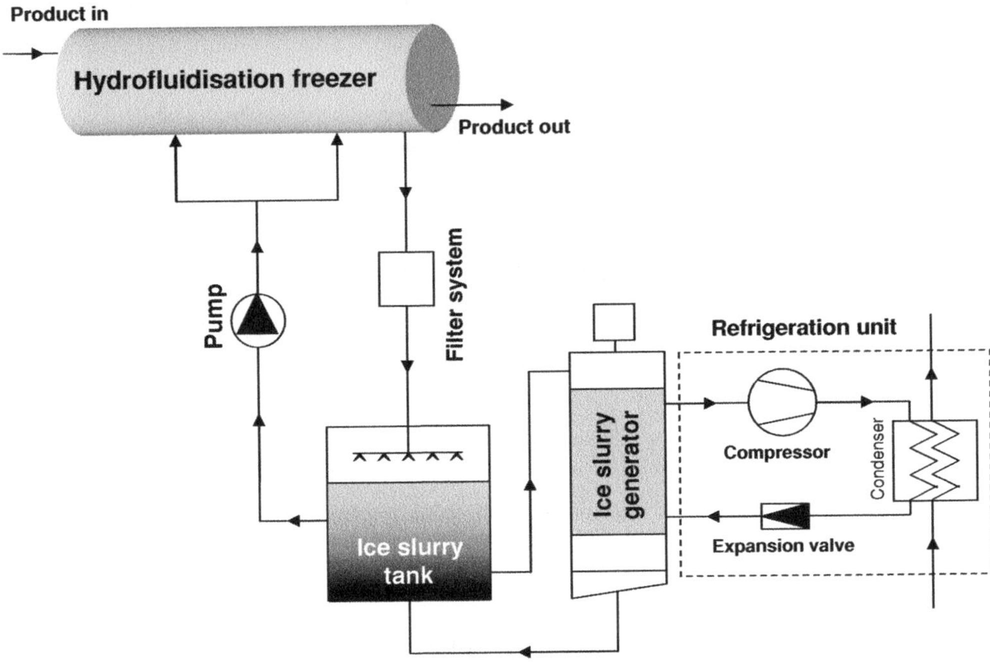

Fig. 5.9 Schematic diagram of an ice-slurry-based hydrofluidisation system HyFloFreeze® (Fikiin and Fikiin, 1998, 1999). Reproduced by permission of the International Institute of Refrigeration: www.iifiir.org.

to be both product- and environment-friendly and to possess a viscosity low enough in terms of pumpability and good hydrofluidisation.

The main advantages of the hydrofluidisation over the conventional freezing modes can be summarised as follows:

- The HFM affords a very high heat transfer rate with a small product–medium temperature difference. The evaporation temperature can be maintained much higher (at −25/−30°C) by a single-stage refrigerating machine with much higher COP and nearly two times lower investments and power costs as compared to the conventional air fluidisation. Cold dissipation through the freezer walls is also lower. The water flow rate or fan power consumption for cooling the condenser decrease as well, due to the reduced mechanical work of the single-stage unit.
- The critical zone of water crystallisation (from −1°C to −8°C) is quickly passed through, that ensures a fine ice crystal structure in foods preventing the cellular tissues from perceptible damage.
- The product surface freezes immediately in a solid crust that hampers the osmotic transfer and gives an excellent appearance. The mass losses tend to zero, while in air freezing tunnels the moisture losses are usually 2–3%.
- New delicious products can easily be formulated by using some selected product-friendly HFM media (for example, fruits frozen in syrup-type sugar solutions turn into dessert products with beneficial effect on colour, flavour and texture). Such media can also include

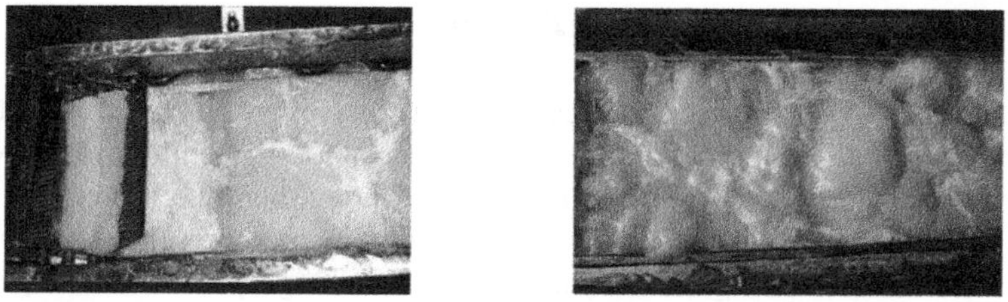

Fig. 5.10 HyFloFreeze® prototype: hydrofluidised bed of highly turbulent ice slurry (Fikiin, 2003). Reproduced by permission of the International Institute of Refrigeration: www.iifiir.org.

appropriate antioxidants, flavourings and micronutrients to extend the shelf-life of the products and to improve their nutritional value and sensory properties.

- The HFM freezers use environmentally-friendly secondary coolants (for instance, syrup-type aqueous solutions and ice slurries) and the refrigerant is closed in a small isolated system, in contrast to the common air fluidisation freezers where large quantities of harmful HCFCs or expensive HFCs circulate to remote evaporators with a much greater risk for emission to the environment.
- Fluidised state is acquired with low velocity and pressure of the fluid jets due to the Archimedes forces and buoyancy of the products, which leads to both energy savings and minimum mechanical action on the foods.
- The operation is continuous, easy to maintain, convenient for automation and the labour costs are substantially low. Further processing or packaging of the HFM-frozen products is considerably easier since they emerge from the freezer in a 'free-flowing' state.
- Ice-slurry-based HFM agents may easily be integrated into systems for thermal energy storage, accumulating ice slurry during the night at cheap electricity charges.

The top view photos in Fig. 5.10 show how a hydrofluidised bed of highly turbulent ice slurry is formed inside the *HyFloFreeze* prototype's freezing compartment.

5.3.4 International research co-operation

Two main innovative aspects of the suggested HFM freezing technique can clearly be distinguished: (i) employment of unfreezable liquids as fluidising agents and (ii) use of pumpable ice slurries as fluidising media. This freezing principle provides an extremely high heat transfer rate, short freezing times, great throughput and better product quality at refrigerating temperatures maintained by a single-stage refrigeration machine. Thus, nearly two times lower investments and power costs are necessary as compared with the popular individual quick freezing methods. Moreover, such hydrofluidisation freezing systems are less hazardous from the environmental viewpoint, since the refrigerant is closed in a small isolated circuit only.

The emerging HFM technology has drawn the attention of a number of academics and industrialists. The identification of optimal design specifications for HFM freezing systems requires an interdisciplinary approach of researchers with complementary skills. The *HyFloFreeze* project was, therefore, funded by the European Commission and performed by an international research consortium of six participating organisations (four universities and two SMEs) from Belgium, Bulgaria, Russia and the UK (Fikiin, 2003).

5.4 HIGH-PRESSURE FREEZING

Non-thermal food processing techniques (e.g. pulse-electric field pasteurisation, high-intensity pulsed lights, high-intensity pulsed-magnetic field, ozone treatment) are presently regarded with special interest by the food industry. Among them, high-pressure processing is gaining in popularity with food processors because of its food preservation capability and potential to achieve interesting functional effects. Under high pressure pathogenic micro-organisms can be inactivated with minimal heat treatment, which results in almost complete retention of nutritional and sensory characteristics of fresh foods, without sacrificing their shelf-life. Other advantages over traditional thermal processing include reduced process times; minimal heat damage problems; retention of freshness, flavour, texture, and colour; lack of vitamin C loss and tangible changes in food during pressure-shift freezing (due to reduced crystal size and multiple ice-phase forms); and minimal undesirable functionality alterations. However, the spore inactivation is a major challenge as methods for full inactivation of spores under pressure are yet to be developed. Hence, another group of research activities worldwide focus on different techniques for treatment of foods by high hydrostatic pressure, including high-pressure-aided freezing and thawing.

A number of products (such as jams and fruit juices) have been processed under high pressure in Japan. There have been 10–15 types of pressurised foods on the Japanese market but several have recently disappeared, while the remaining ones are too specific to excite a substantial commercial interest. Examples of pressurised products in Europe and US are: (i) orange juice (Pernod Ricard Company, France); (ii) acidified avocado purée (Avomex Company, USA and Mexico); and (iii) sliced ham (Espuna Company, Spain). Volumes produced are still very small and some current European food regulations slowed down the launching of new pressurised products because of legislative problems.

The so-called *cryofixation* is a physical method for immobilisation of biological materials by ultra-quick freezing. Unlike the chemical fixation, it preserves thoroughly the ultra-structural morphology, much closer to the natural state of the cell tissue. This results in fast preservation of morphological details without artificial damage, less cross-linking of proteins by aldehyde fixation and reduced masking of the antigenic sites. The water phase diagram (Fig. 5.11) shows that at atmospheric pressure crystalline ice will build up at around 0°C and this water crystallisation leads to some rupture of the biological structures (Fig. 5.1). The cryofixation aims, therefore, to avoid such crystallisation-caused damages. At very high freezing rates particles and large molecules in water serve as cores for a *heterogeneous nucleation*, i.e. water becomes solid in a *vitreous state* and does not show a crystalline structure. The necessary freezing rates can only be achieved for very thin layers of 5–25 μm during freezing at atmospheric pressure. This restriction could be overcome through a depression of the initial freezing (cryoscopic) point of water by adding chemical cryoprotectants or by increasing the ambient pressure. At a pressure of 200 MPa the freezing point drops to about −22°C (see Fig. 5.11), which enables a depth of vitrification of about 200 μm, so that objects with a thickness of up to 0.4–0.6 mm could be well frozen.

Consequently, the *main promising features of high-pressure freezing* are as follows:

- Freezing point depression and reduced latent heat of phase change;
- Short freezing times and resulting benefits (e.g. microcrystalline or vitreous ice);
- Inactivation of micro-organisms and enzymes, and structure modifications with no essential changes of nutritional and sensory quality.

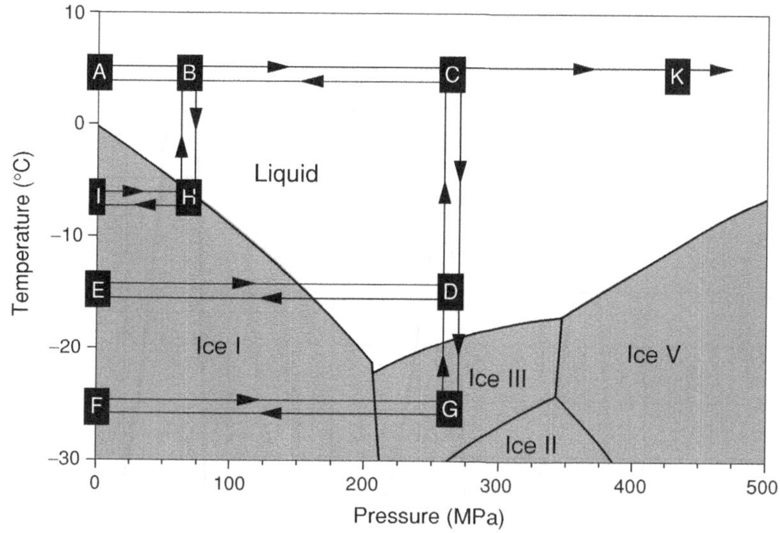

Fig. 5.11 Illustrates various paths of changing the food physical state by external manipulations of temperature or pressure, while Figs. 5.12 and 5.13 show temperature- and pressure-dependent thermal properties of potatoes during pressure-assisted freezing (Schlüter *et al.*, 2000).

Water phase diagram and high pressure effects on the phase transitions:

A – B – C – D – C – B – A	Subzero storage without freezing
A – B – H – I	Pressure-assisted[1] freezing
I – H – B – A	Pressure-assisted[1] thawing
A – B – C – D – E	Pressure-shift[2] freezing
E – D – C – B – A	Pressure-induced[3] thawing
A – B – C – D – G – F	Freezing to ice III
F – G – D – C – B – A	Thawing of ice III
A – B – C – K – ice VI	Freezing above 0°C

[1] assisted: phase transition at constant pressure
[2] shift: phase transition due to pressure change
[3] induced: phase transition initiated with pressure change and continued at constant pressure

The future will reveal soon whether the current achievements in this field are more likely to stay in the laboratories or they could be implemented as a common industrial practice.

5.5 MAGNETIC RESONANCE FREEZING

As already discussed, the conventional refrigeration equipment provides freezing rates which, as a rule, are insufficient to eliminate completely undesirable water migration and mass transfer within a food product undergoing freezing. Realising this circumstance, researchers decided that if water could somehow be retained within the cells while freezing, then the cells would not become dehydrated and foodstuff could keep its original attributes and freshness. A system for magnetic resonance freezing (MRF) preventing such cellular dehydration could be regarded as composed of a common freezer and a special magnetic resonance device. The MRF process (Fig. 5.14) is then applied with the following *two steps* (Mohanty, 2001).

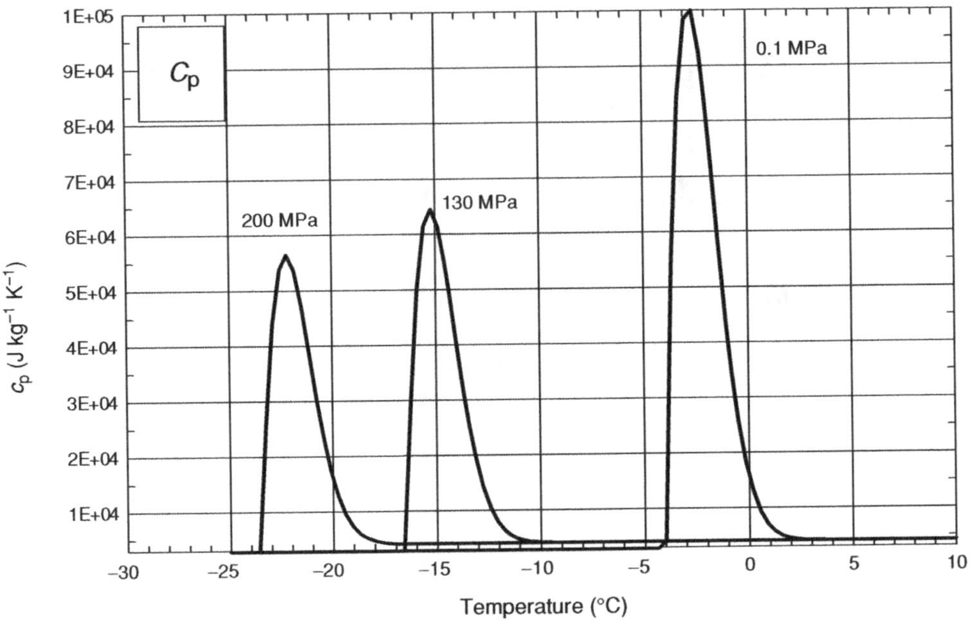

Fig. 5.12 Apparent specific heat capacity of potato tissue at different pressures. Reproduced by permission of the International Institute of Refrigeration: www.iifiir.org.

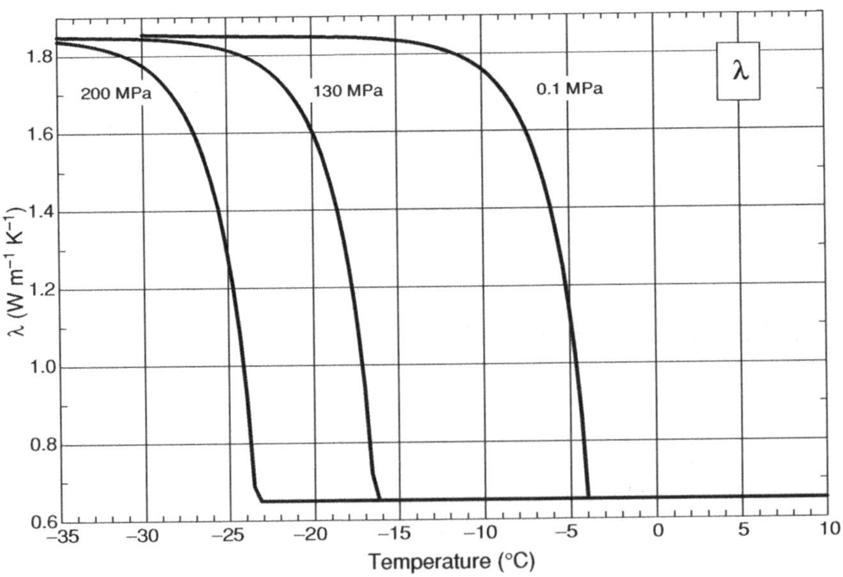

Fig. 5.13 Thermal conductivity of potato tissue at different pressures. Reproduced by permission of the International Institute of Refrigeration: www.iifiir.org.

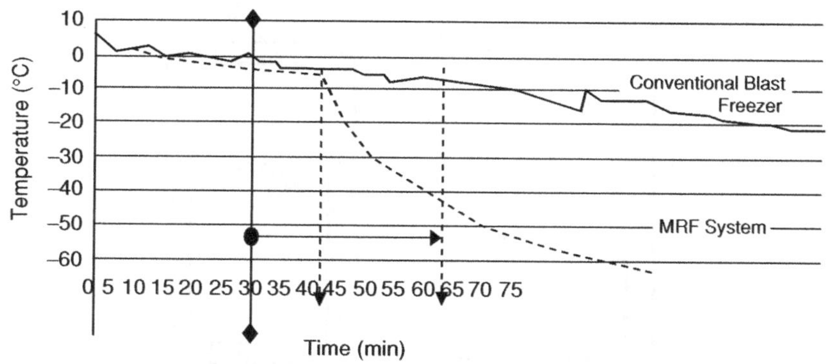

Fig. 5.14 Product freezing curves for conventional and MRF equipment (Mohanty, 2001).

Step 1: Food undergoes *continuous magnetic wave vibrations*, which provide for:

- Impeding the crystallisation;
- Supercooling below the initial freezing point.

Step 2: After a suitable product-specific period of time *the magnetic fields are abruptly removed* with many resulting quality benefits for the frozen end product, e.g.:

- Uniform flash freezing of the entire food volume;
- Quick passing through the critical temperature zone of intense water crystallisation (between −1°C and −6°C);
- Fine ice structure in foods;
- No water migration and undesirable mass transfer phenomena;
- No cellular dehydration;
- Avoiding cracks and related damages;
- Protected integrity of food tissues.

At present MRF data are still kept as a confidential *know-how* of a number of companies, while MRF equipment still needs to prove its claimed advantages and capabilities through extensive tests within a sufficiently representative industrial environment.

5.6 AIR-CYCLE-BASED FREEZING SYSTEMS

The majority of existing food refrigeration equipment has been designed to use halogenated hydrocarbons (CFCs and HCFCs) whose emissions to the environment are damaging the ozone layer and contributing significantly to global warming. Manufacture and import of CFCs is banned in most of the world and many HCFC refrigerants are only short-term replacements, often being more expensive and less efficient. The vapour-compression systems have refrigerant leakage rates to the environment of 15% of the total charge per annum, thus leading to stratospheric ozone depletion and climate change. Some other non-CFC alternatives called *natural refrigerants* (e.g. ammonia, propane, butane, isobutane, carbon dioxide and water) are also being employed or examined. Ammonia is the most common of the alternative refrigerants but is toxic and not always suitable for all refrigeration applications. Natural gas–based refrigerants involve a safety hazard because of their high flammability.

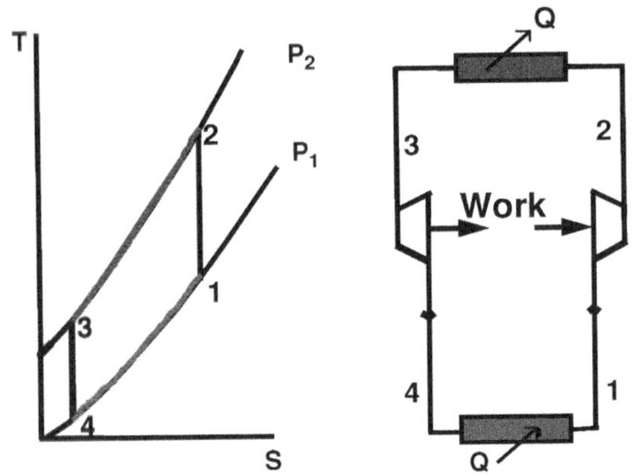

Fig. 5.15 The Air Cycle on a T–S (Temperature–Entropy) diagram and shown diagrammatically.

Transcritical CO_2 cycles needs high pressures and complicated equipment. As a rule, important modifications to the existing refrigerating plants are required to introduce *green* technologies in the food refrigeration sector.

The principle of the air cycle is that when air is compressed its temperature and pressure increase (1–2) as shown in Fig. 5.15. Heat is removed from the compressed air at constant pressure and its temperature is reduced (2–3). Air is then expanded and its temperature decreases (3–4). The air further absorbs heat (gaining temperature) from the process at constant pressure (4–1) where it starts the cycle again (Evans *et al.*, 2005).

Air-cycle refrigeration is an environmentally friendly alternative to the conventional vapour-compression systems. Originally used in the nineteenth century, the technology employed slow-speed reciprocating compressors and low-efficiency expanders. The poor energy efficiency and comparatively high cost of such machinery was a major factor for the replacement of these systems with vapour-compression equipment. However, the remarkable progress in turbine technology since that time (Horlock, 2003), along with the development of air bearings and ceramic components, provided dramatic improvements in efficiency. The air-cycle use for aircraft and railway carriage air-conditioning (which affords equipment simplicity, compactness and robustness) and the development of high-speed rotary compressors and expanders have greatly improved the cycle performance and reliability (Fig. 5.16).

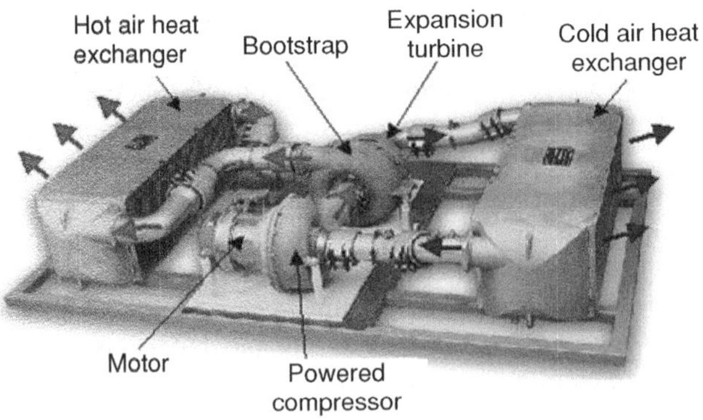

Fig. 5.16 An air-cycle unit for automotive applications (Normalair-Garrett Ltd.)

Combining this with newly available compact heat exchangers with highly enhanced heat transfer characteristics makes competition with existing vapour-compression systems rather attractive.

Air-cycle systems could be viable for many industrial applications (Gigiel *et al.*, 1996; Van Gerwen and Verschoor; 1996; Verschoor, 2001). An air-cycle system can simultaneously produce both heat and cold, the advantage being that heating is provided at higher temperatures as compared to the conventional vapour-cycle heat-pump system. Frequency control of the motor speed permits a smooth variation of the operating capacity depending on the heat loads. On the other hand, mass production techniques make the air systems much cheaper than they were in the past. Air-cycle refrigeration is one of the most attractive ways for solving the environmental problems since the working fluid (air) is completely natural, free and totally environmentally benign. This environmental friendliness, reliability, minimal maintenance and efficiency can successfully be transferred to the food refrigeration industry (Gigiel *et al.*, 1992; Fikiin *et al.*, 1998; Russell *et al.*, 2001, 2001). The energy efficiency can additionally be enhanced by combining the air cycle with systems for thermal energy storage. Feasibility studies have shown that such type of equipment may compete (in terms of energy and investment costs) with vapour-compression systems, coupled with conventional water heating devices (Gigiel *et al.*, 1996; Van Gerwen and Verschoor, 1996).

Both open and closed air-circulation systems can be employed to freeze foods. In the *open cycle* atmospheric air normally enters the system and returns to the atmosphere after being used for heating and cooling purposes (Shaw *et al.*, 1995). The air as a refrigerant is in a direct contact with the refrigerated products, avoiding any intermediate heat exchangers, pipelines, secondary coolants and related energy dissipation. However, the running of such systems is strongly dependent on the atmospheric conditions, which results in more difficult control. Atmospheric air contains water vapours, which may freeze at low temperatures. Obviously, the open systems also require turbo-machines with oil-free air bearings. In the *closed cycle* dry air is circulated around the closed system, similarly to a conventional vapour-compression machine. The advantage of a closed air cycle is that the load on the system can be matched exactly without any change in the efficiency. This is because the efficiency of the rotating machinery used in air-cycle plants depends only on the velocities of the air and not its density. The change of pressure in the system overall results in corresponding changes in the air density and mass flow. Thus the refrigeration and the heating effects from the air-cycle plant can vary in proportion to the mass flow, without change in the efficiency, in contrast to common vapour-compression systems.

Recent developments and improvements in air-cycle equipment enable a wider practical exploitation of this emerging technology for food freezing applications. The benefits are greater reliability, reduced maintenance, simple and compact design, reduced overall labour costs, improved control of product quality, along with using an environmentally benign and free refrigerant unlike the conventional heating and cooling systems. Fast freezing through low-temperature air dramatically reduces freezing times and has many advantages (e.g. flexibility, savings in freezer footprint and higher product quality). With a conventional vapour-compression plant, freezing is limited to refrigerating air temperatures above −40°C. Freezing at lower refrigerating temperatures is limited to cryogens, such as nitrogen or carbon dioxide that are expensive *throw away* liquefied gases. The primary reason for using an air cycle in food freezing is that it can greatly increase the range of operating conditions available (Evans *et al.*, 2005). Lower refrigerating temperatures result in faster freezing, improved food quality and either reduced freezer size or larger throughput through an existing freezer (if the necessary refrigeration capacity is duly provided).

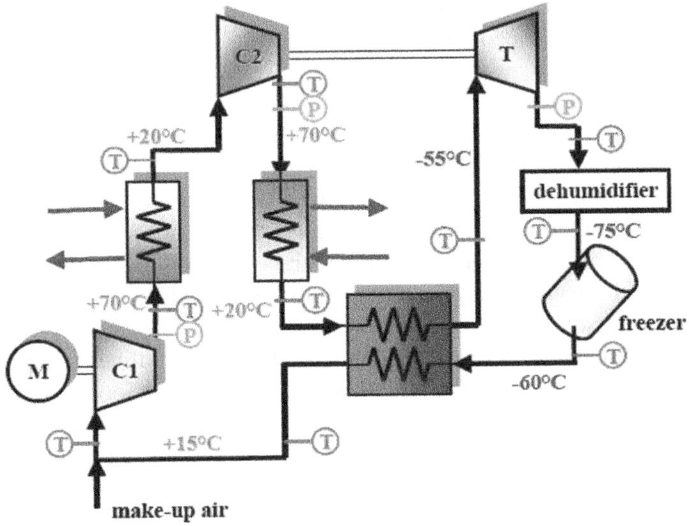

Fig. 5.17 Schematic diagram of an open air-cycle system for quick freezing of foods: C1 = primary compressor, C2 = bootstrap, T = turbine, M = motor (Gigiel *et al.*, 2004). Reproduced by permission of the International Institute of Refrigeration: www.iifiir.org.

An air-cycle system for food freezing is presented schematically in Figure 5.17 (Gigiel *et al.*, 2004), while Figure 5.18 shows how the compressor, expander and driving motor can be assembled in a single compact unit with a powered bootstrap and common power supply (Kulakov *et al.*, 1999; Gigiel *et al.*, 2004; Kikuchi *et al.*, 2005).

A number of theoretical studies have indicated the potential for air cycle in food processing operations (Gigiel *et al.*, 1992; Russell *et al.*, 2000, 2001). With conventional vapour

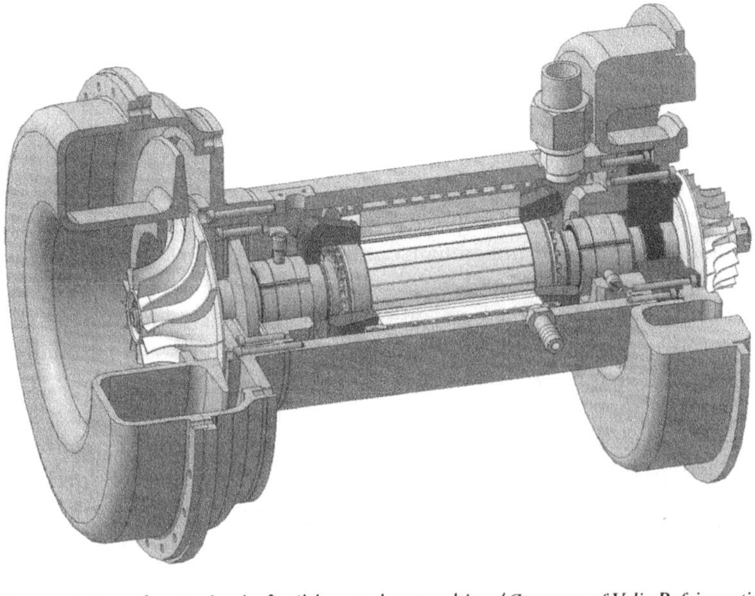

Fig. 5.18 Internal view of a single-shaft oil-less turbo-machine (*Courtesy of Velis Refrigeration, Russia*).

compression plant the refrigerating air temperature ranges between ambient and −40°C. With air cycle, there is no limit on the cold side temperature above the liquefaction point and therefore the processes can be designed to suit the food and not the available equipment (Evans *et al.*, 2005). Theoretically, air temperatures up to 300°C on the hot side can also be obtained at the same time, suitable for direct cooking or the production of steam.

The established public opinion is that the air-cycle is inefficient as it has a low COP. It is fairly mentioned in almost every manual, that air-cycle possesses low thermodynamic efficiency when using it solely for cooling within a temperature range common for the commercial food refrigeration. Nevertheless, thermodynamic evidence shows (Kulakov *et al.*, 1999) that air-cycle refrigeration may reach or exceed the performance of vapour-compression systems in two practically valuable cases: (i) at low refrigerating temperatures (approaching or falling within the cryogenic range), and (ii) when using the air cycle for both cooling and heating, i.e. as a heat pump. In the 1940s Peter Kapitza (Nobel Laureate in physics) introduced for the first-time turbo-technologies for mass production of liquid oxygen at low pressures. His enemies subjected him to malicious attacks at a governmental level because this process has been energy inefficient as compared to common high-pressure cryogenic cycles. However, because of the mass production technologies and the lack of breakdowns, the ‘energy inefficient’ Kapitza process proved to have much higher economic efficiency than alternative cycles (efficient in terms of running costs only). Remarkable progress has recently been achieved as advanced turbo-technologies lie in the basis of the modern power generation and aeronautics – major players, such as Rolls Royce and Boeing, are continuously generating new knowledge in the field (Horlock, 2003).

Consequently, air-cycle food freezing has no intrinsic detrimental ozone depletion and global warming effects. Air cycle is capable of operating at low refrigerating temperatures (close to the cryogenic range), where it outperforms conventional vapour-compression systems (efficient above −40°C) and CO_2-based systems (used above −54°C). Furthermore, the high-velocity airflow exiting the turbo-expander can directly be impinged on the food, thereby ensuring high heat transfer coefficients (over 140 W m^{-2} K^{-1}). Thus, along with *enhanced air-blast freezing*, an open cycle can perfectly serve to freeze foods by *air-cycle-based fluidisation or air impingement*, both associated with high process rates and substantially improved quality of the end product.

5.7 OTHER UNCONVENTIONAL FREEZING METHODS AND CONCLUDING REMARKS

Alongside the emerging and novel freezing techniques mentioned in this chapter (e.g. hydrofluidisation; immersion freezing with smart agitation modes; application of ice slurries or air impingement; air-cycle-based freezing; flash-freezing cryogenic methods; high-pressure shift freezing and magnetic resonance freezing), a number of promising freezing process innovations have also been launched (such as ultrasonic freezing, dehydrofreezing, use of antifreezes and ice nucleation proteins; freeze drying, partial freezing, vacuum and heat pipe applications; solar, thermionic; magnetocaloric, electrocaloric and thermoacoustic refrigeration). Some of these freezing principles are presently small scale only and still unlikely to be implemented for commercial refrigeration in a short-term perspective.

Novel technologies with long-term implementation perspective include magnetic and acoustic Stirling refrigeration. Magnetic refrigeration is based on the magnetocaloric effect

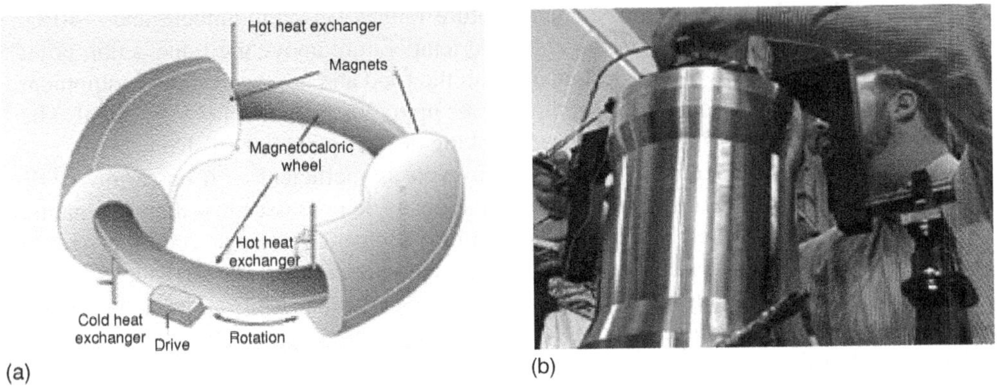

Fig. 5.19 (a) Possible arrangement of a rotary magnetic refrigerator and (b) the world's first commercial thermoacoustic refrigerator for ice cream.

(MCE). Magnetocaloric materials change temperature in response to an applied magnetic field and can therefore be used to cool. Although this is not dissimilar to a traditional vapour-compression cycle (where heat is removed at the condenser and gained at the evaporator), unlike conventional systems, no potentially harmful working fluid is used. The MCE peaks at around the Curie temperature and (due to recent discoveries of materials with high Curie temperatures around ambient temperature and materials with giant MCE) refrigeration devices working at ambient temperatures are being developed (Fig. 5.19a). As a rule, magnetic refrigeration possesses high energy efficiency and, in spite of the comparatively low refrigeration capacities achieved so far, the technology is rapidly developing and might soon become a viable alternative for cooling in small household and retail refrigerators. IIR has recently created a dedicated Working Party on Magnetic Cooling, which holds regular IIR conferences on magnetic refrigeration at room temperature as an important forum to report current findings and a valuable source of up-to-date information (alongside a recent special issue of the *International Journal of Refrigeration*, edited by Auracher and Egolf, 2006).

Backhaus and Swift (1999) developed and reported a regenerator-based thermoacoustic Stirling heat engine, which was more efficient than previous acoustic engines. In thermoacoustic refrigerators sound is used to generate pressure and alternatively compress and expand a gas (usually helium). When the gas compresses it heats up and when it expands it cools. The gas moves backwards and forwards along a tube stopping to reverse direction when the gas reaches either maximum compression or expansion. The Stirling engine principle has been known for almost 200 years but unlike mechanical Stirling machines the acoustic system uses no moving parts. By incorporating plates into the tube the system efficiency can further be enhanced. A small 200-litre prototype of commercial thermoacoustic freezer was developed for the first time by Poese *et al.* (2004) for Ben and Jerry's Homemade (Fig. 5.19b). Further improvements in the design of acoustic refrigerators have been reported extensively by Backhaus and Swift (2000, 2004), So *et al.* (2006), Matveev *et al.* (2007) and Ueda *et al.* (2003).

A number of further publications which feature interesting food freezing innovations in more detail, are readily available for the stakeholders of refrigeration science and industry (see for instance, James *et al.*, 2000; Magnussen *et al.*, 2000; Sun, 2001; 2005).

ACKNOWLEDGEMENTS

Figures 5.1–5.19 include illustrations of R.P. Singh, O. Schlüter, Air Products and Chemicals, Advanced Freezers, Frigoscandia, Normalair-Garrett, Food Refrigeration and Process Engineering Research Centre, Velis Refrigeration, Pennsylvania State University and Ames Laboratory, whose contribution is gratefully acknowledged.

ACRONYMS

CFC	Chlorofluorocarbon
COP	Coefficient of performance
FAO	Food and Agriculture Organisation of the United Nations
HCFC	Hydrochlorofluorocarbon
HFC	Hydrofluorocarbon
HFM	Hydrofluidisation method
IIR	International Institute of Refrigeration
IQF	Individual quick freezing
MCE	Magnetocaloric effect
MRF	Magnetic resonance freezing
SME	Small or medium enterprise
WHO	World Health Organisation

REFERENCES

Auracher, H. and Egolf, P.W., eds. (2006). Magnetic refrigeration at room temperature. *International Journal of Refrigeration* 29(8), 1235–1394.

Backhaus, S. and Swift, G.W. (1999). A thermoacoustic Stirling heat engine. *Nature* 399, 335–338.

Backhaus, S. and Swift, G.W. (2000). A thermoacoustic-Stirling heat engine: detailed study. *Journal of the Acoustical Society of America* 107(6), 3148–3166.

Backhaus, S. and Swift, G.W. (2004). A resonant, self-pumped, circulating thermoacoustic heat exchanger. *Journal of the Acoustical Society of America* 116(5), 2923–2938.

Evans, J.A., Gigiel, A.J. and Brown, T. (2005). Fast freezing using air cycle technologies. In: *Proceedings of EU Workshop 'EUROFREEZE 2005: Individual Quick Freezing of Foods'*, Sofia (Bulgaria), 13–15 January 2005, European Commission.

Fikiin, A.G. (1969). Congélation de fruits et de légumes par fluidisation. *Proceedings Budapest Conference*, IIF/IIR: pp. 197–203.

Fikiin, A.G. (1979). Bases théoriques du procédé de fluidisation lors de l'intensification de la congélation des fruits et des légumes. *Proceedings 15th International Congress Refrigeration*, Venice, 4, pp. 221–230.

Fikiin, A.G. (1980). Physical conditions of fluidized-bed freezing of fruits and vegetables. *Kholodilnaya Tekhnika/Refrigeration Engineering*, 7, 59–61 (in Russian)

Fikiin, A.G. (1985). Method and system for immersion cooling and freezing of foodstuffs by hydrofluidization. *Invention Certificate No. 40164*, Bulgarian Patent Agency INRA.

Fikiin, A.G. (1992). New method and fluidized water system for intensive chilling and freezing of fish. *Food Control* 3(3), 153–160.

Fikiin, A.G. (1994). Quick freezing of vegetables by hydrofluidization. In: *New Applications of Refrigeration to Fruit and Vegetables Processing* – Proceedings of IIR Conference, Istanbul (Turkey), *Refrigeration Science and Technology*, International Institute of Refrigeration, 1994-3, pp. 85–91.

Fikiin, A.G. and Pham, V.H. (1985). System for examination of heat transfer regimes during hydrorefrigeration of foodstuffs. *Invention Certificate No. 39749*, Bulgarian Patent Agency INRA.

Fikiin, A.G., Ditchev, S.P. and Fikiina, I.K. (1966). Principal parameters characterising the fluidisation of fruit and vegetable layers. *Kholodilnaya Tekhnika/Refrigeration Engineering* 11, 33–37 (in Russian)

Fikiin, A.G., Ditchev, S.P. and Karagerov, D.I. (1965). Fluidized bed freezing system for fruits and vegetables with various dimensions. *Invention Certificate No. 10967*, Bulgarian Patent Agency INRA.

Fikiin, A.G., Ditchev, S.P. and Karagerov, D.I. (1970). Fluidized bed freezing apparatus AZF. *Kholodilnaya Tekhnika/Refrigeration Engineering* 7, 55–58 (in Russian).

Fikiin, K.A. (2003). *Novelties of Food Freezing Research in Europe and Beyond.* Flair-Flow Europe Synthetic Brochure for SMEs No.10 (ISBN: 2-7380-1145-4), *INRA: Institut National de la Recherche Agronomique*, Paris (France), p. 55.

Fikiin, K.A. and Fikiin, A.G. (1998). Individual quick freezing of foods by hydrofluidisation and pumpable ice slurries. In: K. Fikiin, ed. *Advances in the Refrigeration Systems, Food Technologies and Cold Chain.* IIR Proceedings Series *'Refrigeration Science and Technology'*, 1998-6: pp. 319–326 (published also in the *AIRAH Journal,* 2001, 55(11), pp. 15–18).

Fikiin, K.A. and Fikiin, A.G. (1999). Novel cost-effective ice-slurry-based technology for individual quick freezing of foods by hydrofluidisation. *CD-Rom Proceedings of the 20th International Congress of Refrigeration*, Sydney (Australia), ICR Paper No. 271.

Fikiin, K.A. and Fikiin, A.G. (2002). Congelación individual rápida de alimentos por hidrofluidificación y compuestos de hielo. *Frío, Calor y Aire Acondicionado* (Madrid) 30(334), 22–27.

Fikiin, K.A. and Fikiin, A.G. (2003a). Quick freezing of foods by hydrofluidisation and pumpable ice suspensions. *Kholodilnaya Tekhnika/Refrigeration Engineering* (1), 22–25 (in Russian).

Fikiin, K.A. and Fikiin, A.G. (2003b). L'Ice Slurry (ghiaccio binario) per una surgelazione veloce degli alimenti – Surgelamento singolo degli alimenti mediante idrofluidizzazione. *Industria & Formazione per il Tecnico della Refrigerazione e Climatizzazione* (Milano) (3), 36–39.

Fikiin, K.A., Fikiin, A.G., Russell, S.L. and Fitt, P.W. (1998). Shelf-life extension of Bulgarian yoghurt by using a novel environment-friendly air-cycle integrated system for thermal and refrigerated processing. In: *Shelf-Life Prediction* – Proceedings of the EU COPERNICUS Workshop, Wageningen (the Netherlands), pp. 159–168.

Fikiin, K.A., Kaloyanov, N.G., Filatova, T.A. and Sokolov, V.N. (2002). Fine-crystalline ice slurries as a basis of advanced industrial technologies: state of the art and future prospects. *Refrigeration Business* (Moscow) (7), 4–11 (in Russian).

Fikiin, K.A., Wang, M.-J., Kauffeld, M. and Hansen, T.M. (2005). Direct contact chilling and freezing of foods in ice slurries – Chapter 9. In: M. Kauffeld, M. Kawaji and P.W. Egolf, eds. *Handbook on Ice Slurries – Fundamentals and Engineering*. International Institute of Refrigeration, pp. 251–271.

Fikiin, K.A., Tsvetkov, O.B., Laptev, Yu.A., Fikiin, A.G. and Kolodyaznaya, V.S. (2001). Thermophysical and engineering issues of the immersion freezing of fruits in ice slurries based on sugar-ethanol aqueous solutions. *Proceedings of the Third IIR Workshop on Ice Slurries*, Lucerne (Switzerland). International Institute of Refrigeration, pp. 147–154 (published also in *EcoLibrium – Journal of AIRAH: Australian Institute of Refrigeration, Air Conditioning and Heating*, 2003, 2(7), 10–15).

Fleshland, O. and Magnussen, O. M. (1990). Chilling of farmed fish. *Proceedings Aberdeen Conference*, IIF/IIR: pp. 185–192.

Gigiel, A.J., Chauveron, S. and Fitt, P. (1992). Air as a replacement for CFC refrigerants. *IIR Congress 'Cold '92'*, Buenos Aires, Argentina, 7–9 Sept.

Gigiel, A., Gibbs, R., Butler, D. and Holder, D. (1996). Air-cycle technologies for building services. *IIR Proceedings Series 'Refrigeration Science and Technology'*, International Institute of Refrigeration – Paris, 1996-3: 687–696.

Gigiel, A., Polonara, F. and Di Nicola, G. (2004). Rapid freezing of food in an open air cycle freezer. In: *Natural Working Fluids 2004* – 6th IIR Gustav Lorentzen Conference. IIR Proceedings Series '*Refrigeration Science and Technology*', International Institute of Refrigeration, 2004–2.

Horlock, J. (2003). *Advanced Gas Turbine Cycles.* Elsevier Science, p. 405.

IIR (1986). *Recommendations for the Processing and Handling of Frozen Foods.* Paris: International Institute of Refrigeration, p. 419.

IIR (1999). *Control of the Cold Chain for Quick-Frozen Foods. Handbook.* Paris: International Institute of Refrigeration, p. 94.

James, C., Ketteringham, L. and James, S.J. (2000). Enhanced heat transfer in food chilling, freezing and thawing using heat pipes. In: K. Fikiin, ed. *Advances in the Refrigeration Systems, Food Technologies and Cold Chain.* IIR Proceedings Series *'Refrigeration Science and Technology'*, 1998/6, pp. 327–333.

Kennedy, C.J., ed. (2000). *Managing Frozen Foods.* Cambridge: Woodhead Publishing, p. 286.

Kikuchi, S., Okuda, S., Igawa, H., Morii, S., Mitsuhashi, M. and Higashimori, H. (2005). Development of an air-cycle system for refrigeration. *Mitsubishi Heavy Industries Technical Review* 42(4), 1–4.

Kulakov, V.M., Vereschagin, M.P., Kulakov, V.V. and Vereschagin, M.M. (1999). Air turbo-refrigerating machines. *Refrigeration Business* 6: 5–10 (in Russian).

Lucas, T. and Raoult-Wack, A.L. (1998). Immersion chilling and freezing in aqueous refrigerating media: review and future trends. *International Journal of Refrigeration* 21(6), 419–429.

Magnussen, O.M., Nordtvedt, T.S. and Torstveit, A.K. (2000). Use of partial freezing in the cold chain. In: K. Fikiin, ed. *Advances in the Refrigeration Systems, Food Technologies and Cold Chain*. IIR Proceedings Series *'Refrigeration Science and Technology'*, 1998/6, pp. 363–370.

Matveev, I.K., Backhaus, S. and Swift, G.W. (2007). Analytical solution for temperature profiles at the ends of thermal buffer tubes. *International Journal of Heat and Mass Transfer* 50, 897–901.

Mohanty, P. (2001). Magnetic resonance freezing system. *AIRAH Journal* 55(6), 28–29.

Pearson, S.F. and Brown, J. (1998). Use of pumpable ice to minimise salt uptake during immersion freezing. *Proceedings Oslo Conference*, IIF/IIR: pp. 712–722.

Poese, M.E., Robert, W.M., Garrett, S.L., van Gerwen, R. and Gosselin, P. (2004). Thermoacoustic refrigeration for ice cream sales. *6th IIR Gustav Lorentzen Conference on Natural Working Substances*, Glasgow, Aug 29–Sept 1, 2004.

Russell, S.L., Gigiel, A.J. and James, S.J. (2001). Progress in the use of air cycle technology in food refrigeration and especially retail display. *AIRAH Journal* 55(11), 20–25.

Schlüter, O., George, S., Heinz, V. and Knorr, D. (2000). Phase transitions in model foods, induced by pressure-assisted freezing and pressure-assisted thawing. In: K. Fikiin, ed. *Advances in the Refrigeration Systems, Food Technologies and Cold Chain*. IIR Proceedings Series *'Refrigeration Science and Technology'*, 1998-6, pp. 240–248.

Shaw, R., Kiczek, E. and Rossman, J. (1995). COLDBLAST air-cycle freezing system. *Proceedings of the 19th International Congress of Refrigeration*, The Hague, 3a: 684–691.

So, J.H., Swift, G.W. and Backhaus, S. (2006). An internal streaming instability in regenerators. *Journal of the Acoustical Society of America* 120(4), 1898–1909.

Sun, Da-Wen, ed. (2001). *Advances in Food Refrigeration*. Surrey: Leatherhead Publishing, p. 482.

Sun, Da-Wen, ed. (2005). *Handbook of Frozen Food Processing and Packaging*. Boca Raton, FL: CRC Press, Taylor & Francis Group, p. 737.

Tressler, D.K. (1968). Food freezing systems. In: D.K. Tressler, W.B. Van Arsdel and M.J. Copley, eds. *The Freezing Preservation of Foods*, Vol. 1. Westport, CT: AVI Publishing, pp. 120–152.

Verschoor, M., ed. (2001). *Guidelines for the Application and Design of Air-cycle Systems for Heating, Ventilating and Air-Conditioning in Buildings*. TNO Environment, Energy and Process Innovation, p. 120.

Van Gerwen, R. and Verschoor, M. (1996). Feasibility of air cycle systems for building air conditioning systems. *IIR Proceedings Series 'Refrigeration Science and Technology'*, International Institute of Refrigeration – Paris, 1996-3: pp. 677–685.

Ueda, U., Biwa, T., Yazaki, T. and Mizutani, U. (2003). Construction of a thermoacoustic Stirling cooler. *Physica B* 329, 1600–1601.

Ure, Z. (1998). Slurry-ice based cooling systems. In: K. Fikiin, ed. *Advances in the Refrigeration Systems, Food Technologies and Cold Chain*. IIR Proceedings Series *'Refrigeration Science and Technology'*, 1998-6: pp. 172–179.

6 Freezing of Meat

Steve James

It is believed that the first meat freezing works were established at Darling Harbour in Sydney, Australia in 1861 (Critchell and Raymond, 1912). In the next decade there were a number of attempts to transport frozen meat. However, the first entirely successful frozen meat shipment was that in the s.s. *Paraguay* from Buenos Aires to Havre in 1877. Due to a collision the ship took 7 months to complete her journey but the 5,500 mutton carcasses were 'in tip top condition' when the ship arrived at Havre. A second voyage was planned but never happened and to quote Bergés (cited in Critchell and Raymond, 1912), "As has often happened in the history of industries, it has been the French who have made the discoveries, and the English have turned them to account to their profit".

The arrival of the s.s. *Strathleven* in London on December 8 1880 with its cargo of 40 tonnes of frozen Australian beef and mutton started the frozen meat trade. It was sold for up to three times its value in Australia and as stated in the Daily Telegraph, "It has been tested by the ordinary method of cooking, and found to be in such good condition that neither by its appearance in the butchers' shops, nor by any peculiarity of flavour when cooked for the table, could it be distinguished from freshly killed English meat." Eighteen months later the s.s. *Dunedin* arrived with a cargo of frozen New Zealand lamb and mutton. By 1892 2,000,000 carcasses were arriving per year and by 1910 602,750 tonnes of frozen meat were being exported from Australia, New Zealand and South America (Critchell and Raymond, 1912).

Currently, meat for industrial processing is usually frozen in the form of carcasses, quarters or boned out primals in 25-kg cartons. It is not unusual for meat to be frozen twice before it reaches the consumer. During industrial processing frozen raw material is often thawed or tempered before being turned into meat-based products, i.e. pies, convenience meals, burgers, etc. or consumer portions, fillets, steaks, and so on. These consumer-sized portions are often refrozen before storage, distribution and sale.

6.1 FREEZING RATE

There are little data in the literature to suggest that, in general, the method of freezing or the rate of freezing has any substantial influence on a meat's subsequent storage life, its quality characteristics or final eating quality. There is some disagreement in the literature about which one has more benefits – fast freezing or slow freezing. Slightly superior chemical and sensory attributes have been found in food cryogenically frozen in a few trials (Sebranek *et al.*, 1978; Dobryzcki *et al.*, 1977; Sebranek, 1980) but other trials did not show any appreciable advantage (Lampitt and Moran, 1933) especially during short-term storage (Hill

Table 6.1 Relationship between freezing rate of beef *M. longissimus dorsi* and weight loss during freezing, thawing and cooking (source: Petrovic *et al.*, 1993).

Freezing rate (cm h^{-1})	**% Weight loss during**		
	Freezing	**Thawing**	**Cooking**
Control	—	—	36.32
0.22	2.83	0.78	38.41
0.39	2.58	0.72	38.00
3.33	1.15	1.21	37.47
3.95	1.05	0.18	37.24
4.92	0.87	0.10	37.15
5.66	0.63	0.03	37.14

and Glew, 1973). In the case of dark, firm and dry (DFD) pork, Kondratowicz *et al.* (2000) found that conventional slow freezing resulted in a better sensory quality than freezing in liquid carbon dioxide. Jackobsson and Bengtson (1973) indicated that there is an interaction between freezing rate and cooking method. Meat that had been cooked from frozen was found to show a favourable effect of faster freezing rates. Mittal and Barbut (1991) showed that freezing rate affected the modulus of rigidity of meat after cooking. Similar values to fresh meat were produced in meat frozen in liquid nitrogen.

In 1980, Añón and Calvelo reported a relationship between the rate of freezing and drip loss. They reported drip loss to reach a maximum when the freezing time from −1°C to −7°C was approximately 17 minutes. Mascheroni (1985) used this relationship to produce a method for determining the rate at which meat had been frozen. However, attempts to replicate the work at Langford (James *et al.*, 1983) were unsuccessful because of the variability in drip loss from meat before freezing. Ngapo *et al.* (1999) reported that drip loss was higher from slow-frozen than from fast-frozen pork samples. However, the difference disappeared after 4 weeks of frozen storage.

Experiments with pork *M. longissimus dorsi* found no difference in drip loss between samples frozen at −20°C or −80°C (Sakata *et al.*, 1995). At −20°C and −80°C, samples took 6 and 3 hours, respectively, to pass from −1°C to approximately −6°C. In the −20°C samples intercellular and intracellular ice was seen but only intracellular at −80°C.

Methods of freezing clearly affect the ultrastructure of muscle. Slow freezing (e.g. 1–2 mm h^{-1}) tends to produce large ice crystals extracellularly, while quick freezing (e.g. 50 mm h^{-1}) gives smaller crystals in and outside cells (Buchmuller, 1986). Obviously, a temperature gradient will occur in large pieces of meat and result in a non-uniform ice morphology (Bevilacqua *et al.*, 1979).

Petrovic *et al.* (1993) found that slowly frozen meat, 0.22 and 0.39 cm h^{-1}, lost more weight during freezing, thawing and cooking than that frozen at 3.95–5.66 cm h^{-1} (Table 6.1). However, higher weight losses during thawing were measured at an intermediate freezing rate of 3.33 cm h^{-1}. Meat frozen at rates of 3.33 cm h^{-1} and faster were rated as tenderer and juicier after cooking than unfrozen controls and slow-frozen samples (Table 6.2). Petrovic *et al.* stated that the optimal conditions for freezing portioned meat are those that achieve freezing rates between 2 and 5 cm h^{-1} to −7°C. Grujic *et al.* (1993) suggest even tighter limits, 3.33–3.95 cm h^{-1}. Slow freezing at up to 0.39 cm h^{-1} resulted in decreased solubility of myofibrillar proteins, increase in weight loss during, freezing thawing and cooking, lower water binding capacity and tougher cooked meat. Very quickly frozen meat (>4.9 cm h^{-1})

Table 6.2 Relationship between freezing rate of beef M. *longissimus dorsi* and texture (1 is extremely rough, 7 is extremely fine), Tenderness (1 is extremely hard, 9 extremely tender) and Juiciness (1 is extremely dry, 9 extremely juicy).

Freezing rate (cm h^{-1})	**Texture**	**Tenderness**	**Juiciness**
Control	7.0	6.8	7.0
0.22	7.0	6.0	6.7
0.39	7.0	6.5	7.0
3.33	7.0	7.5	7.5
3.95	7.0	8.0	8.0
4.92	7.0	8.5	8.5
5.66	6.0	7.0	7.3

Source: Petrovic *et al.*, 1993.

had a somewhat lower solubility of myofibrillar proteins, lower water binding capacity and somewhat tougher and drier meat. The samples were thawed after storage times of 2–3 days at −20°C so the relationship between freezing rates and storage life was not investigated.

Storage times of 48 hours and 2.5 months were used during investigations of the effect of different freezing systems and rates on drip production from small samples of mutton muscle (Sacks *et al.*, 1993). In all cases, drip loss after 2.5 months was at least double the percentage measured after 48 hours (Table 6.3). After 2.5 months, drip loss from samples frozen using cryogenics was >2% less than in those using air freezing.

In 2001, Sundsten *et al.* revealed some commercial advantages of fast freezing, but no quality advantages. The studies compared three different freezing methods: spiral freezing (SF), cryogenic freezing (liquid nitrogen, LN) and impingement freezing (IF). The times required to freeze a 10-mm thick 80-g hamburger from +4°C to −18°C in the SF, LN and IF were 22 minutes, 5 minutes 30 seconds and 2 minutes 40 seconds, respectively. The authors state that dehydration was significantly higher for hamburgers frozen in SF (1.2%) compared to LN (0.4%) and IF (0.4%). No significant difference could be seen in cooking losses, even after 2 months of storage. Ice crystals were significantly larger in hamburgers frozen in SF compared to LN and IF. Sensory analysis revealed no difference in eating quality between the three freezing methods, even after 2 months of storage.

High-pressure freezing and in particular 'pressure shift' freezing is attracting considerable scientific interest (Le Bail *et al.*, 2002). The food is cooled under high pressure to sub-zero

Table 6.3 Drip loss (%) from 77.6 g samples of *longissimus lumborum et thoracis* frozen under different methods and thawed at 4°C.

			Storage time at −20°C	
Freezing conditions	**Freezing time to −2.2°C**	**Freezing rate (cm h^{-1})**	**% weight loss at 48 h**	**% weight loss at 2.5 mo**
Cryogenic −90°C	15 m	6.4	3.34^{a}	9.49^{a}
Cryogenic −65°C	22 m	4.4	4.70^{a,b}	9.72^{a}
Blast freezer −21°C	1.83 h	0.55	5.53^{b}	12.74^{b}
Walk-in-freezer −21°C	1.88 h	0.53	4.71^{a,b}	13.18^{b}
Domestic freezer −25°C	1.96 h	0.51	5.26^{b}	11.72^{b}

Source: Sacks *et al.*, 1993.

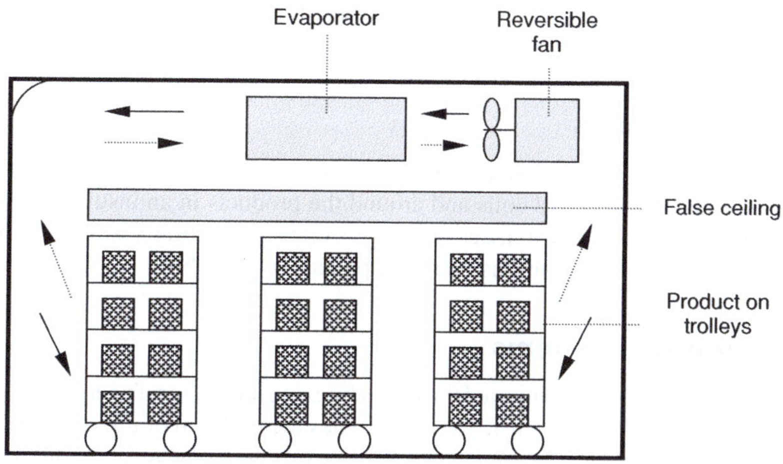

Fig. 6.1 Example of a freezing tunnel with longitudinal air circulation.

temperatures but does not undergo a phase change and freeze until the pressure is released. Rapid nucleation results producing small, even ice crystals. However, studies on pork and beef (Fernandez-Martin *et al.*, 200) and on pork (Zhu *et al.*, 2004) have failed to show any real commercial quality advantages and an increase in toughness in the later study.

Slow freezing from a high initial temperature can provide conditions for microbial growth as compared to a very rapid freezing process. Castell-Perez *et al.* (1989) predicted that slow freezing from an initial product temperature of 30°C could result in an 83% increase in bacterial numbers compared with a 4% increase from 10°C.

6.2 FREEZING SYSTEMS

6.2.1 Air

Air is by far the most widely used method of freezing food, as it is economical, hygienic and relatively non-corrosive to equipment. Systems range from the most basic – in which a fan draws air through a refrigerated coil (evaporator) and blows the cooled air around an insulated room (Fig. 6.1) – to purpose-built, conveyerised blast-freezing tunnels or spirals. Relatively low rates of heat transfer are attained from product surfaces in air systems. The big advantage of air systems is their versatility; especially when there is a requirement to freeze a variety of irregularly shaped products or individual products.

In practice, air distribution is a major problem, often overlooked by the system designer and the operator. The freezing time of the product is reduced as the air speed is increased. An optimum value exists between the decrease in freezing time and the increasing power required to drive the fans to produce higher air speeds. This optimum value can be as low as 1.0 m s^{-1} air speed when freezing beef quarters to 15 m s^{-1} plus for thin products.

Recently, the use of impingement technology to increase the surface heat transfer in air and other freezing systems has received attention (Newman, 2001; Sundsten *et al.*, 2001; Everington, 2001). Impingement is the process of directing a jet or jets of fluid at a solid surface to effect a change. The very high velocity (20–30 m s^{-1}) impingement gas jets "break up" the static surface boundary layer of gas that surrounds a food product. The resulting medium around the product is more turbulent and the heat exchange through this zone becomes much more effective.

6.2.2 Batch Systems

Placing food items in large refrigerated rooms is the most common method of freezing. Fans circulate air through refrigerated coils and around the products in an insulated room. Large individual items such as meat carcasses are hung from overhead rails, smaller products are placed either unwrapped or in cartons on racks, pallets, or large bins.

6.2.3 Continuous systems

In a continuous system meat is conveyed through a freezing tunnel or refrigerated room usually by an overhead conveyor or on a belt. This overcomes the problem of uneven air distribution since each item is subjected to the same velocity/time profile. Some meat products are frozen on racks of trays (2-m high), pulled or pushed through a freezing tunnel by mechanical means. For larger operations, it is more satisfactory to feed meat on a continuous belt through linear tunnels or spiral freezers. Linear tunnels are of simpler construction but are restricted by the length of belt necessary to achieve the cooling time required and on the space available in most factories. Spiral freezers are therefore a more viable alternative.

6.3 CONTACT FREEZERS

Contact freezing methods are based on heat transfer by contact between products and metal surfaces, which in turn are cooled by either primary or secondary refrigerants or direct immersion in a refrigerated liquid. Contact freezing offers several advantages over air-cooling, i.e. much better heat transfer and significant energy savings. However, the need for regularly shaped products with large flat surfaces is a major hindrance with plate systems and the need to wrap and wash off the immersion liquid in immersion systems.

Modern plate cooling systems differ little in principle from the first contact freezer patented in 1929 by Clarence Birdseye. Essentially product is pressed between hollow metal plates containing a circulating refrigerant (Fig. 6.2). A hydraulic cylinder is used to bring the freezing plates into pressure contact with the product. These plates can be either horizontal or vertical.

Good heat transfer is dependent on product thickness, good contact and the conductivity of the product. Plate freezers are often limited to a maximum thickness of 50–70 mm. Good contact is a prime requirement. Air spaces in packaging and fouling of the plates can have a significant effect on cooling time; for example, a water droplet frozen on the plate can lengthen the freezing time by as much as 30–60%.

General advantages of plate freezers over air-blast carton freezers include:

- Freezing is either faster for the same refrigerant evaporating temperature, or can take place at a higher evaporating temperature for a given freezing time.
- Product temperatures are easier to control, especially for smaller cuts.
- Power consumption is significantly reduced – savings of at least 30%, and possibly 50% or more, may be expected because air-circulating fans are not required and because higher evaporating temperatures can be used for the same effective cooling medium temperature.
- In many cases, less building space is required.

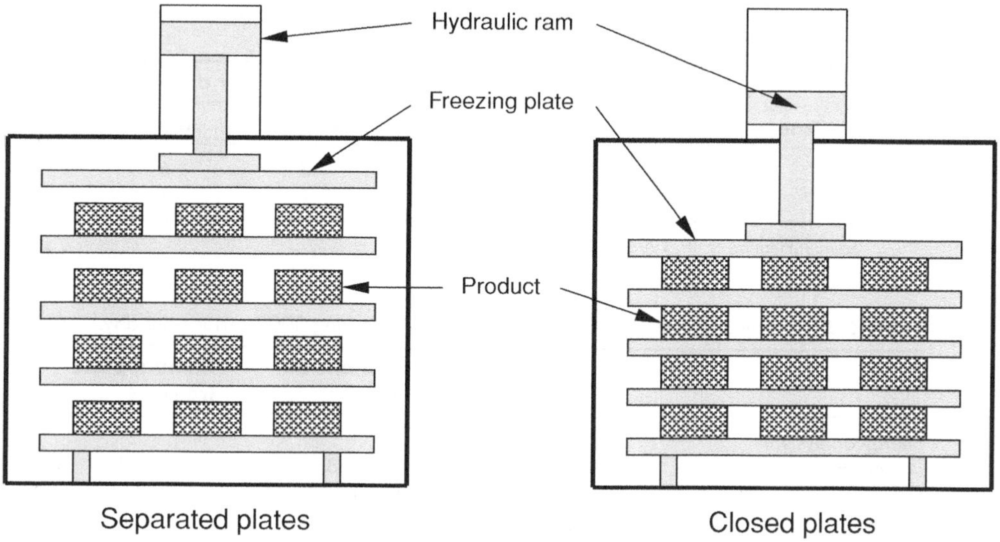

Fig. 6.2 Example of a horizontal plate freezer.

- The product remains uniform and flat after freezing, unlike air-blast frozen cartons, which often bulge. The flat cartons result in stable loads, giving up to 30% higher space utilisation in cold stores. For transport, the stable pallets facilitate unitised loading, and some 8–19% more product can be loaded into a container.

Disadvantages of plate freezers relate mainly to cost aspects:

- Capital costs are significantly higher than for equivalent air-blast freezers. Manually loaded plate freezers are comparable in cost to automatic air-blast tunnel freezers. Fully automatic plate freezers are more expensive.
- A high circulation rate of the liquid refrigerant is required; this results in additional costs for larger accumulators and higher capacity pumps.
- For manual plate freezers, simultaneous loading and unloading may require higher labour input than for a batch air-freezer.
- Each plate must be loaded with product of the same thickness.
- Damp cartons can stick to plates or cause jams when ice forms.
- Air infiltration must be minimised to prevent frost build up on plates.

Freezing unpacked meat has significant advantages because of the substantially shorter freezing times (Fleming *et al.*, 1996). Twice as many freezing cycles per day can be achieved with the bare product (Table 6.4). Overall costs for plate freezing can be comparable to those for air-blast freezing. De Jong (1994) carried out a cost analysis (Table 6.5) for a beef plant using either plate or air-blast freezers in New Zealand which assumed a net electricity cost of \$0.10 kW h^{-1}, a capital recovery over 10 years and an interest rate of 12%.

An immersion freezer is made up of a tank with a cooled freezing liquid that can be any non-toxic salt, sugar or alcohol solution in water and a means of conveying the wrapped meat through the tank. Persson and Löndahl (1993) state that immersion freezing has been commonly used for the surface freezing of turkeys and poultry in markets where a light colour

Table 6.4 Predicted freezing time of meat blocks in a plate freezer operating at −30°C.

	Freezing time (h)		Cycles per day	
Thickness (mm)	**Cartoned**	**Bare**	**Cartoned**	**Bare**
80	6.3	2.5	3	6
160	16.5	8.5	1	2

Source: Fleming *et al.*, 1996.

Table 6.5 Energy and cost requirements for beef plant freezing 3000 cartons per day.

	4 batch air freezers	Automatic air blast	Manual plate freezers
Energy analysis			
Freezing time (h)	38	38	17
Freezer load (kW)	573	477	400
Power consumption (kW)	689	570	462
Economic analysis ($000)			
Capital cost	1,500	2,200	2,000
Annual capital charges	266	389	354
Annual energy costs	463	389	354
Annual labour cost	60	0	90
Total annual cost	789	772	754

Source: de Jong, 1994.

is in demand. The freezing is completed in an air-blast system. Ice slurries are being considered as an alternative to conventional immersion liquids. Such binary systems are described in the scientific literature as flow ice, fluid ice, slush ice or liquid ice. Maria *et al.*, (2005) reported that such systems achieve higher rates of heat transfer than the single state liquids.

6.4 CRYOGENIC FREEZING

Cryogenic freezing normally uses refrigerants, such as liquid nitrogen or solid carbon dioxide, directly. The method of cooling is essentially similar to water-based evaporative cooling. Cooling being brought about by boiling off the refrigerant, the essential difference being the temperature required for boiling. As well as using the latent heat absorbed by the boiling liquid, sensible heat is absorbed by the resulting cold gas.

Due to very low operating temperatures and high surface heat transfer coefficients between product and medium, cooling rates of cryogenic systems are often substantially higher than other refrigeration systems.

Most systems use total loss refrigerants, i.e. the refrigerant is released to the atmosphere and not recovered. Due to environmental and economic factors total loss refrigerants must be both readily available and harmless, which limits the choice to atmospheric air and its components, liquid nitrogen (LN) and liquid or solid carbon dioxide (CO_2).

The particular characteristics of total loss refrigerants that may be regarded as advantages or disadvantages are listed in (Table 6.6).

Table 6.6 Advantages and disadvantages of total loss refrigerants in comparison with mechanical refrigeration.

Advantages	Disadvantages
Low capital investment	High operating cost
High refrigerating capacity	High refrigerating capacity
Low weight when out of use	High weight at start of use
No residual weight (dry ice)	Limited duration without filling
No noise	Poor temperature control
Advantageous storage atmosphere (N_2)	Reduced humidity
Bacteriostatic effect (CO_2)	Suffocation hazard
Low maintenance requirements	Limited availability
Foolproof once installed (dry ice)	

Cryogenic freezing is mainly used for small products such as burgers, ready meals, etc. The most common method is by direct spraying of liquid nitrogen onto a food product while it is conveyed through an insulated tunnel.

Impingement technologies are being used to further increase heat transfer (Newman, 2001). He states that when comparing the overall heat transfer coefficients of cryogenic freezing tunnels, impingement heat transfer is typically three to five times that of a conventional tunnel utilizing axial flow fans. With the increased overall heat transfer coefficient, one can either increase the freezing temperature to increase overall cryogen efficiency or continue to run at very cold temperatures and dramatically increase the overall production rate. Impingement freezing is best suited for products with high surface area to weight ratios, i.e. hamburger patties or products with one small dimension. Testing has shown that products with a thickness less than 20 mm freeze most effectively in an impingement heat transfer environment. When freezing products thicker than 20 mm, the benefits of impingement freezing can still be achieved, however, the surface heat transfer coefficients later in the freezing process should be reduced to balance the overall process efficiency. The process is also very attractive for products that require very rapid surface freezing.

6.5 FREEZING OF SPECIFIC PRODUCTS

6.5.1 Meat blocks

James and Bailey (1979) showed that air temperatures below −30°C and air velocities exceeding 5 m s^{-1} are required to freeze 150-mm thick meat blocks in corrugated cardboard cartons in less than 24 hours (Table 6.7). Creed and James (1981) carried out a survey that indicated that only 58% of industrial throughput is frozen in times within ±20% of the actual freezing time required.

Table 6.7 Average freezing time (hours) from 4 to −7°C at thermal centre of 50, 75 and 100 kg beef quarters.

Weight	50 kg	75 kg	100 kg
Hindquarter	22	29	33
Forequarter	13	20	25

Source: James and Bailey, 1987a.

Table 6.8 Predicted freezing time (hours) from 4°C to −7°C in thermal centre of unwrapped and stockinette wrapped carcasses (source: Creed and James, 1984).

	−30°C, 4 m s^{-1}	−30°C, 0.5 m s^{-1}	−20°C, 0.5 m s^{-1}
Unwrapped			
30 kg	5.5	11.0	16.4
40 kg	8.5	15.6	23.4
Wrapped			
30 kg	8.0	12.5	19.0
40 kg	12.0	17.8	26.5

6.5.2 Beef quarters

James and Bailey (1987) reported that brine spray and liquid nitrogen immersion systems had been used to freeze beef quarters. However, most investigations had used air. Temperature in air systems ranged from −11°C to −40°C and weight loss from 0.3% to 1.19%. In their own investigations beef quarters ranging in weight from 40 to 140 kg were frozen in air at −32°C, 1.5 m s^{-1}. On average, hindquarters below 50 kg and forequarters below 75 kg could be frozen in a 24-hour operation (Table 6.8). There was no statistical difference in bacterial counts before and after freezing.

6.5.3 Mutton carcasses

Since mutton production is seasonal, continuity of supply for processing can be achieved by frozen storage and subsequent thawing and boning (Creed and James, 1984). Data from their investigations were used to verify a predictive program for freezing of mutton carcasses. The predictions indicated that any condition more severe than −20°C, 0.5 m s^{-1} would achieve a 24-hour freezing operation for unwrapped carcasses (Fig. 6.3). To guarantee an overnight (15–16 hours) freezing cycle for wrapped carcasses, conditions more severe than −30°C, 4 m s^{-1} would be required.

6.5.4 Poultry

Poultry meat for further processing is usually frozen in the form of carcasses or bone in and boned out portions in cartons weighing up to 25 kg. Most bulk meat, consumer portions and other poultry products are frozen in air-blast freezers. Some small, individual items, i.e.

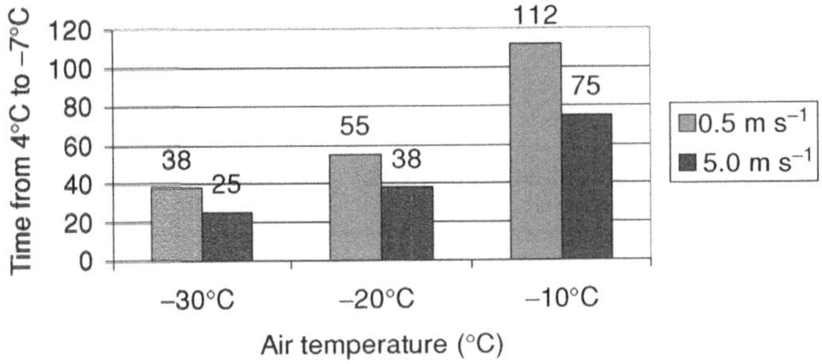

Fig. 6.3 Freezing time of 15-cm-thick boxed blocks (source: James and Bailey, 1979).

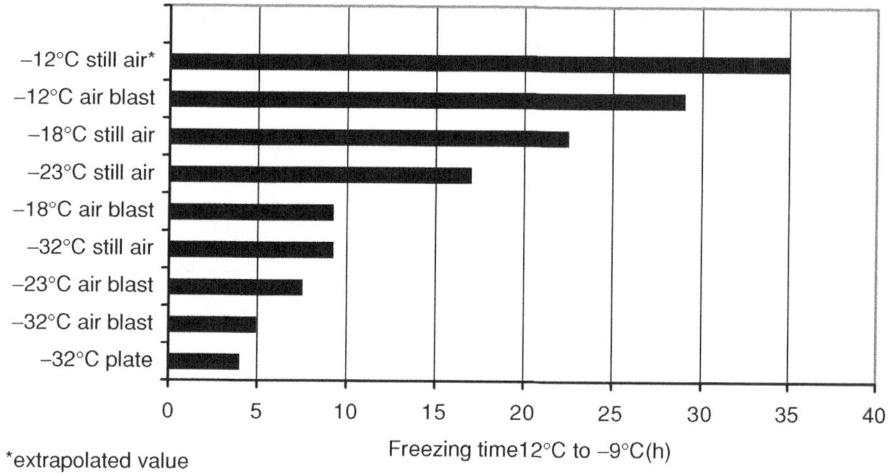

Fig. 6.4 Freezing time of 2.3 kg chicken carcasses under different freezing regimes (DuBois *et al.*, 1942).

chicken burgers may be frozen in cryogenic tunnels and a small amount of offal, mechanically recovered meat (MRM) and other meat is frozen in plate freezers. During industrial processing, frozen raw material is often thawed or tempered before being turned into pies, convenience meals, burgers, etc. or consumer portions, such as breast fillets. These consumer-sized portions are often refrozen before storage, distribution and sale.

Early work (DuBois *et al.*, 1942) on air freezing of poultry carcasses showed that different freezing conditions could result in freezing times from less than 5 to over 35 hours (Fig. 6.4). The different freezing rates resulted in substantial differences in the ice crystal structure in the muscle tissue. However, when the meat was roasted, taste panels detected 'no easily discernible difference in flavour, aroma or texture between the two extremes'. The taste-panel results showed no real differences between any of the freezing treatments and the unfrozen controls. In at least one example the score for aroma, texture, flavour, tenderness, juiciness or overall quality was better from a frozen than an unfrozen treatment.

Immersion freezing is far faster than can be achieved in air-freezing systems (Table 6.9). In air-blast freezing of chicken carcasses both air temperature and air velocity have a substantial effect on freezing time, as shown in Figs. 6.5 and 6.6 (van den Berg and Lentz, 1958). However, the small reduction in freezing time achieved by increasing the air velocity above 4 m s^{-1} would not normally compensate for the increased energy cost. Similarly, use of air

Table 6.9 Freezing times to −9.5°C for chicken and turkeys in different freezing environments.

Type of carcass and average weight	Cooling medium used	Time to freeze to −9.5°C (min)	
		Wrapped	Unwrapped
Broiler, 0.9 kg	−18°C brine	135	55
Broiler, 1.3 kg	−29°C brine	85	35
Turkey Hen, 5.3 kg	−18°C brine	500	280
Turkey Hen, 5.3 kg	−29°C brine	290	140
Turkey Tom, 10.8 kg	−29°C brine	425	—

Source: Esselen *et al.*, 1954.

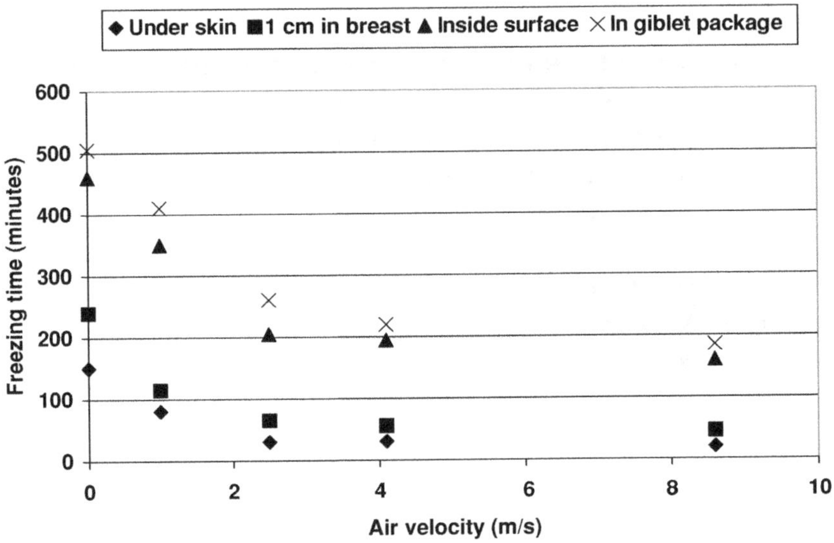

Fig. 6.5 Freezing of 3 kg poultry carcasses from 0°C to –4°C in air at –20°C (van den Berg and Lentz, 1958).

temperatures below –40°C is not commercially viable. Bishop (1972) stated that under the optimum freezing conditions (air at –40°C, velocity 2.5–3.5 m s^{-1}), a 1.5-kg poultry carcass can be frozen in less than 3 hours. The study showed an interesting design for a large-scale air-blast freezer. Cardboard cartons, each containing 10 carcasses were loaded onto racks holding 56 cartons. The racks were suspended on a slightly inclined rail and passed slowly

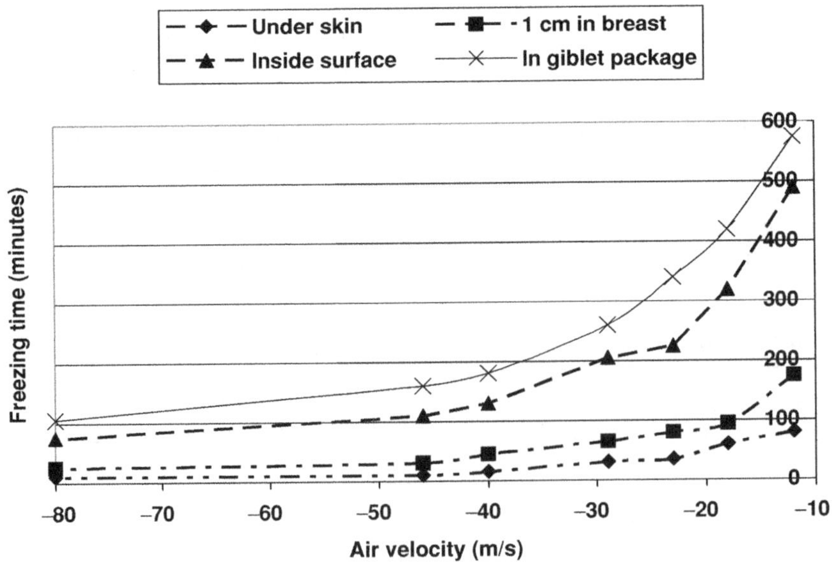

Fig. 6.6 Freezing of 3 kg poultry carcasses from 0 to –4°C in air at 2 m s^{-1} (van den Berg and Lentz, 1958).

Table 6.10 Freezing time equations for liver in plate freezer.

Block thickness (cm)	Equation
7.6	$Y = -0.3547 + 54.4632R + 0.02138I$
10.2	$Y = -0.1917 + 79.9314R + 0.05203I$
15.2	$Y = -0.8020 + 212.119R + 0.08880I$

Source: Creed and James, 1983.

by gravity through a chamber operating at −40°C. After passing through the chamber the racks are unloaded and then slightly elevated. The empty racks were then returned to the starting point by gravity.

6.5.5 Offal

Although edible offal comprises from 3% to 4% of the cold weight of a carcass there is little published data on its refrigeration (Creed and James, 1983). The authors found that liver was amenable to plate freezing and the freezing time to −7°C (Y) was related to the initial temperature (I) and R, where $R = (-1.5°C - \text{refrigerant temperature})^{-1}$ (Table 6.10). The authors extended their studies to examine the effect of different packaging materials on freezing time (Creed and James, 1985).

6.5.6 Small products

The rate of freezing of unwrapped small meat products affects the weight loss with loss increasing as freezing rate decreases. There is also some evidence that it affects losses during cooking and sensory properties. Hanenian *et al.* (1989) have shown that freezing times of individual meat burgers (patties) can range from 22 s to over an hour in different systems (Table 6.11). Freezing in nitrogen and CO_2 substantially reduced the amount of weight loss from unwrapped patties when compared with air systems. However, cooking losses were higher and overall patty quality lower in those frozen in nitrogen. This was mainly due to cracking of patties in the immersion system.

When burgers are stacked and placed in boxes before freezing the freezing times increase by at least an order of magnitude. Studies carried out on packaged burgers looked at freezing rates between 2°C and −18°C ranging from 24 to 96 hours (Berry and Leddy, 1989). Before freezing tenderness scores measured using a taste panel (eight extremely tender to one extremely tough) ranged from 6.6 to 6.1. All the burgers were rated as tougher immediately after freezing and after storage for 18 months (Fig. 6.7). Immediately after freezing burgers

Table 6.11 Freezing of beef burgers from 2°C to −10°C in different systems.

Method	Temperature (°C)	Thickness (cm)	Freezing time (s)	Freezing rate (cm h^{-1})	Weight loss (%)	Cooking loss (%)
Air (23 W m^{-2} K^{-1})	−14	1.10	3939	0.5	2.0	34.4
Air (31 W m^{-2} K^{-1})	−25	1.12	2047	1.0	1.7	34.6
CO_2 snow	−78	1.13	129	16.5	0.1	35.9
N_2 immersion	−196	1.13	22	97.3	0.1	37.7

Source: Hanenian *et al.*, 1989.

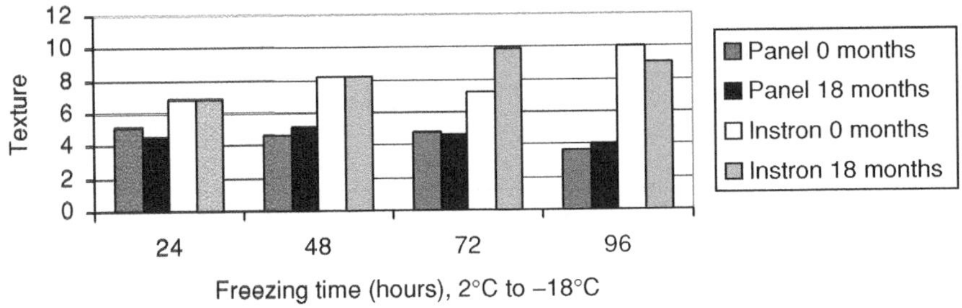

Fig. 6.7 Texture of burgers measured by taste panel (8 is extremely tender, 1 is extremely tough) and Instrom (kg) after 0 and 18 months storage (source: Berry and Leddy, 1989).

frozen in 96 hours were significantly tougher than those frozen in 24 hours; however, the difference was not significant after 18 months of storage. Instrumental texture measurements were in general agreement with those from the taste panel (Fig. 6.7).

Commercial freezing rates can be very slow. Sausages at the centre of a pallet requiring 6–7 days to achieve −15°C from a starting temperature of 7°C (Wanous *et al.*, 1989). However, studies carried out on similar sausages frozen in 9 hours, 2.4 days and 6.8 days showed no effect of freezing rate on TBA (thiobarbituric acid) values during frozen storage of 20 weeks.

Many small meat products such as cubes and strips of ham and poultry meat, poultry pieces, cooked meat balls, slices of salami and mincemeat can be individually quick frozen (IQF) in a rotary cryogenic freezer (Thumel and Gamm, 1994). The product is sprayed with a fine mist of nitrogen to freeze the surface as it enters the drum. As the tilting drum rotates it transports the food through a contra-current flow of cold gas that completes the freezing process.

6.6 TEMPERING AND CRUST FREEZING

There is no exact definition for the word 'tempering' in the meat industry. In practice 'tempering' can be a process in which the temperature of the product is either raised or lowered to a value that is optimal for the next processing stage.

James and James (2002) detail the advantages of tempering and the use of single- and multi-stage systems where the temperature of frozen product is raised as required. With 150-mm-thick meat blocks, single-stage tempering operations can take many days (Brown and James, 2006). However, microwave systems can rapidly temper similar meat blocks in less than 10 minutes (Swain and James, 2005).

Tempering and crust freezing operations are used to produce the optimum texture in a chilled product so that it is suitable for mechanical processing. In this case, the product is semi-frozen so that it is stiff enough to be sliced, cubed etc.

6.6.1 Pork loin chopping

Loins from lamb and pork are often processed into chops with a high-speed guillotining machines. Because of the deformation in this process the yield can be reduced. The yield can be increased by first tempering the meat, providing a stiff outer crust, by freezing with liquid nitrogen or a blast of very cold air. However, the process must be carefully controlled, if too much meat is frozen the subsequent chopped meat will have a large increase in the amount of drip formed, resulting in a loss in yield of some 4–5%. Hence, loin freezing processes must always be carefully controlled.

6.6.2 High-speed ham slicing

Traditional production of ham slices consists in cooling formed ham logs in cold rooms to a core temperature of 2°C. A process that takes between 2 and 7 days (Lammertz and Brixy, 2001). The logs are cut in 1.5-mm-thick slices using standard slicers at rates up to 500 slices per minute. New high-rate slicers operate at rates up to 1000 slices per minute. To produce high-quality slices at this rate the ham logs have to be crust frozen to a temperature of −5°C at a depth of 7 mm. A number of different cryogenic freezers have been developed to perform the crust freezing process (Lammertz and Brixy, 2001).

6.6.3 High-speed bacon slicing

An increasing proportion of bacon is being pre-sliced and packed before it is delivered to wholesalers and retailers. To achieve the throughput required, slicers have to be operated at very high speeds. To maximise the yield of high-quality slices from high-speed slicers the bacon has to be sliced in a semi-frozen, tempered state. The optimum tempering temperature is a function of the bacon and the slicer. Most bacon tempering has been traditionally carried out in a long, single-stage process. However, more efficient two-stage processing systems are now common. The correct design and operation of such systems is critical to the cost effectiveness of the slicing process. Bacon temperature is the critical parameter in a high-speed (typically 800–1400 slices per minute) slicing operation. This operation has more in common with the guillotining of metal than the low-speed slicing normally carried out in a butcher's shop. The bacon must be presented to the blade in a rigid semi-frozen state to minimise distortion and break-up on cutting. Obtaining the correct temperature throughout the bacon middle is crucial for a high yield of undamaged slices (James and Bailey, 1987b). High-speed photography has been used to clearly demonstrate the effect of incorrect slicing temperature. When the temperature was too low the hard bacon shattered and blade wear was excessive; when too high the soft bacon stuck to the blade and the fat was torn away from the lean. The optimum temperature for a particular operation depends on the salt content of the bacon, the maturation time, the type of slicer being used, and the slicing speed.

Experiments carried out in the 1920s (Callow, 1929) showed that there was a near-linear relationship (Fig. 6.8) between the initial freezing point of lean pork and its salt content

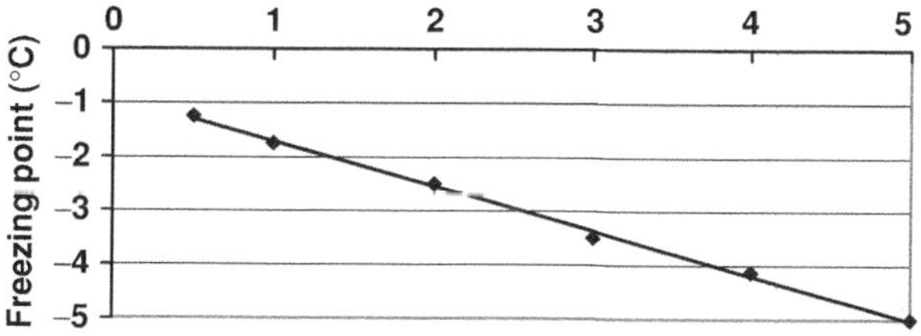

Fig. 6.8 Relationship between initial freezing point of lean pork and its salt content (source: Callow, 1929).

Table 6.12 Salt and moisture contents in bacon middles from different suppliers.

	Salt (g per 100 g water)			Moisture (%)		
Supplier no.	**Maximum**	**Minimum**	**Mean**	**Maximum**	**Minimum**	**Mean**
1	7.1	2.9	5.1	73.5	65.3	69.6
2	6.5	2.7	4.1	74.0	68.0	71.1
3	4.7	2.9	4.0	72.5	68.6	71.0
4	5.9	2.1	4.1	74.6	65.0	71.8
5	6.3	2.2	4.2	74.2	63.9	70.9
6	5.9	2.5	5.0	72.6	64.3	70.3
7	6.5	3.1	4.9	74.4	67.5	71.1
8	4.6	2.2	3.2	74.2	70.0	72.2
9	6.3	2.6	4.7	74.7	66.3	69.7
10	5.2	2.7	4.1	74.8	65.3	70.3

Source: James and Bailey, 1987b.

(expressed in terms of grams of salt per 100 g of water in the meat). Consequently, the initial freezing point, ice content at any temperature and the related slicing temperature depend on both the salt and water content of the bacon. Work has shown that even using very carefully controlled curing methods there are still considerable variations in salt content within individual slices, between slices and between bacon sides in the same batch. This makes it difficult, and in a commercial situation impossible, to carry out analytical tests that will define the optimum slicing temperature for a particular operation. This temperature must therefore be determined experimentally for each slicing operation, and tempering systems developed that will produce a uniform temperature throughout the product in the most efficient way.

6.6.3.1 *Determination of slicing temperature*

There is likely to be substantial variability in the bacon input to a slicing operation. A survey (James and Bailey, 1987b) found that mean percentage salt and water content of bacon supplied to a large slicing plant over a 2-year period varied by 1.9% and 2.6%, respectively (Table 6.12 and Fig. 6.9). The effect on slicing of the variation in salt content, where the overall mean was 4.34%, was much greater than water, which had an overall mean of 70.8%.

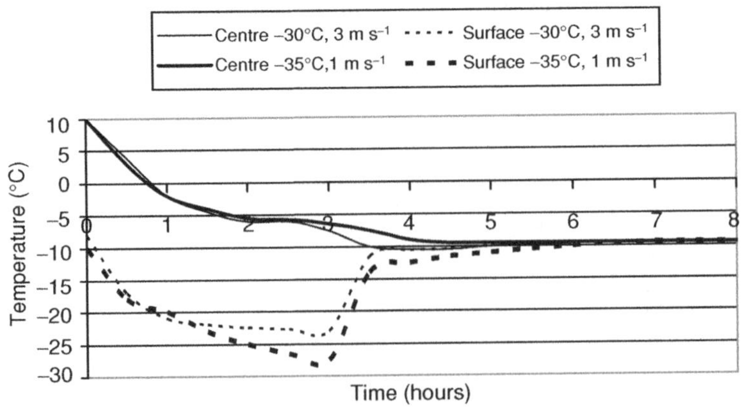

Fig. 6.9 Surface and centre temperature in bacon backs during two-stage tempering operation (source: James and Bailey, 1987b).

Table 6.13 Yield (%) of slices in quality classes from bacon backs from different suppliers sliced at an equalised bacon temperature of −6.5°C, −7.5°C or −9.5°C.

Quality class	'Prime/second'			'Thrifty/catering'			'Bits and pieces'		
Slicing temperature (°C)	−6.5	−7.5	−9.5	−6.5	−7.5	−9.5	−6.5	−7.5	−9.5
Supplier A	—	93.6	94.7	—	3.7	3.0	—	2.7	3.3
Supplier B	79.1	82.0	81.7	12.5	8.3	8.8	8.11	9.7	9.5
Supplier C	—	79.9	75.7	—	10.2	9.7	—	9.9	14.6
Supplier D	84.1	90.8	93.8	10.8	6.0	2.8	5.1	3.2	3.4
Supplier E	—	81.5	78.8	—	11.8	9.6	—	6.7	11.6
Supplier F	—	76.4	78.5	—	10.4	10.3	—	13.2	11.2

Source: James and Bailey, 1987b.

Maximum and minimum values from different suppliers were 7.1% and 2.1% for salt, and 74.8% and 63.9% for water. Initial freezing points, and consequently the optimum slicing temperatures could therefore vary by 5°C or more. Examination of freezing curves of bacon from three of the suppliers showed initial freezing points of −3°C, −3.5°C and −6°C.

Currently, the only method available for determining the optimum bacon temperature for a slicing operation is to carry out slicing trials at different bacon temperatures. Results from such a trial where the yield in each quality grade was determined are shown in Table 6.13. In the specific trial taking into account the relative quantities from each supplier the best slicing temperature was found to be −9.5°C.

6.6.3.2 *Tempering systems*

A small number of operations use plate freezers, liquid immersion systems and cryogenic tunnels to temper bacon for high-speed slicing. However, the majority of industrial systems employ air in a single- or two-stage process.

6.6.3.2.1 *Single-stage tempering*

Single-stage tempering is a very simple process. The bacon middle, back and streaky joints are placed on the shelves of trolleys. The trolleys are then wheeled into a room operating at the desired slicing temperature. The bacon remains in the room until its temperature equalises to that of the room. It is then pressed and sliced.

For example, in a commercial single-stage tempering system backs were held for 18 hours in rooms operating at −9°C to −14°C, 0.2–1.2 m s^{-1} with an average product weight loss of 1.18%. At weekends, the backs remained in the rooms for a total of 64 hours and the average weight loss increased to 1.88%. After 18 hours the surface temperature had reached −10°C but the centre was still above −7°C.

A number of problems are inherent in a single-stage tempering operation. Equalisation times are long, after 18 hours there can still be a 3°C differential across the backs. Using a two-stage system for only half this time resulted in differentials of less than 1°C. The most obvious drawback of single-stage tempering is that to obtain the same throughput systems have to be far larger, probably at least threefold. It is also more difficult to obtain even air distribution and good temperature control in a large room. This problem is exacerbated in that the single-stage system has to fulfil conflicting roles. To remove heat from the bacon a reasonable air/product temperature difference and air movement are required. In contrast, towards the end of the process when all the required heat has been extracted, a very small temperature differential

Table 6.14 Conditions in blast freezer, mean initial weights of bacon backs.

Air conditions		Weight		Equilibrium	No.
Temperature (°C)	Speed (m s^{-1})	Initial (kg)	3 h loss (%)	Temperature (°C)	samples
−30	1.0	5.640 (0.51)	0.71 (0.06)	−8.6	6
−30	3.0	5.150 (0.32)	0.76 (0.08)	−11.6	4
−35	0.5	5.288 (0.13)	0.66 (0.06)	−8.5	6
−35	1.0	5.375 (0.26)	0.55 (0.07)	−9.9	4

Note: % weight losses after 3 hours and equilibrium temperatures; standard deviations are given in parenthses.
Source: James and Bailey, 1987b.

and minimum air movement are desirable to attain an even temperature and a reduced rate of weight loss.

6.6.3.2.2 *Two-stage tempering*

In a two-stage tempering process, an initial blast freezing operation is followed by a separate period of temperature equalisation. It is critical that the desired amount of heat is extracted in the initial blast freezing operation. Examples of conditions and final equalised temperatures are given in Table 6.14. Typical temperature histories at the surface and thermal centre of backs tempered at −30°C, 3.0 m s^{-1} and −35°C, 1.0 m s^{-1} are shown in Fig. 6.10. Surface temperatures tended to be a few degrees lower after 3 hours at −35°C than at −30°C, whilst centre temperatures were very similar, approximately −6°C. In each case the maximum temperature difference across the backs was less than 1°C after 3 hours in the equalisation room and the temperature within any part of the back was within 1°C of the room temperature after 7 hours.

It is critical that the refrigeration system is sized to extract the required amount of heat from the bacon. The energy released per kg of bacon in each half-hour period during one experiment varied by a factor of 2.7 from 0.0139 kW h kg^{-1} to 0.0051 kW h kg^{-1} (Fig. 6.10). The mean total energy extracted per kg of bacon in the 3-hour operation was 0.0535 kW h kg^{-1}.

In a two-stage system there are several practical considerations. In one study a 3-hour blast freeze operation at −35°C, 1.0 m s^{-1} obtained the desired equilibrium temperature of −9.5°C and achieved the lowest weight loss of 0.55%. However, these conditions were very critical and would not allow for a pull down period after loading, and the use of slightly thicker backs or bacon with a higher salt content. Adding heaters to the equalisation room to

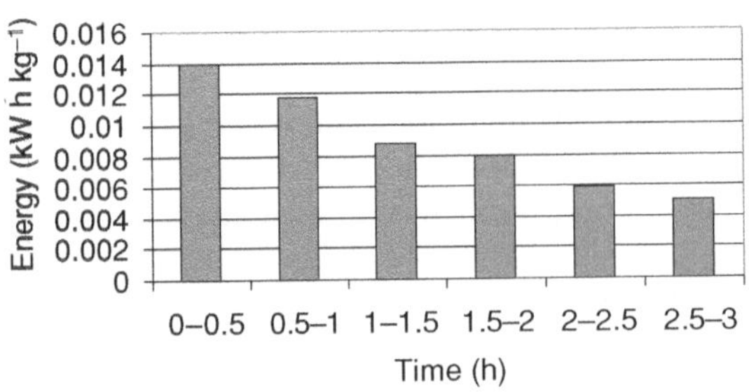

Fig. 6.10 Energy released (kW h kg^{-1}) by bacon backs over 0.5 hour intervals during blast freezing at −30°C, 3.0 m s^{-1} (source: James and Bailey, 1987b).

provide exact temperature control with slightly negative product load was considered more viable than trying to control the refrigeration against the positive load likely in practice. Air distribution and control would also be less exact at 3 m s^{-1} and a variation of ±0.5 m s^{-1} in air velocity over the bacon backs would have far less effect on the final equalised temperature than a similar variation about a mean of 1.0 m s^{-1}. Operating conditions of −30°C, 3.0 m s^{-1} were therefore chosen for this particular industrial plant, because they were less critical and provided a degree of flexibility.

Investigations have showed clearly the need to size the refrigeration system to meet the initial rate of heat release from the warm bacon backs (James 1997). In trials an experimental freezer was unable to maintain the desired set point of −30°C but rose to −27°C immediately after loading and took 1.5 hours to fully recover. The average equalised temperature in this trial was 0.6°C higher than in two successive trials where less bacon was used and the freezer reached −30°C within minutes of loading. Although the average rate of heat release from the bacon backs during the freezing operation was 0.0175 kW kg^{-1} the refrigeration plant had to have twice this capacity to meet the rate of release during the first 0.5 hour. In the industrial situation heat ingress through the open door during loading and the considerable cooling requirement of the supporting racks also have to be taken into consideration. One practical solution is a central refrigeration plant serving a number of separate freezing chambers. These can be loaded and unloaded in a sequence to provide the plant with a nearly constant refrigeration load enabling it to operate at optimum efficiency.

6.7 FROZEN STORAGE OF MEAT AND MEAT PRODUCTS

There are clear differences between the environmental conditions required for cooling, which is a heat removal/temperature reduction process, and those required for storage where the aim is to maintain a set product temperature. Extensive data are available on the optimum storage conditions and attainable frozen storage lives for many products (IIR, 1986; ASHRAE, 2006).

Since the bacteria found on meat will not grow at a temperature below −12°C changes in microbial levels are not a factor in the frozen storage life of meat. Freezing and frozen storage may in fact improve the safety of the meat. Georgsson *et al.* (2006), for example, found that freezing poultry and storing in for 30 days reduced the levels of Campylobacter by up to 2.87log cfu (kg of carcass)$^{-1}$. However, similar reductions were not found with faecal coliforms. While Sandberg *et al.*, (2005) reported a 2log cfu g^{-1} reduction in Campylobacter after 3 weeks at −20°C. On beef trimmings Moorhead and Dykes (2002) reported a 0.6–2.2 log cfu g^{-1} reduction in the first 7 days of storage at −18°C.

The factors that influence the storage life of frozen meat may act in any one of three stages prior to freezing, during the actual freezing process (already discussed) and post-freezing in the storage period itself (James and Evans, 1997).

6.7.1 Pre-freezing treatment

Some pre-freezing factors, i.e. species differences, animal to animal variation or differences between cuts of meat, are inherent in the animal. There are also other factors including feeding and transport that may have an effect on frozen storage.

Species is the main pre-freezing factor that is commonly believed to influence the frozen storage life. Table 6.15 provides data from three sources on the storage life of meat from different species and the average and range from all the publications located.

Table 6.15 Frozen storage life (months) for different species.

Source	1	2	3	Average at	Range at
Temperature(°C)	−20°C	−18°C	−18°C	−18°C	−18°C
Species					
Beef	12	12	18	10.2	2.8–19.4
Pork	6	8	12	17.4	2.8–23.3
Lamb	10	12	24	7.8	2.8–24.3
Chicken	12	10	18	13.6	6.0–23.3

Source: James and Evans, 1997.

There is up to a twofold difference between species in recommended storage times but more importantly the relative ranking, in terms of which can be stored for the longest time, varies between the sources. When all the available data found in the literature are considered the picture becomes even more confusing. Average values for the storage lives of the different species at −18°C (Table 6.15) have a different ranking to that generally accepted and the range of storage lives is very large.

Few publications compare the meat of more than two species under directly comparable conditions. Hiner *et al.* (1951) found that under similar aging and packaging regimes beef was found to store for 69 weeks, pork for 53 weeks and lamb for 44 weeks at −18°C. At −18°C pork remained palatable for longer than lamb but at higher temperatures the lamb was more stable. It seems fair to conclude that most works point towards a difference, but not necessarily a consistent difference, in frozen storage life between species.

To look at animal to animal variation two trials were carried out in New Zealand where lamb was stored at −5°C. In the first trial the lamb was judged rancid after 20 weeks and in a duplicate trial the lamb was found to store for 40 weeks. The only variation that could be determined was that different animals were used in the two trials. There appear to be large variations between animals which cause changes in the storage life of meat but why these differences exist is not completely understood.

Feeding influences frozen storage life. Pork from pigs that had been fed materials containing offal or household refuse had half the practical storage life than that from pigs that had been fed conventional diets (Bailey *et al.*, 1973). Rations with large amounts of highly unsaturated fatty acids tend to produce more unstable meat and fat. The feeding of fish oils or highly unsaturated vegetable oils to poultry is known to produce fishy flavours in the meat but there is some debate as to whether this diet directly affects frozen storage times. The linoleic acid content of meat probably plays a major role in storage. There has been a general trend in the UK in recent years for pigs to be leaner and therefore to have greater proportions of linoleic acid in their tissues. There is a possibility that pork may store less well than might be expected from results dating from 10 or 20 years ago.

Reports of variations in the storage life of different cuts of meat are scarce and primarily show that light meat stored for a longer time than dark meat. This is thought to be due to either higher quantities of haem pigments in the dark muscle, or to higher quantities of phospholipids which are major contributors to oxidised flavour in cooked meat.

Increased stress or exhaustion can produce PSE (pale soft and exudative) or DFD (dark firm and dry) meat, which is not recommended for storage mainly due to its unattractive nature and appearance.

Meat is generally not frozen until rigor is complete and a degree of conditioning has taken place, otherwise toughening and increased drip can occur. In red meat, there is little evidence of any relationship between chilling rates and frozen storage life. However, there is evidence that increasing the time in chilled storage before freezing reduces frozen storage life. Carcasses that have been electrically stimulated have prolonged storage lives and this could be attributed to the shorter interval between slaughter and freezing. In poultry, chilling method does have an effect on storage life. Air-chilled broilers had significant flavour changes after 3 months at −12°C and −20°C, whereas immersion-chilled birds only exhibited changes at −12°C after 6 months and were stable at −20°C. Water chilling of broilers produced a more favourable taste in the leg and breast meat than air chilling. Thielke *et al.*, (2005) found that ageing of poultry carcasses for at least 6 hours after chilling and prior to deboning and freezing improved the texture of cooked breast fillets.

Processing of meat prior to freezing generally results in a longer storage time. Heating prior to freezing can result in a 50% longer PSL (practical storage life) for sausages. However, the heating process could be critical since muscles cooked to higher temperatures are most susceptible to oxidative changes during storage. Heat treatments such as frying tend to produce short storage lives, probably because of the high fat content of the product. Breaded products are often fried and although breading alone may have a protective effect on a product, the addition of oil may have a counteractive effect.

A process such as mincing has been found to effect storage of comminuted products, this is probably due to the induced heating and the increased surface area that results. Addition of fat to mince can lower storage life unless a high-grade wrapping material, which has the ability to exclude air, is used to wrap the product. Smoking is generally advantageous due to the antioxidant properties of the smoke. Smoked broilers and ham store well for over a year without serious quality change.

Additives, such as spices, seasoning, antioxidants and protein concentrates can influence storage life. The use of vegetable extracts such as onion juice, yellow onion peel, hot water extracts of aubergine (egg plant), potatoes and sweet potatoes have been shown to help control rancidity in beef and turkey meat. However, an addition of salt may also reduce the storage life due to increased rancidity. Mechanically recovered meat is used in a range of meat products, but can cause storage problems due to its high fat content and increased rancidity (James and Evans, 1997).

6.7.2 During Frozen Storage

Three factors during storage – the storage temperature, the degree of fluctuation in the storage temperature and the type of wrapping/packaging in which the meat is stored – are commonly believed to have the main influence on frozen storage life.

6.7.2.1 *Storage temperature*

To quote from the IIR Red book, "storage life of nearly all frozen foods is dependent on the temperature of storage . . ." and in the book a table is provided of practical storage lives of different foods at three storage temperatures, extract in Table 6.16. However, few papers have been located where data are presented from experiments on the PSL of meat at different storage temperatures. Many of those that have been located are on products that do not meet the lower temperature–longer storage rule (normal stability).

Table 6.16 Practical storage life (months) at different storage temperatures.

Product	−12°C	−18°C	−24°C
Beef carcasses	8	15	24
Beef steaks/cuts	8	18	24
Ground beef	6	10	15
Veal carcass	6	12	15
Veal steaks/cuts	6	12	15
Lamb carcasses	18	24	>24
Lamb steaks	12	18	24
Pork carcasses	6	10	15
Pork steaks/cuts	6	10	15
Sliced bacon (vac.)	12	12	12
Chicken, whole	9	18	>24
Chicken parts/cuts	9	18	>24
Turkey, whole	8	15	>24
Ducks, Geese, whole	6	12	18
Liver	4	12	18

Source: IIR, 1986.

Experimental data from many different publications have been plotted against the temperature of storage for beef (Fig. 6.11), pork (Fig. 6.12) and lamb (Fig. 6.13). There is a clear effect of temperature on storage life, with lower temperatures resulting in extended storage, but considerable scatter between results at any one temperature.

It has been shown that rancidity in bacon is increased by higher salt content and that the rates of chemical reactions are accelerated as the temperature is lowered when packed in permeable wrap. Cured pork products are known to have an abnormal temperature profile between −5°C and −60°C and store less well between −30°C and −40°C.

Improved aroma scores have been found to be moderately related to lower freezing temperatures, but not related to flavour. Aroma scores for minced beef improved during 6–12 month storage at −12.2°C, −17.8°C or −23.3°C, although a slight increase in rancidity also occurred.

Work in New Zealand (Winger, 1984) has found that consumer panels are often not very sensitive to quality changes and could not tell the difference between samples of lamb stored at −5°C and −35°C. A trained taste panel could differentiate between the two temperatures and scored the samples stored at −5°C as rancid.

6.7.2.2 *Temperature fluctuation*

Generally, fluctuating temperatures in storage are considered to be detrimental to the product. However, it has been reported that repeated freeze–thaw cycles do not cause any essential change in the muscle ultrastructure (Carrol *et al.*, 1981) and that several freeze–thaw cycles during a product's life cause only small quality damage (Wirth, 1979) or possibly no damage at all. In fact, a slight but significant improvement in samples that had been frozen and unfrozen several times was found by a taste panel (Jul, 1982).

Minor temperature fluctuations in a stored product are generally considered unimportant, especially if they are below −18°C and are only of the magnitude of 1–2°C. Well-packed products and those that are tightly packed in palletised cartons are also less likely to show

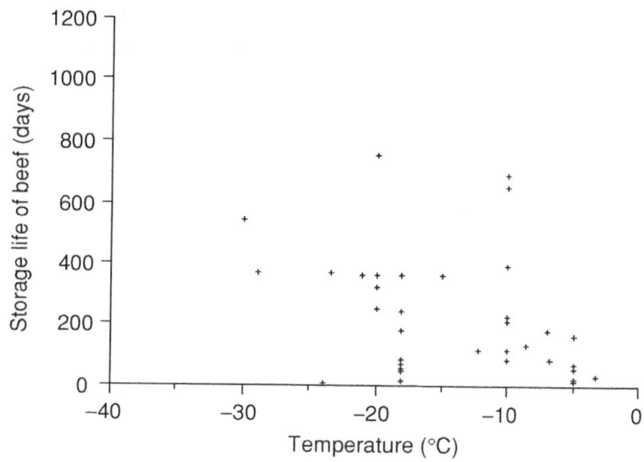

Fig. 6.11 Experimental data on storage life of beef at different temperatures (source: James and Evans, 1997).

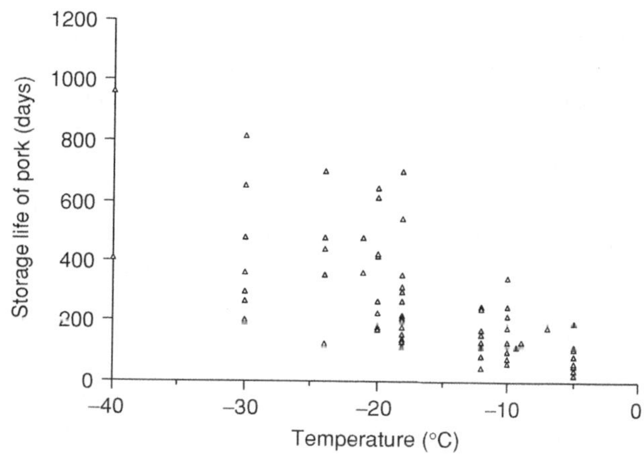

Fig. 6.12 Experimental data on storage life of pork at different temperatures (source: James and Evans, 1997).

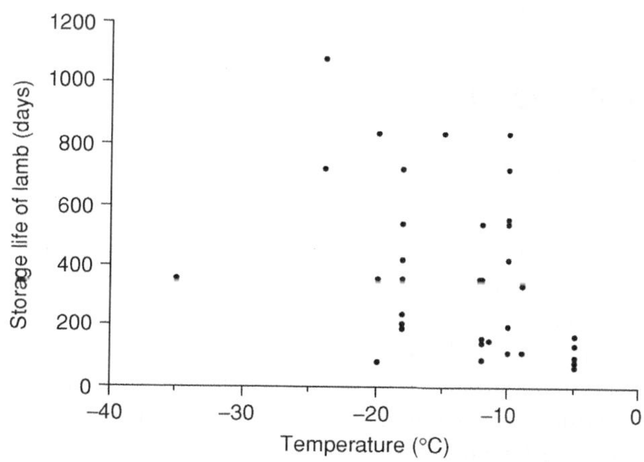

Fig. 6.13 Experimental data on storage life of lamb at different temperatures (source: James and Evans, 1997).

quality loss. However, poorly packed samples are severely affected by the temperature swings. There is disagreement on how much effect larger temperature fluctuations have on a product. Some authors consider temperature fluctuations have the same effect on quality of the product as storage has at an average constant temperature (Dawson, 1969); others consider fluctuations may have an additive effect (Van Arsdel, 1969; Bech-Jacobsen and Bøgh-Sørensen, 1984). There is evidence that exposure to temperatures above −18°C rather than temperature fluctuations may be the major factor influencing quality deterioration (Gortner *et al.*, 1948). Although Reid and Perez Albeta (2006) found that temperature fluctuations of 5°C or 10°C about a −17°C mean increased weight loss from a model system by approximately 0.1% and 0.3%, respectively.

6.7.2.3 *Packaging*

Packaging has a large direct effect on storage life, especially in fatty foods and in extreme cases indirectly due to substantially increasing the freezing time. There are a number of such instances where large pallet loads of warm, boxed meat have been frozen in storage rooms. In these cases, freezing times can be so great that bacterial and enzymic activity results in a reduction of storage life. In most cases, it is the material and type of packaging that influences frozen storage life. Wrapping in a tightly fitting pack having a low water and oxygen permeability (such as a vacuum pack) can more than double the storage life of a product. Waterproof packing also helps us to prevent freezer burn and tight packing helps to prevent an ice build up in the pack. When a product is breaded, packaging appears to have little effect and in a trial where breaded pork chops and breaded ground pork were packed in poor and very good packs, an effect of packing could not be found (Van Arsdel, 1969).

Rancidity occurs in unwrapped meat because its surface dries, allowing oxygen to reach subcutaneous fat. Without wrapping, freezer burn may occur causing extreme toughening and the development of rancidity in the affected area. Packaging can be effective in some cases in reducing discolouration by lessening oxygen penetration into the meat. Lighting, especially ultra-violet, can also increase fat oxidation (Volz *et al.*, 1949; Lentz, 1971). Exposure to the levels of light found in many retail frozen food display areas can cause appreciable colour change within 1–3 days. Development of off flavour can be accelerated and may be noticeable within 1–2 months on display. Products kept in dark or opaque packages may therefore be expected to retain colour longer than those exposed to the light.

6.8 TYPES OF STORAGE ROOM

6.8.1 Bulk storage rooms

Most unwrapped meat and poultry and all types of wrapped foods are stored in large air-circulated rooms. To minimise weight loss and appearance changes associated with desiccation, air movement around the unwrapped product should be the minimum required to maintain a constant temperature. With wrapped products low air velocities are also desirable to minimise energy consumption. However, many storage rooms are designed and constructed with little regard to air distribution and localised velocities over products. Horizontal throw refrigeration coils are often mounted in the free space above the racks or rails of product and no attempt is made to distribute the air around the products.

Using a false ceiling or other form of ducting to distribute the air throughout the storage room can substantially reduce variations in velocity and temperature.

6.8.2 Controlled atmosphere storage rooms

Controlled atmosphere storage rooms were developed for specialised fruit stores, especially those for apples. Interest is growing in the application of this technique to other commodities including meat. In addition to the normal temperature control plant, these stores also include special gas-tight seals to maintain an atmosphere that is normally lower in oxygen and higher in nitrogen and carbon dioxide than air. Additional plant is required to control the CO_2 concentration, generate nitrogen and consume oxygen.

6.8.3 Jacketed cold stores

Cooling the walls, floor and ceiling of a store produces very good temperature control in the enclosed space with the minimum of air movement. It is especially suitable for CA storage and for unwrapped produce that are very sensitive to air movement or temperature fluctuations. The refrigerated jacket can be provided by embedding pipe coils in the structure or utilising a double skin construction through which refrigerated air is circulated.

Although the refrigerated jacket is efficient in absorbing any heat input from the surroundings the lack of air circulation within the enclosed space means that heat removal from the product is very limited. Care must therefore be taken to (a) attain the desired storage temperature throughout the product before storing, (b) minimise any heat loads produced during loading and unloading, and (c) provide the supplementary refrigeration required for any products which respire.

6.9 CONCLUSIONS

(1) Under commercial conditions differences in freezing rates are unlikely to produce noticeable changes in the organoleptic quality of the meat produced. However, current legislation requires a minimum meat temperature of $-12°C$ to be achieved before meat is moved from the freezing system. Freezing time is therefore of considerable economic importance.
(2) Most unprocessed meat is either frozen in batch air systems as bone in carcasses, sides or quarters, or boned out in cartons. Freezing times in such systems are typically 25–72 hours. Some offal is frozen in plate freezers.
(3) Small processed items are typically frozen in continuous belt freezers or in cryogenic tunnels.
(4) Crust freezing and tempering are increasingly being used to allow high-speed mechanical portioning or slicing of meat and meat products. The final temperature distribution produced by the freezing system is critical in such operations.

Although a great deal has been written on the frozen storage life of different meats, the underlying data are backed up by a relatively small number of controlled scientific experiments. Much of the scientific data date back to the time when meat was either stored unwrapped or in wrapping materials that are no longer used. It is not surprising when we consider the changes in packaging and handling methods over the last century that there is a considerable scatter in data on storage lives for similar products.

In recent years energy conservation requirements have caused an increased interest in the possibility of using more efficient storage temperatures than have been used to date. Researchers such as Jul have questioned the wisdom of storage below −20°C and have asked whether there is any real economic advantage in very low temperature preservation. There is a growing realisation that storage lives of several foods can be less dependent on temperature than previously thought. Since research has shown that meat and poultry often produce non-linear time–temperature curves, there is probably an optimum storage temperature for a particular product. Improved packing and preservation of products can also increase storage life and may allow higher storage temperatures to be used. One suggestion is that with storage at −18°C, low stability meats such as mechanically recovered meat should be stored for 8 months or less, medium stability meats such as pork and processed meats should be stored for between 8 and 15 months, and high stability meats, which include all meat and poultry except pork, could be stored for more than 15 months.

REFERENCES

Añón, M.C. and Calvelo, A. (1980). Freezing rate effects on the drip loss of frozen beef. *Meat Science* 4, 1–14.

ASHRAE (2006). *Refrigeration Handbook*, sections 9–29.

Bailey, C., Cutting, C.L., Enser, N. and Rhodes, D.N. (1973). The influence of slaughter weight on the stability of pork sides in frozen storage. *Journal of Science of Food and Agriculture* 24, 1299.

Bech-Jacobsen, K. and Bøgh-Sørensen, L. (1984). TTT-Studies of different retail packed pork products. In: P. Zeuthen, J.C. Cheftel, C. Eriksson, M. Jul, H. Leniger, P. Linko, G. Varela and G. Vos, eds. *Thermal Processing and Quality of Foods*. Elsevier Applied Science Publishers, pp. 716–720.

Berry, B.W. and Leddy, K.F. (1989). Effects of freezing rate, frozen storage temperature and storage time on tenderness values of beef patties. *Journal of Food Science* 54(2), 291–296.

Bevilacqua, A., Zaritzky, N.E. and Cavelo, A. (1979). Histological measurements of ice in frozen beef. *Journal of Food Technology* 14, 237–251.

Bishop, E. (1972). Poultry freezing. *Refrigeration and Air Conditioning* 75, 37–40, 85.

Brown, T. and James, S.J. (2006). The effect of air temperature, velocity and visual lean composition on the tempering time of frozen beef blocks. *International Journal of Refrigeration* 73, 545–552.

Buchmuller, J. (1986). Chilling and freezing meat products with liquid nitrogen. *Fleischwirtschaft* 66(4), 568–570.

Callow, E.H. (1929). The freezing and storage of mild cured bacon. *DSIR Department Food Investigation Board* 67–73.

Carrol, R.J., Cavanaugh, R.J. and Rorer, F.P. (1981). Effects of frozen storage on the ultrastructure of bovine muscle. *Journal of Food Science* 46, 1091–1094, 1102.

Castell-Perez, M.E., Heldman, D.R. and Steffe, J.F. (1989). Computer simulation of microbial growth during freezing and frozen food storage. *Journal of Food Process Engineering* 10, 249–268.

Creed, P.G. and James, S.J. (1981). A survey of commercial meat block freezing in the United Kingdom and Eire. *International Journal of Refrigeration* 4, 348–354.

Creed, P.G. and James, S.J. (1983). The freezing times of liver in a vertical plate freezer. In: *Progress in Refrigeration Science and Technology, Proceedings of the XVI International Congress of Refrigeration*, Tome III. Paris (France). 4, pp. 145–151.

Creed, P.G. and James, S.J. (1984). The prediction of freezing and thawing times of mutton carcasses. In: *Proceedings of the 30th European Meeting of Meat Research Workers*. Bristol. Paper 2:5, pp. 59–60.

Creed, P.G. and James, S.J. (1985). Heat transfer during the freezing of liver in a plate freezer. *Journal of Food Science* 50, 285–288, 294.

Critchell, J.T. and Raymond, J. (1912). *A History of the Frozen Meat Trade*, 2nd edn. London: Constable & Company Ltd.

Dawson, L.E. (1969). Stability of frozen poultry meat and eggs. In: W.B. Van Arsdel, M.J. Copley and R.L. Olson, eds. *Quality and Stability of Frozen Foods – Time-Temperature Tolerance and its Significance*. pp. 143–167.

de Jong, R.S. (1994). Plate freezing technology and economics applied to the meat industry. In: *Proceedings IRHACE Technology Conference*. Paper 11.

Dobryzcki, J., Pietrzak, E. and Hoser, A. (1977). Rheological characteristics of fresh and frozen chicken muscles and their relation to tenderness measured by sensory methods. *Acta Alimentaria* 6(2), 107–111.

DuBois, C.W., Tressler, D.K. and Fenton, F. (1942). The effect of the rate of freezing and temperature of storage on the quality of frozen poultry. *Refrigeration Engineering* 44, 93–99, 111.

Esselen, W.B., Levine, A.S., Pflug, I.J. and Davis, L.L. (1954). Brine immersion cooling and freezing of ready to cook poultry. *Refrigeration Engineering* 62, 61–63.

Everington, D.W. (2001). Development of equipment for rapid freezing. In: *Rapid Cooling of Food*. Meeting of IIR Commission C2. Bristol, UK. Section 2, pp. 173–180.

Fernandez-Martin, F., Otero, L., Solas, M.T. and Sanz, P.D. (2000). Protein denaturation and structural damage during high-pressure-shift freezing of porcine and bovine muscle. *Journal of Food Science* 65, 1002–1008.

Fleming, A.K., Chadderton, T., Amos, N.D. and Cleland, A.C. (1996). Non-traditional refrigeration technologies for the New Zealand meat industry. *MIRINZ publication 961*, ISSN 0465-4390.

Georgsson, F., Porkelsson, A.E., Geirsdottir, M., Reiersen, J. and Stern, N.J. (2006). The influence of freezing and duration of storage on Campylobacter and indicator bacteria in broiler carcasses. *Food Microbiology* 23(7), 677–683.

Gortner, W.A., Fenton, F., Volz, F.E. and Gleim, E. (1948). Effect of fluctuating storage temperatures on quality of frozen food. *Industrial Engineering Chemistry* 40, 1423–1246.

Grujic, R., Petrovic, L., Pikula, B. and Amidzic, L. (1993). Definition of the optimum freezing rate – 1, investigation of structure and ultrastructure of beef *M. longissimus dorsi* frozen at different freezing rates. *Meat Science* 33, 301–318.

Hanenian, R., Mittal, G.S. and Usborne, W.R. (1989). Effects of pre-chilling, freezing rate, and storage time on beef patty quality. *Journal of Food Science* 54(3), 532–535.

Hill, M.A. and Glew, G. (1973). Organoleptic assessment of products frozen two methods: liquid nitrogen and blast freezing. *Journal of Food Technology* 8, 205–210.

Hiner, R.L., Gaddis, A.M. and Hankins, O.G. (1951). Effect of methods of protection, palatability of freezer stored meat. *Food Technology* 223–229.

International Institute of Refrigeration. (1986). *Recommendations for the Processing and Handling of Frozen Foods*, 3rd edn.

Jackobsson, B. and Bengtson, N. (1973). Freezing of raw beef: influence of ageing, freezing rate and cooking method on quality and yield. *Journal of Food Science* 38, 560–565.

James, S.J. and Bailey, C. (1979). The determination of freezing time of boxed meat blocks. In: *The Proceedings of the Institute of Refrigeration*. Session (1978–79), pp. 1–8.

James, S.J. and Bailey, C. (1987). The freezing of beef quarters. In: *Proceedings XVII International Congress of Refrigeration C*, Vienna (Austria), C2-4.

James, S.J. and Bailey, C. (1987b). Bacon tempering for high speed slicing. In: *Proceedings XVII International Congress of Refrigeration C*, Vienna (Austria), C2-1.

James, S.J., Nair, C. and Bailey, C. (1983). The effects of freezing and thawing rates on the drip loss from frozen beef. In: P. Zeuthen, *et al.* eds. *Thermal Processing and the Quality of Foods*, 1st edn. London & New York: Elsevier Applied Science Publishers, pp. 807–813.

James, S.J. (1997). Bacon tempering for high speed slicing. In: *Proc. Meat Refrigeration – Why and How?* EU concerted action programme CT94 1881, Langford, UK.

James, S.J. and Evans, J.A. (1997). Frozen storage of meat and meat products. FAIR Concerted Action PL95-1180.

Jul, M. (1982). The intricacies of the freezer chain. *International Journal of Refrigeration* 5, 226–230.

Kondratowicz, J., Bak, T. and Denaburski, J. (2000). Influence of the freezing procedure – Characteristics of normal pork, as well as pork with PSE- and DFD-traits. *Fleischwirtschaft* 80, 81–83.

Lammertz, M. and Brixy, N. (2001). Continuous process and production improvements by application of refrigeration with cryogenic gases. In: *Rapid Cooling of food*, Meeting of IIR Commission C2, Bristol, UK.

Lampitt, L.H. and Moran, T. (1933). The palatability of rapidly frozen meat. *Journal of the Society of Chemical Industry* 52, L11.21 143t–146t.

Le Bail, A., Chevalier, D., Mussa, D.M. and Ghoul, M. (2002). High pressure freezing and thawing of foods: a review. *International Journal of Refrigeration* 25, 504–513.

Lentz, C.P. (1971). Effect of light and temperature on colour and flavour of pre-packaged frozen beef. *Canadian Institute of Food Science and Technology Journal* 4, 166.
Maria, G.T., Abril, J. and Casp, A. (2005). Surface heat transfer coefficients for refrigeration and freezing of foods immersed in an ice slurry. *International Journal of Refrigeration* 28, 1040–1047.
Mascheroni, R.H. (1985). Generalisation of a method for the characterization of quick frozen beef. *Meat Science* 13, 81–89.
Mittal, G.S. and Barbut, S. (1991). Effects of freezing rate and storage time on the structural properties of minced meat. *Lebensmittel – Wissenschaft und -Technologie* 24, 226–230.
Moorhead, S.M. and Dykes, G.A. (2002). Survival of *Campylobacter jejuni* on beef trimmings during freezing and frozen storage. *Letters in Applied Microbiology* 34, 72–76.
Newman, M. (2001). Cryogenic impingement freezing utilizing atomized liquid nitrogen for the rapid freezing of food products. In: *Rapid Cooling of food*, Meeting of IIR Commission C2, Bristol, UK.
Ngapo, T.M., Babare, I.H., Reynolds, J. and Mawson, R.F. (1999). Freezing and thawing rate effects on drip loss from samples of pork. *Meat Science* 53, 149–158.
Persson, P.O. and Löndahl, G. (1993). Freezing technology. In: C.P. Mallett, ed. *Frozen Food Technology*, Chapter 2. Blackie Academic & Professional.
Petrovic, L., Grujic, R. and Petrovic, M. (1993). Definition of the optimal freezing rate – 2 Investigations of the physico-chemical properties of beef *M. longissimus dorsi* frozen at different freezing rates. *Meat Science* 33, 319–331.
Reid, D.S. and Perez Albeta, L. (2006). The effect of average storage temperature, and temperature fluctuation on the rate of moisture migration in a model frozen food. *IUFoST* 1979–1988.
Sacks, B., Casey, N.H., Boshof, E. and van Zyl, H. (1993). Influence of freezing method on thaw drip and protein loss of low-voltage electrically stimulated and non-stimulated sheep's muscle. *Meat Science* 34, 235–243.
Sakata, R., Oshida, T., Morita, H. and Nagata, Y. (1995). Physico-chemical and processing quality of porcine M. longissimus dorsi frozen at different temperatures. *Meat Science* 39, 277–284.
Sebranek, J.G. (1980). Cryogenic freezing of ground beef patties shows superior organoleptic effects. *Quick Freezing* August, 50–53.
Sandberg, M., Hofshagen, M., Ostensvik, O., Skjerve, E. and Innocent, G. (2005). Survival of Campylobacter on frozen broiler carcasses as a function of time. *Journal of Food Protection* 68, 1600–1605.
Sebranek, J.G., Sang, P.N., Rust, R.E., Topei, D.G. and Kraft, A.A. (1978). Influence of liquid nitrogen, liquid carbon dioxide and mechanical freezing on sensory properties of ground beef patties. *Journal of Food Science* 43, 842–844.
Sundsten, S., Andersson, A. and Tornberg, E. (2001). The effect of the freezing rate on the quality of hamburgers. In: *Rapid Cooling of Food*, Meeting of IIR Commission C2, Bristol, UK.
Swain, M.J. and James, S.J. (2005). Microwave thawing and tempering In: H. Schubert and Regier, M. eds. *Microwave Processing of Food*. Woodhead Publishing Limited.
Thielke, S., Lhafi, S.K. and Kuhne, M. (2005). Effects of aging prior to freezing on poultry meat tenderness. *Poultry Science* 84, 607–612.
Thumel, H. and Gamm, D. (1994). Loose rotary freezing. *Fleischwirtschaft* 74(1), 64–65.
Van Arsdel, W.B. (1969). Estimating quality change from a known temperature history. In: W.B. Van Arsdel, M.J. Copley and R.L. Olson, eds. *Quality and Stability of Frozen Foods – Time-Temperature Tolerance and its Significance*. pp. 237–262.
van den Berg, L. and Lentz, C.P. (1958). Factors affecting freezing rate and appearance of eviscerated poultry frozen in air. *Food Technology* 12, 183–185.
Volz, F.E., Gortner, W.A., Pitz, E.W. and Miller, J.E. (1949). The effect of ultraviolet light in the meat cooler on the keeping quality of frozen pork. *Food Technology* 3, 4.
Wanous, M.P., Olson, D.G. and Kraft, A.A. (1989). Pallet location and freezing rate effect on the oxidation of lipids and myoglobin in commercial fresh pork sausage. *Journal of Food Science* 54(3), 549–552.
Winger, R.J. (1984). Storage life and eating related quality of New Zealand frozen lamb: a compendium of irrepressible longevity. In: P. Zeuthen, J.C. Cheftel, C. Eriksson, M. Jul, H. Leniger, P. Linko, G. Varela and G. Vos, eds. *Thermal Processing and Quality of Foods*. Elsevier Applied Science publishers, pp. 541–552.
Wirth, F. (1979). Chilling, freezing, storage and handling of meat: present state of our knowledge. *Fleischwirtscaft* 59(12), 1857–1861.
Zhu, S.M., Le Bail, A., Chapleau, N., Ramaswamy, H.S. and de Lamballerie-Anton, M. (2004). Pressure shift freezing of pork muscle: effect on color, drip loss, texture, and protein stability. *Biotechnology Progress* 20, 939–945.

7 Freezing of Fish

Ola M. Magnussen, Anne K.T. Hemmingsen, Vidar Hardarsson, Tom S. Nordtvedt and Trygve M. Eikevik

7.1 INTRODUCTION

The term 'fish' covers a wide variety of species with very different characteristics, such as shape, appearance and size. The most common characteristic is that they mostly have gills, tail, and fins and that they are cold blooded. Shellfish is, however, often regarded to be in this group. The most characteristic difference between fish and animal products is the edible muscle. In beef, the fibres are running through the muscle and are strongly bound together by means of connective tissue. Fish mainly live in 'weightless' condition in water and the fish muscle is made up of millions of fibres arranged in short segments or blocks. The short segments are linked together with connective tissue or a thin membrane of collagen material with low material strength. The fish muscle will therefore easily break, or the segments will separate one from another, by mechanical pressure or strain. This is often called gapping and is a common defect for fish products if exposed to tough handling during catch, processing and freezing.

The most common fishes are often divided into two distinct groups, *pelagic* and *demersal*. *Pelagic fish*, like herring, sprat, mackerel and capelin live mainly in the upper layers of the seas. They feed on phytoplankton and zooplankton, which are abundant in spring and summer. Most pelagic fish have a high fat content, but with large variations due to the feeding season. Pelagic fish have thin and soft skin that is easily damaged during handling and freezing. The fat fish muscle has low mechanical strength, and strain must be avoided during processing and freezing. *Demersal fish*, such as cod, haddock and pollock, and flatfishes live near or at the bottom of the seas. They feed mostly on other fish and organisms. The fat reserve is deposited mainly in the liver. The fillet/muscle is lean and a little more capable to resist strain. Spawning generally puts considerable strain on the fish proteins and fat and muscle tissue become watery and soft during the spawning season. The fillet quality and yield then becomes poorer and is not well suited for high-quality products.

Fish farming has lately become a big issue in many countries. Farming gives a high degree of control of quality and fish size and reduces variations in products. The fish are slaughtered and handled in large centralized plants and are often frozen as gutted or as filleted products. Farmed fat fish, such as salmon and halibut, have, unlike pelagic fish, better mechanical strength of muscle and longer shelf-life as frozen products.

7.2 PRE-FREEZING EFFECTS

There is today increasing focus on methods for ethical slaughtering and killing of fish. Anaesthetization before slaughter is recommended to calm down the fish in order to reduce strain and simplify handling and killing. Bleeding should take place immediately after killing and not later than 20–30 minutes after catching. Poor bleeding will result in visible blood vessels in the fish muscle. This occurs in both white and red fish fillets, especially after freezing. Poor handling, which results in local high pressure, blow and impact on the fish, will additionally result in effusions of blood. Gentle handling of fish is therefore required. Gutting and washing should take place immediately after bleeding. The belly flap of ungutted fish will soon become discoloured due to leakage from the urinary bladder. For small fish caught in large numbers, such as herring, sprat and mackerel, immediate gutting is often not possible. This will affect quality and together with other factors will result in low shelf-life for these fish. In fish caught in seasons with abundant access to feed, enzymes activity may be high and can attack the stomach, the belly flap and other organs, resulting in 'belly burst'.

Freezing should take place as soon as possible after the fish is slaughtered, preferable before onset of rigor mortis. All intermediate storage will result in lower quality and reduced shelf-life after freezing. If immediate freezing for some reason, such as transport, filleting and processing, is not possible, chilling should take place immediately after catching and product temperatures below 0°C should quickly be reached. Fish caught on fishing vessels without freezing equipment have to be chilled as soon as possible.

Reducing the product temperatures to the freezing point or 1–2°C below the freezing point before filleting have shown to increase fillet yield and quality, though dependent on fish species. Pre-rigor filleting will result in some fillet shrinkage after rigor onset. Rough handling of in-rigor fillet might cause breakage or gaping of the fillets.

The quality of chilled fish, especially of fat pelagic fish such as herring, is relatively short, see Fig. 7.1.

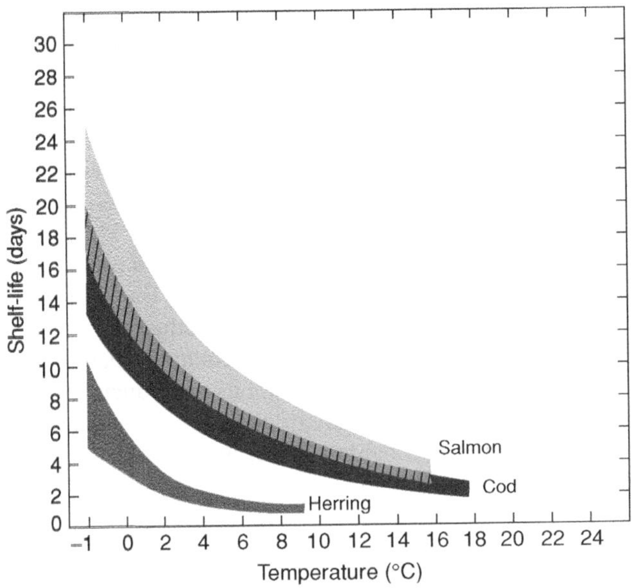

Fig. 7.1 Practical sensory shelf-life of important North Atlantic fish species (Flesland and Magnussen, 1990). Expected shelf-life depends on seasonal variations, like feeding, fat content, spawning seasons and handling (Norwegian commercial standard).

Chill storage of fish before freezing will therefore reduce the shelf-life of the frozen product. Quality losses in fresh products depend to a large extent on micro-organisms present in or around the products. The metabolism of the micro-organisms will mostly terminate in frozen conditions. There is today no appropriate general way to calculate the effect of chill storage on frozen food quality. Effects of handling and chill storage on quality of frozen fish must therefore be established for specific products and species.

7.3 EFFECTS OF FREEZING PARAMETERS

Since the start of microscopy of frozen food in around 1990, the effect of freezing rate and ice crystal size on quality of frozen products have been discussed. Very fast freezing of small tissue parts may give small ice crystals distributed inside the cells. The effect of very fast freezing rate on practical food quality is, however, still discussed. The freezing rate required is very high and can only be achieved with very efficient heat transfer and large temperature differences (cryogenic freezing, N_2 or CO_2). Even then, the freezing rate giving small ice crystals is high enough only on the surface of commercial products. Since the freezing rate is reduced almost proportionally to the square of the ice layer thickness, the ice crystal size and distribution will vary throughout the product. For practical purposes, the effect of freezing rate on quality of most fish products is not proved. However, quality effects of freezing rate and requirements should be tested for new products.

Phase change from liquid to solid water will, different from solidification of most other liquids, give a volume increase and a density reduction of about 5–7%. During freezing of most commercial products and commonly used freezers, the ice formation starts on the surface and heat is transferred through an increasing layer of ice. The internal volume increases due to crystallization of freezeable water. With the most commonly used freezing temperatures, there will still be some liquid inside the muscle after freezing. The volume increase is probably allowed by expansion and movement in unfrozen parts and no visible cracking is seen. Use of very low temperatures will often result in surface/ice temperatures where almost all liquid is frozen and what is termed 'glassy state' exists. The result is little or no expansion possibilities and the build-up of high internal forces until the material cracks in the weakest parts. The limits for low temperatures on product surfaces and expansion possibilities during freezing should therefore be examined when low freezing temperatures are used.

7.4 FREEZING METHODS

In general, the fish processing companies handle and process a number of fish species and products. The products may have different packing, products size and freezing times, which often require flexible freezing equipment and the possibility to handle a variety of items. Selection of freezers is often challenging and demand careful evaluation of the volumes and the requirements for the most important products, seasonal variations and possible use for new products. The result is often general equipment which is labour intensive for most products giving relative high costs. The variations in products, freezing times and volumes give large variations in refrigeration load. There is frequent use of a few large screw compressors with capacity control by slide valve regulation and simple capacity control, which often result in high-energy consumption at part load. Improved systems, for example with such frequency control as should be used to improve energy efficiency. Measurements on a fish processing plant have shown almost constant energy consumption regardless of freezing demand and processing volume. It is therefore strongly recommended to select refrigeration systems capable to cover the variations in load with high energy efficiency. Careful selection of compressors, capacity control and use of frequency control of compressors and fans should be used. Most plants also have large consumption of hot water for cleaning, washing and room heating. To a large extent this should be covered by energy recovery from oil coolers of screw compressors and/or high stage heat pumps and high-pressure compressors.

7.4.1 Air (gas) freezing

Due to the large variety of products to be frozen in most fish processing plants, air-blast freezing is used in most companies. Multi-purpose air freezers can be used for most fishes, finished or semi-finished fish products, with acceptable results due to the air's flow characteristics. On the other hand, the labour requirement, in order to achieve uniform and quick freezing, is often high due to stacking and transport. Desiccation during freezing may cause discolouration of the skin and loss of the characteristic bright surface of high-quality fresh fish. Batch freezing and poor airflow control often results in excessive running times of large fans to ensure freezing of all products. Weight losses of up to 5% have been measured for small consumer products. This will also result in high energy consumption for fans and compressors. Use of improved control and regulation systems is therefore recommended. A large number of types and arrangements of air-blast freezers exist, from the general-purpose types to highly specialised units for one single product. In general, the main challenges are to ensure uniform flow of the fluid over product surfaces. Especially for the *in situ* built freezers, little emphasis has been put on dimensioning, product arrangement and ensuring uniform airflow. Most of the existing air/gas freezer types are found in fish industry; description of the three most common types follows.

7.4.1.1 *Tunnel freezers*

Tunnel freezers are flexible freezers that can be used for fish products of all sizes and geometries, packed or non-packed. Products should be placed on racks with shelf space adapted to product height to allow uniform air flow over the products. Non-packed products will stick to the shelves and tunnel freezers are therefore mainly used for freezing of packed products. The packing material, size and geometry often have strong effects on freezing time. Low thermal conductivity of the packing material and air gaps between packing and product may greatly reduce the heat transer. With the common thickness of products or packing used in the bulk freezing of fish, freezing times of many hours is common. As an example, the freezing of pelagic fish in standard 20-kg packing boxes is shown in Fig. 7.2.

The packing and the air layer between the fish and the lid result in almost no heat flow through the lid. Additionally, with dense stacking on shelves, freezing takes place only from the bottom of the package. Common batch freezing requires 18–20 hours, which when including hours for loading and unloading, results in total freezing times of 24 hours or more. Packing is expensive; substancial cost reductions are possible by use of plastic bags. Freezing time will then also be greatly reduced. Traditionally, air flow over a number of shelves through the tunnel is used, resulting in large temperature increases in the air, especially during the main freezing period. This results in differences in the freezing time and excessive total running of fans and compressors.

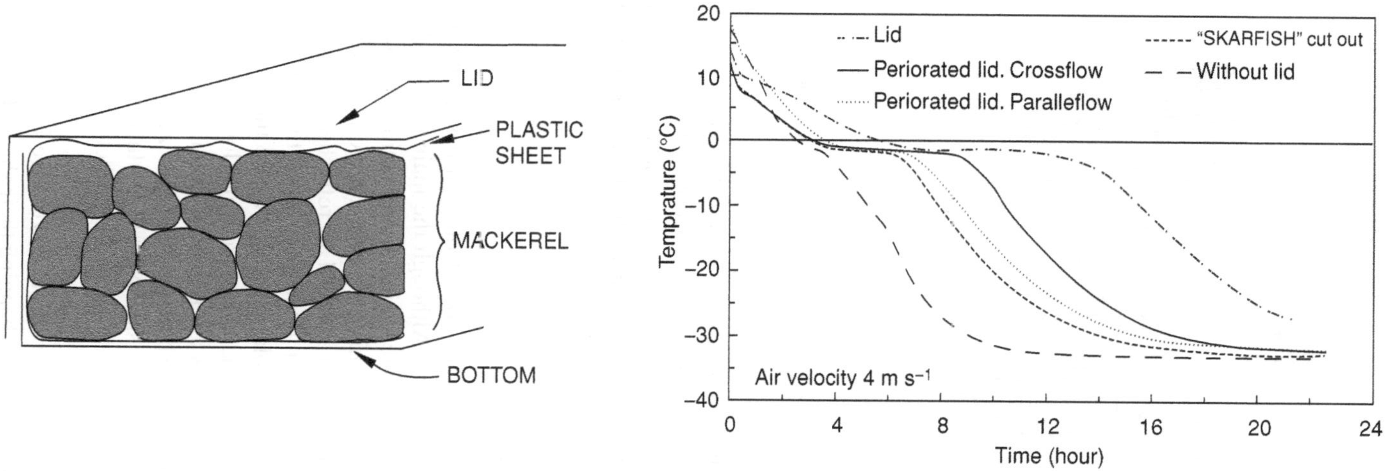

Fig. 7.2 Traditional packing of pelagic fish in 20 kg carton box and effects of perforation and removal of lid. SKAARFISH cut out means that the letters SKARFISH were cut out in the lid.

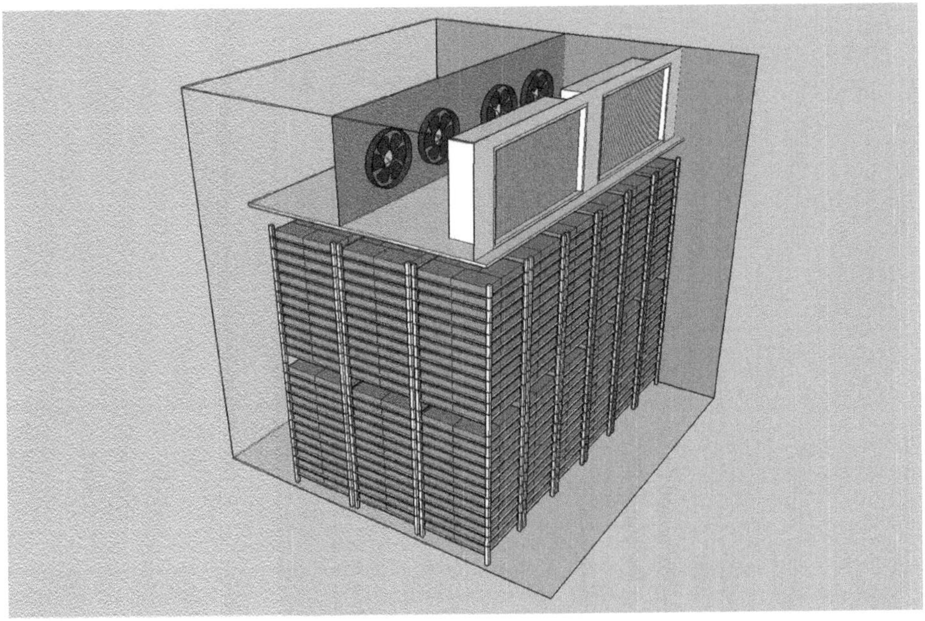

Fig. 7.3 Batch type air-blast freezer for pelagic fish and airflow simulation.

A diagram of a modern air-blast freezer for pelagic fish is shown in Fig. 7.3. The principal design should be chosen from optimisation of airflow, temperatures, required capacity, fan types and structural arrangements. Uniform airflow and freezing time should be designed by use of computer simulation. Energy consumption should be minimized by use of fan diffusers, fan speed control and adjustment of freezing time.

7.4.1.2 *Continuous freezers*

The increased demand for ready-made food from fish has greatly increased the industrial production of single frozen products. Although the main focus today is on fresh products, the industry has also seen a large increase in the production of frozen products, mainly due to the short shelf-life of fresh fish. This has given a large demand for semi-finished frozen products which are thawed, processed and sold as 'refreshed'. Efficient industrial production requires in-line processing and freezing within required times and to avoid discontinuity in the processing line. Continuous air-blast freezers where the products are transported on a belt system are mainly used. To achieve fast freezing, low temperatures and high-velocity air over and/or vertically on the belts are required. Continuous multi-belt air-blast freezers and spiral freezers are often used. The continuous freezers are mainly used for freezing of non-packed products or vacuum or skin-packed products to obtain fast freezing. Weight losses of typical wet fish surfaces of non-packed fish fillets can be reduced by the use of low temperatures.

The advantage of the multi-belt air-blast freezer (see Fig. 7.4) is the wide belts, which permit freezing of large fillets without changing their shape. By using stainless steel plate as the belt marks on the fillets, often found with the use of stainless wire mesh belts or flexible plastic parts, can be avoided. However, this requires enough length of the upper/first belt to

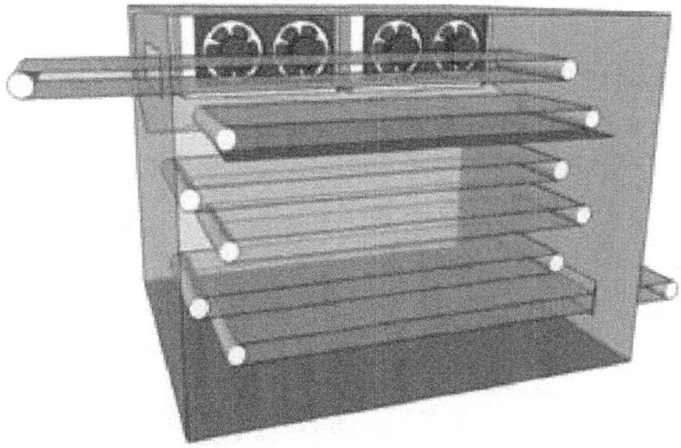

Fig. 7.4 Multi-belt air-blast freezer with horizontal main flow.

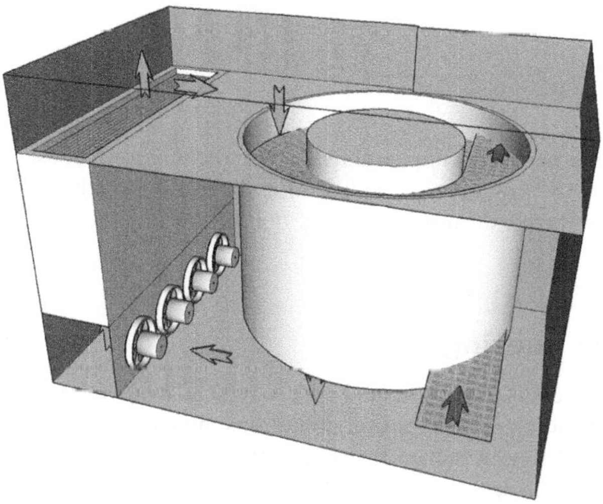

Fig. 7.5 Spiral belt freezer where airflows vertically downwards pass the fish products.

form a rigid body (shell-freezing) so that the products are transferred to the next belt with no changes in shape or quality. The shell-freezing belt length will define the minimum freezer length and thereby the floor space required. In general, the space required for this type of freezer are somewhat large compared to the more common spiral freezers (see Fig. 7.5). The arrangement of the belts can be adjusted to the product height and the belt length will not define floor space. The most efficient freezers have vertical air flow due to better control of air distribution around the products, but are also more expensive.

7.4.1.3 *Fluidised bed freezers*

The main marine product frozen in fluidised bed freezers (see Fig. 7.6) is cooked and peeled shrimps. Due to the size and geometric of the shrimps, the products fluidise satisfactorily. Small parts, such as broken tailpieces and shell parts from peeling, may overflow together with the air stream. Freezing time is short, often less than 10 minutes and water weight loss

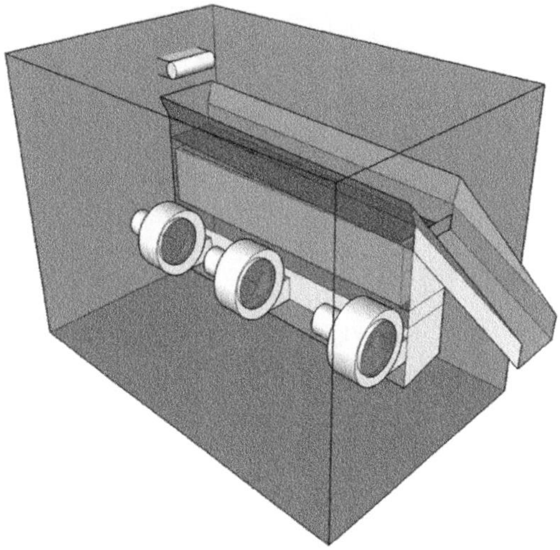

Fig. 7.6 Fluidised bed freezer.

is small. Usually the peeled shrimp is glazed after freezing to increase stability and reduce quality loss. Up to 15% of water is added on a shower belt or by immersion. Controlled sub-cooling of the glazed shrimp is required due to the large temperature increase in the process.

7.4.2 Contact freezing

In contact freezing there is direct contact between a metal evaporator surface and the wet fish surface, giving very efficient heat transfer conditions; no additional energy for the transfer is required. This type of freezer is therefore efficient and compact, requiring little space. Most freezers used in fish freezing today have aluminium or steel plates with channels for direct expansion of refrigerant. Original horizontal plate freezers were introduced in the 1930s for freezing of rectangular fillet blocks, but is today in use for a variety of applications.

7.4.2.1 *Horizontal plate freezers*

This equipment is common in the fillet industry for freezing of the traditional small packed consumer fillets and larger blocks for processing of 'fish fingers', etc. The wet fillet is packed in thin cardboard and stacked on trays in frames to obtain a regular form and placed between the plates (see Fig. 7.7). Good contact between plates and packages is ensured by hydraulic pressure, which also allows for freezing expansion and regular shape. This is especially important for blocks to be cut in processing. Freezing time for consumer packing and industrial blocks is typically between 30 minutes and 2 hours. The filling of product/frames in between the narrow plates and removal of frozen products is time consuming and heavy and the cold product makes this work challenging. Different semi-continuous systems for loading, unloading and stacking of the products are available and are strongly recommended. For small consumer products, automatic semi-continuous horizontal plate freezers are also available for in-line use.

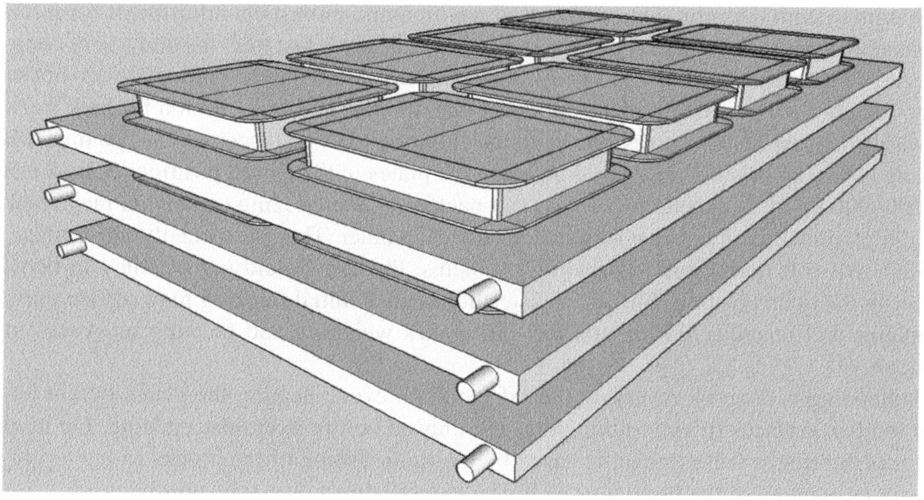

Fig. 7.7 Contact freezer type horizontal plate freezer.

7.4.2.2 *Vertical plate freezers*

Compared to the horizontal plate freezer mainly used for processed fish products, the vertical version developed later is a typical bulk volume freezer. Due to the vertical openings (see Fig. 7.8) the products can be filled in between the plates by weight or volume. For most products, bulk product is required and the products are frozen to the plates. Some products are, however, filled in a plastic bag before freezing. The plate position is controlled by a

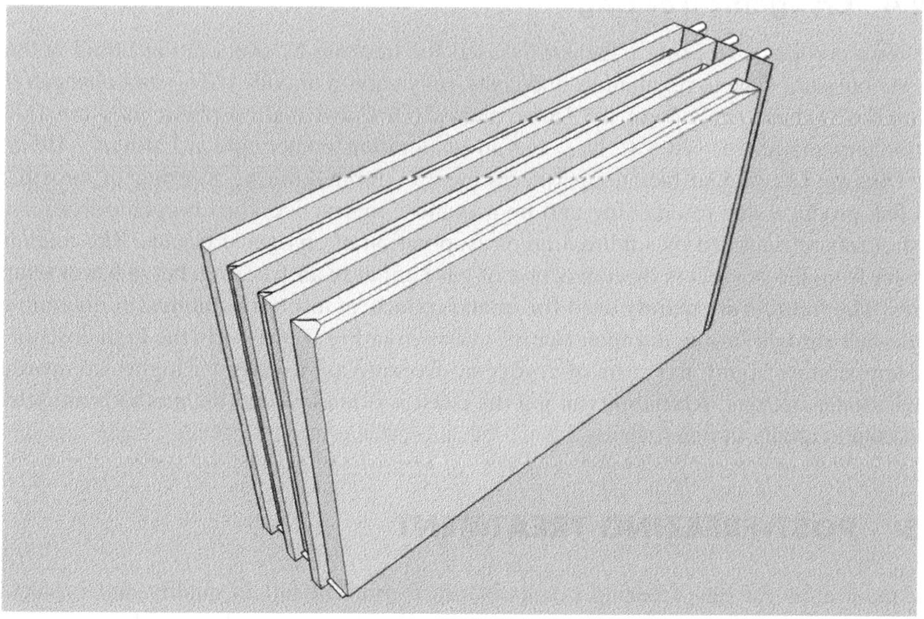

Fig. 7.8 Contact freezing between vertical plates.

hydraulic system between open and closed position and movement adjustment for freezing expansion is permitted. The very efficient heat transfer gives fast freezing and a very compact freezer, which requires a comparatively smaller space. Although the plain plates do not seem to go together with the complex geometry of fish, the freezers are common for bulk freezing of whole fish, especially on vessels. With the common plate distances of 75–100 mm, whole fish up to 5–6 kg or more are placed between the plates in the closed position (bulk densities of 700–900 kg m^{-3}, depending on fish and packing). The freezing expansion may result in local high pressure on fishes positioned across each other. This is especially unfortunate for fat fish and soft flesh. Due to pressing problems, the fish should not be filled in between plates in the open position. Filling of water/liquid after initial freezing may reduce pressure problems and rancidity during storage, but makes weighing and thawing processes more difficult.

Unloading is done by reversing the refrigeration cycle; heating the plates by hot gas so that the fish surfaces frozen to the plates are thawed before hydraulic opening. On modern freezers the frozen fish removal is eased by hydraulic lifting of the blocks to the top of the freezer or opening of the bottom to let the blocks fall down onto transport belts.

7.4.3 Brine freezing

The first method for large-scale industrial fish freezing of pelagic fish was based on brine freezing and introduced by Ottosen and Dahl (Plank, 1954). The most common brines are based on accepted brines, mainly NaCl, and sugar solutions. The freezing temperatures are then limited to −18°C (NaCl) to around −25°C, due to their eutectic points. There are many practical problems in brine freezing, such as corrosion and labour demand, and the method is used only for some fishes and mainly on vessels.

7.4.4 Cryogenic freezing

The term cryogenic freezing is generally used for freezing by evaporating liquid at atmospheric pressure. Due to regulations and costs, only carbon dioxide (CO_2) and nitrogen (N_2) are used. CO_2 has a triple point at 5.18 bar and −56.6°C and at atmospheric pressure a sublimation temperature of −78.8°C. $N_{2liq.}$ has an evaporation temperature at 1 atm of −195.8°C. The freezing rate and surface temperatures must be limited during freezing of most fishes and fish products due to cracking and poor quality. In practice, the cryogenic freezers are gas freezers refrigerated by sublimation or evaporation of cryogenic liquids. The main heat transfer from the product is by convection of gas similar to air freezing, but at lower temperatures. The freezers are mainly used for small products at low temperatures. A diagram of a single-belt straight in-line nitrogen freezer is shown in Fig. 7.9. Due to the high costs of the low-temperature liquid, the costs of cryogenic freezing are usually far higher compared to conventional freezers. Availability along the coast is often low and the methods are seldom used commercially in fish freezing.

7.5 POST-FREEZING TREATMENT

The main aims for post-freezing processes are to prevent fall in quality and protect the products during storage and distribution. For some fish products this will have a great effect on shelf-life, especially on fatty fish. Correct post-freezing treatment will also reduce

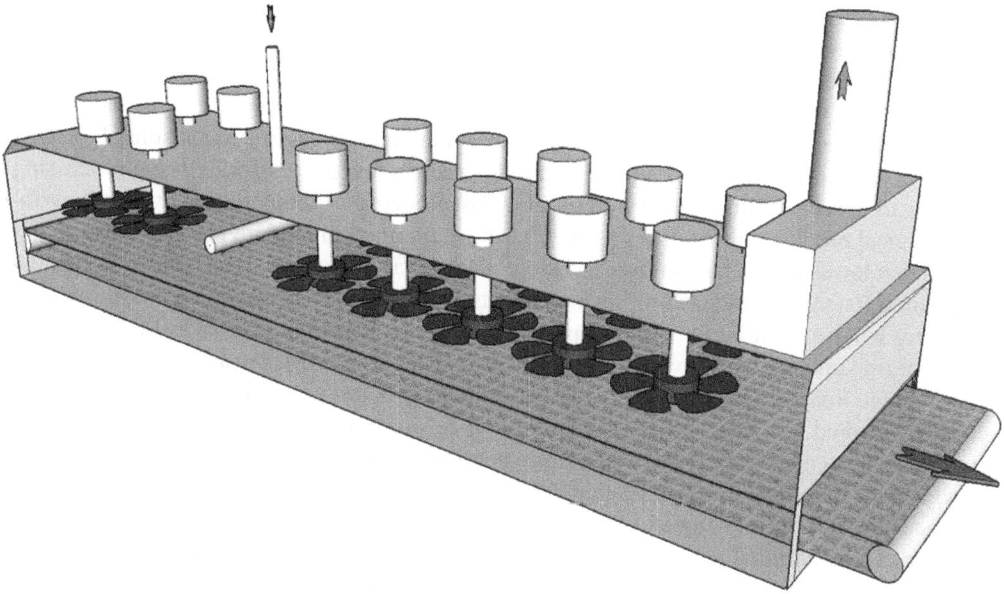

Fig. 7.9 Single belt straight in-line nitrogen freezer.

decolourisation of the fish flesh. Freezing of a thin layer of water or water solution with antioxidants, polyphosphates or alginates on the product surfaces is usually called *glazing*. The film protects the product from losses of oxygen and water/ice from the product ('freezer burn'). Industrial glazing is usually done by spraying the product surfaces or short dipping of the products in a water solution. The main challenge is to obtain a uniform film on the rough fish surface and to control the process. As fish have a high surface-to-volume ratio the products may experience a large temperature rise during glazing and may require refreezing after glazing.

Efficient *packing* is essential to reducing microbial and chemical contamination, dehydration and mechanical damage from the surroundings. The typical fish smell will be reduced by packaging, and liquid loss from the product, especially after thawing, will also be reduced. The large assortment of different packing materials makes selection challenging and a systematic approach taking into account quality factors, protection and distribution needs, costs and effect on processing is important.

7.6 TRENDS IN TECHNOLOGY DEVELOPMENTS

As in most of the food industry, consumer requirements and trends in eating habits are changing rapidly. Busy consumers in the modern society have little time for food preparation and demand food that requires little preparation and cooking. The focus on healthy food and the positive effect of both fat fish (unsaturated fat) and fish/marine food in general is important. Many consumer groups also require sustainability and ethics in production and processing of fish products. In addition, the increased willingness of consumers to pay for high-quality and healthy fish has changed the industrial production trends. Increasing volumes have given more rational and effective production and processing plants. Use of automation and improved control systems are frequent. At the same time, flexibility is required to increase variation in products and packing.

The increased demand for high-quality products makes temperature control of product from harvest to the consumers especially important. Immediate chilling of caught and farmed fish down to the freezing point and minimal storage before processing are absolutely necessary. It is further important to keep product temperatures low and uniform during processing in order to ensure optimal quality and yield. Fast freezing to temperatures well below today's typical product temperatures (−30°C) is now common in modern plants. Research has shown that lowering the temperature to below −45°C has a great effect on quality, and that it is required for some products (i.e. sushi products). Also, lower storage and distribution temperatures will be required for a new improved cold chain. Thawing is often more demanding than freezing and new improved thawing processes and equipment must be used.

The large floating fish factories process the catch on board producing consumer products or semi-finished products. Fillets are mainly frozen in horizontal plate freezers (compact equipment), but also tunnels and continuous-belt freezers are used. An increasing number of trawlers and long-liners now have changed from fresh fish production to freezing of gutted unprocessed fish for thawing and processing on land. Vertical plate freezers or bulk tunnels are used. The traditional fish processing industry is based on fish from smaller coastal vessels and trawlers. The seasonal variation in volume of fish requires freezing of raw materials. The raw material is thawed before processing. Knowledge of thawing procedures and equipment is therefore also required and is essential to retain the high yield and quality of fish products.

7.7 LEGISLATIVE ASPECTS OF FREEZING

The legislative requirements of freezing of fish varies with region and have lately been changed somewhat in Europe. In Europe the EU regulations are the most common standard. The requirements are mainly recommendations. In Norway the directions are given in 'The description of quality of fish and fish products' and are summarized below.

Fish and fish products must be frozen in freezing equipment with sufficient capacity and in a way so that the temperature region for maximal crystallisation is passed as soon as possible for the actual product. The freezing shall reduce the temperature in the warmest place of the product to −18°C or lower. When choosing the freezing method, the freezing rate and the properties of the raw material and the finished product must be taken into consideration. Contact freezing, air freezing and evaporation of freezing medium can be used direct on the products. The only freezing media that can be directly in contact with the products are air, nitrogen and carbon dioxide. Stowing of products in a freezing tunnel must be done in a way so that the freezing medium can flow along the whole surface of every product unit.

Freezing of fish and fish products must be done as soon as possible after all necessary processing operations are finished. The time from when the raw material is put into production until it is placed in a freezer should not be more than 4 hours. Fish that are exempted from the directions of gutting can be frozen round. Fish that are frozen round should mainly be free from parasites and have low content of feed in the stomach. Fish that are frozen for use as baith should during freezing and freeze storage be labelled and stored such that confusion with human food does not happen.

With contact freezing, the freezing time shall be set so that the temperature in the warmest point of the product is −18°C or lower after a time in hours equivalent to the product's half thickness in centimetre (for example, 5 cm thickness gives maximum 2.5-h freezing time).

With air freezing, the freezing time for round fish (single frozen, in blocks or packaging) shall be maximum 24 hours (the temperature in the warmest point of the product should be $-18°C$ or lower within this time limit). For whole fish with thickness above 16 cm, the freezing time can be prolonged to 72 hours. Glazing of fish products shall be performed hygienically, with pure water and in a way so that the temperature rise is minimal.

Freezing storage rooms shall be insulated and have refrigeration equipment so that the fish products after thermal stabilization are stored so that the core temperature is $-18°C$ or lower in the whole product. Transfer of products from freezing equipment to freezing storage equipment shall not be done before the products have a core temperature of minimum $-18°C$. The products shall be transferred to freezing storage equipment immediately after departing the freezing equipment. Frozen raw materiel and finished products shall, if nothing else is decided, be labelled, stored, transported and presented as frozen products and the freezing chain shall be maintained throughout every point.

7.8 FOOD QUALITY AND FREEZING RATE

Literature on freezing of fish is often confusing about what happens to the fish as it freezes. The difference between slow and quick freezing is not well understood and only in recent years freezing processing knowledge has advanced sufficiently to explain the differences between slow and quick freezing.

A common understanding has been that rapid freezing of fish (small products: 50–100 mm h^{-1}) is unsatisfactory due to disruption of muscle tissue during sudden cooling. Another theory is that the cell walls during rapid freezing can burst under the induced pressure caused by expansion of water during rapid freezing. It has also been assumed that slow freezing (in blast freezer room: 2 mm h^{-1}) results in formation of large ice crystals that can cause damage to the cell walls, which would result in considerable drip loss during thawing. It has, however, been shown that the differences in the size of the ice crystals does not provide a full explanation as the wall of the fish muscle cells are sufficiently elastic to adapt to the larger ice crystals without excessive damage. Additionally, the water in fish muscle is bound to the protein in the form of a gel so that little fluid will be lost during thawing even if the damage of the cell walls has occurred. Slow bulk freezing does, however, result in products with poorer quality and this is now explained mainly by temperature-dependent denaturation of the proteins. The dentauration processes are reduced as the temperature is reduced and at slow freezing a longer time is spent at higher temperatures (Johnston *et al.*, 1994).

Many of the advantages of rapid freezing can be lost during subsequent storage (Fennema, 1975) and often too much effort has been placed on high freezing rates in process freezers. Achieving complete product freezing is probably more critical because rates of temperature reduction in frozen storage is much lower leading to significant loss of product quality (Valentas *et al.*, 1997). Very fast freezing (100–1000 mm h^{-1}) in liquefied gases such as nitrogen or carbon dioxide may cause stresses resulting in splitting or cracking of the tissue.

In general, the lower the temperature of frozen storage, the lower is the rate of microbial and biochemical changes. Freezing and frozen storage has a variable effect on micro-organisms and does not inactivate enzymes. Different types of micro-organisms vary in their resistance to low temperatures. At normal storage temperatures of $-18°C$, there is a slow loss of quality (Fellows, 1997) which for fish is mainly due to oxidation of lipids.

Factors influencing protein denaturation during freezing include salt concentration, pH, ionic strength, lipid oxidation, enzymatic reaction, surface tension and the physical effects of ice and dehydration. For many proteins, the temperature between −2°C and −10°C is critical. Freezing rate and temperature control can be used to maintain protein integrity during freezing. The rate of freezing should be high enough to prevent formation of large ice crystals in the extra-cellular spaces. The effect of frozen storage on colour, pH and texture seem to have been studied much more than the effect of freezing rate on these properties (Sun, 2006).

REFERENCES

Flesland, O. and Magnussen, O.M. (1990). Chilling of farmed fish. In: *Chilling and Freezing of New Fish Products, Preprints*. Aberdeen, September 18–20, pp. 159–164.

Plank, R. (1954). Handbuch der kältetechnik, Erster band, Entwiklung wirtschaftliche bedeutung werkstoffe, Springer-Verlag, Berlin/Göttingen/Heidelberg, p. 568.

Johnston, W.A., Nicholson, F.J., Roger, A., Stroud, G.D. (1994). *Freezing and refrigerated storage in fisheries*. FAO Fisheries Technical Paper. No. 340. Rome, FAO. p. 143.

Fennema, O.R. (1975). Freezing preservation. In: *Principles of Food Science-Part II Physical Principles of Food Preservation*. New York: Marcel Dekker, pp. 173–215.

Valentas, K.J., Rotstein, E., Singh, P.R. (1997). *Handbook of Food Engineering Practice*. Boca Raton, New York: CRC Press, p. 718.

Fellows, P.J. (1997). *Food Processing Technology, Principles and Practice*. Camebridge, England: Woodhead Publishing Limited, p. 505.

Sun, Da Wen (2006). *Handbook of Frozen Food Processing and Packaging*. Boca Raton, London, New York: CRC, Taylor & Francis, p. 737.

8 Freezing of Fruits and Vegetables

Cristina L. M. Silva, Elsa M. Gonçalves and
Teresa R. S. Brandão

8.1 INTRODUCTION

The benefits of eating fruits and vegetables are well recognised by the general consumers. Fruits and vegetables are key parts of a proper daily diet, resulting from being great sources of vitamins, mineral salts and dietary fibres with health-promoting or disease-preventing properties.

Fruits and vegetables are perishable foods with extremely rapid deterioration; this means that their stability after harvesting and during sub-sequent storage is critical (Canet, 1989). Preservation of foods usually involves technologies that prevent microbial growth as well as retard quality degradation reactions. Among such processes, freezing is a proven and efficient method. For vegetables, freezing is widely recognised as the most satisfactory method for long-term preservation, and it is an important segment of the frozen food market. The market for frozen fruits is rising slowly (Wisnow, 2006). They may suffer from the image of being second best, when compared to fresh or chilled produces. However, the growing consumer demand for healthy products that ease busy lifestyles, with little meal preparation and shopping, makes frozen fruits a fashionable market.

Raw fruits and vegetables contain large quantities of water in proportion to their weight and, consequently, the water phase change occurring in freezing makes these products more susceptible to ice crystal formation and thawing than other types of food. Due to their cellular structural characteristics, fruits are less resistant to the freezing process than vegetables. Adding to these characteristics, the fact that frozen vegetables are most often consumed cooked, with thawing and cooking operations occurring simultaneously, make quality aspects less relevant for frozen vegetables than for frozen fruits. Consumers expect more from a frozen fruit. The attractive aroma, colour, texture and freshness are strong characteristics that are difficult to dissociate from the raw product, and hence the negative impact of freezing on fruit quality is huge.

Some novel methods for rapid freezing and thawing of foods, aiming at improving quality, have been applied in frozen food technology (Li and Sun, 2002). In the particular case of fruits and vegetables, dehydrofreezing has been successfully used, since part of the water is removed before freezing, thus being an advantage for plant texture preservation. Other techniques, yet expensive and challenging, such as addition of antifreeze proteins, may be applied aiming at controlling the growth of ice crystals and recrystallisation, and thus improving textural properties.

The impact of the freezing process on product quality is enormous. However, a large number of other factors may also contribute to the final frozen product characteristics. Factors such as product type and variety, degree of ripening, raw product quality, harvesting methods, elapsed time between harvesting and processing and pre-freezing treatments are some examples (Parrenõ and Torres, 2006; Torreggiani and Maestrelli, 2006). All these factors, and/or their combinations, combined with packaging and storage conditions during the distribution chain, lead to difficulties in accurate prediction of frozen fruit and horticultural product quality.

In terms of safety, consumers have confidence in frozen products. Obviously, those safety records depend directly on the quality of the raw products prior to the freezing process, and on hygiene practices and standards along all process steps, storage and distribution.

Food industries want to diversify and to innovate whilst maintaining high quality and safety levels. Currently, the challenge for the frozen food industry is to maintain a role for frozen products in a market where the consumer prefers fresh food.

Frozen food fits well with changing consumer trends. This type of product allows great flexibility in meal preparation, and enables consumers to select healthy foods. Consumers have not only expected frozen food suppliers to meet their needs, for convenient and healthy food, but also to expand variety and improve its quality.

Quality and safety aspects and traditional and innovative freezing technologies applied to fruits and vegetables will be examined further in the following sections.

8.2 QUALITY AND SAFETY OF FROZEN FRUITS AND VEGETABLES

Quality and safety are two major food issues. Quality is a broad concept, which embraces both sensory attributes readily perceived by the human senses (such as general appearance, texture, taste and aroma), and hidden attributes (such as nutritional value, chemical constituents, and mechanical and functional properties) that may involve quantification and instrumental measurements (Shewfelt *et al.*, 1997; Abbott, 1999). Safety is related to chemical and microbiological characteristics of the product. Obviously, more than characteristics, quality and safety are requirements and standards of product excellence.

A huge variety of products are commonly named as 'vegetables'. This is not a botanical term and refers to the most diverse plant parts, such as roots (carrots), stems (asparagus), leaves (lettuce), flowers (broccoli), seeds (peas) and also fruits such as cucumbers and pumpkins. This great diversity of plant tissues and structures is one important influencing factor for the final frozen fruit and vegetable quality. Most vegetables, however, have fibrous structures that allow structural retention when they are thawed after freezing. On the other hand, fruits present softer structures, being physically much more susceptible to firmness loss (Sterling, 1968; Alonso *et al.*, 1997).

Quality and stability of frozen fruits and vegetables is influenced markedly by the Product itself, by the applied freezing Process and by Packaging. These factors are commonly referred to as PPP factors. These aspects are well documented in several handbooks of frozen foods (Deitrich *et al.*, 1977 and Torreggiani and Maestrelli, 2006 are two examples).

8.2.1 Product influence

The raw material used in the preparation of frozen fruits and vegetables is an important influencing factor, affecting both physical characteristics and nutritive value of the final product. The suitability for freezing varies greatly with agronomic practices and conditions, the involved species and varieties, the degree of ripening, and the time elapsed between harvesting and processing (Fennema, 1966; Munoz-Delgado, 1977; Canet, 1989).

It is also obvious that only raw material that is clean and sound should be selected for freezing, but this is not enough. The agronomic factors, including soil type, climatic conditions, irrigation, methods of cultivation, fertiliser composition and application, can and do have significant effect on the nutrient content and texture of the frozen product.

Many authors, such as Buss (1977), Karel (1979) and Reid (1990), have studied the influence of different cultivars of the same variety on the nutritive value, texture and flavour of the final frozen product. On the other hand, it seems that differences in cultivars can also influence the processing conditions, especially the required thermal treatment (blanching) time before freezing (Mundt and McCarty, 1960).

Other authors studied the influence of different maturity stages on composition and freezing conditions (Lisiewska *et al.*, 1999; Robertson and Sissins, 1966; Thompson *et al.*, 1983). It should be pointed out that most fruits and some vegetables intended for freezing are harvested at the same stage of maturity as for fresh consumption. However, the determination of the optimum maturity stage is very important for maximum quality retention of some species and varieties during frozen storage, particularly for fruits.

Another obvious quality loss source is the time elapsed between harvesting and freezing. Raw materials change with time and, unless careful handling, transportation, and storage procedures are used, the initial quality of the raw materials fed into a process may be lost. Products that undergo some rapid changes in chemical composition, sensory attributes and nutritional value and the ones with intense metabolism, directly related to high respiration heat (Maestrelli, 2000), should be processed as quickly as possible. Sweet corn, green peas and asparagus are some of the most susceptible vegetable products (Reid, 1990), since a severe decrease in quality may occur if the time between harvesting and processing is long.

Vegetables such as peas and lima beans, which are vined mechanically and badly bruised during the threshing process, may suffer rapid deterioration. Bruising during the vining operation brings about abnormal respiration, which is responsible for off-flavour production, just as if these vegetables had been under a low-oxygen atmosphere. Holding under refrigeration temperatures retards spoilage but, even so, they will not keep longer than 5 or 6 hours in good conditions, unless they are blanched (Tressler and Evers, 1957).

In spite of the above-mentioned studies, little has been done to relate the structure and chemical composition of the raw tissue, to the organoleptic properties of the frozen fruit or vegetable products. Research into the specific mechanisms that produce these quality losses in individual products is required, as well as in the selection and breeding of varieties more adapted to the freezing process. Research workers should turn greater attention to these aspects, with the aim of expanding the number of available varieties suitable for the freezing process.

8.2.2 Process influence

Processing often results in a marked influence on the finished product. The following subtopics consider the stages of preparatory or pre-freezing treatments, where blanching is included, and the freezing process itself.

8.2.2.1 *Pre-freezing treatments*

Preparatory or pre-freezing treatments are usually necessary to obtain ready-to-use products and to provide the best preservation conditions (Munoz-Delgado, 1977).

Grading, cleaning, sorting, removal of defective produce and inspecting, and in some cases peeling, shelling, trimming, chopping and slicing are the main common operations for fruits and vegetables (Cioubanu and Niculescu, 1976). These operations reduce microbial load, remove foreign material and minimize product variation, but may also destroy the protective barrier provided by cellular compartmentation and allow oxygen access. The resulting effects include leaching of nutrients, browning, desiccation and interaction of enzymes, which can lead to loss of nutrients, and texture and colour changes (Canet, 1989; Cano *et al.*, 1990; Reid, 1990).

Some endogenous enzymes are responsible for undesirable changes, such as off-flavours and odours, and colour and nutritive alterations, during frozen storage. The main food quality-related enzymes have been discussed by Svensson (1977). According to this author, the enzymes can be separated into four groups, related to changes in flavour, colour, texture/consistency and nutritional value. However, which are the enzymes responsible for quality deterioration in frozen vegetables, still remains a question. The amount of research work carried out trying to find a correlation between frozen vegetable's quality losses and its enzyme content is large. These deteriorative enzymes may have to be inactivated by applying particular treatments before the freezing step. Therefore, blanching is the most important pre-freezing treatment for vegetable tissues stabilisation (Fennema, 1966). Blanching is a thermal treatment designed to inactivate a target enzyme to a given extent. Normally, peroxidase, followed by catalase, and in some cases polygalacturonase or lipoxygenase, are used as criteria, due to their higher resistance to thermal treatment (Fennema, 1985). However, as mentioned above, there is no evidence that these enzymes are the main factors responsible for quality deterioration.

Blanching also affords a series of secondary benefits, which result from the complementary functions of washing, destroying vegetative cells of micro-organisms present on the surface, eliminating any remaining insecticide residues, enhancing the colour of green vegetables and eliminating off-flavours produced by gases and other volatile substances, that may have formed during the time between harvesting and processing (Fennema, 1966; Shams and Thompson, 1987).

However, since blanching is a thermal treatment, detrimental effects exist, like alteration of plant tissue and consequent texture change, solubilisation and destruction of nutrients and vitamins in the blanching medium, loss of weight and colour changes, resulting in quality loss.

Due to these changes, research has also been carried out into possible alternatives that can replace blanching, without producing adverse effects on product quality. As a result of such research, some vegetables, with a high natural flavour, low enzymatic activity or when shorter frozen storage periods are envisaged, can be preserved without blanching. This is the case for onions, green peppers, parsley, leeks and cucumbers (Leino, 1992; Munoz-Delgado, 1977). Except for these few products, blanching remains an essential step in the freezing process for vegetables, and consequently research in this area should be continued, in order to optimize procedures and reduce the adverse effects on the final product quality.

In general terms, the optimisation of the blanching process (Mundt and McCarty, 1960; Steinbuch, 1983; Selman, 1987; Reid, 1990) implies:

- a careful characterisation of the raw material, since enzyme level depends on variety, maturity, and other factors;

- product analysis prior to blanching, to identify the enzyme levels within the tissue and estimate the required heat treatment to inactivate those enzymes responsible for deleterious changes in freezing and frozen storage;
- the selection of the most adequate blanching system, and;
- the understanding and quantification of the degree to which the physico-chemical and sensorial enzymatic changes occur during the process.

Fruits cannot be subjected to a blanching treatment, due to their tissue sensitivity. Therefore, alternative pre-treatments have to used, such as chemical treatments or use of additives, to inactivate deteriorative enzymes. Barbosa-Cánovas *et al.* (2005) compiled a freezing guide for fruits, but there is still a great need for research in this field, in order to obtain better quality products.

8.2.2.2 *Freezing process*

The process of freezing requires the controlled removal of heat from the product, at a steady uniform rate, until the heat remaining in the product is equal to its equilibrium after stabilisation (Fennema, 1975).

Besides the great number of species and varieties included in the category of fruits and vegetables, these products have in common a considerable amount of water in their constitution. This characteristic makes them very sensitive to water phase changes and, consequently, to quality deterioration.

The ice crystal formation that occurs during the freezing processes tends to disrupt cellular structure. Ice crystals begin to form in the extracellular medium and progress towards the cytoplasm, after the cell membrane has lost its permeability. The ice crystals' growth causes cells to decompartmentalise, which does not allow the return of the water to the intracellular medium during thawing. Consequently, the turgidity of the cells is affected and texture may suffer pronounced damage. These alterations may also promote drip loss while thawing. Fruit tissues, being more delicate, are particularly susceptible, and the impact of freezing on cell turgidity and firmness can be disastrous (Sousa *et al.*, 2006).

In conclusion, the colour, flavour, taste, texture and aroma *bouquet* of fresh fruits and vegetables is affected strongly by the freezing technology, and these characteristics determine the product quality excellence and therefore they should be preserved.

The freezing process involves several stages (Fig. 8.1). The first stage – pre-freezing stage (a) – corresponds to the removal of heat from the food during the cooling period, when the temperature is reduced to the freezing point. This initial freezing temperature varies with product, depending on the moisture content. During this initial period sensible heat is removed from the product.

The second stage – super-cooling (b) – when the temperature falls below the freezing point, is essentially transitory and not always observed.

The third stage – freezing stage (c) – is the period of transformation of water into ice, throughout the whole mass of food. During this stage, the temperature remains constant in an ideal system, but in real situations falls slowly and continually while latent heat is extracted. The transformation of water into ice is an example of crystallisation. Crystallisation is the formation of a systematically organised phase from a solution. The crystallisation process consists of nucleation and crystal growth. Nucleation is the association of molecules into a tiny ordered particle of a size sufficient to survive and serves as a site for crystal growth.

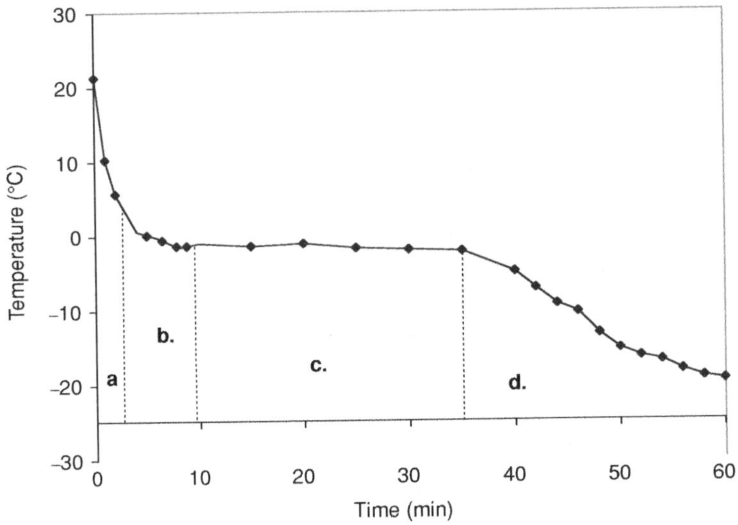

Fig. 8.1 Typical temperature history of a product throughout the freezing process.

Crystal growth is simply the enlargement of the nucleus by the orderly addition of molecules (Fennema, 1975).

The last stage – sub-freezing stage (d) – is the period where the product temperature is lowered to the end temperature, which should be the intended storage temperature. In this part of the process, mostly sensible heat is removed.

The freezing time is affected by the product size (particularly thickness) and shape, composition of the fruit or vegetable, and by the parameters of the heat transfer process and the temperature of the cooling medium. Many attempts have been made to mathematically model the freezing process, therefore the freezing time can be theoretically predicted from the system physical parameters (Holdsworth, 1968; Bakal and Hayakama, 1973; Cleland and Earle, 1984a, 1984b; Mannapperuma and Singh, 1988).

From a physical point of view, foods may be considered as dilute aqueous solutions, with a freezing point below 0°C. The freezing point depression is 1.86°C mol^{-1} L^{-1}, which means that the freezing point depends on the concentration of dissolved molecules in the water phase, and not only on the water content.

In general, the temperature range which causes most irreversible changes is from about 1°C to −5°C. Therefore, during freezing, foods should pass this temperature range reasonably quickly (Boegh-Sorensen and Jul, 1985). Tressler and Evers (1957) suggested that the solidification, the zone of the maximum crystal formation between 0°C and −3.9°C, must be passed in less than 30 minutes.

It is a characteristic of frozen foods that a high proportion of the water content is ice, and it is also well-known that the quicker the cooling process is, the smaller the ice crystals will be. During the freezing of foods, ice crystals begin to form in the liquid between the cells, and the main reason is assumed to be the higher freezing point of this extracellular liquid, compared with the intracellular liquid (Boegh-Sorensen and Jul, 1985). During slow freezing, ice crystals grow between the cells making the extracellular liquid more concentrated. The cells will lose water by osmosis, and this leads to an extensive dehydration and contraction of the cells. The result is, relatively few large ice crystals in between shrunken cells. On the other hand, during rapid freezing, heat is removed so quickly that there is little time for dehydration of the cells, and ice crystals will also be formed in the cells. The distribution and size of frozen food's ice crystals has been the subject of many investigations (Luyet, 1968; Fennema, 1975). It is easy to imagine that a slowly frozen product will lose a certain amount of water during thawing, simply because the water may be unable to return to its original position. It is equally easy to conclude that no such problems arise in a rapidly frozen product. Therefore, quick freezing leads to small ice crystals and superior quality of the frozen foods (Reid, 1983; Boegh-Sorensen and Jul, 1985; Reid, 1990), specially frozen plant materials.

During the freezing process, water is converted into ice crystals with a high degree of purity, leading to a concentrated solution of salts, minerals, and other substances. The concentration extent depends on the product, the end temperature, and also on the freezing rate. This increased solutes concentration often causes a pH change, usually towards the acid side, that can influence product quality (van den Berg, 1968).

Freezing may also cause physico-chemical changes, such as loss of water-binding capacity, resulting in drip loss; protein changes, leading to toughening or dryness; and loss of turgour. Many of these changes increase with increasing water-phase solute concentration, but may at the same time decrease with colder freezing temperature (Boegh-Sorensen and Jul, 1985).

Different kinds of cell damage can occur in the freezing processes, depending on the rate of heat removal and the water permeability. The pre-freezing stage brings an increase in the permeability of the tissue membranes, with a final loss of intracellular pressure, but the irreversible adverse effects of freezing on quality are the result of crystallisation. Fennema (1975) stated that these changes are due to ice formation, rather than to the decrease in temperature *per se*.

The water volume increase, caused by the change of state of water into ice, depends mainly on the free water content in tissues and the amount of gas in their intercellular spaces. Through the expansion, caused by freezing of water, cell tissues will be exposed to strong mechanical forces. Due to the volume increase, intercellular ice development forces the cells apart, rupturing the middle lamellae and tearing the cell wall. Therefore, freezing causes disruption of the cell membranes, considerable cell disorganisation, the major result being usually a loss in tissue firmness (Reeve, 1970; Rahman *et al.*, 1971; Fuchigami *et al.*, 1995a; Khan and Vincent, 1996).

Related to the freezing rate, as an example, Khan and Vincent (1996) concluded that for potato low freezing rates promoted more mechanical damage than fast freezing. But confusion exists about the real influence of freezing rate on quality of frozen foods. It seems that for many products the effects, which may be caused by different freezing rates, are not big enough (Fuchigami *et al.*, 1995a, 1995b; Bartolome *et al.*, 1996). In most vegetables, where freezing follows blanching, the freezing rate is of minor importance, because blanching induces marked structural changes.

The number of studies available in the literature about the effect of freezing *per se*, on the nutrient components and changes in fruits and vegetable pigments is very limited. Lopez and Williams (1985) determined the effect of freezing on the concentration of essential elements, such as cadmium and lead in frozen green beans, and concluded that there were fewer changes. Jansen (1969) reported little or no effect of freezing rate on ascorbic acid and vitamin B in peas and snap beans.

8.2.3 Packaging influence

The packaging has a determinant role on frozen fruits and vegetables quality preservation, by protecting the food from external contamination or deterioration that may occur along the distribution chain from producer to consumer. Therefore, the main packaging function is product preservation. In addition to that, an attractive package with a good design projects an image of quality.

At the same time, the packaging material cannot affect and/or contaminate the food, and the choice of the material has to be in line with the existing legislation.

The materials employed in packages for frozen foods should, first of all, possess all the usual features normally required for food packaging, namely: they should be chemically inert and stable; odour-free and not permeable to odour; free of toxic substances, that could be absorbed; impermeable to water vapour, volatile substances, and external odours; mechanically transformable into the appropriate size and shape for display in retail sales; easy to open; attractive; and able to afford protection against microbial contamination (Feinberg and Hartzell, 1968).

In addition to those requirements, packaging materials should be shaped in such a way as to promote rapid freezing of the product inside, yet resistant to food products expansion during freezing. They should also be: impermeable to liquids; resistant to moisture, weak acids and low temperatures; reflective and as opaque as possible; and permeable and resistant to microwave energy, in those cases in which reheating or cooking may be done in microwave ovens (Villalvilla, 1988).

Frozen fruits and vegetables have their own special requirements for preparatory treatments and packaging. Certain products are particularly fragile, calling for packages that can withstand the compression and shocks during production. Ultraviolet radiation, to a wavelength of 500 nm, can also catalyse certain chemical reactions that may give rise to significant colour changes, in the case of chlorophyll-containing vegetables, and makes the use of opaque packaging materials essential.

Frozen fruits and vegetables can undergo significant dehydration during storage, as a result of storage temperatures fluctuations, and water permeability of packaging. Such dehydration is irreversible, giving rise to ice formation inside the package and exerting detrimental effects on quality (changes in colour and flavour, freezer burn, increased risk of oxidation and structural deterioration). Consequently, during storage packages should ideally be air-tight, totally impermeable to water vapour, and effective as thermal insulators to limit possible temperatures fluctuations within the product (Ahvenainen and Malkki, 1984).

There are numerous papers dealing with the mechanical and physical properties of packaging material, but relatively few with the effect of packaging type on the quality and stability of frozen fruits and vegetables (Zhuang *et al.*, 1994; Orunã-Concha *et al.*, 1998).

8.2.4 Safety

Nowadays, due to the global market and consumer behaviour, the occurrence of microbiological outbreaks is common. Nevertheless, frozen foods are generally recognised as safe (Barbosa-Cánovas *et al.*, 2005). The freezing and frozen storage under proper conditions do not affect significantly the microbial level, and the frozen foodstuff's final safety record depends mainly on the quality of the raw materials, thawing conditions and final handling by consumer.

The spoilage of fresh fruits and vegetables can be caused by microbial activity. The rich nutrient composition and high water activity of these products make them attractive substrates for micro-organisms to grow. Soil and improper irrigation waters are prime sources of contamination. If the contaminants are pathogenic bacteria, viruses or parasites, consumers' health will certainly be at risk, since serious foodborne illnesses are associated with those micro-organisms.

In terms of spoilage, from a quality point of view, the genera *Bacillus*, *Clostridium*, *Corynebacterium*, *Cytophaga*, *Erwinia*, *Pseudomonas* and *Xanthomonas* are the most common bacteria. Fungi also adversely affect some produces. *Salmonella*, *Shigella*, *Escherichia coli*, *Listeria monocytogenes* and *Aeromonas* are the major threats to fruits and vegetables safety (Sumner and Peters, 1997).

For maximum hazard control and consumer defence, this microbial flora should be inactivated before the freezing stage. Disinfectant washings, thermal (blanching) or alternative non-thermal treatments (such as ozonation, ultrasonication, UV-C radiation) may be applied to reduce the risk of final frozen product contamination.

In most processing plants, conveyer belts and other equipment are the chief source of microbial contamination. Therefore, continuous cleaning of the equipment and environment is a requirement.

The freezing process *per se* affects microbial activity in foods as unfavourable conditions for microbial survival are involved (e.g. the low temperatures and the water phase change). However, this impact varies greatly with the type of micro-organisms and its physiological state, the type of fruit and vegetable and its composition, and the rates of freezing and thawing.

The level of nitrates in foods has caused concern, because of potential toxicity, and high levels of nitrites may cause methaemoglobinaemia in infants. Several authors have demonstrated that the blanching treatment can reduce plant material nitrate content significantly (Bodiphala and Ormrod, 1971; Kmiecik and Lisiewska, 1999). Blanching was shown to remove other contaminants, such as DDT (dichlorodiphenyltrichloroethane) and carbonyl residue (Elkins *et al.*, 1968), di-syston (Kleinschmidt, 1971), adrin, heptaclor epoxide, and endrin (Solar *et al.*, 1971).

8.2.5 Legislative aspects of frozen fruits and vegetables

Existing specific legislative and legal standards for frozen fruits and vegetables are mainly regulatory advisory establishments. These specific documents are guidelines for proper practice of processing and handling.

In terms of hygiene, environmental issues, working conditions, labelling, HACCP, packaging, storage and distribution, the EU legislation for frozen stuffs is common with general food production (Sørensen, 2002). The Council Directive 89/108/EEC defined the approximation of the laws of the Member States relating to quick-frozen foodstuffs for human consumption, which was then implemented through the measures Commission Directive 92/1/EEC and 92/e/EEC, respectively on the monitoring of temperatures in transport, warehousing and storage, and on laying down the sampling procedure and method of analysis for the official control of the temperatures.

Recommended International Codes of Practice for the Processing and Handling (FAO/WHO Food Standards, 2006) are for quick frozen foods in general, and deal with raw materials and preparation, freezing process, storage, transport and distribution, retail display, packaging and labelling, and hygiene topics. Only sound and wholesome raw materials should be used and should be in prime condition just before processing. After preparation, the product should be quick frozen without delay, using appropriate equipment to minimise physical, biochemical and microbial changes. Furthermore, the range of temperatures for maximum crystallisation should be passed quickly and the process is achieved only when the product temperature reaches $-18°C$ after thermal stabilisation. Thereafter, the temperature has to be kept constant and repackaging can be done under controlled conditions. Sørensen (2002) presented a critical analysis on the available EU legislation, mentioning also the US and Australian laws, and concluded that it is limited due to the enormous amount of different products, give few guidelines to producers, and more detailed information can be found in some countries.

Particularly for frozen fruits and vegetables, several Codex Standards (FAO/WHO Food Standards, 2006) exist for a good number of quick-frozen products, such as strawberries, peaches and different berries, and spinach, broccoli, brussels sprouts, peas, leek, cauliflower, green beans and potatoes. These Codex Standards emphasise good codes of practice for maximising quality aspects, and mention the allowed optional ingredients and additives.

Finally, several import-export regulations exist (e.g. the Canadian Food Inspection Agency Liaison – Preparedness and Policy Coordination, 2006).

8.3 TRADITIONAL FREEZING TECHNOLOGIES

As described before, the freezing process can be divided into two main phases, which are the pre-freezing treatments that include blanching processes, and freezing. The next subtopics summarise the traditional freezing technologies and give indications for innovative techniques, aiming at maximising the final frozen fruits and vegetables products quality.

8.3.1 Traditional preparatory treatments

The most common preparatory procedures for freezing fruits and vegetables are grading, selection, washing, peeling and, in some cases, shelling, trimming, chopping and slicing (Cioubanu and Niculescu, 1976).

Washing cleans the product of dirt and impurities and removes pesticide residues. Beuchat (2000) reported that wash-water with about 5 ppm chlorine reduced microbial populations by more than 90%, from an initial population of 10^4–10^6 CFU g^{-1}. However, the efficacy of this operation depends on pH, temperature, type of product and diversity of micro-organisms. Garg *et al.* (1990), for example, observed that dipping lettuce in water containing 300 ppm chlorine, reduced total microbial counts, by about 1000-fold, but had no effect on microbial counts on carrots.

Peeling, one of the most delicate pre-treatments, is performed industrially by abrasion, high-pressure steam or treatment with sodium hydroxide solution. The disadvantage of all these methods is the substantial raw material losses (Canet, 1989).

After washing and peeling, the product may, depending upon the product type and variety, be subjected to other procedures such as shelling, chopping and slicing. All these operations must be carried out with the utmost care, under the most stringent hygienic conditions, in order to prevent product contamination and mechanical damage. The varying degree of complexity, and automation of the preparatory treatments, requires a thorough understanding of the mechanisms according to the product type, and further research studies are desirable, in order to improve quality and optimise procedures.

Blanching is a thermal treatment, commonly applied to a variety of vegetables, with different objectives, the most important being to preserve and stabilise the products through enzyme inactivation. Blanching may be carried out in boiling water, steam, or in a combination of both. The product is heated typically by brief immersion in water, to a temperature between

Table 8.1 Freezing rate of different freezing methods.

Freezing method	Freezing rate (cm h^{-1})
Ultra rapid freezing	Over 10
Rapid freezing	1–10
Normal freezing	0.3–1
Slow freezing	0.1–0.3
Very slow freezing	less than 0.1

85°C and 100°C and for times of about 1–10 minutes, depending on the product requirements, or in steam up to 100°C.

Mechanical arrangements to obtain uniform heating of all pieces, and control of piece size, is important, to shorten the heating time for adequate blanching. The relative merits of steam and water blanching have been widely studied (Odland and Eheart, 1975; Fellows, 1988; Howard *et al.*, 1999). Although some disagreement exists, the consensus seems to be that retention of soluble nutrients is higher in steam-blanched than in water-blanched vegetables, due to less leaching losses (Fitz, 1979; Selman, 1987; Fellows, 1988; Howard *et al.*, 1999).

As described before, being a thermal treatment, blanching always presents a negative impact, and actual research is directed towards innovative technologies that try to maximise final product quality.

8.3.2 Traditional freezing methods

Because it is not very meaningful to compare freezing times for products of vastly different size, the concept of freezing rate has been introduced. Freezing rate can be expressed in several ways. Temperature change per time unit, e.g. °C s^{-1}, is sometimes used, but the temperature change will vary considerably from the surface to the centre, thus making this approach less useful for characterising certain freezing processes. Freezing rate is normally expressed, therefore, as the average velocity at which the ice front advances from the surface to the thermal centre. When depth is measured in centimetres and freezing time in hours, the freezing rate is expressed in cm h^{-1}.

The freezing methods may be characterised by the freezing rate (Table 8.1), and the usual categories are:

- Ultra-rapid freezing may be achieved by freezing small-sized products in liquid nitrogen, or carbon dioxide (cryogenic freezing).
- Rapid freezing can take place in fluidised-bed freezers and plate freezers.
- Normal freezing is found in most air-blast freezers.
- Slow freezing is found during air-blast freezing of foods in cartons.
- Very slow freezing can result when the air speed around the product, in a blast freezer, is too low, or when the air temperature is not much low, or a product is frozen in too large units.

Barbosa-Cánovas *et al.* (2005) presented a systematic review on the available equipment for freezing of fruits and vegetables.

Air-blast and multi-plate freezers are most widespread, while air fluidising systems are used for IQF (individual quick freezing) of small products. Cryogenic IQF is more restricted, because of the high price of the liquefied gases. However, the need for longer shelf-life and improved product taste and quality has motivated the development of this type of equipment (Fellows, 1988; Ramaswamy and Marcotte, 2005).

Air fluidisation (IQF) was studied extensively and has been increasingly used commercially, during the last 40 years. This freezing process has many attractive features, including:

- High freezing rate due to the small product size and thermal resistance of the IQF products, and high surface heat transfer coefficients;
- Good quality of the frozen products that have an attractive appearance and do not stick together;
- Continuity and possibilities for complete automation of the freezing process.

Notwithstanding these advantages, fluidisation freezing by air has some drawbacks as well, such as:

- Lower surface heat transfer coefficients and freezing rates in comparison to immersion methods;
- Need for a high-speed and -pressure air flow, that results in high fan energy consumption;
- Some moisture losses from the product surface and a rapid frosting of the air coolers (evaporator), caused by the great temperature differential between the products and the evaporating refrigerant;
- Excessive sensitivity of the process parameters to the product shape, mass and sizes, which requires careful control, specific for each food commodity.

The immersion freezing in non-boiling liquid refrigerating media is a well-known method, having several important advantages: high heat transfer rate, fine ice crystal system in foods, great throughput, low investments and operational costs. The immersion applications have been limited, because of the uncontrolled solute uptake by the refrigerated products and operational problems with the immersion liquids (high viscosity at low temperatures, difficult maintaining the medium at a definite constant concentration and free from organic contaminants). Recent achievements in heat and mass transfer, physical chemistry, fluid dynamics and automatic process control make it possible to solve these problems and to develop advanced innovative immersion IQF systems (Fikiin, 2003).

The selection of the most adequate system for a given product must be a balance between costs, quality and feasibility (Bebilacqua *et al.*, 2004).

8.4 INNOVATIVE TECHNOLOGIES

Freezing is unquestionably the most satisfactory method currently available for long-term preservation of fruits and vegetables. The nutrient content is largely retained and the product resembles the fresh material more closely than thermally processed foods. However, as it has been described, some changes occur that are mainly textural, and to a smaller extent losses of nutrients, colour and odour.

Therefore, there is a strong driving force to develop innovative research mainly dedicated to overcome the drawbacks of the required pre-treatments, specially blanching, and freezing process.

8.4.1 Innovative preparatory treatments

Developments have been mainly on systems that try to minimise thermal pre-treatments, or even avoid it, in order to better retain the original quality of the raw fruits and vegetables.

Several patents exist for microwave blanching, both at atmospheric and higher pressures (Cryodry Corporatio, 1969; Jeppson, 1970; Smith and Williams, 1970). The use of microwaves may not be generally advantageous in blanching, partly because of its characteristic non-uniform heating. It has been reported that microwaved vegetables were worse in colour than water or steam blanched green vegetables (Drake *et al.*, 1981). Moreover, the use of microwave may not reduce significantly the operation time. However, some advantage is seen in the use of this system for vegetables of large cross-section, such as potatoes and brussels sprouts (Selman, 1987).

It seems that the optimisation of the system can involve the joint use of microwaves and steam, to raise the product temperature rapidly, followed by a holding time controlled by steam alone (Decareau, 1985). However, there is still little information on the consequences of this alternative blanching on quality (Begum and Brewer, 2001; Ramesh *et al.*, 2002).

The use of new blanching techniques, such as electro-conductive, is still very limited (Vigerstrom, 1976; Garrote *et al.*, 1988). This may be due to the high capital cost needed to replace current water blanchers, and to the fact that more work is required to show that product quality is higher, compared to conventionally blanched products.

The increasing consumer demand for high quality standards has spurred the search for new and gentle processing technologies that prolong shelf-life without the detrimental effects caused by blanching. This fact promoted the search and the development of other methods, as efficient as blanching, to reduce the enzyme activity and microbial load on products (Piyasena *et al.*, 2003; Knorr *et al.*, 2004). Non-thermal methods have emerged as attractive alternatives to conventional methods of thermal processing, and constitute challenging methods aiming at reducing pernicious effects of thermal methods, by preserving quality and nutritional attributes of fruits and vegetables, and yielding safe and less-perishable products.

The application of ozone, ultrasounds and ultraviolet (UV-C) irradiation are examples of non-thermal technologies that may have potential applications in the food industry. Because they do not use heat, these technologies are commonly designated as non-thermal technologies. Moreover, they consume small amount of energy, and appear much more economical and environment-friendly technologies (Piyasena *et al.*, 2003).

Ozone is a gas molecules of which are formed by three oxygen atoms. In nature, this triatomic molecule is formed by the UV part (185 nm) of sunlight. Commercially, this molecule is obtained by submitting oxygen molecules to electrical discharges. This molecule is very unstable, and it rapidly dissociates, returning to its former oxygen form (Butz and Tauscher, 2002).

Ozone is considered as a potent disinfecting agent, due to its high oxidation power (Guzel-Seydim *et al.*, 2004). Studies show that ozone enables a fast inactivation of micro-organisms through the reaction with intracellular enzymes, nucleic material or membrane components, destroying the coating of spores or viral capsules (Kim *et al.*, 1999).

Disinfection is the most usual and known application of ozone. Some applications of ozone in the food industry include food preservation, surface hygienisation, sanitation, water disinfection and wastewater reutilisation (Graham *et al.*, 1969; Schneider *et al.*, 1991; Sheldon and Brown, 1986).

Ozone, applied as a gas or in dissolved water, has been tested for the post-harvest treatments of fruits and vegetables, such as apples, oranges, berries, grapes, onions, lettuce and spices

(Beuchat, 1992; Zao and Cranston, 1995; Kim *et al.*, 1999; Pérez *et al.*, 1999; Suslow, 2004). However, its use still presents some controversies, and its efficacy, concerning its application in foods, still needs to be further studied.

Ultraviolet (UV) light occupies a wide band of wavelengths in the non-ionising region of the electromagnetic spectrum, between X-rays (200 nm) and violet part of visible light (400 nm). The germicidal range is in the region of short-wave UV (UV-C), with wavelengths between 200 and 280 nm, with 254 nm being the most lethal. Exposure to low doses of UV-C has been reported to reduce post-harvest decay of fruits and vegetables (Lu *et al.*, 1987; Erkán *et al.*, 2001; Marquenie *et al.*, 2002; Allende and Artés, 2003), with potential applications in the post-harvest industry.

Ultrasound is defined as pressure waves with a frequency of 20 kHz or more (Butz and Tauscher, 2002). Ultrasound may be used at frequencies between 20 kHz and 10 MHz. Higher-power ultrasound, at lower frequencies (20–100 kHz), has the ability to cause cavitation, which has the capacity to inactivate microbes and enzymes (Knorr *et al.*, 2004; Piyasena *et al.*, 2003).

Ultrasounds technology has been increasingly used in the food industry, either for analysis or for the modification of foods. Low-intensity ultrasound provides information about physical and chemical properties, and high-intensity ultrasound is normally used to physically and chemically change the properties of foods, such as emulsification, cell disruption, chemical reaction promotion, enzyme inhibition, meat softening and modification of crystallisation processes (McClements, 1995).

The single use of ultrasounds seems not to be effective in inactivating micro-organisms in foods. However, the conjoint application of mild temperatures may enhance the ultrasonication effect (thermosonication) (Mason *et al.*, 1996; Lopéz-Malo *et al.*, 2005; Cruz *et al.*, 2006), with minor changes in terms of quality parameters, when compared to conventional thermal methods. Other combinations that also seem to be successful in terms of microbial and enzymatic inactivation are the manosonication and thermomanosonication, which combine the use of pressure, and pressure and temperature together with ultrasound, respectively (McClements, 1995; Mason *et al.*, 1996).

8.4.2 Innovative freezing methods

Recent developments in freezing technology have been characterised mainly by improvements in process control to increase freezing rate and reduce costs (Barbosa-Cánovas *et al.*, 2005).

As explained, ice crystal nucleation and growth are one of the main causes of degradation of fruits and vegetables quality during freezing. Therefore, a series of strategies for controlling ice crystals in food have been attempted, and are listed below:

- *The inhibition of nucleation.* There is an attempt to reduce freezing temperature, with the benefits of minimising chemical and physical processes, without the deleterious effects of freezing and freeze concentration. This approach has been pursued by the Rich Corporation in a series of patents and products, where the freezing point has been lowered by introducing massive quantities of osmotically active materials (Blanshard and Franks, 1987).
- *The control of nucleation.* Since ice nucleation and growth are temperature-dependent rate processes, with optima at different temperatures, the relative rates of nucleation and growth of ice crystals may be exploited, by the appropriate manipulation of the rates of heat transfer

(Diller, 1985). Examples of novel techniques are high-pressure freezing and impingement jet freezing. High-pressure freezing promotes instantaneous and homogeneous formation of ice throughout the whole volume of the product, and is one promising innovative technology to improve product's quality (Li and Sun, 2002; Van Buggenhout *et al.*, 2005). Impingement jet freezing, characterised by its high turbulence, results in a very rapid freezing process and less expensive method, compared to cryogenic freezing (Soto and Bórquez, 2001).

- *The control of ice crystal growth.* Crystals are desired, but of the right size. One approach may be to inhibit partially the accretion of water molecules at the ice interface. Alternatively, the presence and accumulation of micro- and macro-molecular additives may modify the diffusion/colligative properties at the ice crystal–water interface and, thereby, limit extensive ice crystal growth (Blanshard and Franks, 1987). Antifreeze proteins may be directly added to the product to lower the freezing temperature and slow recrystallisation rate during frozen storage (Griffith and Ewart, 1995; Li and Sun, 2002).
- *Exploitation of the glassy state.* This area has been the subject of a number of investigations. Boutron (1986) has observed that, depending on the rate of cooling in a system, the fraction of water in the ratio of vitrified to crystalline state will vary in a systematic fashion. Therefore, there is an obvious and exploitable correlation between the kinetics of ice crystallisation and the degree of cell damage, depending on whether the ice crystallises inside or outside the cells. An even more interesting development has been the conclusion of Levine and Slade (1989), who have demonstrated that glassy temperature (T_g) is a function of molecular mass, and have discussed how the use of appropriate raw materials, in a formulated food product, allow us to manipulate T_g and, therefore, promote product stability.
- *Bacterial ice nucleation.* Bacterial ice nucleation, by strains of *Pseudomonas*, *Erwinia* and *Xanthomonas*, has been both detected and investigated since the 1970s, and it has been recognised as one of the major causes of frost injury in plants. These micro-organisms are very useful in such processes where ice nucleation is a limiting step (Margaritis and Bassi, 1991; Watanabe and Arai, 1994). The application of these bacterial ice nucleators to the freezing of some model food systems, elevating nucleation temperature, may reduce freezing times and improve the quality (e.g. the flavour and textural properties) of frozen foods (Li and Lee, 1995).
- *Reduction of water content.* By reducing the product's water content, especially of fruits, the negative impact of freezing may be minimised. Dehydrofreezing is a promising new technology to reduce or avoid the negative phenomena that occurs in most frozen fruits and vegetables after storage and thawing (Maestrelli *et al.*, 2001; Li and Sun, 2002; Senesi, 2003).

REFERENCES

Abbott, J.A. (1999). Quality measurement of fruits and vegetables. *Postharvest Biology and Technology* 15(3), 207–225.

Ahvenainen, P. and Malkki, Y. (1984). Influence of packaging on the shelf life of frozen carrot, fish and ice cream. In: O. Zeuthen, J.C. Cheftel, C. Eriksson, M. Jul, H. Leniger, P. Linko, G. Varela and G. Vos, eds. *Thermal Processing and Quality of Foods*. New York: Elsevier Applied Science Publishers, pp. 528–532.

Allende, A. and Artés, F. (2003). UV-C radiation as a novel technique for keeping quality of fresh processed 'Lollo rosso' lettuce. *Food Research International* 36(7), 739–756.

Alonso, J., Canet, W. and Rodriguez, M.T. (1997). Thermal and calcium pre-treatment affects texture, pectinesterase and pectic substances of frozen sweet cherries. *Journal of Food Science* 62(3), 511–515.

Bakal, A. and Hayakama, K.I. (1973). Heat transfer in the freezing and thawing of foods. *Advance Food Research* 20, 217–256.

Barbosa-Cánovas, G.V., Altunakar, B. and Mejía-Lorío, D.J. (2005). Freezing fruits and vegetables. *FAO Agricultural Services Bulletin* 158, Rome.

Bartolome, A.P., Ruperez, P. and Fuster, C. (1996). Freezing rate and frozen storage effects on color and sensory characteristics of pineapple fruit slices. *Journal of Food Science* 61(1), 154–156.

Bebilacqua, M., D'Amore, A. and Polonara, F. (2004). A multi-criteria decision approach to choosing the optimal blanching-freezing system. *Journal of Food Engineering* 63(3), 253–263.

Begum, S. and Brewer, M.S. (2001). Physical, chemical and sensory quality of microwave-blanched snow peas. *Journal of Food Quality* 24(6), 479–493.

Beuchat, L.R. (1992). Surface desinfection of raw produce. *Dairy, Food and Environmental Sanitation* 12(1), 6–9.

Beuchat, L.R. (2000). Surface decontamination of fruits and vegetables eaten raw: a review. Food Safety & Hygiene, Food Science Australia. Available from: http://www.who.int/fsf/fos982~1.pdf [2001].

Blanshard, J.M.V. and Franks, F. (1987). Ice crystallization and its control in frozen-food systems. In: J.M.V. Blanshard and P. Lillford, eds. *Food Structure and Behaviour.* Orlando, FL: Academic Press, pp. 51–65.

Bodiphala, T. and Ormrod, D.P. (1971). Factors affecting the nitrate content of vegetables and fruit foods. *Journal of Canadian Institute of Food Technology* 4(1), 6–8.

Boegh-Sorensen, L. and Jul, M. (1985). Effects of freezing/thawing on foods. In: R.K. Robinson, ed. *Microbiology of Frozen Foods.* London and New York: Elsevier Applied Science Publishers, pp. 41–82.

Boutron, P. (1986). Comparison with the theory of the kinetics and extent of ice crystallization and of the glass-forming tendency in aqueous cryoprotective solutions. *Cryobiology* 23, 88–102.

Buss, D.H. (1977). Nutritional priorities, with particular reference to frozen foods. In: W.K. Downey, ed. *Food Quality and Nutrition.* London: Applied Science Publishers.

Butz., P. and Tauscher, B. (2002). Emerging technologies: chemical aspects. *Food Research International* 35, 279–284.

Canadian Food Inspection Agency Liaison – Preparedness and Policy Coordination. (2006). Available from: http://www.inspection.gc.ca/english/reg/rege.shtml (December 2006).

Cano, P., Marin, M.A. and Fúster, C. (1990). Freezing of banana slices. Influence of maturity, level and thermal treatment prior to freezing. *Journal of Food Science* 55(4), 1070–1073.

Canet, W. (1989). Quality and stability of frozen vegetables. In: S. Thorne, ed. *Developments in Food Preservation,*), Vol. 5. New York: Elsevier Science Publishing, pp. 1–50.

Cioubanu, A. and Niculescu, L. (1976). *Cooling Technology in the Food Industry.* In: A. Cioubanu, G. Lascu, V. Bercescu and L. Niculescu, eds. Kent: Abacus Press.

Cleland, A.C. and Earle, R.L.J. (1984a). Assessment of freezing time prediction methods. *Journal of Food Science* 49, 1034–1042.

Cleland, A.C. and Earle, R.L.J. (1984b). Freezing time predictions for different final products *Journal of Food Science* 49, 1230–1232.

Cruz, R.M.S., Vieira, M.C. and Silva, C.L.M. (2006). Effect of heat and thermosonication treatments on peroxidase inactivation kinetics in watercress (*Nasturtium officinale*). *Journal of Food Engineering* 72(1), 8–15.

Cryodry Corporatio (1969). *Microwave Heating Apparatus.* British Patent no 1: 143 272.

Decareau, R.V. (1985). *Microwave in the Food Processing Industry.* New York: Academic Press.

Deitrich, W.C., Feinberg, B., Olson, R.L., Roth, T.L. and Winter, F.H. (1977). Freezing vegetables. In: N.W. Desroier and D.K. Tressler, eds. *Fundamentals of Food Freezing.* Westport, Conneticut: AVI Publishing Company, pp. 81–134.

Diller, K.R. (1985). The influence of controlled ice nucleation on regulating the thermal history during freezing. *Cryobiology* 22, 268–281.

Drake S.R., Spayd, S.E., and Thompson, J.B. (1981). The influence of blanch and freezing methods on the quality of selected vegetables. *Journal of Food Quality* 4(4), 271–278.

Elkins, E.R., Lamb, F.C., Farrow, R.P., Kimball, J.R. and Cook, R.W. (1968). Removal of DDT, parathion, and carbaryl from spinach by commercial and home preparative methods. *Journal of Agricultural and Food Chemistry* 16(6), 967–973.

Erkán, M., Wang, C.Y. and Krizek, D.T. (2001). UV-C irradiation reduces microbial populations and deterioration in *Cucurbita pepo* fruit tissue. *Environmental and Experimental Botany* 45, 1–9.

FAO/WHO Food Standards. (2006). Available from: http://www.codexalimentarius.net/web/standard_list.jsp (December 2006).

Feinberg, B. and Hartzell, R.P. (1968). Factors affecting quality in frozen foods. In: D.K. Tressler, W.B. van Arsdel and M.J. Copley, eds. *The Freezing Preservation of Foods*, Vol. 2. Wesport, CT: AVI Publishing, pp. 260.

Fellows, P. (1988). *Food Processing Technology – Principles and Practice*. Chichester: Ellis Horwood Ltd.

Fennema, O. (1966). An over-all view of low temperature food preservation. *Cryobiology* 3(3), 197–213.

Fennema, O. (1975). Effects of freeze-preservation on nutrients. In: Roberts S. Harris and Endel Karmas, eds. *Nutritional Evaluation of Food Processing*, 2nd edn. Westport, CT: AVI Publishing, pp. 244–288.

Fennema, O.R. (1985). *Food Chemistry*. New York: Marcel Dekker.

Fikiin, K. (2003). *Novelties in Food Freezing Research in Europe and Beyond*. Technical Bulletin of the Flair Flow 4, Technical University of Sofia, Bulgaria.

Fitz, V.L. (1979). The vibratory spiral blancher-cooler. *Quick Frozen Foods International* 21(1), 124–125.

Fuchigami, M., Hyakumoto, N. and Miyazaki, K. (1995a). Programmed freezing effects on texture, pectic composition and electron microscopic structure of carrots. *Journal of Food Science* 60(1), 137–141.

Fuchigami, M., Miyazaki, K. and Hyakumoto, N. (1995b). Frozen carrots texture and pectic components as affected by low-temperature-blanching and quick freezing. *Journal of Food Science* 60(1), 132–136.

Garg, N., Churey J.J. and Splittstoesser, D.F. (1990). Effect of processing conditions on the microflora of fresh-cut vegetables. *Journal of Food Protection* 53, 701–703.

Garrote, R.L., Silva, E.R. and Bertone, R.A. (1988). Blanching of small whole potatoes in boiling water and by electroconductive heating. *Lebensmittel-wissenschaft und Tecnologie* 21(1), 41–45.

Graham, H.N., Struder, V.V. and Gurkin, M. (1969). *US patent*, 3, 484, 247.

Griffith, M., and Ewart, K.V. (1995). Antifreeze proteins and their potential use in frozen foods. *Biotechnology Advances* 13(3), 375–402.

Guzel_Seydim, Z.B., Greene, A.K. and Seydim, A.C. (2004). Use of ozone in food industry. *Lebensmittle Wissenschaft und Technologie* 37, 453–460.

Holdsworth, S.D. (1968). *Mathematical Procedures for Calculating Freezing Times for Foodstuffs and Methods of Determining Temperature Distributions in Frozen Products*. Technical Bulletion No 16, Campden Food Preservation Research Association, Chipping Campden, Glos., UK.

Howard, L.A., Wong, A.D., Perry, A.K. and Klein, B.P. (1999). β-Carotene and ascorbic acid retention in fresh and processed vegetables. *Journal of Food Science* 64(5), 929–936.

Jansen, E.J. (1969). Quality related chemical and physical changes in frozen foods. In: W.B. Van Arsdel, M.J. Copley and R.L. Olson, eds. *Quality and Stability of Frozen Foods*. New York, London, pp. 19–42.

Jeppson, M.R. (1970). Method and device for treating substances with microwave energy. West German Patent Application no 1 492 600.

Karel, M. (1979). Effect of storage on nutrient retention of foods. *Food Technology* 33(2), 26–37.

Khan, A.A and Vincent, J.F.V. (1996). Mechanical damage induced by controlled freezing in apple and potato. *Journal of Texture Studies* 27, 143–157.

Kim, J.G., Yousef, A.E. and Dave, S. (1999). Application of ozone for enhancing the microbiological safety and quality of foods: a review. *Journal of Food Protection* 62, 1071–1087.

Kleinschmidt, M.G. (1971). Rate of Di-syston (0,0-diethyl *S*-[2-(ethylthio)ethyl] phosphorodithioate) in potatoes during processing. *Journal of Agriculture and Food Chemistry* 19(6), 1196–1197.

Kmiecik, W. and Lisiewska, Z. (1999). Effect of pretreatment conditions and period of storage on some quality indices of frozen chive (*Allium schoenoprasum* L.). *Food Chemistry* 67, 61–66.

Knorr, D., Zenker, M., Heinz, V. and Lee, D.-U. (2004). Applications and potential of ultrasonics in food processing. *Trends Food Science and Technology* 15, 261–266.

Leino, M.E. (1992). Effect of freezing, freeze-drying and air-drying on odor of chive characterized by headspace gas chromatography and sensory analyses. *Journal of Agricultural and Food Chemistry* 40, 1379–1384.

Levine, H. and Slade, L. (1989). A food polymer science approach to the practice of cryostabilization technology. *Comments-on-Agricultural-and-Food-Chemistry* 1(6), 315–396.

Li, J. and Lee, T.C. (1995). Bacterial ice nucleation and its potential application in the food industry. *Trends in Food Science and Technology* (6), 259–265.

Li, B. and Sun, D.-W. (2002). Novel methods for rapid freezing and thawing of foods – a review. *Journal of Food Engineering* 54(3), 175–182.

Lisiewska, Z., Kmiecik, W. and Gebczynski, P. (1999). Effect of maturity stages on the content of ash components in raw, frozen, and canned broad beans. *Food Chemistry* 67(2), 155–162.

Lopez, A. and Williams, H.L. (1985). Essential elements and cadmium and lead in fresh, canned, and frozen green beans (*Phaseolus vulgaris* L.). *Journal of Food Science* 50, 1152–1157.

Lopéz-Malo, A., Palou, E., Jiménez-Fernández, M., Alzamora, S.M. and Guerrero, S. (2005). Multifactorial fungal inactivation combining thermosonication and antimicrobials. *Journal of Food Engineering* 67, 87–93.

Lu, J.Y., Stevens, C., Yakabu, P., Loretan, P.A. and Eakin, D. (1987). Gamma, electron beam and ultraviolet radiation on control of storage rots and quality of Walla Walla onions. *Journal of Food Processing and Preservation* 12, 53–62.

Luyet, B. (1968). Basic physical phenomena in the freezing and thawing of animal and plant tissue. In: D.K. Tressler, W.B. van Arsdel and M.J. Copley, eds. *The Freezing Preservation of Foods*, Vol. 2. Westport, CT: AVI Publishing, pp. 1–25.

Maestrelli, A. (2000). Fruit and vegetables: the quality of raw material in relation to freezing. In: C.J. Kennedy, ed. *Managing Frozen Foods*. Cambridge: Woodhead Publishing Ltd., pp. 27–55.

Maestrelli, A., Lo Scalzo, R., Lupi, D., Bertolo, G. and Torreggiani, D. (2001). Partial removal of water before freezing: cultivar and pre-treatments as quality factors of frozen muskmelon (*Cucumis melo*, cv *reticulatus* Naud.). *Journal of Food Engineering* 49(2–3), 255–260.

Mannapperuma, J.D. and Singh, R.P. (1988). Prediction of freezing and thawing times of foods using a numerical method based on enthalpy formulation. *Journal of Food Science* 53(2), 626–630.

Margaritis, A. and Bassi, A.S. (1991). Principles and biotechnological applications of bacterial ice nucleation. *Critical Revue of Biotechnology* 11(3), 277–295.

Marquenie, D., Michiels, C.W., Geeraerd, A.H., Schenk, A., Soontjen, C., Van Impe, J.F. and Nicolăi, B.M. (2002). Inactivation of conidia of *Botrytis cinerea* and *Monilinia fructigena* using UV-C and heat treatment. *International Journal of Food Microbiology* 73, 187–196.

Mason, T.J., Paniwnyk, L. and Lorimer, J.P. (1996). The uses of ultrasound in food technology. *Ultrasonication and Sonochememistry* 3, S253–S260.

McClements, D.J. (1995). Advances in the application of ultrasound in food analysis and processing. *Trends in Food Science and Technology* 6, 239–299.

Mundt, J.O. and McCarty, I.E. (1960). Factors affecting the blanching of green beans. *Food Technology* 309–311.

Munoz-Delgado, J.A. (1977). Effect of freezing, storage and distribution on quality and nutrition attributes of foods, in particular of fruits and vegetables. In: W.K. Downey, ed. *Food Quality and Nutrition: Research Priorities*. London: Applied Sciences Publishers, pp. 353–384.

Odland, D. and Eheart, M.S. (1975). Ascorbic acid, mineral and quality retention in frozen broccoli blanched in water, steam and ammonia-steam. *Journal of Food Science* 40(5), 1004–1007.

Orunã-Concha, M.J., Gonzalez-Castro, M.J., Lopez-Hernandez, J. and Simal-Lozano, J. (1998). Monitoring of the vitamin C content of frozen green beans and padrón peppers by HPLC. *Journal of the Science of Food and Agriculture* 76, 477–480.

Parrenõ, W.C. and Torres, M.D.A. (2006). Quality and safety of frozen vegetables. In: D.-W. Sun, ed. *Handbook of Frozen Food Processing and Packaging*. Boca Raton, FL: CRC Press, pp. 377–415.

Pérez, A.G., Sanz, C., Rios, J.J. and Olías, R. (1999). Effects of ozone treatment on postharvest strawberry quality. *Journal of Agricultural Food Chemistry* 47, 1652–1656.

Piyasena, P., Mohareb, E. and McKellar, R.C. (2003). Inactivation of microbes using ultrasound: a review. *International Journal of Food Microbiology* 87, 207–216.

Rahman, A.R., Henning, W.L. and Wescott, D.E. (1971). Histological and physical changes in carrots as affect by blanching, cooking, freezing, freeze drying and compression. *Journal of Food Science* 36, 500–502.

Ramaswamy, H. and Marcotte, M. (2005). *Food Processing – Principles and Application*. Boca Raton, FL: CRC Press, p. 420.

Ramesh, M.N., Wolf, W., Tevini, D. and Bognar, A. (2002). Microwave blanching of vegetables. *Journal of Food Science* 67(1), 390–398.

Reeve, R.M. (1970). Relationships of histological structure to texture of fresh and processed fruit and vegetables. *Journal of Texture Studies* 1, 247–284.

Reid, D.S. (1983). Fundamental physicochemical aspects of freezing. *Food Technology* 110–115.

Reid, D.S. (1990). Optimizing the quality of frozen foods. *Food Technology* 44(7), 78–82.

Robertson, J. and Sissins, D.J. (1966). The effect of maturity, processing, storage in the pod and cooking on the vitamin C content of fresh peas. *Nutrition* 20(1), 21–24.

Schneider, K.R., Steslow, F.S., Sierra, F.S., Rodrick, G.E. and Noss, C.I. (1991). Ozone depuration of *Vibrio vulnificus* from the southern quahog clam, *Mercenaria campechiensis*. *Journal of Invertebrate Patholology* 57(2), 184–190.

Selman, J.D. (1987). The blanching process. In: Stuart Thorme, ed. *Developments in Food Preservation*, Vol. 4. London and New York: Elsevier Applied Science Publishers, pp. 205–250.

Senesi, E. (2003). Applications of mild technology to horticultural products. Experiences from the IVTPA. *Food, Agriculture & Environment* 1(2), 150–159.

Shams, M.A. and Thompson, D.R. (1987). Qualitative determination of pea losses as affected by conventional water blanching. *Journal of Food Science* 52(4), 1006–1109.

Sheldon, B.W. and Brown, A.L. (1986). Efficacy of ozone as a disinfectant for poultry carcasses and chill water. *Journal of Food Science* 51(2), 305–309.

Shewfelt, R.L., Erickson, M.C., Hung, Y.C. and Malundo, T.M.M. (1997). Applying quality concepts in frozen food development. *Food Technology* 51(2), 56–59.

Smith, T.J. and Williams, L.G.US. (1970). Microwave blanching. Patent no 3 578 463.

Solar, J.M., Liuzzo, J.A. and Novak, A.F. (1971). Removal of aldrin, heptachlor epoxide, and endrin from potatoes during processing. *Journal of Agriculture and Food Chemistry* 19(5), 1008–1010.

Sørensen, L.B. (2002). *Frozen Food Legislation*. Bulletin of the International Institute of Refrigeration 2002–4.

Soto, V. and Bórquez, R. (2001). Impingement jet freezing of biomaterials. *Food Control* 12, 515–522.

Sousa, M.B., Canet, W., Alvarez, M.D. and Tortosa, M.E. (2006). Effect of processing on the texture and structure of raspberry (cv. *Heritage*) and blackberry (cv. *Thornfree*). *European Food Research Technology* 223, 517–532.

Steinbuch, E. (1983). Techniques for blanching of vegetables for freezing cost, energy and waste aspects. *Revue Generale du Froid* 73(1), 27–38.

Sterling, C. (1968). Effect of low temperature on structure and firmness of apple tissue. *Journal of Food Science* 33(6), 577–580.

Sumner, S.S. and Peters, D.L. (1997). Microbiology of vegetables. In: D.S. Smith, J.N. Cash, W.-K. Nip and Y.H. Hui, eds. *Processing Vegetables, Science and Technology*. Lancaster, PA: Technomic Publishing Company, pp. 87–116.

Suslow, T.V. (2004). *Ozone Applications for Post Harvest Disinfection of Edible Horticultural Crops, Division of Agriculture and Natural Resources. University of California*, Publication 8133.

Svensson, S. (1977). Inactivation of enzymes during thermal processing. In: T. Hoyem and O. Kvale, eds. *Physical, Chemical and Biological Changes in Food Caused by Thermal Processing*. London: Applied Science Publishers, pp. 202–217.

Tressler, D.K. and Evers, C.F. (1957). *The Freezing Preservation of Foods - Freezing of Fresh Foods*, Vol. I. The AVI Publishing Company, Inc. Westport, CT.

Thompson, D.R., Hung, Y.C. and Norwing, J.F. (1983). Influence of raw material properties on the freezing of sweet corn. In: B.M. McKenna, ed. *Engineering and Food*, Vol. 1. Elsevier Applied Science Publishers, pp. 299–309.

Torreggiani, D. and Maestrelli, A. (2006). Quality and safety of frozen fruits. In: D.-W. Sun, ed. *Handbook of Frozen Food Processing and Packaging*. Boca Raton, Florida: CRC Press, pp. 417–440.

Van Buggenhout, S., Messagie, I., Van Loey, A. and Hendrickx, M. (2005). Influence of low-temperature blanching combined with high-pressure shift freezing on the texture of frozen carrots. *Journal of Food Science* 70(4), S304–S308.

van den Berg, L. (1968). Physicochemical changes in foods during freezing and subsequent storage. In: J. Hawthorn and E.J. Rolfe, eds. *Recent Advances in Food Science. Vol. 4 – Low Temperature Biology of Foodstuffs*. Oxford: Pergamon Press, pp. 205–219.

Villalvilla, C. (1988). Packaging. More than protection!. *Surgelation* 26, 39–45.

Vigerstrom, K.B. (1976). Deep-fried potato products. US Patent no 3 997 678.

Watanabe, M. and Arai, S. (1994). Bacterial ice-nucleation activity and its application to freeze concentration of fresh foods for modification of their properties. *Journal of Food Engineering* 22, 453–473.

Wisnow, S.J. (2006). Frozen vegetables and fruits are good: So why aren't more of them being eaten?. *Quick Frozen Foods Intenational –*

Available from: http://www.qffintl.com/pdf/october_2006/ FrozenVegetablesandFruitsAreGood.cfm

Zao, J. and Cranston, P.M. (1995). Microbial decontamination of black pepper by ozone and the effect of the treatment on volatile oil constituents of the spice. *Journal of the Science of Food and Agriculture* 68(1), 11–18.

Zhuang, H., Barth, H. and Hildebrand, D. (1994). Packaging influenced total chlorophyll, soluble protein, fatty acid composition and lipoxygenase activity in broccoli florets. *Journal of Food Science* 59, 1171–1174.

9 Freezing of Bakery and Dessert Products

A. LeBail and H.D. Goff

9.1 BAKERY PRODUCTS

9.1.1 Bake off technology and bakery products

Bakery products (termed BCG) such as bread, viennoiserie, and many other specialities remain an important input in a well-balanced diet, being at the base of the pyramid of nutrition (http://www.usda.gov/cnpp/DietGd.pdf). Consumers prefer freshly baked products, which are traditionally prepared by baking from scratch. In addition, consumers are searching for a larger variety of products including ethnic products, health promoting products, non-allergenic products (i.e. gluten-free bread) with a high level of 'freshness' ('freshly baked') and organoleptic qualities. In parallel, the reduced availability of skilled bakers and the cost of shop space mean that traditional bakers have difficulties to match production with demand. The food industry has been proposing several alternatives to solve this problem. The corresponding technologies called 'Bake Off Technologies' (BOT) encompass several baking processes. The spirit of BOT is to keep the technological and time-consuming operations at the industrial level and the final preparation of products before retailing is carried out in small shops equipped with a minimum of equipment (thawing–proofing chamber, oven) and with minimally skilled personnel.

The market share of BOT is growing at a very high rate (more or less around 10% per year), whereas the overall bread consumption is almost constant; there is thus a transfer from traditional baking (scratch baking) towards BOT.

The main technologies of bread making are listed in Table 9.1. The three key technologies of BOT are the *Unfermented Frozen dough* (UFD), the *Partially Baked Frozen Bread* (PBF) and the *Partially Baked Unfrozen Bread* (PBUF).

UFD requires skilled personnel but can also produce products with a quality comparable to scratch baking. At least 3 hours are needed to prepare bread with UFD technology. UFD products are the most preferred frozen bakery products in France. PBF or PBUF do not require highly skilled personnel. Usually PBF and PBUF, breads are thawed before being baked in a convection oven. Partially baked (PBF and PBUF) products can be prepared in less than 1 hour but their production is energy demanding. The fermented frozen dough (F-FD) technology has not been extensively developed due to quality problems. This technology, also called 'ready to bake', is promising; however, if temperature abused in the cold chain, the surface of the dough pieces might reach the melt point and stick together, resulting in an unacceptable crust after baking.

In 2003, the European bread market was divided into 47% for traditional scratch baking, 6% for bread production in supermarket (traditional scratch baking), 36% for industrial fresh

Table 9.1 Terminology and acronyms of the conventional and Bake Off Technology bread-making processes.

Name	Acronym	Brief description of the process
Fully baked unfrozen or conventional direct bread making	FB-U	Main steps are mixing – rest – dividing – shaping – fermentation – baking – refrigeration
Fully baked frozen	FB-F	The bread is obtained from FB-U process and is frozen
Partially baked un-frozen	PB-UF	The bread is obtained from FB-U except that baking is stopped before crust colouration. Bread is cooled and is stored in packaging at room temperature
Partially baked frozen	PB-F	The bread is obtained from FB-U except that baking is stopped before crust colouration. Bread is cooled and is frozen in packaging at freezing conditions (i.e.; −20°C)
Unfermented frozen dough	U-FD	Dough is prepared as for FB-U; the rest period is shortened and care is taken to prevent excessive fermentation before freezing, which is done as soon as possible after shaping. A slow freezing is usually recommended.
Fermented frozen dough	F-FD	Dough is prepared as for FB-U; the fermentation is started and is stopped before full development. The fermented dough is then frozen. Rapid freezing is usually recommended. Baking is usually carried out in one process (frozen – ready to bake) or after thawing

Source: LeBail *et al.*, 2006. Reproduced with permission from Russell Publishing.

breads and 11% for industrial baked and part-baked products. France, Italy and Germany represent 75% of the European bread market. Concerning frozen breads, the market share in volume at the European level in 2003 was 63% for FPB, 32% for UFD and 5% for FBF. The production was divided into 6% for traditional scratch bakers, 19% for out-of-home consumption, 20% for affiliated bakers and downtown vending shops and 55% for supermarkets. Among these mean values, differences exist from one country to another. For example, in France, only 24% of fresh bread comes from industrial bakeries, whereas in UK 51% of the production of fresh bread comes from industrial bakeries (in total 88% of the overall bread production is industrial in the UK).

9.1.2 Frozen dough

Breads made from frozen dough are exposed to a global loss of volume in the final product. This problem has been abundantly mentioned and studied in the literature (Anon *et al.*, 2004). The final volume of a bread is a function of the fermentation step and from the oven rise. The expansion during fermentation, which is one of the most important phases of the bread making process, is linked to three main properties: the activity of the yeast, the rheology of the dough and the presence of miniature bubbles embedded in the dough during mixing. The expansion during baking is linked to the following key points: the condition of the dough at the end of fermentation, the baking conditions and the rheology of the dough during the baking process (dough–crumb transition and dough–crust transition at surface). Two types of studies are available in the literature focussing either on the formulation and/or on the process.

9.1.2.1 *Process conditions*

Process parameters that have been examined are, for example, the mixing conditions (Rouillé *et al.*, 2000), degree of fermentation prior to freezing (Kline and Sugihara, 1968; Zounis *et al.*, 2002), freezing temperature (Stecchini *et al.*, 2002), freezing rate (Havet *et al.*, 2000; Hsu *et al.*, 1979), storage conditions LeBail *et al.*, 1999; Ribotta *et al.*, 2004; Giannou and Tzia, 2007), thawing rate (Lucas *et al.*, 2005).

A higher bread volume was observed with a prolonged mixing time in a study that combined the impact of selected improvers (ascorbic acid and α-amylase) (Rouillé, 2000). The impact of prefreezing fermentation has been studied by numerous authors. Industrial practice has confirmed that freezing must be achieved as fast as possible after shaping of the dough to minimise the start of fermentation. Dough temperature must be kept in the range of 20°C during mixing; the higher the temperature of the dough before freezing, the lower the baking score. Zounis *et al.* (2002) found that a dough frozen at a lower temperature resulted in a better freeze tolerance. Stecchini *et al.* (2002) focus on the relationship between the freeze tolerance of a baker's yeast and dough water content, freezing, storage temperatures, and prefermentation before freezing. They observed that yeast viability was mostly affected by the freezing temperature and by the storage temperature. The freezing rate being linked to the freezing temperature (Plank, 1941), there was no clear evidence to suggest whether the freezing temperature or the freezing rate was the governing parameter to consider. The same authors observed that storage at temperatures below the glass transition temperature (T_g) gave the highest survival ratios; this result is logical, in that water mobility is strongly reduced below the glass transition temperature. Nevertheless, T_g of a dough, which is in the range, or lower than, −30°C (Ribotta and LeBail, 2007a) depends on its water content and on the formulation. Therefore, storage below T_g might appear out of scope for most sub-zero industrial mass storage.

Freezing also affects dough rheology as shown by Havet *et al.* (2000), Lu and Grant (1999) and Inoue and Bushuk (1991). This is attributed to a modification of the free:bound water ratio. Indeed, it has been shown that during storage the amount of frozen water (free water) increased with increasing storage duration (Lu and Grant, 1999). This phenomenon resulted in dehydration of the matrix and therefore in a change of the dough rheology which became more 'elastic'.

The impact of the freezing rate is often mentioned as an important parameter. A general trend confirmed by several authors is that baker's yeast is better preserved with a slower freezing rate. Neyreneuf and Delpuech (1993) used cryogenic freezing conditions and clearly demonstrated this fact. Similar results were obtained by Havet *et al.* (2000) and LeBail *et al.* (1998) using dough sticks and freezing conditions closer to industrial freezers.

One of the major inaccuracies encountered in the literature lies in the evaluation of the 'freezing rate' or 'freezing velocity'. The disparity in geometry and in mode used for the evaluation of the freezing rate do not permit direct comparisons between the available studies. The term 'freezing rate' is generic and is rather used to compare the freezing processes on a relative basis. When a numerical value has to be given to the freezing rate, authors consider it as a global value that applies to the whole geometry of the sample. In fact, the freezing rate or freezing velocity is dependent on the location considered in a given geometry. The freezing rate can be numerically represented as the velocity of the phase change front (Plank,

1941) with distance/time as a unit or as a cooling velocity during phase change with °C/time as a unit. Most authors use the latter approach using an equation (equation (9.1)) in which T is the temperature and t the time. The SI unit of equation (9.1) is K s^{-1} but most authors use K min^{-1}.

$$F_T(r) = \frac{T_1 - T_2}{t_1 - t_2}, \text{ in K min}^{-1} \tag{9.1}$$

A study on the evaluation of the freezing rate has been carried out by LeBail Apace and Cornier (1994) and has been applied to yeast activity in frozen dough (LeBail *et al.*, 1998). Results showed that the initial and final criterion of equation (9.1) should be a temperature 1°C lower than the initial freezing point and a temperature ca. 10°C below the initial freezing point, respectively, for levels '1' and '2' of equation (9.1). In the case of a cylindrical or spherical geometry, one can easily demonstrate that the velocity of the freezing front (or freezing rate) is increasing when the freezing front is travelling toward the centre of the geometry (Plank, 1941). This has an impact on the freezing process applied to dough systems. When yeast is more tolerant to slower freezing rates, then the freezing process has to be slowed down when the freezing front is inside the geometry. A specific study has been carried out by LeBail *et al.* (2001). A comparison on the impact of the duration of freezing on the bread volume was done with dough sticks (French baguettes). A higher bread volume was obtained by withdrawing dough sticks from the freezer when its centre temperature was −5°C in comparison to withdrawing the bread when its centre temperature reached −12°C. Freezing was completed by placing dough sticks in a static freezer (slow freezing). This result was explained by the fact that a slower freezing rate was applied in the first case. Grenier (2003) used magnetic resonance imaging as a tool to assess internal porosity of a dough stick during fermentation (Grenier *et al.*, 2003). The local porosity was observed in a dough stick (5 cm diameter) during fermentation. In the case of a higher freezing rate, a significant difference was observed between the dough porosity at surface and at centre. The higher density was observed at the centre location, which underwent the highest freezing rate, confirming the results observed by most authors.

Storage of frozen dough seems to have a significant impact on the yeast performance and in the baking performance of the frozen dough. Both storage temperature and storage duration are influential. Storage at a temperature lower than the glass transition of the dough seems to be preferable, even though the corresponding temperature level might be outside the scope of most industrial storage conditions (Stecchini *et al.*, 2002) as the glass transition of a dough system is in the range of −30°C. Several studies are available on the impact of storage duration (Hsu *et al.*, 1979; LeBail *et al.*, 1999; Ribotta and LeBail, 2007b; Inoue and Bushuk, 1991; LeBail *et al.*, 1998; Inoue and Bushuk, 1991; Wang and Ponte, 1995; Wang and Ponte, 1994; Takasaki and Karasawa, 1992). During storage, ice clusters affect the uniformity of the dough. During prolonged storage, water diffuses from the matrix towards ice clusters resulting in a decrease in the extensibility of the dough (Havet and LeBail, 1999) and in an increase of the amount of frozen water (Lu and Grant, 1999). The use of hydrocolloids is sometimes proposed to minimise this water diffusion. Temperature fluctuation might have an important role as shown by LeBail *et al.* (1999) who compared stable storage with storage in an interrupted cold chain. Three weeks of stable conditions were found comparable to 1 week of storage in an interrupted cold chain in terms of bread volume.

9.1.2.2 *Formulation*

Several studies are available on the impact of formulation on the freeze tolerance of dough. The basic formulations can be quite different from one study to the other, using or not, for example, shortening. The quality of the basic ingredients first seems to be an important issue. The quality of the gluten can be as important as its amount, as shown by Inoue and Bushuk (1992). Flour with higher gluten is usually recommended for frozen dough (i.e. higher then 12%). This can be obtained by using specific blends of flour or by adding vital gluten (Wang and Ponte, 1995, 1994).

Another point of interest is the yeast. Yeast amount is usually higher for frozen dough; often multiplied by a factor of 2. Nevertheless, specific studies have focused on the development of freeze tolerant yeast (Stecchini *et al.*, 2002; Baguena *et al.*, 1991; Gelinas *et al.*, 1994; Oda *et al.*, 1986). Specific properties of the yeast appear to be important such as trehalose content (Meric *et al.*, 1995) or lipid content (Murakami *et al.*, 1996; Gelinas *et al.*, 1991) for example. Improvers for baking are often used and are made of a mix of selected ingredients such as specific amylases, gluten and ascorbic acid. Such mixes are developed and adjusted constantly by specialised companies and they vary from one country to another depending on the quality of the flour (for example, humid weather during growing might develop a higher amylasic activity, dryer weather may result in more damaged starch and in possibly better 'strength' of the dough due to a modified amount and quality of gluten). Mixes are also developed for specific applications such as frozen dough and part-baked applications. Amylases facilitate the release of free sugars for the fermentation. Gluten and ascorbic acid make the rheology of the dough more tolerant to freezing. A specific study (Rouillé *et al.*, 2000) showed the combined effect of mixing time, ascorbic acid and α-amylases on selected quality parameters (fermentation rate, bread volume). They observed that the bread volume was positively correlated not only to ascorbic acid amount but also to the mixing time. Other technical parameters, such as the length of the dough stick at the end of the moulding machine, show that the dough is more elastic with a higher amount of ascorbic acid.

Shortening and emulsifiers have attracted a lot of interest (Matuda *et al.*, 2005). The most encountered emulsifiers are diacetyl tartrate ester of monoglycerides (DATEM) and sucrose esters. Such substances are most of the time combined with the use of shortening (El-Hady *et al.*, 1999). They have a combined effect and help in making the gas bubbles that are embedded in the dough during mixing, and that develop during fermentation, smaller and more numerous (Zounis *et al.*, 2002), giving the dough more tolerance to freezing. (Ribotta *et al.*, 2004, pp. 37, 38). This results also in a foamier crumb texture (numerous cells of uniform diameter). Such a 'foamy' crumb has not been researched for specific products such as French baguette (crispy rolls) for which the consumers expect a crumb with irregular alveolar structure.

Recently, the use of gums and hydrocolloids have attracted research efforts. One of the expected roles of hydrocolloids is to immobilise water in order to reduce the 'freezable' water (or free frozen water) in the dough, resulting in an improved tolerance to storage. Some researchers found that hydrocolloids can also have an effect of the retrogradation of starch (Ribotta *et al.*, 2004, p. 39). The use of hydrocolloids in frozen dough tends to improve their stability during storage. Ribotta *et al.* (2004) observed that guar gum and DATEM improved bread volume and texture and resulted in a reduced staling rate. Nevertheless, they were not able to counterbalance the negative effect of frozen storage on dough microstructure. Hydrocolloids are therefore thought mainly to interact with water distribution in the dough and with dough rheology, offering in principle a better tolerance to frozen storage.

9.1.3 Frozen part-baked bread

9.1.3.1 *Introduction*

The market share for frozen part-baked bread is growing and this product can be considered as the success story of BOT. This is probably because of its convenience and because the technology is kept under control at an industrial level. Therefore, its production can be managed with well-defined protocols and quality targets, and its final transformation requires minimally trained personnel. One may also mention that this concept of 'frozen and partially baked' applies to several other products such as frozen pizza, quiches and pies.

Literature references related to frozen part-baked bread are scarce. There are some French references related to this product but the information available remains very general (Anonymous, 1988, 2002). A French patent (LeDuff, 1985) concerns the process of partial baking. This patent is mainly related to the baking conditions; it describes an optimised baking temperature profile adapted to the production of part-baked bread. Monzie (2003) presents specific data on special breads (ethnic breads). Part-baked breads can indeed offer shops the possibility to sell several types of breads, without having the constraints (as in the case of scratch baking) of adjusting the production to the demand of consumers.

Specific problems (detachment) might arise when grains or seeds are applied onto the surface of ethnic breads. A technological study was carried out by Bonnardel and Maitre (1988) who demonstrated that partial baking process induced higher water loss than conventional direct baking (4% versus 2.5%). These authors compared different baking conditions (10 min/210°C, 15 min/190°C, 20 min/170°C) and observed that the water loss increased with the duration of the baking process. More recently, Leuschner *et al.* (1998) and Leuschner *et al.* (1999) studied the impact of storage on the microbial quality and on the moisture distribution of part-baked breads.

Quality loss during frozen storage has been assessed by Vulicevic *et al.* (2004) in a 9-month trial. The change in quality during the first month of storage was acceptable but it degraded further subsequently. The impact of frozen storage was also studied by Barcenas *et al.* (2003). Miniature dough samples were set in DSC pans and were stored at −18°C for 7, 15 and 30 days. They were baked *in situ* and were kept as 'bread' samples and stored at 4°C for 2, 4 and 7 days. Amylopectin retrogradation, which is an indicator of bread staling, increased during storage at 4°C after final baking and then stabilised after 7 days of storage. It was observed that this staling phenomenon increased with increasing storage duration. Carr and Tadini (2003) studied the impact of the amount of yeast and of vegetable shortening on the organoleptic quality and on the texture of frozen part-baked bread. Increasing the amount of yeast resulted in a more aerated crumb, whereas increasing the shortening resulted in a softer crumb. Leuschner *et al.* (1997) studied the optimisation of the baking parameters for partially baked bread.

9.1.3.2 *Crust flaking of part-baked bread*

One of the quality problems for frozen partially baked products is the quality of the crust in terms of colour, glossiness and crispiness, Sometimes, flaking off of the crust (crust flaking also called crust peeling) is mentioned by industry. There is very limited literature available on this problem. The crust is a very specific part of a crispy roll (typically a French Baguette) and may contain up to one-third of the dry matter. The hydrothermal treatment undergone by the bread during baking is applied to a crust dry area where non-gelatinised starch is still

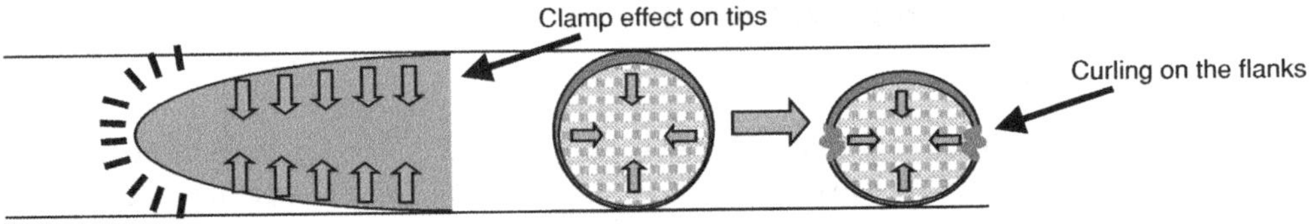

Fig. 9.1 Scheme showing the impact of crumb contraction during bread chilling and freezing. The clamp effect is developing on the tip of a crispy roll. In parallel, crumb contraction induces curling of the surface of the flank of the product, which has the thinner and therefore less mechanically resistant crust.

present. This affects the organoleptic perception of the consumer and also the nutritional value of the bread.

The impact of process conditions on crust flaking has been studied by LeBail *et al.* (2005). The impact of fermentation condition (50% and 90% air humidity at 27°C), chilling condition after partial baking (50–55% and 90–95% air humidity at 20°C) and freezing condition (blast air freezing, bread centre temperature 35°C, 45°C or 55°C at the freezer entrance) on crust flaking of part-baked bread was investigated. Crust flaking was evaluated by crushing a baguette on the top and on the sides and collecting the crust flakes. Crust flaking was quantified from the mass ratio of crust to bread. An experimental design procedure was used to compare the relative effect of the selected parameters. It was found that the chilling condition after partial baking was the most influential parameter on crust flaking. This showed that chilling is only second in order of importance as the reason for crust flaking. Other parameters were not significantly influential; nevertheless, flaking was minimised by placing the product in the freezer at −35°C rather than at a higher temperature. This study showed that crust flaking was reduced if higher air humidities were used during fermentation and chilling. The authors observed that the size of the crust flakes increase with increasing crust flaking ratio. An important observation is that even though crust flaking was only fully revealed by the final baking process, the phenomenon may be predicted and was obviously visible at the end of the freezing process. Two hypotheses are proposed by the authors to explain crust flaking;

Assumption (a): Ice concentration below the crust
Assumption (b): A thermo-mechanical problem between the crust and crumb.

The first assumption is driven by the fact that during freezing, the centre of the bread remains hot for a certain duration while the freezing front is travelling from the surface of the bread. This situation induces a strong vapour pressure gradient, which results in moisture diffusion from the warmer locations (centre) toward the cooler location (freezing front). The accumulation of ice on the freezing front might expose the crumb–crust interface to mechanical stress. This assumption has been investigated by Hamdami *et al.* (2004a, 2004b, 2004c); they found that the small amount of water collected by the freezing front was not significant enough to be considered the cause of crust flaking.

The second assumption is driven by the fact that bread contracts during chilling and freezing. This fact can be visible on baked bread especially on the flanks of a baguette (curling of the surface). Crust flaking mostly occurs on the flanks and on the tip of a baguette. The scheme proposed in Fig. 9.1 explains the clamp effect that develops on the tips and the curling of the flanks with thinner crust thickness and therefore these two result in deformation caused by crumb contraction.

The evidence of crumb contraction was first assessed using magnetic resonance imaging (MRI) by Lucas *et al.* (2005). They found MRI signal amplitude to increase below the crust during chilling and freezing; they attributed it to densification of the crumb–crust material. More recently, the impact of selected enzymes on crumb contraction was studied by Ribotta and LeBail (2007c). Relative strains corresponding to crumb contraction can reach a few percent during the chilling and freezing phases. The strain (contraction of the crumb) seems to be related to the retrogradation phenomenon which corresponds to the recrystallisation of macromolecules that are gelatinised during baking. The contraction of the crumb exposes the crust to stresses. Contraction stops once the crumb reaches the glass transition temperature. In this study, crumb contraction was studied using a Dynamic Mechanical Analyser operated as a dilatometer. A piece of crumb (8-mm thick) was exposed to a temperature scan (3°C per min) from 20°C to −70°C while crumb contraction was continuously logged.

Additional work (Ribotta and LeBail, 2007c) showed that crumb contraction seems to be related to amylose recrystallisation, which occurs during the post-baking chilling process. Specific technological aids such as enzymes or adapted processing conditions may reduce the contraction of the crumb; work on this is in progress. It also seems that the milder the baking process, the greater the contraction of crumb. Optimisation of the partial baking condition may therefore also be an objective of further research.

9.1.3.3 *Conclusion on frozen part-baked bread*

One can easily see that the production of high-quality frozen part-baked bread remains a challenge for the food industry. The concept of quality varies from one country to another. Texture of the crust is obviously a central characteristic that is difficult to control; possibly each step of the process has its own role in the final crust quality. The quality of the flour, the mixing conditions, the fermentation conditions, condition of the batter at the end of fermentation, the partial baking conditions, the chilling conditions, the freezing conditions, the storage conditions, the final baking conditions are process parameters that may impact the final quality of the crust. The formulation is also very important and may need adaptation from traditional formulations for frozen part-baked production.

9.1.4 Freezing of partially fermented dough

Partially fermented dough (also called ‘ready to bake’ or ‘ovenrise’) has appeared in the market over the last decade and has become especially popular over the last 5 years. The objective of using partially fermented dough is to make a product that may ideally be transferred directly from the freezer to the oven.

Usually conventional blast air freezing equipment is used for freezing of prefermented dough. A high freezing rate is recommended in the literature, and so an efficient freezer should be used. Cryogenic freezing may be used, even though the operating cost of such equipment is hardly justified by the low cost of the bakery products.

The degree of fermentation before freezing plays an important role. The structure of the dough, which is made of closed cells will collapse due to gas contraction during freezing and. the volume that is lost has to be recovered during the thawing and baking phases. Hanneforth *et al.* (1994) recommend a fermentation ratio of 1/3 to 1/2 in comparison to full fermentation (volume ratio between non-frozen dough and fermented dough). Research work carried out by Rasanen (1998) investigated the frozen storage stability of pre-fermented frozen bread dough during frozen storage for 14 days at −20°C. Shorter fermentation time (25 min versus

Table 9.2 Experimental conditions used to assess the impact of preproving time on bread volume in the case of partially fermented frozen bread dough.

Experiment #	Preproving time	Freezing temperature (°C)	Time to reach −18°C at centre (min)
1	1 h	−20	86
2	1 h	−40	32
3	2 h	−20	103
4	2 h	−40	39
5	1 h 30 min	−30	45
6	1 h 30 min	−30	43
7	1 h 30 min	−30	48

Source: LeBail A., unpublished.

40 min) prior to freezing improved the freeze–thaw stability of frozen dough with a 20% increase in loaf volume, a more uniform pore structure and a thicker network of gluten around gas bubbles, making them more resistant to freezing stress. No differences in storage stability of frozen dough were observed immediately after 25- and 40-minute prefermentation with most of the damage (expressed as the impact on the final bread volume) occurring during the first week of storage. A reduced moisture content (minus 2% versus optimum content in fresh dough) resulted in a higher volume. The use of hydrocolloids was also evaluated as a solution to minimise the dehydration of the dough by ice coarsening (increase in the mean size of ice crystals) during frozen storage. For further details of this work, see Rasanen *et al.* (1995) and Rasanen *et al.* (1997).

LeBail (unpublished) carried out a set of specific experiments to assess process conditions for partly fermented dough. The experimental processing conditions are presented in Table 9.2. Preproving was between 1 hours (half of optimal proving time) and 2 hours (full optimal proving time). Dough formulation was 100 g flour (type 55), 58 g water, 2.2 g salt and 3 g compressed yeast. Fermentation ('proving') was carried out in a fermentation cabinet at 30°C with a relative humidity of 80%. Pre-fermentation was either 1, 1.5 or 2 hours and the corresponding final fermentation after frozen storage was 1 hour, 30 minutes or 0 minute so that the total fermentation time (including prior to freezing and post-frozen storage) was 2 hours. This total fermentation duration of 2 hours did not include thawing time after frozen storage. Displacement transducers were used to measure the expansion of the dough during fermentation (see Fig. 9.2). A typical plot of dough expansion is presented in Fig. 9.3. Freezing was carried out in a blast air tunnel (air velocity 4 m s^{-1}) at −20°C or −40°C; dough sticks were withdrawn once the centre temperature became −20°C. The dough was thawed in a fermentation cabinet and was considered as completed once the centre of the dough stick (baguette 3.2-cm diameter after shaping–modelling) was 20°C. Storage was for 1 day at −20°C to focus on the effect of freezing *per se*. Baking was done in a ventilated oven at 230°C. Total baking duration was 15 minutes with steam injection at the beginning of baking and opening of the oven extraction after 5 minutes.

The experimental results showed that the lower the freezing rate the higher the dough collapse during freezing. In parallel, the higher the expansion ratio during the fermentation prior to freezing, the higher the dough collapse during freezing. These results are logical as rapid freezing permits the dough stick to stabilise and result in a much better preserved dough volume. A higher expansion ratio prior to freezing result in a more fragile dough with a higher gas ratio. A higher contraction is thus expected during freezing. An experimental design has been used to assess the interaction of the selected parameters. The highest bread volume was obtained with the highest freezing rate and with the lowest pre-fermentation time.

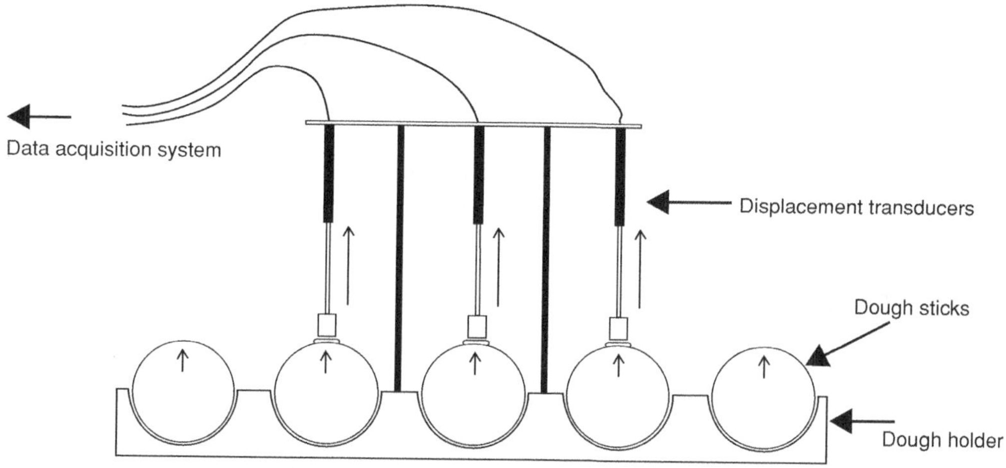

Fig. 9.2 Scheme showing a cross-section of the dough sticks installed in the dough holder. Displacement transducers were used to measure the vertical expansion of the dough stick continuously.

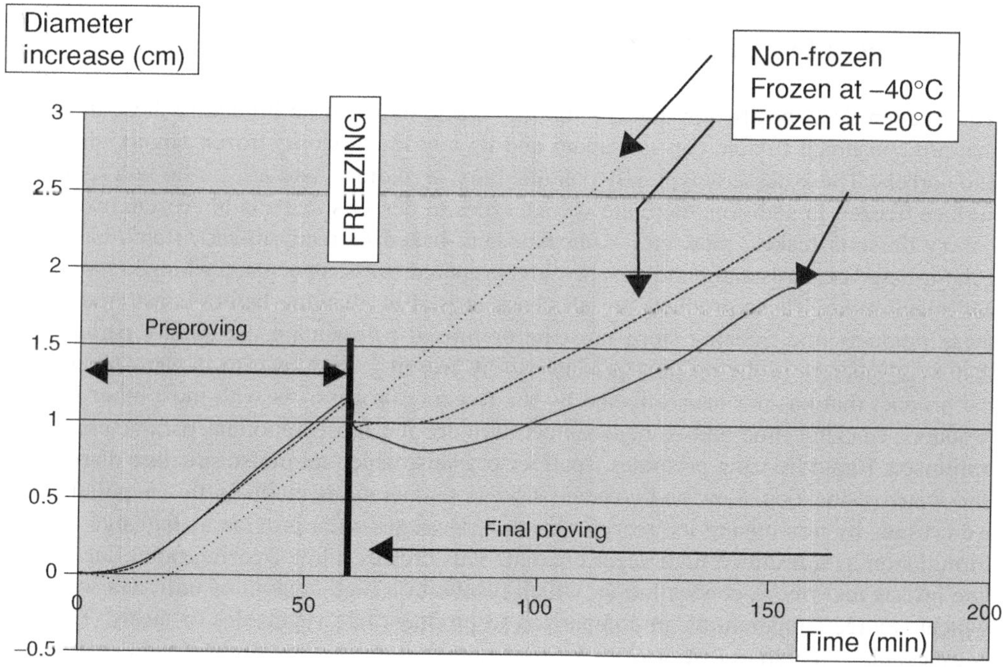

Fig. 9.3 Expansion (vertical rise) of a dough stick in the case of a prefermentation of 60 minutes followed by frozen storage of 1 day at –20°C and second fermentation of 60 minutes. The graph does not show the thawing period before the second final proving. Fermentation was stopped when the diameter increase was around 20 mm (overall volume expansion ratio was 2.5 for an initial diameter of 3.2 cm). A higher final volume at the end of the 2-hour fermentation was observed for the fastest freezing rate. A significant volume reduction was observed in comparison to direct fermentation (non-frozen conditions).

One of the other problems with pre-fermented frozen dough is that during fermentation, the dough tends to dehydrate, resulting in a reduction in the initial freezing temperature. The initial freezing temperature of a dough is around −3°C/−4°C; therefore, the initial freezing temperature of the surface of the dough stick might be reduced by a few °C due to dehydration. During frozen storage, a rupture of the cold chain might induce a partial thawing of the surface of the dough sticks, which might stick them together. Such an incident would result in unacceptable crust in the finished product.

9.1.5 Conclusion

The industry has developed several bakery technologies to keep as much technology and know-how at the industrial level as possible, so as to ensure good control of the final quality of the product. This 'Bake Off Technology' is growing by about 10% every year whereas the share of scratch baking is reducing. Nevertheless, this technology still has quality problems and has a high energy demand, approximately double that required for scratch baking. New products such as pre-fermented frozen dough have therefore appeared in the market in the last decade. Nevertheless, new innovative technologies are still expected to enhance the convenience, organoleptic quality and to reduce the energy demand of these products.

9.2 FROZEN DESSERTS

This section describes frozen desserts, but the discussion will be limited to those desserts that are consumed frozen, e.g. ice cream and its low-fat versions: frozen yogurt, sherbets and sorbets. These desserts typically contain dairy ingredients and are frozen in a scraped-surface freezer. In addition, there are several types of desserts that can be frozen, including bakery desserts (cakes, pies, etc. – unbaked, pre-baked or ready-to-eat), starch-based or gelatin-based custards and puddings, raw and prepared fruits, sweetened whipped toppings, and many more. These products are all characterised by thawing before consumption. In these products, the freezing steps must be optimised to maintain the quality parameters that were inherent in the original product before freezing, to deliver to the consumer a final product that has not been affected by the freezing process. As with most other frozen products, freezing time and storage temperature are the most important parameters to be optimised. Rapid freezing promotes small ice crystals, which minimise structure disruption and water dislocation. Low and constant storage temperatures maintain this population of ice crystals by minimising ice recrystallisation. Desserts often provide a challenge due to formulation as a result of high sugar content. This creates a low freezing point due to solute effects and freeze-concentration, which results in a high content of unfrozen water at typical storage temperatures, in comparison to products like vegetables or meats. Thus, it is especially imperative that desserts be maintained at the lowest storage and distribution temperatures possible. Functional ingredients, such as polysaccharide stabilisers or modified starches, may be used to help control water redistribution and ice recrystallisation. The rest of this section will focus on desserts that are consumed frozen, with the discussion specific to ice cream, unless otherwise noted. The reader is referred to Goff (2002, 2003, 2006), Marshall *et al.* (2003) and Goff and Hartel (2006) for further details than those provided here.

Table 9.3 A typical compositional range for the components used in various frozen dessert mix formulations.

Component	Premium ice cream	Ice cream	Light ice cream	Low fat ice cream	Sherbet	Frozen yogurt
Milk fat	14–16	10–14	6–10	2–6	0–2	0–3
Milk solids-not-fat	8–10	10–11	11–12	11–12	2–5	9–12
Sucrose	10–16	10–16	10–16	10–16	20–25	10–16
Corn syrup solids	0–4	4–6	4–6	4–6	0–10	4–6
Stabilisers/emulsifiers	0–0.3	0–0.5	0–0.5	0–0.5	0–0.5	0–0.5
Total solids (100 – water content)	40–45	36–40	32–36	28–32	25–32	28–32

9.2.1 Ingredients

The range of compositional variables found in most frozen dairy dessert mix formulations is shown in Table 9.3. The composition of frozen dairy dessert products in most countries is standardised and regulated.

9.2.1.1 *Fat*

The fat content is an indicator of the perceived quality and/or value of the ice cream. The fat component of the mix increases the richness of flavour of ice cream, produces a characteristic smooth texture by lubricating the palate, contributes to making the structure and aids in producing desirable melting properties. Milk fat – from cream, sweet (unsalted) butter, frozen cream, condensed milk blends or whey cream – is the principal and only fat source for dairy ice cream formulations. Vegetable fats can also be used as fat sources in non-dairy ice cream. Blends of oils are often used in ice cream manufacture to gain on physical characteristics, flavour, availability and cost. Palm kernel oil, coconut oil, palm oil, sunflower oil, peanut oil, fractions thereof, and their hydrogenated counterparts are all used to some extent. Regardless of the source, it is important that a considerable quantity of the fat be crystallised at the time of whipping/freezing. During freezing of ice cream, the fat emulsion that exists in the mix will partially coalesce or destabilise as a result of the presence of crystals of fat, emulsifier action, air incorporation, ice crystallisation and high shear in the freezer, thus establishing a three-dimensional network of fat throughout the product that entraps air and provides shape retention and meltdown resistance.

9.2.1.2 *Milk solids-not-fat*

The milk solids-not-fat (SNF) or serum solids contain the lactose, caseins, whey proteins, minerals (ash), vitamins, acids, enzymes and gases of the milk or milk products from which they were derived. The proteins are essential for their functional contributions of emulsification, aeration and water holding capacity/viscosity enhancement. An excess of lactose, however, may lead to problems due to excessive freezing point depression or lactose crystallisation, leading to a textural defect. Thus, SNF sources should be chosen to optimise protein functionality but limit lactose content. Traditionally, the best sources of milk SNF for high-quality products have been fresh concentrated skimmed milk or spray dried low-heat skimmed milk powder. Others include those containing whole milk protein (e.g. condensed or sweetened condensed whole milk, milk protein concentrates, dry or condensed buttermilk), those containing casein (e.g. sodium caseinate), or those containing whey proteins (e.g. dried or condensed whey, whey protein concentrate, whey protein isolate).

9.2.1.3 *Sweeteners*

Sweeteners improve the texture and palatability of ice cream and enhance flavours. Their ability to lower the freezing point of a solution imparts a measure of control over the temperature–hardness relationship. The most common sweetening agent is sucrose, alone or in combination with other sugars. In determining the appropriate blend of sweeteners to use, the sweetness, freezing point depression and contribution to total solids – all have to be considered. Sucrose and lactose are most commonly present in ice cream in the supersaturated or glassy state, with few crystals being present. In many ice cream formulations, sweeteners derived from corn syrup are substituted for all or a portion of the sucrose. The use of corn starch hydrolysis products (corn syrups or glucose solids) in ice cream is generally perceived to provide greater smoothness by contributing to a firmer and more chewy body, to provide better meltdown characteristics, to reduce heat shock potential, which improves the shelf-life of the finished product, and to provide an economical source of solids.

9.2.1.4 *Functional additives (stabilisers, emulsifiers)*

Ice cream stabilisers are a group of hydrocolloid ingredients (usually polysaccharides) used in ice cream formulations to produce smoothness in body and texture and retard or reduce the growth of ice and lactose crystals during storage, especially during periods of temperature fluctuation, known as heat shock. Thus, by physical means, they effectively increase the shelf-life of ice cream. They also increase the viscosity of the mix, aid in suspension of flavouring particles in the semi-frozen ice cream, produce a stable foam with easy cut-off and stiffness at the barrel freezer for packaging, slow down moisture migration from the product to the package or the air in frozen product, and contributes to preventing shrinkage of the product volume during storage. Stabilisers commonly used include: locust bean (carob) gum, guar gum, carboxymethyl cellulose, sodium alginate, xanthan, and gelatin. Each stabiliser has its own characteristics and often two or more of these stabilisers are used in combination to lend synergistic properties to each other and improve their overall effectiveness. Carrageenan is a secondary hydrocolloid used to prevent serum separation in the mix, which is usually promoted by one of the other stabilisers. Hence, it is found in most stabiliser blends

Small-molecule surfactants (emulsifiers) are usually integrated with the stabilisers in proprietary blends but their function and action are very different from those of the stabilisers. They are used to: (a) improve the aeration properties of the mix; (b) produce an ice cream at extrusion with good shape retention properties (referred to as 'dryness') to facilitate moulding, fancy extrusion and novelty product manufacture; (c) produce a smoother body and texture in the finished product; and (d) produce a product with good shape-retention properties during melting. Their mechanism of action can be summarised as follows: they lower the fat–water interfacial tension in the mix, resulting in protein displacement from the fat globule surface, which in turn reduces the stability of the fat globule to partial coalescence that occurs during the whipping and freezing process, leading to the formation of an aggregated fat structure in the frozen product which contributes greatly to texture and melt-down properties. Emulsifiers used in ice cream manufacture are of two main types: mono- and di-glycerides and sorbitan esters. Of the latter, polysorbate 80 is a very strong promoter of fat destabilisation in ice cream and is used in many commercial stabiliser/emulsifier blends.

9.2.1.5 *Flavours*

Liquid flavours and colours, those that are to be homogeneous throughout the frozen product, are added to the flavour tank prior to scraped-surface freezing while particulate flavours are added post-freezing. While there are numerous flavours that exist for ice cream, there are a few limitations, such as the effect of the flavour addition on protein stability in the mix (ethanol, for example), the effect on freezing point depression (in the case of high sugar flavours), the water activity balance between the frozen ice cream and the ingredient (as it affects water migration), hardness of the flavour ingredient at product consumption temperatures, etc.

9.2.2 Processes

The manufacturing process for most of these products is similar. It involves the preparation of a liquid mix; whipping and freezing this mix dynamically under high shear to a soft, semi-frozen slurry; incorporation of flavouring ingredients to this partially frozen mix; packaging the product, and further freezing (hardening) of the product under static, quiescent conditions. Swept (scraped)-surface freezers are used for the first freezing step, while forced convection freezers, such as air blast tunnels or rooms, or plate-type conduction freezers are used for the second freezing step.

9.2.2.1 *Mixing, pasteurisation, homogenisation and aging*

Ice cream processing operations can be divided into two distinct stages: mix manufacture and freezing operations (Fig. 9.4). The manufacture of ice cream mix involves the following unit operations: combination and blending of ingredients, batch or continuous pasteurisation, homogenisation and ageing. Ingredients are chosen to supply the desired components, e.g. cream or butter to supply fat, on the basis of availability, ease of handling, desired quality and cost. An algebraic solution of the formulation is required, since many of the ingredients supply more than one component.

Pasteurisation is designed to kill pathogenic bacteria. In addition, it serves a useful role in reducing the total bacterial load and in solubilising some of the components (proteins and stabilisers). Both batch (>~69°C for ~30 min) and continuous (high temperature–short time, HTST, >~80°C for ~15–25 s) systems are in common use.

Homogenisation is responsible for the formation of fat emulsion by forcing the hot mix through a small orifice under pressures of 15.5–18.9 MPa (2000–3000 psi gauge), depending on the composition of the mix. Small fat globules are needed to establish the appropriate degree of fat partial coalescence. A large increase in the surface area of the fat globules is responsible in part for the formation of the fat globule membrane, comprised of adsorbed materials that lower the interfacial free energy of the fat globules. With single-stage homogenisers, fat globules tend to cluster as bare fat surfaces come together or adsorbed molecules shared. Therefore, a second homogenising valve is frequently placed immediately after the first with applied back pressures of 3.4 MPa (500 psi gauge), allowing more time for surface adsorption to occur.

An ageing time of 4 h or longer following mix processing prior to freezing is recommended to produce a smoother texture and better quality product. The temperature of the mix should

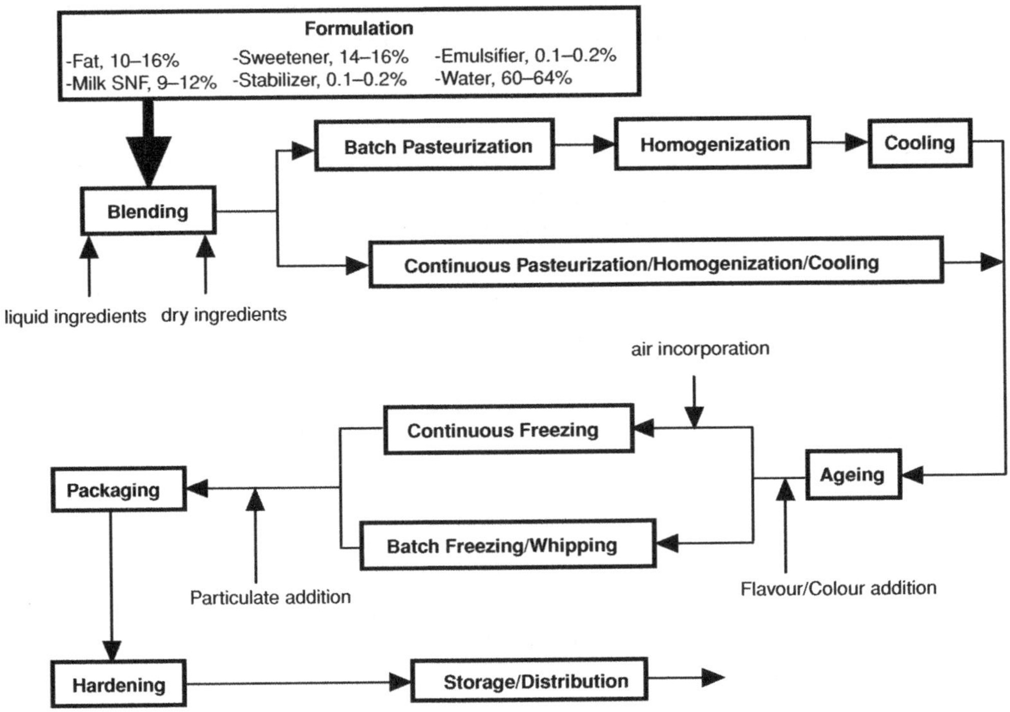

Fig. 9.4 Flow diagram of the ice cream manufacturing process.

be maintained as low as possible without freezing (≤4°C). Ageing allows for hydration of milk proteins and stabilisers (some increase in viscosity occurs during the ageing period), crystallisation of the fat globules and membrane rearrangement. The appropriate ratio of solid:liquid fat must be attained at this stage, a function of temperature and the triglyceride composition of the fat used, as a partially crystalline emulsion is needed for partial coalescence during the whipping and freezing steps. Emulsifiers generally displace milk proteins from the fat surface during the ageing period.

9.2.2.2 *Dynamic freezing*

Ice cream freezing also consists of two distinct stages: passing mix through a swept (scraped)-surface heat exchanger under high shear conditions to promote extensive ice crystallisation and air incorporation; and freezing the packaged ice cream under conditions that promote rapid freezing and small ice crystal sizes. The concomitant freezing and whipping process is one of the most important unit operations for the development of quality, palatability, and yield of the finished product. Flavouring and colouring can be added as desired to the mix prior to passing through the barrel freezer, and particulate flavouring ingredients, such as nuts, fruits, candy pieces, or ripple sauces, can be added to the semi-frozen product at the exit from the barrel freezer prior to packaging and hardening.

Continuous freezers dominate the ice cream industry. In this type of process, mix is drawn from the flavouring tank into a swept-surface heat exchanger (Fig. 9.5), which is jacketed with a liquid, boiling refrigerant (e.g. ammonia). Following the incorporation of air into the

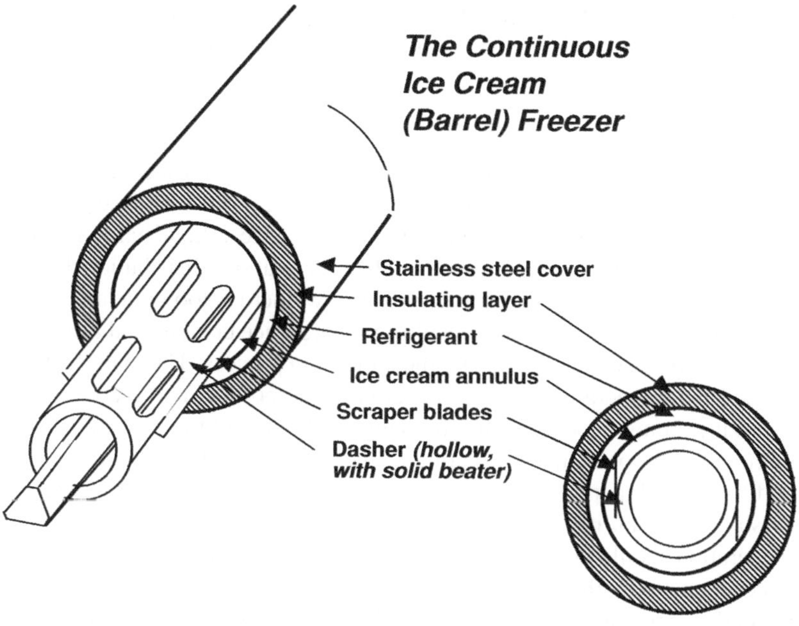

Fig. 9.5 Schematic diagram of a continuous dynamic ice cream freezer. Mix enters at the rear and freezes on the cold wall as it passes through the ice cream annulus. Continuous scraping of the wall mixes the ice crystals from the frozen layer into the bulk, where they grow. Agitation from the rotating dasher also disperses the air into small foam bubbles.

mix, the water in the mix is partially frozen as the mix and air combination passes through the barrel of the scraped-surface heat exchanger. Rotating knife blades scrape the ice layer off the surface and dashers keep the product agitated, which incorporates the air phase as tiny bubbles and maintains discrete ice crystals as they grow in the bulk liquid. Residence time for mix through the annulus of the freezer varies from 0.4 to 2 minutes, freezing rates can vary from 5 to 27°C min^{-1}, and draw temperatures of −6°C can easily be achieved.

Incorporation of air into ice cream, defined by its 'overrun' or increase in volume, as a percent of mix volume, is a necessity to produce a desirable body and texture. As a guide, maximum overrun should be 2.5–3 times the total solids content to avoid possible defects in the finished ice cream. In older systems, the pump configuration resulted in a vacuum either at the pump itself or on the mix line entering the pump. Air was then incorporated through a spring-loaded, controllable needle valve. Newer types of freezers utilise compressed air, which is injected into the mix. This type of air handling system allows for filtration of the air prior to entering into the mix and for better control of overrun.

Batch freezing processes differ slightly from the continuous systems just described. The barrel of a batch-swept surface heat exchanger is jacketed with refrigerant and contains a set of dashers and scraper blades inside the barrel. It is filled to about one-half volume with the liquid mix. Barrel volumes usually range from 2 to 12 L. The freezing unit and agitators are then activated and the product remains in the barrel for sufficient time to achieve the desired degree of overrun and stiffness. Whipping increases with time and cannot exceed the amount that will fill the barrel with product (i.e. 100% overrun when starting half full). Batch freezers are used in smaller operations where it is desirable to run individual flavoured mixes on a small scale or to retain an element of the 'homemade'-style manufacturing process. They are also operated in a semi-continuous mode for the production of soft-serve type desserts. A hopper containing the mix feeds the barrel as product is removed.

9.2.2.3 *Static freezing*

Following dynamic freezing, ingredient addition, and packaging, the ice cream is immediately transferred to a hardening chamber (−30°C or colder, either forced convection or plate-type conduction freezers) where most of the remaining water freezes. Rapid hardening is necessary for product quality, as it helps to maintain the small ice crystal size distribution that was created in the scraped-surface freezer. When hardening is slow, there is too great an opportunity for small ice nuclei formed to recrystallise, resulting in larger ice crystals and a coarser product. Many factors need to be considered during the hardening process. The main factors affecting heat transfer are the temperature difference between the product and the freezing medium, the area of product being exposed to the freezing medium, and the heat transfer coefficient for the particular operation. The temperature of the ice cream when placed in the hardening room should be as cold as possible. Draw temperatures from the barrel freezer are limited by the necessity of flowability for packaging the product. The addition of ingredients and the packaging operation should not increase the temperature of the ice cream as it is drawn from the barrel freezer any more than necessary. The temperature of the hardening chamber is also critical for rapid freezing and smooth product. The surface area of the ice cream also needs to be considered, and is especially important when packaging in large packages or in shrink-wrapping product bundles. Palletising or stacking of product should not interfere with rapid air circulation and fast freezing. Convective heat transfer coefficients are greatly increased through the use of forced convection systems. The evaporator should be free of frost from the outside and oil from the inside of the coils as these act to reduce the heat transfer coefficients. Following rapid hardening, ice cream storage should occur at low, constant temperatures, usually at −25°C.

9.2.3 Structure and stability

The whipping and freezing processes occurring concomitantly in the scraped-surface freezer together account for the formation of the freeze-concentrated, unfrozen, continuous phase and all of the discrete phases of ice cream structure: the ice crystals, the air bubbles and the partially coalesced fat globule structure (Fig. 9.6) (Goff *et al.*, 1999; Muse and Hartel, 2004). This structure, in turn, dictates the final texture as consumed. The ice phase is of critical importance to the quality and shelf-life of frozen products. The objective of ice cream manufacturers is to produce ice crystals of size that is below, or at least not significantly above, the threshold of sensory detection at the time of consumption: 40–50 μm. Consequently the freezing steps of the manufacturing process and the temperature profile throughout the distribution system are the critical factors in meeting this objective.

The dissolved sugars, lactose, and salts, result in an initial freezing temperature in the mix of about −2.5°C, depending on their concentration. From Raoult's law, the freezing point is a function of both the concentration and the molecular weight of the solutes. As water freezes out of solution in the form of pure ice crystals, solutes are excluded from the growing ice and concentrate in an ever-decreasing amount of water. Water and its dissolved components are referred to as the serum or matrix of the mix. Because the freezing point of the serum is a function of the concentration of dissolved solids, the formation of more ice concentrates the serum and results in an ever-decreasing freezing temperature for the remaining serum.

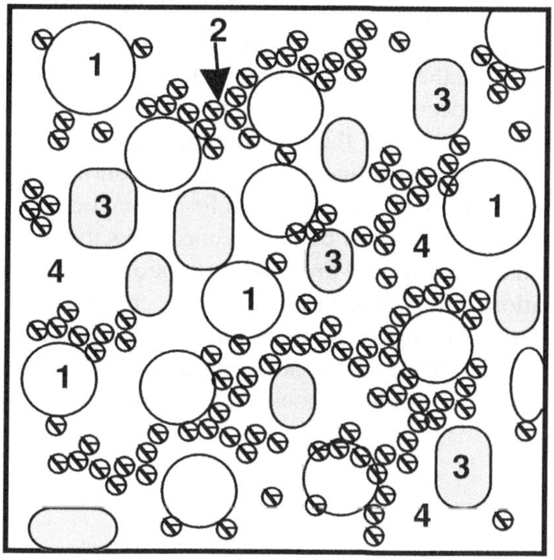

Fig. 9.6 Schematic diagram of the physical structure of ice cream: (1) air cells, (2) fat globules and partially coalesced fat surrounding air cells, (3) ice crystals, (4) unfrozen, freeze-concentrated solution of dissolved and suspended solids.

Thus at temperatures several degrees below the initial freezing temperature, there is always an unfrozen phase present. Ice cream hardness is a function of temperature due to its effect on this conversion of unfrozen water to ice and further concentration of the serum phase surrounding the ice crystals, which gives ice cream its property that it can be scooped and chewed at freezer temperatures. Composition-based freezing curves for ice cream can readily be calculated (Marshall *et al.*, 2003; Livney *et al.*, 2003).

The dynamic whipping and freezing process is also responsible for the formation of a fat network or structure in the product. Ice cream is both an emulsion and a foam. The milk fat exists in tiny globules that have been formed by the homogeniser. There are many proteins that act as emulsifiers and give the fat emulsion its needed stability. The emulsifiers actually reduce the stability of this fat emulsion, because they replace proteins on the fat surface. When the mix is subjected to the whipping action of the scraped-surface freezer, the fat emulsion begins to partially coalesce and the fat globules begin to flocculate (Fig. 9.6). The air bubbles that are being beaten into the mix are stabilised by this partially coalesced fat. This process is similar to that which occurs during the whipping of heavy cream as the liquid is converted to a semi-solid with desirable stand-up qualities and mouthfeel. In ice cream, the emulsion destabilisation phenomenon also results in desirable textural qualities both at the time of draw from the barrel freezer and during consumption, resulting in a smoother, creamier product with a slower meltdown.

During storage, the ice crystals that were created during manufacture undergo morphological changes in number, size, and shape. This is known collectively as recrystallisation and leads to a coarse, icy texture and the defect of iciness. Recrystallisation is probably the most important change producing quality losses and limitations in shelf-life. It also probably accounts for countless lost sales through customer dissatisfaction with quality. The majority of recrystallisation is stimulated by fluctuating temperatures (known as heat shock) and can be minimised by maintaining a low, constant storage temperature. Heat shock occurs readily in ice cream products. If the temperature during the frozen storage increases, some of the ice crystals melt, particularly the smaller ones as they have the highest free energy and lowest melting point. Consequently, the amount of unfrozen water in the serum phase increases. Conversely, as temperatures decrease, water does not renucleate but rather refreezes on the surface of larger crystals, resulting in the total number of crystals diminish and the mean crystal size increase. Temperature fluctuations are common during frozen storage, due to the cyclic nature of refrigeration systems and the need for automatic defrost. However, mishandling of product is probably the biggest culprit. If one tracks the temperature history of ice cream during distribution, retailing, and finally consumption, one would find a great number of temperature fluctuations. Each time the temperature changes, the ice to serum content changes, and the smaller ice crystals decrease in numbers or disappear while the larger ones grow even larger, leading to a change in the mean crystal size distribution. Proper formulation with stabilisers designed to combat heat shock is an essential defence against the inevitable growth of ice crystals.

REFERENCES

Anon, M.C., Le-Bail, A. and Leon, A.E. (2004). Effect of freezing on dough ingredients. In: Y.H. Hui, *et al.* eds. *Handbook of Frozen Foods*. New York: Marcel Dekker., pp. 571–580.

Anonymous. (2002). Le Prêt à cuire, une révolution prometteuse. *Filière Gourmande* 93, 24–26.

Anonymous. (1988). Précuit surgelé-précuit frais. *Stratégies Gourmandes* June 1988: pp. 62–68.

Baguena, R., Soriano, M.D., Martinez, A.M.A. and Barber, C.B.D. (1991). Viability and performance of pure yeast strains in frozen wheat dough. *Journal of Food Science* 56(6), 1690–1694.

Barcenas, M.E., Haros, M., Benedito, C. and Rosell, C.M. (2003). Effect of freezing and frozen storage on the staling of partbaked bread. *Food Research International* 36, 863–869.

Bonnardel, P. and Maitre, H. (1988). La technologie boulangère du pain français précuit. *Industries des Céréales* 51(février 88), 31–37.

Carr, L.G. and Tadini, C.C. (2003). Influence of yeast and vegetable shortening on physical and textural parameters of frozen part baked French bread. *Lebensmittel-Wissenschaft und Technologie* 36, 609–614.

El-Hady, E.A., El-Samahy, S.K. and Brümmer, J.M. (1999). Effect of oxidants, sodium-stearoyl-2-lactylate and their mixtures on rheological and baking properties of nonprefermented frozen doughs. *Lebensmittel-Wissenschaft und Technologie* 32(7), 446–454.

Fanta, G.F. and Christianson, D.D. (1996). Starch Hydrocolloids composites prepared by steam jet cooking. *Food Hydrocolloids* 10(2), 173–178.

Gelinas, P., Fiset, G., Willemot, C. and Goulet, J. (1991). Lipid content and cryotolerance of baker's yeast in frozen dough. *Applied and Environmental Microbiology* 57(2), 463–468.

Gelinas, P., Lagimoniere, M. and Rodrigue, N. (1994). Performance of cream or compressed yeast in frozen and nonfrozen doughs. *Cereal Chemistry* 71(2), 183–186.

Giannou, V. and Tzia, C. (2007). Frozen dough bread: Quality and textural behavior during prolonged storage – Prediction of final product characteristics. *Journal of Food Engineering* 79(3), 929–934.

Goff, H.D. (2006). Ice cream. In: P.F. Fox and P.L.H. McSweeney, eds. *Advanced Dairy Chemistry – 2. Lipids*, 3rd edn. New York: Kluwer Academic, pp. 441–450.

Goff, H.D. and Hartel, R.W. (2006). Ice cream and frozen desserts. In: Y. H. Hui, ed. *Handbook of Food Science, Technology and Engineering*, Vol. 4 Boca Raton. FL: CRC/Taylor & Francis, pp. 154–1 to 154–48.

Goff, H.D. (2002). Formation and stabilization of structure in ice cream and related products. *Current Opinion Colloid Interface Science* 7, 432–437.

Goff, H.D. (2003). Ice cream. In: P.F. Fox and P.L.H. McSweeney, eds. *Advanced Dairy Chemistry – 1. Proteins*, 3rd edn. New York: Kluwer Academic, pp. 1063–1082.

Goff, H.D., Verespej, E. and Smith, A.K. (1999). A study of fat and air structures in ice cream. *International Dairy Journal* 9, 817–829.

Grenier, A. (2003). *Caractérisation par IRM et modélisation de transferts thermiques dans une matrice à structure alvéolée évolutive. Application à la congélation et décongélation-pousse d'un pâton de pain*, in *Sciences pour l'Ingénieur*. Nantes: University of Nantes, p. 150.

Grenier, A., Lucas, T., Collewet, G. and Le Bail, A. (2003). Assessment by MRI of local porosity in dough during proving. theoretical considerations and experimental validation using a spin-echo sequence. *Magnetic Resonance Imaging* 21(9), 1071–1086.

Hamdami, N., Monteau, J.Y. and Le-Bail, A. (2004b). Heat and mass transfer in par-baked bread during freezing. *Food Research International* 37, 477–488.

Hamdami, N., Monteau, J.Y. and Le-Bail, A. (2004a). Thermophysical properties evolution of French partly baked bread during freezing. *Food Research International* 37(7), 703–713.

Hamdami, N., Monteau, N. and Le-Bail, A. (2004c). Heat and moisture transfer simulation during freezing of bread. In: *ICEF9 – 2004 International Conference Engineering and Food*. Montpellier, France.

Hanneforth, U., Brack, G. and Valerius, U. (1994). *Herstellung von Hefefeingebaeck aus vorgegaerten Tiefkuehlteigen* (in German).Title in English :*Manufacture of fine yeasted products from prefermented frozen dough.* Getreide, Mehl und Brot 48(4), 36–42.

Havet, M. and LeBail, A. (1999). Frozen bread dough: impact of the freezing rate and the storage duration on gassing power. In: *Workshop 'Process Engineering of Cereals'*. Montpellier, France.

Havet, M., Mankai, M. and Le, B.A. (2000). Influence of the freezing condition on the baking performances of French frozen dough. *Journal of Food Engineering* 45(3), 139–145.

Hsu, K.H., Hoseney, R.C. and Seib, P.A. (1979). Frozen Dough. III. Effects of Freezing and Storing Conditions on the Stability of Yeasted Doughs. *Cereal Chemistry* 56(5), 424–426.

Inoue, Y. and Bushuk, W. (1991). Studies on frozen bread doughs. I: Effects of frozen storage and freeze-thaw cycles on baking and rheological properties. *Cereal Chemistry* 68, 627–631.

Inoue, Y. and Bushuk, W. (1991). Studies on frozen doughs. I. Effects of frozen storage and freeze-thaw cycles on baking and rheological properties. *Cereal Chemistry* 68(6), 627–631.

Inoue, Y. and Bushuk, W. (1992). Studies on frozen doughs. II. Flour quality requirements for bread production from frozen dough. *Cereal Chemistry* 69(4), 423–428.

Inoue, Y., Sapirstein, H.D. and Bushuk, W. (1995). Studies on frozen doughs. IV. Effect of shortening systems on baking and rheological properties. *Cereal Chemistry* 72(2), 221–225.

Kline, L. and Sugihara, T.F. (1968). Frozen bread doughs prepared by the straight dough method. *The Bakers Digest* 42(5), 45–50.

LeBail, A. and Cornier, G. (1994). Control of the cooling rate during freezing : which cooling rate and which control parameters. *Application to bread dough*. In: *ACOFOP III*. Paris.

LeBail, A., Grinand, C., Le-Cleach, S., Martinez, S. and Quilin, E. (1999). Influence of storage conditions on frozen French bread dough. *Journal of Food Engineering* 39, 289–291.

LeBail, A., Monteau, J.Y., Margerie, F., Lucas, T., Chargelegue, A. and Reverdy, Y. (2005). Impact of selected process parameters on crust flaking of frozen part baked bread. *Journal of Food Engineering* 69, 503–509.

LeBail, A., Havet, M. and Hayert, M. (2001). Improvement of the baking performance of frozen dough by short freezing method. Application to French baguette. In: *Rapid Cooling of Foods*. Bristol, UK: International Institute of Refrigeration.

LeBail, A., Havet, M. and Pasco, M. (1998). Influence of the freezing rate and of storage duration on the gassing power of frozen bread dough. In: *Symposium of the International Institute of Refrigeration*. Nantes, France: International Institute of Refrigeration.

LeBail, A., Hamdami, N., Jury, V., Monteau, J.Y., Davenel, A., Lucas, T. and Ribotta, P. (2006). *Examining Crust Problems Resulting from Processing Conditions with Respect to Freezing*, Vol. 3. *New Food Magazine* – Russell Publishing pp. 10–14.

LeDuff, L. (1985). *Pain Français précuit congelé et son procédé de fabrication*.

Leuschner, R.G.K., O'Callaghan, M.J.A. and Arendt, E.K. (1998). Bacili spoilage in part-baked and rebaked brown soda bread. *Journal of Food Science* 63(5), 915–918.

Leuschner, R.G.K., O'Callaghan, M.J.A. and Arendt, E.K. (1999). Moisture distribution and microbial quality of partbaked breads as related to storage and rebaking conditions. *Journal of Food Science* 64(3), 543 546.

Leuschner, R.G.K., O'Callaghan, M.J.A. and Arendt, E.K. (1997). Optimization of baking parameters of part-baked and rebaked Irish brown soda bread by evaluation of some quality characteristics. *International Journal of Food Science and Technology* 32, 487–493.

Livney, T., Verespej, E. and Goff, H.D. (2003). On the calculation of ice cream freezing curves. *Milchwissenschaft* 58, 640–642.

Lu, W. and Grant, L.A. (1999). Effects of prolonged storage at freezing temperatures on starch and baking quality of frozen doughs. *Cereal Chemistry* 76(5), 656–662.

Lucas, T., Quellec, S., Le-Bail, A. and Davenel, A. (2005). Chilling and freezing of part-baked breads. II. Experimental assessment of water phase changes and of structure collapse. *Journal of Food Engineering* 70(2), 151–164.

Marshall, R.T., Goff, H.D. and Hartel, R.W. (2003). *Ice Cream*, 6th edn. New York: Kluwer Academic/Plenum Publishers.

Matuda, T.G., Parra, D.F., Lugao, A.B. and Tadini, C. (2005). Influence of vegetable shortening and emulsifiers on the onfrozen water content and textural properties of frozen French bread dough. *Lebensmittel-Wissenshaft und Technologie*, 38, 275–280.

Meric, L., Lambert, G.S., Neyreneuf, O. and Richard, M.D. (1995). Cryoresistance of baker's yeast *Saccharomyces cerevisiae* in frozen dough: contribution of cellular trehalose. *Cereal Chemistry* 72(6), 609–615.

Monzie, S. (2003). Course à l'innovation. *Filière gourmande* 95(Février-Mars 2003), 32–33.

Murakami, Y., Yokoigawa, K., Kawai, F. and Kawai, H. (1996). Lipid composition of commercial baker's yeasts having different freeze-tolerance in frozen dough. *Bioscience, Biotechnology, and Biochemistry* 60(11), 1874–1876.

Muse, M.R. and Hartel, R.W. (2004). Ice cream structural elements that affect melting rate and hardness. *Journal Dairy Science* 87, 1–10.

Neyreneuf, O. and Delpuech, B. (1993). Freezing experiments on yeasted dough slabs. Effects of cryogenic temperatures on the baking performance. *Cereal Chemistry* 70(1), 109–111.

Oda, Y., Uno, K. and Otha, S. (1986). Selection of yeasts for breadmaking by the frozen-dough method. *Applied Environmental Microbiology Review* 11, 339–356.

Plank, R. (1941). Bietrage zur berechnung und bewertung der gefriergeschwindigkeit von lebensmittel. *Beihelft zur Zeitschrift für die gesamte Kalte-industrie* 3(10), 1–16.

Rasanen, J., Harkonen, H. and Autio, K. (1995). Freeze-thaw stability of prefermented frozen lean wheat doughs: effect of flour and fermentation time. *Cereal Chemistry* 72(6), 637–642.

Rasanen, J. (1998). *Prefermented Frozen Lean Wheat Doughs*. VTT Publications p. 352.

Rasanen, J., Laurikainen, T. and Autio, K. (1997). Fermentation stability and pore size distribution of fozen prefermented lean wheat doughs. *Cereal Chemistry* 74(1), 56–62.

Ribotta, P. and LeBail, A. (2007a). Effect of additives on the thermo-mechanical behaviour of dough systems at sub-freezing temperatures. *European Food Research and Technology* 224, 519–524.

Ribotta, P. and LeBail, A. (2007b). Thermo-physical assessment of bread during staling. *Lebensmittel-Wissenshaft und Technologie* (in press).

Ribotta, P. and Le-Bail, A. (2007c). Thermophysical and thermomechanical assessment of partially baked bread during chilling and freezing process. Impact of selected enzymes on crumb contraction to prevent crust flaking. *Journal of Food Engineering* 78(3), 913–921.

Ribotta, P.D., Perez, G.T., Leon, A.E. and Anon, M.C. (2004). Effect of emulsifier and guar gum on micro structural, rheological and baking performance of frozen bread dough. *Food Hydrocolloids* 18(2), 305–313.

Rouillé, J., Le-Bail, A. and Courcoux, P. (2000). Influence of formulation and mixing conditions on breadmaking qualities of French frozen dough. *Journal of Food Engineering* 43(4), 197–203.

Stecchini, M.L., Maltini, E., Venir, E., Del-Torre, M. and Prospero, L. (2002). Properties of wheat dough at sub-zero temperatures and freeze tolerance of a baker's yeast (*Saccharomyces cerevisiae*). *Journal of Food Science* 67(6), 2196–2201.

Takasaki, S. and Karasawa, K. (1992). Effect of storage period on survival of freeze tolerant yeast and on the rheological properties of frozen dough. *Journal of Japanese Society of Food Science and Technology [Nippon Shokuhin Kogyo Gakkaishi]* 39(9), 813–820.

Vulicevic, I.R., Abdel-Aal, E.-S.M., Mittal, G.S. and Lu, X. (2004). Quality and storage life of partbaked frozen breads. *Lebensmittel-Wissenshaft und Technologie*, 37, 205–213.

Wang, Z.J. and Ponte, J.G. Jr. (1995). Storage stability of gluten-fortified frozen dough. *Cereal Foods World* 40(11), 827–831.

Wang, Z.J. and Ponte, J.G.J. (1994). Improving frozen dough qualities with the addition of vital wheat gluten. *Cereal Foods World* 39(7), 500–503.

Wijnans, G.J.M. and Baal, H.C.I. (1997). Emulsifiers: sucrose esters of fatty acids. *Food Technology International Europe* 35–37.

Zounis, S., Quail, K.J., Wootton, M. and Dickson, M.R. (2002). Effect of final dough temperature on the microstructure of frozen bread dough. *Journal of Cereal Science* 36(2), 135–146.

10 Developing Frozen Products for the Market and the Freezing of Ready-Prepared Meals

Ronan Gormley

SUMMARY

Frozen ready-meals are a major sector of the frozen foods market and a range of topics regarding the market, preparation/assembly, health and safety, and the importance of product–process–package (PPP) and time–temperature–tolerance (TTT) elements is addressed. There is particular emphasis on freeze-chilling and *sous vide*-freezing of ready-meals, and ready-meal components, as the application of these processes to ready-meals is relatively new. The conclusion is that freeze-chilling and *sous-vide*-freezing are suitable technologies for many ready-meals and ready-meal components, including sauces, and offer logistic and shelf-life benefits during distribution and retailing.

10.1 INTRODUCTION

The value of the ready-meal market in Europe is forecast to reach €13.4 billion by 2010 (Anon, 2005a), which presents a major opportunity for processors and retailers. The diversity of the ready-meals available continues to broaden with convenience products aimed at children, adolescents, working couples, families and the aged. This chapter addresses frozen ready-meals under the headings preparation and assembly, health and safety, product–process–package (PPP) and time–temperature–tolerance (TTT) considerations, freezing processes, freeze-chill and *sous vide*-freezing technologies. The application of freeze-chilling and *sous vide*-freezing to ready-meals is a relatively recent development and has been the subject of considerable research activity at Ashtown Food Research Centre. For this reason, freeze-chilling and *sous vide*-freezing of ready-meal components and ready-meals are considered in depth.

10.2 MARKET FOR READY-MEALS

The European (EU-15) ready-meal market is increasing steadily (Anon, 2005a) and frozen ready-meal sales were valued in Europe at €600 million in 2005 (Anon, 2005b). The total market (chilled and frozen) totalled nearly 2.3 million tonnes in 2005, with France accounting for 592 and the UK 497 kilotonnes. Frozen ready-meals accounted for 857 kilotonnes in 2005. In the UK, the demand is for high-quality authentic-tasting ready-meals (Birks, 1999) including ethnic dishes, fish dishes, vegetarian meals, pasta-based ready-meals, and larger family-sized portions. Investment in process technology to ensure maintenance of food quality during cooking, freezing and microwaving is a feature of these developments. Chilled ready-meals are often perceived by consumers as of higher quality than frozen (Vickers, 1999) and this has acted against sales of the latter. In France, sales of complete frozen meals were worth €308 million in 2004 but were down 4.8% by value and up 1.2% by volume compared to 2003. Lynn (2000) also showed a decline in frozen meal popularity in the UK in comparison to the chilled. In contrast, the frozen dinners and entrees market rose 3.2% in the 12 weeks to September 2005 in US with multi-serve meals (up 10.8%) and hand-held entrees (up 6.6%) being the key growth elements (Anon, 2005c).

10.3 PREPARATION AND ASSEMBLY

Ready-meals usually comprise protein (beef, pork, chicken, fish), carbohydrate (rice, pasta, potato), vegetable (carrots, peas, beans, sweet corn, onions, peppers) and sauce components. Cheese is often included, e.g. in lasagne, or as a topping. The sauces can be water (stocks, flavours) or oil (hollandaise, mayonnaise) based. The ratio of the components can vary considerably but a general figure is protein (15–20%), carbohydrate (40–50%), vegetable (15–20%) and sauce (20–30%). Some of these totals exceed 100% but the sauce often incorporates the vegetable and spice components. Meals vary considerably around this basic formula as there is now a large range of ready-meals including ethnic and vegetarian items. In preparation, all the ingredients may be cooked on-site, or alternatively some may be cooked on-site and others bought-in as frozen product and then added to the meal during assembly. In on-site cooking, the protein components are cooked or par-cooked and are placed in the tray 'as is' or can be individually quick frozen (IQF) for inclusion in the trays at a later stage during assembly. Carbohydrates are usually blast chilled after cooking/par-cooking while the vegetables are often delivered frozen to the trays. Trays may be compartmented for the different components and in some cases all the ingredients may be in a vacuum bag as in *sous vide* cooking (see below). In the case of frozen assembly, the filled trays and product are given a final freeze before the product goes into frozen storage. In warm assembly, the sealed trays proceed to a freezer (blast or spiral freezer) for a full freezing process and are then sent to frozen storage. Tasses (2001) described a new ready-meal tray that delivers a meal of similar quality to restaurant meals. The sterilised meal is placed in a plastic tray, which is then subjected to freezing and frozen storage. A serving dish is also provided whose shape matches that of the carrier tray such that the frozen food can be transferred onto the dish prior to thawing. This can then be reheated and the ready-meal served.

10.4 HEALTH AND SAFETY

Ready-meals offer great potential for the inclusion of nutraceuticals and bioactive compounds but this has not been explored extensively to date. This has particular application in ready-meals for the elderly who may require supplements and are reluctant to consume them in tablet form. Similarly, ready-meals for small children, pregnant women and others with particular nutritional requirements represent a commercial opportunity that will be realised in the relative short term. Khan (2000) has discussed ethnicity and health as related to health and pleasure from foods and the growing consumer awareness of health issues. There is emphasis on ready-meals with lower fat and calorie contents and this trend will continue. Mitchell (personal communication, 2006) is working on reduced salt ethnic ready-meals with emphasis on sensory properties. Salt levels *circa* 33% of conventional levels are being explored and, on occasions, panellists are finding difficulty discriminating between spiciness and saltiness. Ethnic ready-meals have major potential in areas with high ethnic populations, e.g. the Indian population in regions of the UK (Darrington, 1999).

Safety is of paramount importance for all foods including frozen ready-meals. Safety is also a relative concept and in this context frozen ready-meals are potentially safer than chilled because of the low storage temperature and ice content of the former. Fabrega Fernandez and Forcadell Berenguer (1992) found that 69 of 353 ready-meals exceeded the legal maximum number of Lancefield group D *Streptococcus*, six exceeded the limit for *Staphylococcus aureus*, and nine contained *Salmonella* spp. An overview of the microbiological hazards and controls relating to raw materials and processing of ready-meals has been prepared by Kyriakides and Batchford (2002) and they point to a number of potential black spots. Microbiological hazards relating to freeze-chilled (Redmond *et al.*, 2004c) and *sous vide*-processed (Tansey *et al.*, 2005) ready-meal components/meals have also been considered (see below). Process validation is a key component of ready-meal production as is the use of equipment that is designed with efficient cleaning in mind (Holah, 2000).

10.5 PPP AND TTT CONSIDERATIONS

The successful marriage of the product–process–package (PPP) elements (Bøgh Sørensen, 1990) is key in any food production process as are time–temperature–tolerance (TTT) considerations (van Arsdel *et al.*, 1969). For ready-meals, most products are relatively robust for cooking and freezing, with the exception of sauces, which need to be freeze–thaw stable. Some of the components of ready-meals are suited to IQF freezing such as peas or diced carrot. However, with most ready-meals the product in-tray needs to be frozen and this is a large unit size and so requires a system such as a spiral freezer to achieve a reasonably fast freezing rate (i.e. core temperature of −20°C in 0.5 h). Cryogenic freezing can also be used for delicate product but this process is usually too expensive for conventional ready-meals where cost margins are tight. Many frozen ready-meals are produced on a 'just-in-time basis' and so deleterious changes during prolonged frozen storage are not a feature. However, temperatures fluctuating below the freezing point can cause icing and other changes in the product even over short (3 month) periods of frozen storage (Gormley *et al.*, 2002).

Packaging trays for ready-meals must withstand the extremes of cold and hot as in freezing and frozen storage of the product at −30°C followed by a heating step in a microwave or convection oven. Crystalline polyethylene terephthalate (CPET) trays are best for this application. The trays must also be consumer friendly in terms of attractiveness and easy-opening. Ready-meals are also being processed by *sous vide* and, in some cases, this is being followed by freezing rather than chilling (see below). The *sous vide* bags and seals must withstand vacuum conditions, pressure during the cooking phase, and low temperature during frozen storage.

Reheating is the final process before consumption and is normally conducted directly from frozen in a microwave, convection or household oven depending on where the ready-meal is going to be consumed. Swain *et al.* (1995) studied the effect of the initial temperature of a food, and continuous use of a microwave oven, on the temperature of the food after heating. As expected, low initial temperatures in the product (mashed potato) gave lower temperatures after microwaving. Lower-powered microwave ovens were better in that they gave more uniform heating than higher powered versions when heating foods directly from frozen. Two of five microwave ovens tested maintained a mean lowest temperature of >90°C in the mashed potato samples while in the other three the mean lowest temperature fell from 81°C to 71–76°C after continuous use on a sequence of samples for 1.5 hours.

10.6 FREEZE-CHILL TECHNOLOGY

Chilled ready-meals are perceived to be of better quality than frozen. However, the former have a relatively short shelf-life while the latter offer better manufacturing and distribution flexibility, food safety and extended storage time (Kobs, 1997). Freeze-chilling is a dual process consisting of freezing and frozen storage followed by thawing and chilled retail display (O'Leary *et al.*, 2000). It has logistic advantages over chilling as it (i) allows bulk preparation of frozen products followed by controlled batch release of thawed product into the chill chain, (ii) enables 'chilled' products to reach distant markets more easily, and (iii) reduces the level of product recalls as it enables routine microbiological tests to be completed before the product is released from the factory. The reason for conducting the freeze-chilling trials described below arose from the concern that freeze-chilling might predispose products to more rapid quality deterioration in the chill phase of the process compared to chilled foods that had not been previously frozen. For example, the thawed food could be more susceptible to microbial growth due to nutrients in the drip, and because freezing may open up the cell structure of vegetables and meats thereby adversely affecting colour, texture, vitamin retention and sensory acceptability. The research addressed the freeze-chilling of mashed potatoes (three cultivars), carrots, green beans, lasagne, white sauces, and commercial ready-meals supplied by manufacturers. The elements of temperature abuse, tempering (thawing), re-freezing and modified atmosphere packaging were also investigated.

10.6.1 Procedures used

The process treatments, testing schedules and gas flushing details for the different products are given in the text below. Texture measurement (probe penetration in mashed potato), colour (Hunter Lab), centrifugal drip, vitamin C content, taste panel acceptability, and total viable counts (TVCs) were conducted, as appropriate to the product range, as described by Redmond *et al.* (2004c).

10.6.2 Freeze-chilling of mashed potato: different cultivars

The objective was to examine the effect of freeze-chilling on the quality of mash from Rooster, Golden Wonder and Maris Piper with a view to their use in ready-meals. The texture of mashed potato is critically important for acceptability and a soft mealy texture is preferred to one that is sticky, cohesive or gluey (Longree, 1950). The mashed potato (in plastic cups) was subjected to one of the following treatments: (i) *Freeze-chill*: blast freeze at −30°C (air velocity 3.75 m s^{-1}) for 2.5 hours; store at −25°C for 4 days; thaw overnight at 4°C; store at 4°C for 4 days; test; (ii) *Freeze:* blast freeze at −30°C for 2.5 hours; store at −25°C for 4 days; thaw overnight at 4°C; test; (iii) *Chill:* store at 4°C for 4 days; test; (iv) *Fresh:* cook, chill to 4°C and then test.

Freeze-chilling and freezing led to less firm, less adhesive potato mash compared to chilling and preparing fresh (Fig. 10.1) ($P < 0.001$). This was most likely due to cell wall damage caused by ice crystal formation during the freezing process (Fennema, 1993). Of

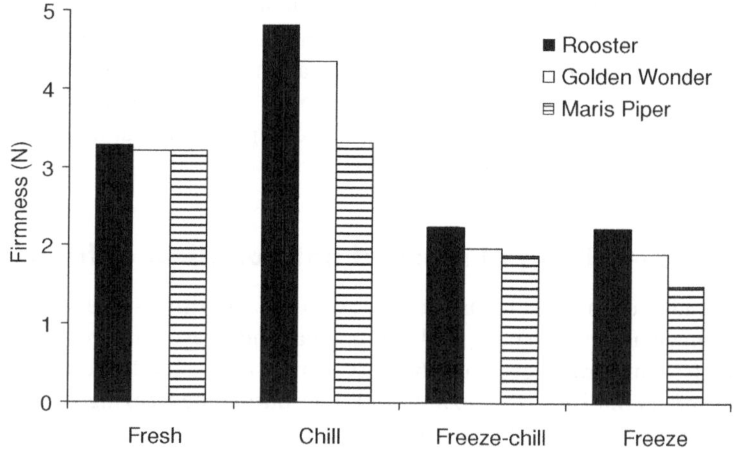

Fig. 10.1 Firmness (penetration force) of fresh or processed mashed potato. Reprinted from Redmond, G.A., Gormley T.R. and Butler, F. (2003a), The effect of short- and long-term freeze-chilling on the quality of mashed potato, *Innovative Food Science and Emerging Technologies* 4(1), 85–97, with permission from Elsevier.

the cultivars, Rooster produced a firmer, more adhesive mash than Golden Wonder or Maris Piper ($P < 0.001$). Chilling generally led to a brighter coloured mash (higher L/b value) than any of the other treatments. This finding was similar to that of O'Leary *et al.* (2000) for reconstituted potato flakes. Rooster mash had a lower L/b ($P < 0.001$) value than mash from Golden Wonder or Maris Piper. Freezing and freeze-chilling led to a higher centrifugal drip than chilling or preparing fresh (Fig. 10.2). This may be due to the breakdown in cell structure as a result of freezing with a consequent loss of fluids (IIR, 1986) on thawing. Mash from Rooster had a higher vitamin C content than that from Golden Wonder or Maris Piper

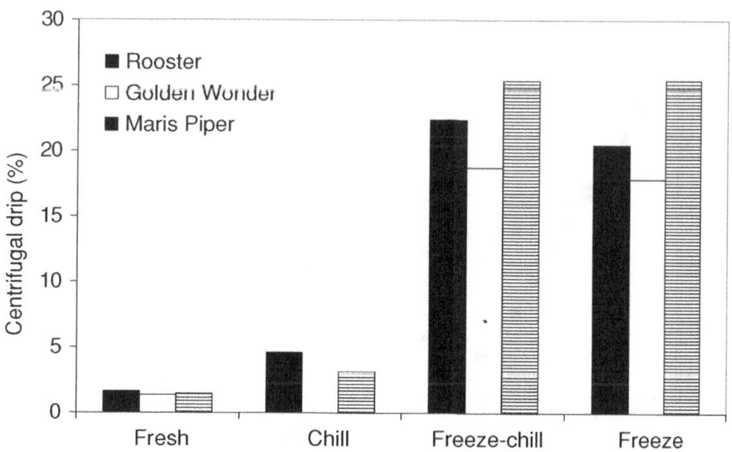

Fig. 10.2 Centrifugal drip (%) from fresh or processed mashed potato of cultivars. Reprinted from Redmond, G.A., Gormley T.R. and Butler, F. (2003a), The effect of short- and long-term freeze-chilling on the quality of mashed potato, *Innovative Food Science and Emerging Technologies* 4(1), 85–97, with permission from Elsevier.

although all values were low. As expected, fresh potato mash had a slightly higher vitamin C content than the other three treatments and chilled and freeze-chilled potatoes had the lowest content (Redmond *et al.*, 2003a). It is stressed that all mashed potato has an inherently low vitamin C content irrespective of the cultivar due to the cooking and mashing process. For this reason tests were also conducted on the use of encapsulated vitamin C for maintaining vitamin status in freeze-chilled mashed potato (see below).

10.6.3 Freeze-chilling of mashed potato: effect of freezing rate

Freezing can lead to changes in texture, colour, drip loss, nutritive value and microbial load (IIR, 1986; Fennema, 1993). There is also a gradual and irreversible reduction in product quality during frozen storage, even at −60°C (Gormley *et al.*, 2002). Tests were conducted to examine the effect of different freezing conditions during the freezing stage of the freeze-chilling process on the quality of mashed potato. Reconstituted potato flakes (termed reconstituted) and mash made from freshly boiled potatoes (termed fresh) were used. The freezing stage of the freeze-chill process was carried out using a liquid nitrogen cryogenic environmental chamber (CM-2000, Carburos Metalicos, Madrid, Spain). The mashed potato was frozen at −30 (air speed 3.75 m s^{-1}), −60 (0.5 m s^{-1}) or −90°C (0.5 m s^{-1}) to an internal temperature of −25°C; stored at −25°C for 1 week; thawed at 4°C; stored at 4°C for 1 week; then tested. As expected, the lowest freezing temperature (−90°C) led to the fastest reduction in temperature ($P < 0.001$) (Fig. 10.3). The pots of potato treated at −90°C reached an internal temperature of −25°C within 24 minutes compared to 78 minutes for samples frozen at −30°C. Similar results were obtained for fresh mashed potato.

A textural problem associated with reconstituted mashed potato is stickiness/firmness, presumably due to excessive extracellular 'free starch' produced by diffusion of starch through the cell walls during cooking, and by rupture of the cooked cell walls during mashing and mixing (Faulks and Griffiths, 1983). Ice crystal formation during freezing leads to further destruction of cell walls and, therefore, more 'free starch'. This may explain why frozen

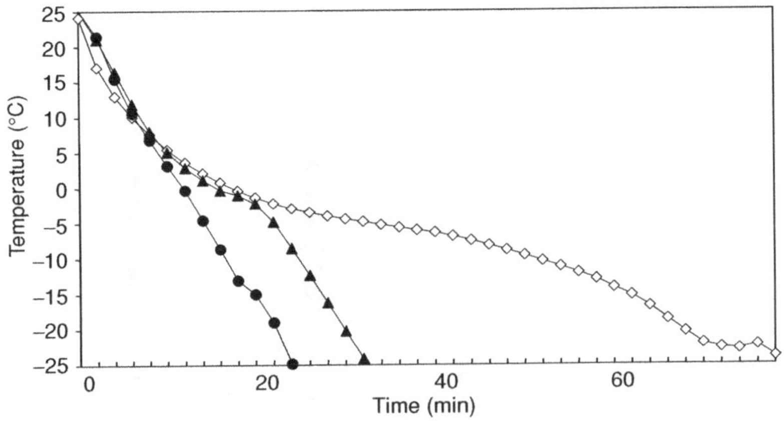

Fig. 10.3 Average core temperatures for reconstituted mashed potato during freezing at −30°C (◇), −60°C (▲) and −90°C (•). Reprinted from Redmond, G.A., Gormley T.R. and Butler, F. (2002), The effect of freezing conditions on the quality of freeze-chilled reconstituted mashed potato, *Lebensmittel-Wissenschaft und- Technologie* 35, 201–204, with permission from Elsevier.

reconstituted mashed potato was firmer in the current study than chilled. Lowering the temperature of freezing resulted in a softer mash for rehydrated potato but not for fresh mashed potato. Lowering the freezing temperature had no effect on vitamin C content, colour or sensory acceptability of mashed potato (fresh and reconstituted), but drip loss decreased linearly ($P < 0.001$) with temperature of freezing.

10.6.4 Vitamin C retention in freeze-chilled mashed potato

Vitamin C is a labile nutrient and losses occur in cooked, and then chilled vegetables (Williams *et al.*, 1995), and also in freeze-chilled mashed potato (Decazes *et al.*, 2001; O'Leary *et al.*, 2000). To overcome this, the addition of encapsulated vitamin C to freeze-chilled mashed potato was investigated. Cooled mashed potato sample sets (20 g per pot; cultivar Rooster) were prepared containing (a) ordinary vitamin C (OVC) (33 mg per 100 g), (b) encapsulated vitamin C (EVC) (50 mg per 100 g), and (c) no vitamin C. All samples underwent one of the following process treatments: (i) *Fresh*: cook and test, (ii) *Chill:* store at 4°C for 8 days, (iii) *Freeze-chill:* blast freeze at −30°C for 2.5 hours; store at −25°C for 7 days; thaw overnight at 4°C; store at 4°C for 8 days; (iv) *Freeze:* blast freeze at −30°C for 2.5 hours; store at −25°C for 1.5 days; thaw overnight at 4°C. The vitamin C content of the not-supplemented mashed potato (NVC) was low (*circa* 2 mg per 100 g) in the fresh and in the processed products (Fig. 10.4). This may have been due to the cooking and mashing of the potato. The chill and freeze-chill treatments caused a dramatic reduction ($P < 0.001$) in the vitamin C content of the samples with added OVC. In contrast, the chilled and freeze-chilled samples with added EVC had *circa* 15 mg per 100 g (Fig. 10.4) indicating that encapsulation was exerting a protective effect during chilled storage. The freeze-only treatment resulted in a good retention of both the OVC and EVC (Fig. 10.4) (Redmond *et al.*, 2003b). This result was expected as vitamin C is normally well retained in frozen foods (Favell, 1998).

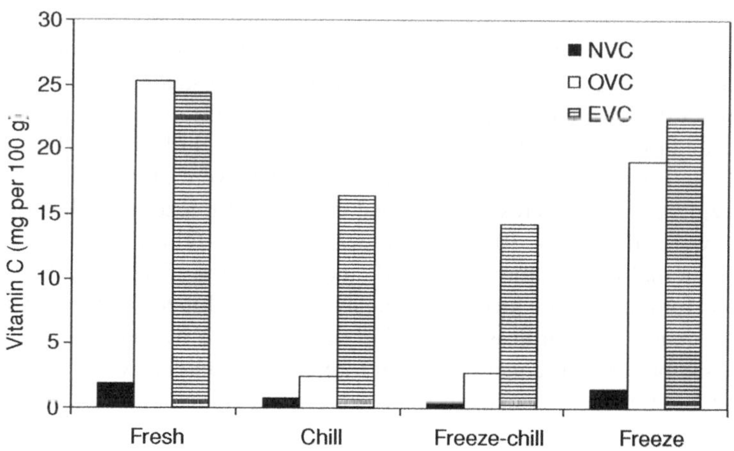

Fig. 10.4 Effect of process treatments (and fresh) on the vitamin C content of fresh and processed mashed potato without added vitamin C (NVC), with added vitamin C (OVC) and with added encapsulated vitamin C (EVC). Reprinted from Redmond, G.A., Decazes, A.M., Gormley T.R. and Butler, F. (2003b), The vitamin C status of freeze-chilled mashed potato, *Journal of Food Engineering*, 56(2–3), 219–221, with permission from Elsevier.

10.6.5 Freeze-chilling of steamed carrots and green beans

Steamed carrots and green beans are often used as components in ready-meals and the effect of short- and long-term freeze-chilling on their quality was investigated in a recent study (Redmond *et al.*, 2004a). Freeze-chilling and freezing led to less firm (lower shear values) carrots than chilling or preparing fresh but this was not the case for green beans. Freeze-chilling and freezing led to a higher drip loss than the fresh or chill treatments for both carrots and green beans. For carrots, drip loss from the fresh and chill treatments was low (average = 2.1%) compared to the freeze-chill and freeze treatments (average = 19.2%), whereas drip loss from the fresh and chilled green beans was much higher (average = 22.1%), although significantly lower than the freeze-chilled or frozen green beans (average = 31.4%).

No difference was found in cooked carrot colour between the four treatments but differences were found in cooked green bean colour, in that freeze-chilling and chilling led to lower hue angles (less green) than freezing or preparing fresh. No difference was found in total viable count (TVC) values between the treatments for either carrots (average = 2.2 $\log_{10}$ cfu g^{-1}) or green beans (average = 3.2 $\log_{10}$ cfu g^{-1}), and there was no difference in sensory acceptability between the process treatments for steamed carrots. However, sensory acceptability was lower for the fresh green beans than for processed beans. These data suggest that freeze-chilling is a technology suitable for steamed carrots and green beans. (Redmond *et al.*, 2004a).

10.6.6 Freeze-chilling of white sauces

Starch-based white sauces are frequent constituents of ready-meals. The sauce enhances appearance and imparts a flavour sensation to the protein food which complements the eating experience. It also prevents dehydration and freezer-burn during storage. However, freezing traditional white sauce releases water (syneresis) after thawing, which decreases product acceptability. It is important, therefore, that sauces for frozen ready-meals are freeze–thaw stable and trials were conducted on white sauces made with two modified starches, i.e. Purity W, Thermflo (modified waxy maize starches), and National Frigex CL (modified tapioca starch). Each sauce was subjected to both freeze-chilling and chilling. The amount of liquid separation (syneresis) in the freeze-chilled and chilled sauces was low (less than 4%) throughout the 8-day chilled display period. There was no difference in colour or viscosity of the chilled and freeze-chilled sauces during the period of chill storage (Dempsey, 2003).

Thermoflo and National Frigex CL modified starches were also used to investigate the effect of freezing rates on syneresis, colour and rheology of freeze-chilled white sauces. The white sauces were frozen at −30 (air speed 3.75 m s^{-1}), −60 (0.5 m s^{-1}) or −90°C (0.5 m s^{-1}) to an internal temperature of −25°C, and stored at −25°C for 2 days followed by holding at 4°C for 8 days. The level of syneresis in freeze-chilled sauces frozen at different rates was less than 2%, regardless of freezing rate and there was no effect on viscosity or colour of the samples. It was concluded, therefore, that the starches were freeze–thaw stable and were suitable for use in sauces in freeze-chill applications in ready-meals. (Dempsey, 2003).

10.6.7 Freeze-chilling of lasagne

Lasagne is an increasingly popular ready-meal with average consumption per capita in Ireland of 0.32 kg annually, compared to 0.2 kg in the UK and 1.0 kg in Belgium (Anon, 2001a).

Table 10.1 Effect of short-term frozen storage on the quality of freeze-chilled lasagne.

	Fresh	Chill	Freeze	Freeze-chill	*F* test	LSD[a]
Unheated lasagne						
Centrifugal drip (%)	0.84	1.56	10.07	3.41	$P < 0.001$	1.99
Colour (L/b)	5.59	5.17	5.37	5.55	$P < 0.05$	0.29
Shear value (N)	274	258	289	256	NS	48
TVC ($\log_{10}$ cfu g^{-1})	1.76	1.95	3.62	4.03	$P < 0.001$	0.64
Heated lasagne						
Centrifugal drip (%)	—	2.15	2.77	2.29	NS	1.57
Colour (L/b)	—	2.49	4.01	3.17	$P < 0.01$	0.7
Shear value (N)	—	216	229	232	NS	30
Sensory acceptability	3.75	3.64	3.38	3.60	NS	0.54

Note: [a]LSD, Least significant difference.

Reprinted from Redmond, G.A., Gormley, T.R. and Butler, F. (2004b), The effect of short- and long-term frozen storage with MAP on the quality of freeze-chilled lasagne, *Lebensmittel-Wissenschaft und- Technologie* **38**(1), 81–87, with permission from Elsevier.

Trials were conducted to examine the effect of freeze-chilling on the quality of beef lasagnes (unit size 300 g) (Redmond *et al.*, 2004b). A fresh sample was used as control.

Freeze-chilled lasagnes had a brighter colour (L/b) than chilled lasagnes before heating, but this was not evident after heating (Table 10.1). Frozen and freeze-chilled lasagne had higher drip loss than fresh or chilled before heating but again this was not evident after heating. No difference was found in shear values or sensory acceptability between the treatments but freeze-chilled and frozen lasagne had higher TVCs than chilled or fresh. This result was unexpected as freezing usually gives a reduction in TVC values (Redmond *et al.*, 2004b). All microbial counts were below the upper limit of log 5 cfu g^{-1} set out by the Food Safety Authority of Ireland for ready-to-eat meals at point of sale (Anon, 2001b).

Frozen and freeze-chilled lasagne had significantly higher drip loss values than fresh or chilled before heating but not after heating. This could be due to structural damage to components as a result of freezing (Khan and Vincent, 1996). However, it is likely that moisture redistribution took place during reheating thereby offsetting the adverse effects of freezing. This increase in drip loss for the unheated lasagne is not of major concern for manufacturers as the lasagnes are reheated before consumption. No difference was found in texture between freeze-chilled, frozen, chilled or fresh lasagne, either before or after reheating suggesting that storage conditions had little effect on the texture of the final heated product (Redmond *et al.*, 2004b).

10.6.8 Long-term freezing of products

One of the advantages of freeze-chilling is that products can be stored frozen prior to chilled retail. Tests were conducted, therefore, to examine the effects of long-term frozen storage (up to 12 mo) prior to thawing and chilling, on the quality of mashed potato, steamed carrots, steamed green beans and lasagne. The products were subjected to one of the following treatments: (i) *Freeze – chill*: blast freeze at −30°C for 2.5 hours; store at −25°C for 0, 3, 6, 9 or 12 months; thaw overnight at 4°C; store at 4°C for 6 or 7 days; (ii) *Freeze:* blast freeze at −30°C for 2.5 hours; store at −25°C for 0, 3, 6, 9 or 12 months; thaw overnight at 4°C; test.

Frozen storage for up to 12 months followed by chilling led to a firmer product than freezing alone for mashed potato and green beans. Length of time in frozen storage had

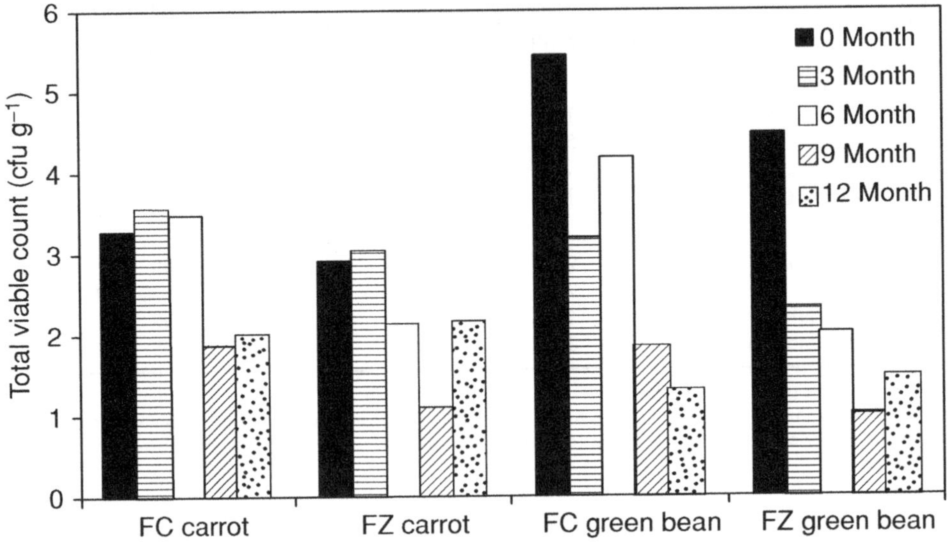

Fig. 10.5 Effect of long-term frozen storage on the total viable count (TVC) values of frozen and freeze-chilled cooked carrots and green beans: frozen carrots (FZ carrot); frozen green beans (FZ green bean); freeze-chilled carrots (FC carrot); freeze-chilled green beans (FC green bean).

no effect on drip loss values of freeze-chilled or frozen mashed potato (average = 21%), cooked carrots (average = 26%) or green beans (average = 31%). Time in frozen storage reduced ($P < 0.001$) the brightness (i.e. lower Hunter L/b values) of freeze-chilled and frozen mashed potato. Process treatments were also different ($P < 0.001$) with freeze-chilled mash having a brighter colour (L/b value of 3.20) than frozen mash (2.98). Frozen storage time had no effect on the colour (hue angle) of carrots (average = 45°) or green beans (average = 118°), and no difference was found between freeze-chilling and freezing. Length of time in frozen storage had no effect on the vitamin C content of mashed potato but freeze-chilling led to a reduction in vitamin C content compared to freezing (0.64 versus 1.68 mg per 100 g, respectively) (Redmond *et al.*, 2003a, 2003b). The length of time in frozen storage had no effect on the sensory acceptability of the three products and no differences were found in sensory acceptability between freeze-chilled and frozen carrots or green beans. However, freeze-chilling led to a lower ($P < 0.01$) sensory score (2.52) than freezing (2.75) for mashed potato on a 5-point hedonic scale.

Length of time in frozen storage had no effect on the TVC values of freeze-chilled and frozen mashed potato but did have an effect on values for both green beans and carrots (Fig. 10.5). For green beans, TVC values before frozen storage were higher than at any other storage time, while carrots stored for 9 months had the lowest TVC values. Freeze-chilling also led to significantly higher TVC values than freezing for the products, which agrees with the findings of O'Leary *et al.* (2000) for a range of products. However, all values for the processed samples were below the limit (log 5 cfu g^{-1}) set by the Food Safety Authority of Ireland for cooked vegetable products at point of sale (Anon, 2001b). The overall findings suggest that long-term frozen storage does not result in an increased microbial load in the follow-on chill phase of the freeze-chill process for mashed potato, steamed carrots or green beans.

In tests with freeze-chilled lasagne (commercially produced) the samples were stored at −30°C for up to 12 months followed by thawing and chilling at 4°C for 6 days. Frozen lasagne was used for comparison. The term 'heated' refers to samples microwaved at 700 W as per the manufacturer's instructions followed by cooling to 4°C before testing. Drip losses tended to decrease with length of frozen storage time especially for the freeze-chilled samples. This may reflect increased water binding by the meat or pasta components of the lasagne. However, this was not reflected in product softness as lasagne stored frozen for 12 months had the highest shear values. Sensory tests indicated a non-significant deteriorative trend in sensory acceptability with increasing length of time in frozen storage as indicated by sensory scores of 3.9 (6 mo), 3.6 (9 mo) and 3.4 (12 mo). Several panellists commented on the pasta component of the meal being tough or leathery. However, even after 12 months of frozen storage, lasagnes were still acceptable (3.4 on a scale of 0 (unacceptable) to 5 (very acceptable)) (Redmond *et al.*, 2004b).

10.6.9 Modified atmosphere packaging (MAP) with freeze-chilling

Trials were conducted to assess the synergistic effect, if any, of combining MAP with freeze-chilling on the quality and shelf-life of mashed potato, steamed carrots, steamed green beans and lasagne. The products were packed in: (i) air; (ii) CO_2/N_2 (40:60%); (iii) $CO_2/O_2/N_2$ (40:30:30%); (iv) CO_2 (100%); (v) N_2 (100%); then blast frozen at −30°C for 2.5 hours; stored at −25°C for 3 weeks; chilled at 4°C for 9 days; tested. The treatments had minimal effects on colour, texture and centrifugal drip of the product range (Redmond *et al.*, 2004c). TVCs were not influenced by the treatments with the exception of lasagne where an interaction ($P < 0.05$) was found between process treatment and atmosphere, i.e. TVCs increased in the order air, CO_2/N_2 (40:60%), $CO_2/O_2/N_2$ (40:30:30%) for the frozen lasagnes while the order was the opposite for the freeze-chilled samples (Redmond *et al.*, 2004b).

10.6.10 Temperature abuse of freeze-chilled products

Chilled and freeze-chilled foods may be more at risk of temperature abuse than frozen products (O'Leary *et al.*, 2000) and a model study was conducted by Oxley (2003) to assess the effects of different temperatures (4°C, 7°C, 10°C) in the final chill phase of the freeze-chill process on TVCs (mashed potato) and psychrotrophic counts (mashed potato and steamed carrots) over an 8-day period. No significant difference was found for TVCs (potato) or for psychrotrophic counts (potato or cooked carrot) between freeze-chilling and chilling on any of the test days indicating that the freezing and thawing of the mashed potato and cooked carrot did not make the products more susceptible to microbial growth over the chill storage period. However, increasing chilled storage temperatures in the chill phase of the freeze-chill process gave an increased microbial growth for both carrots and potatoes. For example, mashed potato stored at 10°C had a TVC of 6.5 compared to 3.1 and 1.3 $\log_{10}$ cfu g^{-1} for potato at 7°C and 4°C, respectively, on day 8. Similarly, psychrotrophic counts for cooked carrots after 8 days were 8.6 (10°C), 5.7 (7°C) and 3.4 $\log_{10}$ cfu g^{-1} (4°C). These data show that the products stored at 10°C for 8-day post-thawing had unacceptable microbial counts and indicate the need for careful temperature control in the final chill phase of the freeze-chill process (Oxley, 2003). While TVCs and psychrotrophic counts do not necessarily signify a safe or unsafe product, they are an index of the potential for growth of pathogens if such contamination was to occur.

10.6.11 Tempering of freeze-chilled products

Tempering (thawing) of freeze-chilled foods is of paramount importance to food companies and a series of trials was conducted by Gerety (2004). The main trial involved tempering in a commercial unit (SB10, six-pallet unit; cubic capacity 20 m^3) supplied by Dawson Rentals, which blows cycles of warm and cold air over the product being thawed. In the tests, 216 commercially produced lasagne ready-meals were placed in 27 boxes (outers) (2 rows of 4 in each box). T-type wire thermocouples were inserted in the geometric centre (core) and between the meal and plastic seal (surface) of 12 meals at random before freezing. The frozen meals in the boxes were placed in the tempering unit in one of three box configurations: (a) separated, (air space between every box), (b) three together, (air space between groups of three boxes), and (c) stacked (no air spaces).

When the boxes were stacked in threes, the temperatures in the centremost boxes were *circa* 10°C lower at each testing time than the temperature in the more exposed boxes, and the product surface and core temperatures were still *circa* −4°C after 24 hours. This shows that tight stacking is not an option and the best practice for efficient tempering is a separated box configuration.

10.6.12 Re-freezing of freeze-chilled products

Many prepared food products have 'suitable for freezing' stamped on the pack and consumers freeze them for convenience and to extend shelf-life. However, with freeze-chilled foods home freezing or refreezing means a second freezing, which may raise issues of product quality and safety. A trial to address this involved freeze-chilled mashed potato, sliced steamed carrots and commercially produced lasagne ready-meals. Frozen samples of the three products were prepared (Redmond *et al.*, 2004a, 2004b), held at −25°C for 7 days, thawed overnight at 4°C, stored at 4°C for 7 days followed by testing (to determine baseline values, i.e. thaw 1 tests). The freeze-chilled samples were then refrozen by fast (blast freezing at −30°C for 2.5 hours (air speed 3.75 $m\ s^{-1}$) or slow (cabinet; still air at −25°C) procedures and held at −25°C for 7 days. They were thawed at 4°C overnight and tested again (i.e. thaw 2 tests). The colour, shear values, centrifugal drip loss and vitamin C (potatoes only) content of the freeze-chilled mashed potato, steamed sliced carrot and heated lasagne were not affected by refreezing (both methods) or thawing. However, refreezing/thawing did reduce the adhesiveness of the freeze-chilled mashed potato but the effect was small in practical terms. The microbial status of the freeze-chilled (unheated) lasagne was not affected by refreezing/thawing and TVC values were below the upper limit of $\log_{10} 5$ cfu g^{-1} specified by the Food Safety Authority of Ireland for ready-to-eat meals at point of sale (Anon, 2001b). These data suggest that freeze-chilled mashed potato, steamed sliced carrots, or lasagne can be refrozen/thawed without impairing quality or safety. However, it is imperative that no temperature abuse occurs during the chill phase of the freeze-chill process.

10.6.13 Overall outcomes on freeze-chilling

Freeze-chilling is a technology suitable for many ready-meals and ready-meal components, including sauces, and offers logistic and shelf-life benefits during distribution and retailing. Extended frozen storage (up to 1 yr) prior to chilling had little adverse effect on the quality of a range of freeze-chilled products. Modified atmosphere packaging (MAP) did not benefit the quality and shelf-life of freeze-chilled mashed potato, steamed carrots or lasagne.

Freeze-chilled mashed potato, steamed sliced carrots and lasagne can be successfully refrozen provided temperature abuse does not occur during the final chilled phase of the freeze-chilled process. Results from a commercial tempering unit showed that boxed ready-meals should be stacked with air spaces between them to ensure adequate air thawing. Good manufacturing practice (GMP) and HACCP are imperative throughout the freeze-chilling process. Freeze-chilled products should be labelled as 'previously frozen' for reasons of consumer information. A use-by date should also be employed and this label should be attached at the start of the thawing process.

10.7 *SOUS VIDE*-FREEZING TECHNOLOGY

Sous vide processing is an interrupted catering system in which raw or par-cooked food is sealed under vacuum in a laminated plastic pouch or container, heat-treated by controlled cooking, rapidly chilled and then reheated for service after a period of chilled storage (SVAC, 1991). The chilled storage period is up to 21 days at 0–3°C. The minimum recommended thermal process for *sous vide* products is 90°C for 10 minutes, or its time–temperature equivalent (SVAC, 1991). This process ensures a minimum 6-log reduction in psychrotrophic *Clostridium botulinum* spores and in vegetative pathogens such as *Listeria*, *Salmonella* and *Escherichia coli*.

Sous vide has been used mostly in the catering and food service sectors (Creed and Reeve, 1998), but more recently it is being adopted by the ready-meals sector. Concerns about the safety of *sous vide* products, mainly due to the potential for temperature abuse in the chill chain, has prevented its widespread use (Betts, 1992). For this reason, the use of *sous vide*-freezing as an alternative to *sous vide*-chilling for ready-meal components has been investigated in a major study (Tansey and Gormley, 2005; Tansey *et al.*, 2003a, 2005) and selected findings are reported here. *Sous vide*-freezing has at least two advantages: (i) it minimizes the risk of growth of *Clostridium botulinum* spores, and (ii) it extends product shelf-life. A potential disadvantage is that it could offset some of the quality advantages of the *sous vide* process, especially product texture, due to structural damage by ice crystals. Ten different foods were studied in three groups: carbohydrates (potatoes, pasta and rice); vegetables (carrots and broccoli); and muscle foods (salmon, cod, chicken, beef and lamb). A P-SV-F system for *sous vide* processing the 10 ready-meal components embracing pretreatments (P), *sous vide* cook time/temperature (SV), and freezing (F) post-*sous vide* cooking was developed and validated, and there was particular focus on the effects of the P-SV-F system on product texture and overall quality.

10.7.1 Procedures used

Sous vide/freezing tests were conducted on carbohydrates (sliced potatoes, pasta, rice), vegetables (sliced carrots, broccoli florets) and muscle foods (salmon portions, cod portions, chicken, beef and lamb pieces/cubes) as described by Tansey *et al.* (2005). All samples received pretreatments (Table 10.3) prior to *sous vide* cooking. Cooking time (Barriquand Steriflow cooker) varied from food to food due to variation in heat transfer rates and the size of the food pieces (Table 10.3). Pasteurisation values (*P* values) were recorded using an Ellab time–temperature recorder. The *sous vide* products were blast frozen (2 h at −30°C) or chilled (+4°C) and tests conducted for sensory acceptability/comparison, shear, colour, gravity drip, centrifugal drip and moisture as well as specific tests such as β-carotene levels

Table 10.2 Ideal texture range (shear values; kN) for 10 foods with corresponding steaming/boiling times.

Component	Ideal texture (hot shear) (kN)	Ideal texture (cold shear) (kN)	Steaming/boiling time to ideal texture (min)
Carrots	0.8–2.2	1.3–2.6	25–30
Broccoli	0.6–1.0	0.8–1.2	15
Potatoes	0.3–0.5	0.5–0.8	20–25
Pasta	0.8–1.2	1.3–2.0	10
Rice	0.4–0.7	0.9–1.5	15–20
Salmon	1.8–2.1	1.9–2.2	25–35
Cod	1.9–3.5	2.4–4.2	20–50
Chicken	2.0–3.0	3.0–4.0	40–120
Beef	4.0–4.6	4.8–6.2	120–180
Lamb	3.7–4.3	4.5–5.5	60–90

(carrots), vitamin C levels (broccoli) and thiamine levels (salmon and cod) (Tansey *et al.*, 2005).

10.7.2 Determining ideal texture

Good textural properties in food products are a key requirement for acceptance. Hence, the importance of determining ideal texture ranges for the 10 products. These were determined by simultaneous sensory analyses and shear tests on hot, and also on cold, steamed/boiled samples (samples removed at different time intervals) of the product range (Tansey *et al.*, 2003c; 2005) and the outcome for the 10 foods is shown in Table 10.2. These texture ranges became the target textures for the products following pretreatment, *sous vide* cooking, and freezing. The vegetable and carbohydrate components started hard but gradually became very soft on cooking. However, the muscle foods toughened slightly and either gradually softened or remained consistently tough with cooking. The ideal texture for pasta was achieved quickly after 10 minutes boiling whereas that for beef was achieved after 120 minutes and was maintained at 180 minutes (Table 10.2).

10.7.3 Pretreatments, process optimisation, and freezing

Pretreatments were used to firm the texture in carrots, broccoli and sliced potatoes in order to counter the texture-softening effects of *sous vide* processing followed by blast freezing. A range of other pretreatments was used for the other seven products ranging from chilled versus frozen starting material, to water soaking or to pan searing (Table 10.3). The influence of pretreatments on selected quality attributes of the 10 *sous vide* products have been published (Tansey *et al.*, 2005) as have additional aspects of the effects on carrots (Tansey and Gormley, 2002; Gormley and Tansey, 2004), broccoli (Tansey *et al.*, 2003a) and fish (Gormley *et al.*, 2003; Tansey *et al.*, 2003b).

The ideal texture ranges for the 10 products (Table 10.2) became the target textures for the optimised *sous vide* process for each product. This embraced a combination of pretreatment, *sous vide* cook time, and freezing or chilling post-*sous vide* cooking. The optimised process, based on these factors, for each product is given in Table 10.3 together with the time–temperature values for each thermal process. All were in excess of 10 minutes at 90°C (i.e. $P_{90} > 10$) with the exception of carrots processed for 10 minutes at 90°C which had a value

Table 10.3 Optimised pretreatments and *sous vide* cooking process for ready-meal components, which were blast frozen or chilled (4°C) and were within the ideal texture range.[a]

Component (150 g in *sous vide* bags)	Pre-treatment	Cooking time and temperature[b]	Pasteurisation values (*P* values, min)	Storage	Sensory tests[c] (frozen vs. chill)
Carrot slices (5 mm)	Blanch at 50°C for 15 min, then blanch at 90°C for 2 min with 0.1% salt and cool	10 min, 90°C	$P_{70} > 2$	Frozen	10/10
		30 min, 90°C	$P_{90} > 10$	Chilled	Not significant
Broccoli florets (10–50 g heads)	Blanch at 50°C for 15 min, then blanch at 90°C for 2 min with 0.1% salt and cool	30 min, 90°C	$P_{90} > 10$	Frozen	6/14
		40 min, 90°C	$P_{90} > 10$	Chilled	Not significant
Potato slices (5 mm)	Blanch at 50°C for 15 min, then blanch at 90°C for 2 min with 0.1% salt and cool	20 min, 90°C	$P_{90} > 10$	Frozen	10/10
		20 min, 90°C	$P_{90} > 10$	Chilled	Not significant
Salmon fillet (150 g portions)	None	20 min, 90°C	$P_{90} > 10$	Frozen	8/12
		40 min, 90°C	$P_{90} > 10$	Chilled	Not significant
Cod fillet (150 g portions)	None	20 min, 90°C	$P_{90} > 10$	Frozen	8/12
		30 min, 90°C	$P_{90} > 10$	Chilled	Not significant
Pasta shells (Roma)	Simmer for 4 min in water with 2% vegetable oil, drain and wash. Soak in water with 2% vegetable oil and drain	15 min, 90°C	$P_{90} > 10$	Frozen	11/9
		20 min, 90°C	$P_{90} > 10$	Chilled	Not significant
Long grain rice (Uncle Ben's long grain)	Soak 3 parts water: 1 part rice for 3 h, add 2% vegetable oil and drain	40 min, 90°C	$P_{90} > 10$	Frozen	11/9
	Soak 2 parts water: 1 part rice for 3 h, add 2% vegetable oil and drain	30 min, 90°C	$P_{90} > 10$	Chilled	Not significant
Beef (diced shoulder)	Sear for 5 min in frying pan with vegetable oil	200 min, 90°C	$P_{90} > 10$	Frozen	13/7
		200 min, 90°C	$P_{90} > 10$	Chilled	Not significant
Lamb (diced shoulder)	Sear for 5 min in frying pan with vegetable oil	200 min, 90°C	$P_{90} > 10$	Frozen	8/12
		200 min, 90°C	$P_{90} > 10$	Chilled	Not significant
Chicken (diced breast)	Sear for 5 min in frying pan with vegetable oil.	90 min, 90°C	$P_{90} > 10$	Frozen	9/11
		90 min 90°C	$P_{90} > 10$	Chilled	Not significant

Note: [a]See Table 10.2.
[b]2nd phase of Barriquand retort cycle.
[c]Paired comparison; 20 tasters.

of $P_{70} > 2$. The cook time–temperature values (Table 10.3) are not product core temperatures but refer to the holding times/temperatures in the middle (2nd) phase of the Barriquand retort cycle; the other phases are come-up (1st) and cooling (3rd). Beef and lamb required much longer cooking times than the other products. Samples to be frozen after *sous vide* cooking received a shorter process than those to be chilled in the case of carrots, broccoli florets, salmon/cod fillets and pasta shells (Table 10.3).

Paired comparison taste panel tests indicated no statistically significant preference for frozen versus chilled *sous vide* products (Table 10.3). This is a key finding as freezing post-*sous vide* cooking gives additional product safety without seriously impairing product quality. Freezing versus chilling post-*sous vide* cooking had a variable effect on the physico-chemical properties of the product range (Tansey *et al.*, 2005). Some of these were statistically significant but the differences were small in practical terms as indicated by the paired comparison taste panel data (Table 10.3).

10.7.4 Effect of freezing rate and length of frozen storage

Rapid freezing is usually associated with high product quality and in this regard *sous vide* samples of the 10 products were frozen by cabinet (−20°C; still air), air blast (−30°C for 2 h; air speed 3.75 m s^{-1}) and liquid nitrogen (LN) (−196°C) freezing methods. These procedures represent slow, intermediate and fast freezing rates, respectively, and the effect on the texture (shear values), colour and gravity/centrifugal drip values of *sous vide* samples was evaluated. The impact of the different freezing methods on the quality parameters was small and the efficacy of LN as a rapid freezing medium was lessened due to the large unit size (200 g) of the *sous vide* packs (Tansey *et al.*, 2005). In further tests, Carbonell and Oliveira (see Tansey *et al.*, 2005) showed that high-pressure shift freezing gave frozen cooked potatoes similar to just-cooked ones, particularly in terms of texture. This result is applicable to vegetables in general. However, the technology is expensive to acquire and to run, and may lack robustness. A simpler low-cost fast-freezing method, which has application to *sous vide* packs, is immersion in an ethylene glycol solution.

Long-term frozen storage has the potential to adversely affect product quality due to ice crystal growth, moisture migration and oxidation (Gormley *et al.*, 2002). The *sous vide* packs for the 10 products in the trial were blast frozen (−35°C for 2.5 h), stored at −20°C and tested after 0, 3, 6 and 9 months. Length of storage time at −20°C had a statistically significant effect on the colour parameters (Hunter Lab) of some of the 10 *sous vide* products. However, the magnitude of these effects was small in practical terms and would not have a major effect on product visual appearance as sensed by the consumer. Length of time at −20°C had no effect on the texture (shear values) of *sous vide* cooked broccoli, cod, salmon, chicken, beef or lamb. Pasta and rice samples became progressively firmer with storage time as indicated by shear values of 3.11, 3.35, 3.44 and 3.66 (pasta), and 4.11, 4.67, 5.13 and 5.60 kN (rice) at the 0, 3, 6 and 9 month test dates. There was no pattern in the shear data for carrots or potatoes (Tansey *et al.*, 2005).

10.7.5 Safety aspects of *sous vide* processing

The traditional microbiological concern associated with *sous vide* processing is that a mild heat treatment in conjunction with vacuum packaging and an extended shelf-life (in chill) may pose a risk from the survival and growth of *Clostridium botulinum* and from toxin production. However, the introduction of a freezing step to the process, if properly controlled, will overcome this risk. Extensive microbiological trials were conducted as part of the P-SV-F system on the effects of *sous vide* processing and post-processing storage conditions on the survival of food pathogens (Bourke and O'Beirne, 2003a, 2003b). The research focused on the effects of different thermal treatments, pretreatments and preparation steps, stress conditions and combined thermal, freezing and frozen storage on the survival of a range of pathogens. Specific aspects included: (a) strain effects on thermal resistance properties of *Listeria* spp.; (b) comparison of thermal resistance of *E. coli O157:H7* and *Salmonella* spp.; (c) effects of stress conditions on the thermal resistance of *L. innocua* (Tansey *et al.*, 2005).

10.7.6 Overall outcomes on *sous vide* freezing

A P-SV-F system for the *sous vide* cooking of ready-meal components (10 products) embracing pretreatments (P), *sous vide* cook times/temperatures (SV), and freezing (F) post-*sous vide* cooking has been tested, optimised and validated. Taste panellists were unable to detect a difference between *sous vide* products that were frozen versus chilled after *sous vide* cooking. This is a key result as the former ensures safety and is easily managed (i.e. frozen storage), while the latter requires a carefully controlled chill chain. Rate of freezing, or long-term frozen storage had a minimal effect on the quality attributes of the 10 *sous vide* cooked products.

Acknowledgements

Thanks are extended to Karen Hussey, Thomas Walshe, Lisa Darley, Anne Decazes, Alicia Marcilla, Sophie Noiret, Anne Jestin, Claudia Duelk, Anne Sainlot and Tanja Budde who conducted elements of the freeze-chilling and *sous vide* work. Thanks are also extended to the Irish Department of Agriculture and Food for funding the freeze-chilling and *sous vide* research under the Food Institutional Research Measure (FIRM) as part of the Irish National Development Plan.

REFERENCES

Anonymous (2001a). UK consumption of lasagne meals looks set to double. *Frozen and Chilled Foods Europe* April, 2.

Anonymous (2001b). *Guidelines for the Interpretation of Results of Microbiological Analysis of Some Ready-to-eat Foods at Point of Sale. Guidance Note No. 3*, Published by the Food Safety Authority of Ireland, Dublin, Ireland.

Anonymous (2005a). Opportunities still exist: EU ready-meals/ingredients market overview. *Food Ingredients and Analysis International* 27(3), 4–6.

Anonymous (2005b). Disposals on the way?. *Frozen Food Europe* 17(6), 28–30.

Anonymous (2005c). Scrumptious! (US frozen ready-meals market). *Frozen Food Age* 54(5), 14–17.

Betts, G. (1992). *The Microbiological Safety of* Sous Vide *Processing, Technical Manual No. 39*, Campden and Chorleywood Food Research Association [CCFRA], Chipping Campden, Gloucestershire, UK.

Birks, S. (1999). The 'eat go' generation. *Food Manufacture* 74(2), 32–34.

Bøgh-Sørensen, L. (1990). The TTT-PPP concept for chilled foods. In: P. Zeuthen, J.C., Cheftel, C., Eriksson, T.R., Gormley, Linko, P. and K. Paulus, eds. *Processing and Quality of Foods*, Vol. 3: *Chilled Foods: The Revolution in Freshness.* London and New York: Elsevier Applied Science, pp. 3.42–3.52.

Bourke, P. and O'Beirne, D. (2003a). Effects of *sous vide* processing and post-processing storage conditions on the survival of *L. innocua* in salmon and beef. In: *Proceedings of the 33rd Annual Food Science and Technology Research Conference.* University College Cork, Cork, Ireland, p. 97.

Bourke, P. and O'Beirne, D. (2003b). Thermal inactivation of *L. innocua* in a range of *sous vide* model products: some effects of prior heat stress and acid adaptation. In: *Proceedings of the 1st FEMS Conference of European Microbiologists*. Ljubljana, Slovenia, p. 136.

Creed, P.G. and Reeve, W. (1998). Principles and applications of *sous vide* processed foods. In: S. Gazala, ed. *Sous Vide and Cook-Chill Processing for the Food Industry*. Gaithersburg, MD: Aspen Publications, pp. 25–26.

Darrington, H. (1999). High noon at noon. *Food Manufacture* 74(7), 30–31.

Decazes, A.M., Redmond, G.A., Gormley T.R. and Butler, F. (2001). Vitamin C retention in mashed potato. In: *Proceeding of 31st Annual Food Science and Technology Research Conference*. University College Cork, Cork, Ireland, p. 45.

Dempsey, A. (2003). *The Effect of Starch Type and Freezing Rates on Freeze-Chilled Starch-based Sauces*. MSc thesis, University College Dublin, Dublin, Ireland.

Fabrega Fernandez, A. and Forcadell Berenguer, M.L. (1992). Microbiological study of frozen ready-meals for bulk sale. *Alimentaria* No. 229, 61–68.

Faulks, R.M. and Griffiths, N.M. (1983). Influence of variety, site and storage on physical, sensory and compositional aspects of mashed potato. *Journal of the Science of Food and Agriculture* 34, 979–986.

Favell, D.J. (1998). A comparison of the vitamin C content of fresh and frozen vegetables. *Food Chemistry* 62, 59–64.

Fennema, O. (1993). Frozen foods: challenges for the future. *Food Australia* 45, 374–380.

Gerety, A. (2004). *Investigation of the Thawing Process for Selected Ready-Meals and their Components*. MSc thesis, University College Dublin, Dublin, Ireland.

Gormley, T.R. and Tansey, F.S. (2004). Emerging freeze-chill and *sous vide* technologies in the production of ready-meals. *European Food and Drink Review* 1, 28–30.

Gormley, T.R., Duelk, C. and Tansey, F.S. (2003). *Sous vide*/freezing of cod and salmon portions. *Proceedings of the Transatlantic Fisheries Technology Conference (TAFT)*. The Icelandic Fisheries Laboratories, Reykjavik, Iceland, pp. 66–67.

Gormley, T.R., Walshe, T., Hussey, K. and Butler, F. (2002). The effect of fluctuating vs constant frozen storage temperature regime on some quality parameters of selected food products. *Lebensmittel-Wissenschaft und-Technologie* 35(2), 190–200.

Holah, J. (2000). Food processing equipment design and cleanability. *FLAIR-FLOW EUROPE Technical Manual No. F-FE 377A/00*. Ashtown Food Research Centre, Teagasc, Dublin, Ireland, p. 47.

IIR (International Institute of Refrigeration). (1986). *Recommendations for the Processing and Handling of Frozen Foods*, 3rd edn. Paris: IIR.

Khan, A.A. and Vincent, J.F.V. (1996). Mechanical damage induced by controlled freezing in apple and potato. *Journal of Texture Studies* 27, 143–157.

Khan, Y. (2000). Product design: ready-meals. *Food Ingredients and Analysis International*, January/February, pp. 21–22, 25–26.

Kobs, L.(1997). Designing frozen foods. *Food Product Design* January, 27–43.

Kyriakides, A. and Batchford, S. (2002). The microbiological safety and quality of ready-meals. In: *Ready-Meal Technology*. Leatherhead Publishing, pp. 192–228.

Longree, K. (1950). Quality problems in cooked, frozen potatoes. *Food Technology* 4, 98–104.

Lynn, S. (2000). New packaging products for convenience foods. *Packaging News* June, 15–16.

Mitchell, M. (2006). Ashtown Food Research Centre (Teagasc), Dublin, Ireland. Personal communication.

O'Leary, E., Gormley, T.R., Butler, F. and Shilton, N. (2000). The effect of freeze-chilling on the quality of ready-meal components. *Lebensmittel-Wissenschaft und- Technologie* 33(3), 217–224.

Oxley, E. (2003). *The Effect of Freeze-Chilling on the Quality and Shelf Life of Selected Ready-Meal Components Under Varying Storage Temperature Conditions*. MSc thesis, University College Dublin, Dublin, Ireland.

Redmond, G.A., Gormley T.R. and Butler, F. (2003a). The effect of short- and long-term freeze-chilling on the quality of mashed potato. *Innovative Food Science and Emerging Technologies* 4(1), 85–97.

Redmond, G.A., Decazes, A.M., Gormley T.R. and Butler, F. (2003b). The vitamin C status of freeze-chilled mashed potato. *Journal of Food Engineering* 56(2–3), 219–221.

Redmond, G.A., Gormley, T.R. and Butler, F. (2004a). The effect of short- and long-term freeze-chilling on the quality of cooked green beans and carrots. *Innovative Food Science and Emerging Technologies* 5(1), 65–72.

Redmond, G.A., Gormley, T.R. and Butler, F. (2004b). The effect of short- and long-term frozen storage with MAP on the quality of freeze-chilled lasagne. *Lebensmittel-Wissensehaft und- Technologie* 38(1), 81–87.

Redmond, G.A., Gormley, T.R., Butler, F., Dempsey, A., Oxley, E. and Gerety, A. (2004c). *Freeze-Chilling of Ready-to-Eat Meal Components. End-of-Project Report.* Ashtown Food Research Centre, Teagasc, Dublin, Ireland, p. 32.

SVAC (*Sous Vide* Advisory Board). (1991). *Codes of Practice for Sous Vide Catering Systems.* Tetbury, Gloucestershire, UK.

Swain, M.J., Foster, A.M., Phillips, I.C., Wilkins, T.J., James, S.J., Radhakrishnan, C.D. and Burfoot, D. (1995). *Effect of Initial Temperature of Food and Continuous use of a Microwave Oven on the Temperature of Food after Heating.* UK: Ministry of Agriculture, Fisheries and Food Publications.

Tansey, F.S. and Gormley, T.R. (2002). *Sous vide*/freezing of ready-meals. *Farm and Food Research* 12(1), 18–22.

Tansey, F.S. and Gormley, T.R. (2005). Sous vide/freezing technology for ready-meals. In: G.V. Barbosa M.S. Tapia and M.P. Cano, eds. *Novel Food Processing Technologies.* Boca Raton, FL: Marcel Decker/CRC Press, pp. 477–490.

Tansey, F.S., Budde, T. and Gormley, T.R. (2003a). The effect of *sous vide*/freezing technology on the quality of broccoli florets. In: *Proceedings of the 33rd Annual Food Science and Technology Research Conference.* University College Cork, Cork, Ireland, p. 44.

Tansey, F.S., Duelk, C. and Gormley, T.R. (2003b). The effect of *sous vide*/freezing technology on the texture of salmon portions. *Proceedings of the 12th World IUFOST Congress.* Chicago, IL, www.iufost.org.

Tansey, F.S., Gormley, T.R., Bourke, P., O'Beirne, D. and Oliveira, J.C. (2003c). Texture, quality and safety of *sous vide*/frozen foods. In: J.S.A. Edwards and I.B. Gustafsson, eds. *Culinary Arts and Sciences IV: Global and National Perspectives.* Bournemouth University, Poole, UK, pp. 199–207.

Tansey, F., Gormley, T.R., Carbonell. S., Oliveira, J., Bourke, P. and O'Beirne, D. (2005). *Developing sous vide/freezing systems for ready-meal components. End of Project Report.* Ashtown Food Research Centre, Teagasc, Dublin, Ireland, p. 36.

Tasses, P. (2001). *Process and Apparatus for Preparing Meals.* European Patent Application, Patent No. 1248537 A1.

van Arsdel, W.B., Copley, M.J. and Olson, R.L. (1969). *Quality and Stability of Frozen Foods.* New York: Wiley Interscience.

Vickers, R. (1999). Could the ice be melting for frozen ready-meals?. *Food Processing (UK)* 68(8), 17–18.

Williams, P.G., Ross, H. and Brand Miller, J.C. (1995). Ascorbic acid and 5-methyltetrahydrofolate losses in vegetables with cook/chill or cook/hot-hold in foodservice systems. *Journal of Food Science* 60(3), 541–546.

11 Frozen Storage

Noemi E. Zaritzky

The market for frozen food has expanded worldwide over the last 50 years to include a variety of products. Even if a food product is adequately frozen, physico-chemical and biochemical changes during storage can lead to a degradation in its quality. Quality of frozen food is highly dependent on storage temperature and there is a need for a constant and systematic control on maintaining the required temperature throughout the frozen foods distribution in the cold chain, from production to final consumption. Storage and transport conditions have great influence on the quality of frozen foods.

11.1 QUALITY LOSSES IN FROZEN FOOD

Ice formation involves a series of modifications that deteriorate food quality. Although these modifications slow down at low temperatures, they continue during frozen storage conditions. Physical and chemical changes can be described as deterioration mechanisms that are produced during frozen storage causing quality losses in frozen foods.

11.1.1 Physical changes during frozen storage

The main physical changes during storage of frozen food are moisture migration and ice recrystallisation. Both phenomena are related to the stability of frozen water inside and on the surface of the product (Fennema *et al.*, 1973; Zaritzky, 2000).

11.1.1.1 *Moisture migration*

A slow freezing process of tissues can allow sufficient time for water migration due to osmotic forces from the inner region of a cell to the freeze-concentrated intercellular region. This can result in cell desiccation, cell wall disruption, loss of turgor and crushing of the dried cell by the large intercellular ice mass. These phenomena affect not only the texture of the frozen product, but also a significant drip loss during thawing and cooking can occur, leading to a loss of nutrients.

During frozen storage the existence of temperature gradients within a product creates water vapour pressure profiles resulting in moisture migration and relocation of the water, both within and from the product. There is an overall tendency for moisture to move into the void spaces around the foodstuff and to accumulate on the product surface and on the internal package surface. In packaged frozen food, moisture migration leads to ice formation inside the package (Pham and Mawson, 1997). Temperature fluctuations (cooling–warming cycles) produce a net migration of moisture from the interior towards the surface of the foodstuff, or to the wrapping. The temperature of the packaging material follows the temperature fluctuations in the storage room faster than the product itself. As the surrounding temperature decreases, moisture inside the pores sublimes and diffuses to the packaging film; when ambient temperature increases, the ice on the wrapping tends to diffuse back to the surface of the food; however, reabsorption of water in the original location is impossible, and the process can be considered irreversible, producing undesirable weight losses. Moisture migration can be minimised by maintaining small temperature fluctuations and small internal temperature gradients and by the inclusion of internal barriers within the product and within the packaging.

Weight losses during freezing and frozen storage have economic consequences, unless the product is packaged in films of low water vapour permeability.

According to Pham and Mawson (1997) typical weight losses during meat processing are 1–2% during chilling, 1% during freezing, and about 0.5–1% per month during storage and transport, unless the product is packaged in an impervious film; the rate of sublimation doubles for every 10°C rise in temperature.

Freezer burn is a surface desiccation defect caused by sublimation that can occur when frozen tissues are stored without an adequate moisture barrier packaging. It manifests itself as an opaque dehydrated surface, produced by moisture losses in frozen foods. An excessive desiccation can accelerate oxidative alterations at the surface of the product. Freezer burn increases oxygen contact with the food surface area and raises oxidative reactions, which irreversibly alter colour, texture and flavour. It is caused by the sublimation of ice on the surface region of the tissue where the water pressure of the ice is higher than the vapour pressure of the environment. In cold storage rooms the temperature of the freezing coil (evaporator) is always lower than the surrounding air, therefore ice forms and accumulates on the coil. As moisture is removed, the relative humidity of the air in the cold room drops. Since the water vapour pressure over the surface of the frozen product is higher than that of the air, a constant loss of water in the form of vapour is produced from unprotected materials; sublimation continues as long as this vapour pressure difference continues. Glazing, dipping or spraying a thin layer of ice on the surface of an unwrapped frozen product helps to prevent drying. Freezer burn is prevented if a product is packed in tight-fitting, water-proof, vapour proof material, avoiding evaporation. Surface coating of prepared meals reduces the effect of this quality loss and may even add value to the product.

11.1.1.2 *Recrystallisation of Ice*

Slow freezing results in a low rate of nucleation and the production of a small number of large ice crystals, whereas fast freezing causes a high rate of nucleation leading to the formation of a large number of small ice crystals. However, during frozen storage ice crystals undergo metamorphic changes. Recrystallisation reduces the advantages of fast freezing and includes any change in the number, size, shape, orientation, or perfection of crystals following completion of initial solidification (Fennema *et al.*, 1973). In frozen aqueous solutions recrystallisation is the process by which the average ice crystal size increases with time. Small ice crystals are thermodynamically unstable, having a high surface/volume ratio and therefore a large excess of surface free energy. The net result of minimising free energy is that the number of crystals decreases at constant ice phase volume but their mean size increases (Sutton *et al.*, 1996). Recrystallisation basically involves small crystals disappearing, large crystals growing and crystals fusing together, and affects the quality of the product because small ice crystals make the product quality better; large crystals often cause damage during freezing. As the temperature of an aqueous specimen increases within the subfreezing range, the rate of recrystallisation also increases (Rapatz and Luyet, 1959; Mac Kenzie and Luyet, 1967; Reid *et al.*, 1987).

There are different types of recrystallisation processes described in literature (Fennema *et al.*, 1973; Hartel, 2001; Zaritzky, 2006): (a) iso-mass, (b) migratory, (c) accretive, (d) pressure-induced, and (e) irruptive.

11.1.1.2.1 *Surface iso-mass recrystallisation*

This includes changes in the shape or internal structure of a crystal and reduction of the defects as the crystal tends to a lower energy level maintaining a constant mass of ice. This 'rounding off' process may be produced by surface diffusion of the water molecules. Ice crystals of irregular shape and large surface-to-volume ratio (dendritic crystals) adopt a more compact configuration with a smaller surface-to-volume ratio and a lower surface energy. Sharper surfaces are less stable than flatter ones and show a tendency to become smoother over time.

11.1.1.2.2 *Migratory recrystallisation or grain growth*

This refers to the tendency of large crystals in a polycrystal system to grow at the expense of the smaller ones. Ostwald ripening refers to migratory recrystallisation that occurs at constant temperature and pressure due to differences in surface energy between crystals, which is proportional to the crystal curvature. Melting–diffusion–refreezing or sublimation–diffusion–condensation are possible mechanisms leading to an increase in average crystal size, a decrease in the number of crystals, and a decrease in surface energy of the entire crystalline phase. At constant temperature and pressure, migratory recrystallisation is the result of differences in the surface energies of large and small crystals. The small crystals, with a very small radii of curvature, cannot bind the surface molecules as firmly as larger crystals, thus, small crystals exhibit lower melting points than large ones. Migratory recrystallisation is enhanced by temperature fluctuations inducing a melt–refreeze behaviour due to ice content changes induced by temperature variations. Melt–refreeze behaviour can lead to complete disappearance of smaller crystals during warming and growth of larger crystals during cooling, or to a decrease in size of crystals during partial melting and regrowth of existing crystals during cooling. Melt–refreeze should occur to a greater extent at higher temperatures and more rapidly for smaller crystals.

11.1.1.2.3 *Accretive recrystallisation*

This is produced when contacting crystals join together increasing crystal size and decreasing the number of crystals and surface energy of the crystalline phase. The proposed mechanism of crystal aggregation is surface diffusion. Accretion refers to a natural tendency of crystals in close proximity to fuse together; the concentration gradients in the areas between them are high, thus, material is transported to the point of contact between crystals and a neck is formed. Further 'rounding off' occurs because a high-curvature surface like this has a natural tendency to become planar. The number of molecules leaving a curved surface is larger than the number of molecules arriving on that surface. The continued exchange of molecules at the interface serves to reduce the curvature of a single crystal (forming a sphere) or to reduce the number of small crystals by adding to the larger crystals.

11.1.1.2.4 *Pressure-induced recrystallisation*
If force is applied to a group of crystals, those crystals that have their basal planes aligned to the direction of force grow at the expense of those in other orientations. This type of recrystallisation is found not very frequently in foods.

11.1.1.2.5 *Irruptive recrystallisation*
Under conditions of very fast freezing, aqueous specimens solidify in a partially non-crystalline state and not all the freezable water is converted to ice. Upon warming to some critical temperature, crystallisation of ice occurs abruptly. This phenomenon is called 'irruptive recrystallisation'; however 'devitrification' is also used when the frozen specimen is totally non-crystalline after initial solidification.

11.1.1.2.6 *Rates of ice recrystallisation*
Rates of ice recrystallisation in frozen solutions and in frozen muscle tissue were reported by Bevilacqua and Zaritzky (1982), Martino and Zaritzky (1987, 1988, 1989). In these works it was proposed that the driving force for recrystallisation of ice is the difference in the surface energy of two adjacent crystals, with this energy being proportional to the crystal curvature. Ice crystal size distributions were measured from micrographs, and a direct relationship between crystal size and the number of crystal sides was established. Small crystals with three or four sides show concave surfaces, and tend to disappear because the crystal boundaries move towards the centre of curvature. Ice crystals with six sides have planar surfaces and are stable, and those with a higher number of sides tend to grow. Histograms of the relative frequencies of crystal diameters as a function of equivalent diameter were obtained for different freezing rates and storage conditions. Crystal enlargement occurs at a constant temperature but it is accelerated by fluctuations and thermal steps. As temperature increases, the small ice crystals melt; then when temperature decreases again, since new nuclei cannot be formed, water is converted to ice on the existing crystals hence increasing their size. Martino and Zaritzky (1988) working on meat tissues, reported that ice crystal size reaches a limiting value D_l that was related to the tissue matrix characteristics. The following equation was proposed considering that the driving force of this phenomenon is the difference between the instantaneous curvature of the system and the limit curvature:

$$\frac{dD}{dt} = k\left(\frac{1}{D} - \frac{1}{D_l}\right) \tag{11.1}$$

where D is the mean equivalent ice crystal diameter at time t, D_l is the limit equivalent diameter and k the kinetic constant. Integration of equation (11.1) leads to the following expression:

$$\ln\left(\frac{D_l - D_o}{D_l - D}\right) + \frac{1}{D_l}(D_o - D) = \frac{k}{D_l^2}t \tag{11.2}$$

where D_0 is the mean initial equivalent diameter. This model fits experimental data at short and long storage times satisfactorily (Martino and Zaritzky, 1989).

Donhowe and Hartel (1996) described the recrystallisation process by the following equation:

$$D = D_o + kt^{1/n} \tag{11.3}$$

where D is the mean crystalline diameter, D_0 is the initial diameter, k is the recrystallisation rate and n is the power law exponent. Recrystallisation was studied in either model sugar systems or in ice cream (Harper and Shoemaker, 1983; Sutton *et al.*, 1996; Sutton *et al.*, 1996) and results showed that ice crystals increased in size as a function of time to a power $(1/n)$ between 0.33 and 0.5 (Donhowe and Hartel, 1996).

The addition of hydrocolloids (locust bean gum, guar gum, carrageenan, xanthan gum) is important in the case of ice cream to control ice recrystallisation (Regand and Goff, 2003). Hydrocolloids are also added in frozen gelatinised starch-based systems that are used as sauces in precooked foods. Ferrero *et al.* (1993a, 1993b, 1994) performed recrystallisation studies on corn starch pastes (10% w/w wet basis) and observed that the addition of hydrocolloids such as xanthan gum (0.3% w/w wet basis) to these pastes modified neither the size of ice crystals at different freezing rates nor ice recrystallisation rates compared to pastes without gum addition. Hydrocolloids are recommended as ice crystal inhibitors, but different studies show that the stabilisation character of hydrocolloids should be explained on another basis, such as their capability to undergo molecular entanglement in the freeze concentrated matrix surrounding ice crystals (Ferrero *et al.*, 1993a).

Ice crystal size influences the stable shelf-life of the product mainly by its effect on the texture. This is important in products that are consumed in the frozen state (e.g. ice cream) in these cases a coarse or sandy texture is normally observed when large ice crystals are present.

11.1.2 Chemical changes produced by freezing and frozen storage

During the freezing of food, water is transferred into ice crystals and solutes concentrate in the unfrozen matrix. Slow freezing results in a maximum ice crystal purity and maximum concentration of solutes in the unfrozen phase, leading to equilibrium conditions (Fennema *et al.*, 1973).

In contrast, rapid freezing results in a considerable entrapment of solutes by growing crystals and a lower concentration of solutes in the unfrozen phase. The increasing concentration of solutes in the unfrozen matrix increases the ionic strength and can produce changes affecting the biopolymer structures. Water structure and water–solute interactions may be altered and interactions between macromolecules such as proteins may increase. Formation of ice crystals can produce the release of contents of food tissues; reactions that normally do not occur in intact cells may occur as a consequence of the freezing process. The possibility that enzymes come in contact with different substrates increases, leading to quality deterioration during frozen storage. Most enzymes exhibit substantial activity after freezing and thawing and many enzymes show significant activity in partially frozen systems. Freezing can give unusual effects on chemical reactions; temperature and concentration of the reactants in the unfrozen phase (freeze concentration effects) are the main factors responsible for changes in the reaction rates in frozen products. In many frozen systems, reaction rates as a function of temperature go through a maximum at some temperature below the initial freezing point. This is a consequence of opposing factors: low temperatures that bring reaction rates down, and increasing solute concentration in the unfrozen phase that may increase these reaction rates. For example, oxidation of myoglobin (meat pigment) was accelerated at temperatures close to −5°C (Lanari *et al.*, 1990; Lanari and Zaritzky, 1991).

The most important chemical changes that can proceed during freezing and frozen storage are: enzymatic reactions, protein denaturation, lipid oxidation, degradation of pigments and vitamins, and flavour deterioration (Zaritzky, 2006).

11.1.2.1 *Enzymatic reactions*

Storage at low temperatures can decrease the activity of enzymes in tissues but cannot inactivate them. In raw products, hydrolases (enzymes catalysing hydrolytic cleavage) such as lipases (carboxylic ester hydrolases), phospholipases (phosphatid cleaving enzymes), proteases (hydrolases cleaving peptide bonds) etc., may remain active during frozen storage. Hydrolytic enzymes can produce quality deterioration in products that are not submitted to thermal treatments before freezing; however, blanching of vegetables or cooking of meat inactivates these enzymes (Sista *et al.*, 1997). Lipases and phospholipases, hydrolyse ester linkages of triacylglycerols and phospholipids, respectively; the hydrolysis of lipids can lead to undesirable flavour and textural changes. Certain lipases can remain active in frozen food systems stored even at −29°C. Lipase activity is evident in the accumulation of free fatty acids. Freezing may accentuate lipolysis by disrupting the lysosomal membrane that releases hydrolytic enzymes, especially at low freezing rates and under fluctuating temperatures. During storage, the release of short-chain free fatty acids can lead to hydrolytic rancidity, producing off flavours and may interact with proteins forming complexes that affect texture.

Proteases catalyse the hydrolysis of proteins to peptides and amino acids; in meat this endogenous enzymes are considered beneficial, tenderising the muscles during *rigor mortis* (Sista *et al.*, 1997). Conditioned meat on freezing not only retains the texture quality, but also has a smaller tendency to drip on thawing.

The browning of plant tissue is caused by enzymatic oxidation of phenolic compounds in the presence of oxygen. Disruption of cells by ice crystals can start enzymatic browning by facilitating contact between *o*-diphenol oxidase and its substrate. The oxido-reductases are of primary importance because their action leads to off-flavour and pigment bleaching in vegetables, and to browning in some fruits.

In vegetable and fruit tissues, endogenous pectin methyl esterases catalyses the removal of the methoxyl groups from pectins. In the case of frozen strawberries, these enzymes produce gelation during storage. Hydrolytic enzymes, such as chlorophylases and anthocyanases present in plants, may catalyse destruction of pigments in frozen tissues affecting the colour, if they are not inactivated by blanching.

11.1.2.2 *Protein denaturation*

The main causes of freeze-induced damage to proteins are ice formation and recrystallisation, dehydration, salt concentration, oxidation, changes in lipid groups and the release of certain cellular metabolites. Freeze-induced protein denaturation, and related functionality losses are commonly observed in frozen fish, meat, poultry, egg products and dough.

During freezing, proteins are exposed to an increased concentration of salts in the unfrozen phase; the high ionic strength can produce competition with existing electrostatic bonds, modifying the native protein structure. Losses in functional properties of proteins are commonly analysed by comparing water-holding capacity, viscosity, gelation, emulsification, foaming and whipping properties. Freezing has an important effect in decreasing water-holding capacity of muscle systems on thawing, producing changes also in protein solubility. This decrease occurs during freezing, because water–protein associations are replaced by protein–protein associations or other interactions (Xiong, 1997). Proteins exposed to the aqueous medium of the biological tissues have a hydrophobic interior, and charged (or polar) side chains in the surface. The migration of water molecules from the interior of the tissue during extracellular freezing leads to a more dehydrated state disrupting protein–solvent interactions. Protein molecules exposed to a less polar medium have a greater exposure of hydrophobic chains, modifying protein conformation. To maintain the minimum free energy, protein–protein interactions via hydrophobic and ionic interactions occur, resulting in protein denaturation and the formation of aggregates (Wagner and Añón, 1986).

Oxidative processes during frozen storage can also contribute to protein denaturation; oxidizing agents (enzymes, haem and transition metals) can react with proteins.

11.1.2.3 *Lipid oxidation*

Lipid oxidation is another reaction that severely limits the shelf-life of a frozen product, leading to loss of quality (flavour, appearance, nutritional value, and protein functionality). Lipid oxidation is a complex process that proceeds upon a free radical process (Erickson, 1997). During the initiation stage a hydrogen atom is removed from a fatty acid, leaving a fatty acid alkyl radical that is converted in the presence of oxygen to a fatty acid peroxyl radical. In the next step, the peroxyl radical subtracts a hydrogen from an adjacent fatty acid forming a hydroperoxide molecule and a new fatty acid alkyl radical. Breakdown of the hydroperoxide is responsible for further propagation of the free radical process. Decomposition of hydroperoxides of fatty acids to aldehydes and ketones is responsible for the characteristic flavours and aromas (rancidity).

Redox-active transition metals are major factors catalysing lipid oxidation in biological systems; iron in particular is a well-known catalyst.

Both enzymatic and non-enzymatic pathways can initiate lipid oxidation. One of the enzymes considered important in lipid oxidation is lipoxygenase that is present in many plants and animals and can generate offensive flavours and also loss of pigment colours. Lipid oxidation is particularly important in meats (including poultry) and seafood. Fatty meats and fish, in particular, suffer from this adverse reaction during long-term frozen storage.

Oxidative flavour deterioration is produced in both plant and animal products. It is identified more with frozen muscle than with frozen vegetable products, because blanching is typically applied to vegetables prior to freezing.

Pigment degradation and colour quality deterioration are also related to lipid oxidation. Haem pigments in red meats, and carotenoid-fading in salmonid flesh, are subjected to oxidative degradation during storage. Chlorophyll is also capable of serving as a secondary substrate in lipid oxidation.

11.1.3 Effect of freezing and frozen storage on food quality

11.1.3.1 *Plant tissues*

The epidermis of plant tissues which is structurally adapted to provide protection against biological or physical stress consists of tightly packed cells containing waxy material. The parenchymatous tissue performs much of the metabolic activities of the plant and is constituted by semi-rigid, polyhedral cells containing cellulosic cell walls bounded by pectinaceous middle lamella and often including a network of air spaces. Mature plant cells contain starch granules and a number of organelles such as chloroplasts, chromoplasts, large vacuoles, protein bodies, amyloplasts. The vacuole, which may account for most of the space inside a mature plant cell, contains organic acids, phenols and hydrolytic enzymes that can be released when the fragile membranes are disrupted by freezing. Firmness and crispiness (textural properties associated with fruits and vegetables) are attributed to the osmotic pressure developed within the cell when pressure is exerted on the rigid cell walls. Exposure of cell wall to hydrolytic enzymes that attack pectins, hemicelluloses, and non-cellulosic carbohydrate material constituents would dissipate the osmotic pressure (Sista *et al.*, 1997).

During freezing, the formation of ice in the tissues of fruits and vegetables results in undesirable changes in texture, such as loss of turgour. Preservation by freezing is applicable only to those vegetables that are cooked for consumption. Generally fruits do not require blanching and can be packaged in sugar or syrup before freezing. Most vegetables benefit from quick freezing, which is better to maintain the textural quality of the tissue. During slow freezing of fruit tissues, extracellular ice crystals can damage cell walls and middle lamellae to such an extent that the thawed product is much softer than the fresh fruit. The texture damage often found in frozen thawed plant tissues is attributed to the semi-rigid nature of the cells. Loss of membrane semi-permeability and disruption of cellular compartments in fruits and vegetables can be minimised using rapid freezing rates, low storage temperatures, and slow thawing. Softening caused by freeze–thawing can sometimes be minimised by pretreatment of the tissue with calcium chloride and/or sucrose.

The most common chemical changes related to quality deterioration in frozen fruits and vegetables are reactions that produce off-odours and off-flavours, pigment degradation, enzymatic browning and autoxidation of ascorbic acid. Certain frozen fruits undergo enzymatic oxidative discolouration due to the action of the polyphenoloxidases on naturally phenolic constituents. Antioxidants such as ascorbic acid can be used to inhibit enzymatic reactions. Vegetables undergo enzymatic browning if they are not blanched. Another cause of change in colour is the partial loss of anthocyanin pigments, such as in frozen berries. Improvement of colour and flavour can be achieved by packaging the fruits with sugar or syrups, and by decreasing storage temperature to −18°C. Chlorophylls and carotenes are usually well retained in frozen vegetables, but some degradation occurs during storage. During frozen storage of blanched green vegetables at −18°C or above, the bright green colour of the recently frozen product (*a* and *b* chlorophyll pigments) slowly changes to brownish green (pheophytin). The rate of pigment degradation depends on the amount of tissue damaged prior to freezing.

Unblanched (or underblanched) vegetables change flavour due to the action of lipases and lipoxygenases. Volatile compounds such as carbonyl compounds and ethanol accumulate in the tissue, producing off-odours.

11.1.3.2 *Muscle tissues*

Muscle cells (myofibrils) are long, parallel bundles of contractile proteins (myosin and actin). These flexible and elongated fibres are aligned in a parallel arrangement, with minimal air spaces and separated by an extracellular matrix rich in glycoprotein. A large portion of hydrolytic enzymes is located in the lysosome (an organelle similar to the vacuole in plant cells). After death of the animal, meat is left in a contraction state until hydrolytic enzymes present in the cytoplasm can disrupt the proteins and tenderise the meat.

Meats have excellent frozen storage life (Marsdon and Hendrickson, 1993). However, freezing and thawing of myosystems decrease the water-holding capacity of the tissues, resulting in drip losses. Some muscles are susceptible to cold shortening and thaw rigor; in general, allowing the muscle to undergo *rigor mortis* prior to freezing is recommended. The two important causes of quality loss in frozen stored meat are lipid oxidation and protein denaturation. The development of oxidative rancidity in frozen muscles is caused by the accumulation of carbonyl compounds formed during autoxidation of muscle lipids. Enzymatic hydrolysis of lipids, with the liberation of free fatty acids, occurs during the frozen storage of meats. Fish and pork, which contain a higher proportion of more reactive polyunsaturated fatty acids, are more susceptible to the development of rancidity.

In red meats, colour is determined by the relative concentration of purple myoglobin, bright red oxymyoglobin and brown metmyoglobin. Freezing–thawing accelerates pigment oxidation and the production of metmyoglobin. The colour of meats may become unattractive to the consumer; this may be due either to desiccation of the meat surface with the consequent development of grey areas (attributed to light scattering effects without ice crystals) or to the darker colour of myoglobin compared to oxymyoglobin.

Metmyoglobin formation in red meats and carotenoid bleaching in fish and poultry tend towards parallel fat oxidation (Haard, 1997). In the case of fish, the major problems found during freeze processing are oxidative deterioration, dehydration, toughening, loss of juiciness, and excessive drip.

11.1.3.3 *Ice cream*

During its manufacture it is very important to produce a large number of small ice crystals to get an optimal texture. Recrystallisation is the most important change during frozen storage that produces quality losses and limits the storage life. Storage temperature fluctuations must be avoided (Kennedy, 2000; Goff, 2006).

11.1.3.4 *Cheese*

During the freezing and frozen storage of cheese, the body and texture become more crumbly and mealy. An increase in the unordered structure of proteins in frozen cheese is consistent with the greater proteolysis and damage to the microstructure, affecting the viscoelastic properties (Bertola *et al.*, 1996; Graiver *et al.*, 2004; Goff, 2006).

11.1.3.5 *Frozen eggs*

Liquid egg products are pasteurised to eliminate *Salmonella* before freezing. Appropriate freezing processes generally cause minor changes in raw egg white; however, freezing egg yolk at a temperature below −6°C causes gelation that is an irreversible change in its fluid texture leading to an undesirable product. Addition of cryoprotectants improves the quality and extends the shelf-life. For frozen egg products sodium chloride and sucrose at a level of 10% are commonly added to the yolk to prevent gelation. Syrup, glycerin, phosphates and sugars can also be used. Fast freezing results in less gelation both of egg yolk and of whole egg due to lesser damage to protein structure (Lai, 2006).

11.1.3.6 *Frozen ready meals*

They are complex multi-component products with a wide variety of ingredients that are cooked and then frozen (Creed, 2006). The modes of deterioration in frozen convenience foods during frozen storage are: rancidity in meat portions, weeping and curdling of sauces and discolouration. They normally include sauces, gravies which act as protective agents for the solid elements, minimising dehydration and reducing rancidity development. Lipid oxidation leads to flavour deterioration; colour changes can also result from pigment degradation in meat and vegetables because of lipid oxidation. Sauces based on gelatinised starch are composed of an amylose matrix filled with granules of different degrees of fragmentation. Starch gels are metastable and non-equilibrium systems, and therefore undergo structural transformation during storage and processing. Upon ageing, starch retrogradation is produced, which involves partial crystallisation of starch components. Starch molecules reassociate, depending on the affinity of hydroxyl groups and the attractive forces or hydrogen bonding between hydroxyl groups on adjacent chains. The process induces an increase in paste rigidity in the viscoelastic system. Starch retrogradation consists of two distinct processes: a rapid gelation of amylose via formation of double helical chain segments that is followed by helix–helix aggregation, and a slow recrystallisation of short amylopectin chain segments (Miles *et al.*, 1985; Morris, 1990). Starch systems undergo freezing damage such as rheological changes and syneresis after thawing, which may alter the desired characteristics of the products, reducing consumer acceptability. Freezing rate has an important effect on exudate production in starch pastes (Ferrero *et al.*, 1993b). High freezing rates (>100 cm h^{-1}) lead to lower exudate values. During frozen storage at $-5°C$ a spongy matrix is formed. This structure was not observed when samples were frozen in liquid nitrogen. The spongy structure is attributed to the water release caused by slow freezing, producing local high-starch polymer concentrations and interaction between molecular chains. Amylopectin retrogradation measured by differential scanning calorimetry (DSC) was only detected for corn starch pastes frozen at low rates (<1 cm h^{-1}). Storage temperature also has an effect on starch retrogradation. Amylopectin retrogradation was detected at $-1°C$ and $-5°C$, but not at lower temperatures, (Ferrero *et al.*, 1993a, 1994). Textural characteristics of sponginess observed at low freezing rates, and at high frozen storage temperatures, can be attributed to amylose retrogradation, described as the coarsening of the network.

The use of hydrocolloids is highly recommended in the food industry, in order to restrict syneresis and ice crystal growth besides their traditional role of texturisers and emulsion stabilisers. The addition of xanthan gum to starch gelatinised systems decreases exudate production and avoided the formation of the spongy structure, even at low freezing rates and at high storage temperatures. These effects can be explained, considering that amylose–hydrocolloid interaction competes with amylose–amylose aggregation, decreasing the probability of amylose retrogradation occurrence (Ferrero *et al.*, 1993a, 1993b).

11.1.4 Nutritional aspects of freezing and frozen storage

In general, freezing preservation is considered to be a method that delivers a product comparable in nutritional quality to the fresh product. Available experimental data tend to show that this method is less destructive than other processing methods. The degradation of vitamins during the freezing process, in contrast to lipid and protein degradation, generally has a more significant impact on nutritional value. The main adverse effect of extended frozen storage on nutritive value may be the losses of the more labile vitamins, such as some of the water soluble B vitamins (B1, thiamin; B2, riboflavin) and vitamin C (ascorbic acid), that are frequently used as indicators of the food processing effect (Jul, 1984). Ascorbic acid losses have been studied in fruits and vegetables and are attributed to oxidative mechanisms during frozen storage; however, blanching is an important contributor to vitamin degradation. In the presence of dissolved oxygen, ascorbic acid in aqueous solution is oxidised to dehydroascorbic acid and other oxidised products. Ascorbate oxidase exists naturally in many plant tissues, and if it is not inactivated, it catalyses ascorbic acid oxidation during the freezing process.

11.1.5 Microbiology of frozen products

Microbial deterioration mechanism is not a problem in frozen foods because they are stored at temperatures below the lower limits of microbial growth (approximately −10°C). However, with temperature fluctuations during storage and distribution, these may become significant. The major objective of freezing as a method for food preservation is to prolong storage life of the products by retarding or inhibiting microbial growth. Freezing (and the subsequent frozen storage) can produce certain lethality in some micro-organisms but this process is very slow and variable, depending on the type of foodstuff. Freezing cannot be regarded as a method to reduce microbial contamination; for this reason, hygienic and sanitary conditions prior to processing are very important. Storage temperatures below −10°C inhibit bacterial growth, whereas yeasts and moulds cannot multiply below −12°C and −18°C, respectively (Zaritzky, 2000).

11.1.6 Glass transition temperature and stability of frozen food

Most food materials can be considered biopolymers and exist in an amorphous state. Melting of crystalline polymers results in the formation of an amorphous system, which can be supercooled to a viscoelastic, rubbery state or to a solid glassy state. The glass transition takes place when a completely amorphous polymer undergoes a transition from a vitreous state to a high-mobility rubbery state where diffusion-controlled changes occur. The effect of water as a plasticiser of food systems is manifested as a depression of the glass transition temperature of the amorphous components. Below the glass transition temperature (T_g), polymer material becomes glassy, and the molecular motion is so slow, that crystallisation does not occur in a finite period of time; thus, deterioration governed by diffusion is inhibited. At temperatures above T_g and below the crystal melting temperature (T_m), the material is rubbery and sufficient motion of polymers occur to allow crystallisation. Ice formation in food materials results in freeze concentration of the solutes; non-equilibrium ice formation is a typical phenomenon of rapidly cooled biological materials at low temperatures. Freeze concentration and lowering of temperature increase the viscosity of the unfrozen phase until this concentrated solution becomes a glass (Roos, 1995). The glass transition temperature of a slowly frozen sample is higher than that of a rapidly cooled sample, and is considered to be the glass transition temperature of the freeze-concentrated solute matrix surrounding the ice crystals in a maximally frozen solution (T_g') (Levine and Slade, 1986). At this temperature the unfrozen water in the matrix is unable to freeze and then ice formation ceases within the time-scale of normal measurement (Franks, 1985; Goff, 1997; Levine and Slade, 1988, 1991; Roos and Karel, 1991; Le Meste *et al.*, 2002).

Below T_g' the unfrozen matrix takes on solid properties because of reduced molecular motion, which is responsible for the marked reduction in translational, not rotational, mobility (Slade and Levine, 1995; Roos and Karel, 1991). Solutions cooled rapidly to temperatures lower than T_g' show non-equilibrium ice formation. In rapidly cooled systems being in the glassy state, ice-formation can occur during rewarming (exothermic devitrification) at a temperature above T_g' (Roos and Karel, 1991a, 1991b) and ice is produced by crystallisation of the immobilised water, before the onset of ice melting. Roos and Karel (1991) suggested that formation of such maximally frozen solutions in the unfrozen matrix requires annealing slightly below the initial ice melting temperature within the maximally frozen solution. The cryostabilisation of frozen foods is related to the possibility of maintaining the product below the glass transition temperature of the freeze concentrated matrix (T_g'), or to modify the formulation of the food to increase glass transition temperatures to above normal storage temperatures. If a product is stored at a temperature below T_g' it may be expected to be composed of ice and a freeze-concentrated phase in the glassy state and long-term stability may be predicted. If the storage temperature is between T_g' and T_m, the freeze-concentrated phase is not in the glassy state, it is more diluted (rubbery) and processes governed by diffusion are not inhibited. These processes can lead to deterioration during food storage. Frozen foods stored below T_g' are stable to ice recrystallisation and other physical changes (Fennema, 1996).

Food systems based on gelatinised starch may undergo important textural changes related to amylose and amylopectin retrogradation (Miles *et al.*, 1985; Morris, 1990); they can show syneresis (exudate production) and a spongy texture due to slow freezing and frozen storage at relatively high temperatures. These changes may make such products unacceptable for the consumer. As starch retrogradation is a recrystallisation process, it is controlled by diffusion and depends on solute mobility in the system.

Working with a frozen gelatinised starch system, Ferrero *et al.* (1996) demonstrated that storage at temperatures below T_g' minimises physical changes in the concentrated matrix. Using DSC, they reported T_g' onset values ranging between −4.5°C and −5.5°C for annealed frozen starch pastes; Slade and Levine (1991) reported similar temperature ranges. Rapid frozen samples stored at −10°C and −20°C had a homogeneous structure without a spongy network; they showed no evidence of starch retrogradation because at temperatures below T_g' amylose and amylopectin chains in concentrated matrix have a reduced mobility that limits the molecular association responsible for the crystallisation (retrogradation) of starch molecules. The addition of hydrocolloids such as xanthan gum helps to maintain in the frozen samples the rheological characteristics of unfrozen starch pastes, even under low freezing rate conditions. Xanthan gum does not prevent amylopectin retrogradation, because amylopectin remains in the starch granule but inhibits the development of a spongy structure produced by amylose retrogradation. Ferrero *et al.* (1996) demonstrated that the addition of small quantities of hydrocolloids (in the ranges used in common formulations) did not change T_g' of the samples, but had an important role in minimising structural damage in frozen starch systems; they improved rheological properties related to amylose retrogradation. Additionally, hydrocolloids can make systems in the rubbery state more viscous, decreasing molecular mobility and avoiding retrogradation related to sponge formation (Ferrero *et al.*, 1994).

Frequently, low molecular solutes such as sucrose are added to starch–water systems; the addition of sucrose shifts T_g' from −5°C towards a lower value of −23°C. By measuring viscoelastic properties, Ferrero and Zaritzky (2000) reported amylose retrogradation (spongy structure and high values of the dynamic complex modulus G*[1]) during common storage temperatures such as −18°C. This behaviour was not observed in quickly frozen samples stored at −80°C whose G* values coincided practically with those of the just quickly frozen sample. Results can be explained again considering that sample storage temperature (−18°C) lay above T_g' (−23°C) allowing molecular mobility.

Kasapis (2006) reported that the glass transition measured by calorimetry (DSC) remains unaltered by the presence of low levels of polysaccharide; however, the mechanical profile of the rubber-to-glass transition is strongly influenced by the polysaccharide, particularly if

[1] G* is the complex viscoelastic shear modulus of a material ($G^* = G' + iG''$) and represents the contributions of G′ (the shear storage modulus or elastic component) and G″ (the shear loss modulus or viscous component).

its network forming is affecting the rheological properties of the system. Reduction in the diffusion kinetics and increase in the relaxation time of the unfrozen phase in the presence of polysaccharide as monitored by stress relaxation studies, confirmed the utility of a three-dimensional network in the stabilisation at sub-zero temperatures.

Reid *et al.* (2003) have reported the following T_g' values (mobility temperatures) using DSC and a capacitance cell: −20°C (potato), −23°C (green been), −24°C (broccoli), −29°C (spinach), −20°C (apricot), −25°C (plum), −30°C (pear), −27°C (apple), −33°C (strawberry), −27°C (salmon), −30°C (cod), −24°C and −34°C (beef), −23°C (chicken). Knowledge of the influence of factors such as freezing rate and composition of the system on glass transition temperature would help to determine adequate formulations, processing or storage conditions in order to enhance frozen foods shelf-life.

11.2 TEMPERATURE REQUIREMENTS DURING FROZEN STORAGE

Frozen foods deteriorate during storage by different mechanisms and low temperatures are required to maintain the food quality. Deterioration is accentuated by the fluctuating time–temperature environment during storage.

The required temperature conditions need to be maintained in the cold chain from the producer to the consumer assuring a temperature of −18°C – a limit set by the majority of international regulations.

Monitoring and control of the cold chain is a prerequisite for reliable quality management and optimisation (Giannakourou *et al.*, 2006). According to the EU Directive 89/108 (Quick Frozen Food Directive, QFF), after quick freezing, the product should be maintained at −18°C or colder after thermal stabilisation (Anonymous, 1989). Some frozen foods (broilers, beef, butter) have a fairly long storage life even at −12°C whereas foods such as lean fish require storage temperatures around −28°C. In the United States a temperature of −18°C or colder is recommended; however, ice creams require −23°C or colder. The EU Directive 92/1 (Anonymous, 1992) requires that a temperature-recording device must be installed in each storage facility to register and store for at least a year the temperature data of air surrounding the perishable food.

Legislation on control of transport equipment and temperatures during transport has been getting more and more strict. The Quick Frozen Food Directive requires that the temperature of quick frozen food must be maintained at −18°C or colder at all points of the product, with possible brief upward fluctuations of no more than 3°C during transport.

Storage temperatures can fluctuate significantly over the complete cold chain. Typically, −23°C has been the average for the manufacturer's cold store with a maximum target of no more than −18°C during fluctuations. This may rise to −18°C during distribution to either the wholesaler or the retailer. A further rise in the mean is quoted at retail level. Although −18°C is the target, the norm is to maintain it at least at −15°C. The fluctuations become even greater when the consumer enters the chain, because this transport can start surface thawing (McKenna, 2006). The domestic freezer will probably be close to that in the retail outlet, but can have large variations. In addition, temperature changes during the defrost cycles at both retail and domestic levels can have a significant impact. However, it is generally admitted that fluctuations of 3–5°C may occur in any one part of the chain.

Fluctuations in storage temperature significantly affect the shelf-life of the product. The amplitude of the ambient thermal fluctuation is reduced inside the package; besides it was mathematically demonstrated that ambient temperature fluctuations with large oscillation periods cause higher amplitude fluctuations inside the product, this situation being more detrimental to the quality of frozen food, than short oscillation periods (Zaritzky, 1982).

11.3 SHELF-LIFE OF FROZEN FOOD

The shelf-life of a frozen food is a complex concept that depends on the characteristics of the food product and the environmental conditions to which the food is exposed after being subjected to the freezing process. Packaging also plays an important role in maintaining the quality of foods. The International Institute of Refrigeration (IIR, 1986) has recommended two definitions: practical storage life (PSL) and high-quality life (HQL).

The *practical storage life* (PSL) or acceptability time is the period of proper frozen storage after freezing of an initial high-quality product, through which the frozen food retains its quality characteristics and is suitable for consumption or for use in further process.

High-quality life parameter (HQL) is defined as the storage period through which the initial quality was maintained from the time of freezing up to the point where 70% of the trained taste panel members are capable of detecting a noticeable difference between the frozen food stored at different temperatures and the corresponding controls stored at −40°C in a triangular sensory test; therefore this parameter is also known as Just Noticeable Difference (JND).

The three main factors that affect product quality of any given frozen food are: initial quality of the original foodstuff, processing and packaging of the product, and temperature and duration of storage. For any specific frozen product, which mode determines its shelf-life, depends on the product characteristics (raw materials, ingredients, formulation), prefreezing treatment, freezing process, packaging film, and storage temperature. Besides, the shelf-life of a frozen food can be extended through ingredient selection, process modification and change of package or storage conditions.

Data on practical storage and high-quality life of different frozen foods have been reported in literature (Van Arsdel *et al.*, 1969; Jul, 1984). Table 11.1 shows values of PSL adapted from IIR (Anonymous, 1986).

The ratios between practical storage life (PSL) and HQL for frozen foods range between 2 and 3. Reported values are the following: 1.9–2.0 for lean meat, 2.0–2.4 for fatty meat, 1.9–2.2 for lean fish, 2.4–2.7 for fatty fish, 3.1–3.5 for vegetables, 2.8–3.1 for fruit and 2.8–3 for precooked foods. Considering that PSL values are greater than the HQL values, the determination of HQL in frozen foods can be used to shorten shelf-life testing times (Fu and Labuza, 1997).

11.3.1 Shelf-life testing

Shelf-life testing consists basically of selecting the quality characteristics, which deteriorate most rapidly in time and the mathematical modelling of the change. Quality characteristics, which depend on the specific product, usually require professional judgement.

The criterion for the end of shelf-life may be variable depending on the definition of product quality grade, so the shelf-life of a product may also be variable. Chemical oxidations that cause an off-flavour development and loss of colour are readily recognisable by consumers.

Several frozen foods deteriorate mainly by slow chemical reactions such as loss of nutritional value. For example, the vitamin C content of some frozen fruits and vegetables may

Table 11.1 Effect of storage temperature on practical storage life (in months) of frozen foods.

Product	Storage temperature °C		
	−12°C	**−18°C**	**−24°C**
Fruits			
Peaches, apricots, cherries (raw)	4	18	>24
Raspberries, strawberries (raw)	5	24	>24
Raspberries, strawberries (in sugar)	3	24	>24
Concentrated fruit juice	—	24	>24
Vegetables			
Asparagus (with green spears)	3	12	>24
Beans (green)	4	15	>24
Broccoli	—	15	24
Brussels sprouts	6	15	>24
Carrots	10	18	>24
Cauliflower	4	12	24
Cut corn	4	15	18
Mushrooms	2	8	>24
Peas	6	24	>24
Peppers (red and green)	—	6	12
French fried potatoes	9	24	>24
Spinach	4	18	>24
Onions	—	10	15
Meat and meat products			
Beef, ground/minced	6	10	15
Beef steaks	8	18	24
Veal steaks	6	12	15
Lamb steaks	12	18	24
Pork steaks	6	10	15
Bacon (sliced, vacuum packed)	12	12	12
Chicken (whole or cuts)	9	18	>24
Turkey (whole)	8	15	>24
Liver	4	12	18
Seafood			
Fatty fish (lazed)	3	5	>9
Lean fish	4	9	>12
Shrimps (cooked/peeled)	—	2	5
Eggs			
Whole egg	—	12	>24
Milk products			
Butter (lactic, unsalted, pH 4.7)	15	18	20
Butter (lactic, salted, pH 4.7)	8	12	14
Cream	—	12	15
Ice cream	1	6	24
Bakery and confectionery			
Cakes (cheese, chocolate, fruit)	—	15	24
Breads	—	3	—
Raw dough	—	12	18

Source: IIR Recommendations for the Processing and Handling of Frozen Foods. International Institute of Refrigeration, Paris, 1986.

fall below the required standard before sensory quality becomes inadequate. The criteria for shelf-life may also vary depending on the sensitivity of the consumer. For consumers, taste, odour, and appearance are the most obvious criteria; in academia and in the industry, both sensory evaluation and instrumental measurements of a given quality index (e.g. vitamin C level) are usually conducted.

Since 1950 different time–temperature–tolerance (TTT) experiments have been performed by the USDA Western Regional Research Center in Albany, California to analyse the stability of a great number of frozen foods (fruits, vegetables and meats) stored at different temperatures for various periods of time (Van Arsdel *et al.*, 1969; Jul, 1984). The quality was measured by organoleptic testing carried out by taste panels, and by various objective measurements such as ascorbic acid deterioration, the change of chlorophyll to pheophytin, etc. Results were expressed as straight lines in a semi-logarithmic diagram as stability time (number of days taken to undergo a certain amount of quality deterioration) versus storage temperature. The PSL of frozen foods depends not only on time and temperature of storage, but also to a large extent on product, process and packaging; these are the so-called PPP factors introduced by Jul (1984). In general, the changes in quality during frozen storage at different temperatures are cumulative and irreversible; these changes are normally smaller at lower temperatures, thus storage temperature is the determinant factor that governs quality and shelf-life.

Reid *et al.* (2003) have proposed a new method for accelerated shelf-life prediction for frozen foods which requires around 60 days and involves direct determination of shelf-lives at the more elevated storage temperature (where the change is more rapid), coupled with information on the mobility temperature (T_g'). The authors measured mobility temperatures of typical foods, chlorophyll and ascorbic acid retentions in plant tissues, myofibrillar protein solubility in meat muscle and dimethylamine content in fish muscle. The role of T_g' in diffusion-controlled processes is clear; however, its influence on chemical reactions has to be investigated more deeply.

11.3.2 Modelling the loss of quality in frozen food

The knowledge of shelf-life is very important in the development of any new frozen food product, or when changes of packaging or storage/distribution conditions are introduced. This life must at least exceed the minimum distribution time required from the processor to the consumer.

Quality is a complex attribute of food, which influences the degree of its acceptability to the consumer. In the case of frozen food, physical phenomena that lead to quality changes and kinetics of chemical reactions, play an important role in food shelf-life. The change of food quality can be represented by the loss of one or more quality index (e.g. flavour, colour, vitamin C) and by the formation of undesirable products (e.g. peroxide value).

These quality factors can be sensory parameters or can be measured using instrumental analysis. Chemical and physical analysis such as moisture, nutrient loss, free fatty acids, peroxides, oxidative volatiles (e.g. hexanals) and colour that correlates closely to sensory attributes can supplement sensory techniques increasing the confidence level of the results (Fu and Labuza, 1997). Once a frozen product starts its distribution from the manufacturer's plant to warehouse, distribution centre, retail store and finally consumer freezer, the rate of quality loss is primarily temperature dependent.

Considering that the representative quality factor (P) changes with time (t), the following kinetic equation can be proposed, where k is the apparent reaction rate constant and n the apparent order of the reaction:

$$\frac{dP}{dt} = kP^n. \tag{11.4}$$

Most published data related to quality deterioration do not give rate constants as a function of temperature; they are in the form of an overall storage life (end-point analysis) as a function of storage temperature. The storage life θ can be considered determined by a defined value of the property P_θ . Integrating equation (11.4) with $n = 0$ (zero order reaction) the following is obtained:

$$f_0(P_\theta) = P_\theta - P_0 = k\theta \tag{11.5}$$

where P_0 is the initial quality value, P_θ corresponds to the quality value at the end of shelf-life θ and $f_0(P_\theta)$ is the quality function for a zero order reaction, evaluated at the end of the storage life. Equation (11.5) implies that the rate of quality loss is constant at constant storage temperature.

For a first order reaction, the change in quality follows an exponential decay with storage time then:

$$f_1(P_\theta) = \ln\left(\frac{P_\theta}{P_0}\right) = k\theta \tag{11.6}$$

where $f_1(P_\theta)$ is the quality function of the first order reaction evaluated at θ.

As can be seen in equations (11.5) and (11.6), the shelf-life (θ) of a frozen food depends on storage temperature and is determined by a defined quality change given by the function $f(P_\theta)$. The product $k\theta$ will be constant at different storage temperatures and equal to the change in the property defined by the test conducted to determine the storage life. The shelf-life θ of the frozen food stored at a constant temperature is then inversely proportional to the rate constant k. However, when it is necessary to determine, at a given temperature, the effect of time on a quality property, knowledge of the reaction kinetics is necessary. The majority of food reactions that have been studied have been characterized as pseudo-zero or pseudo-first order. When a reaction has less than 50% conversion, the zero and first orders might be indistinguishable in goodness of fit (Labuza, 1979).

If the end of shelf-life is within less than 20% conversion, for practical purposes either a pseudo-zero or pseudo-first order model can be used.

The temperature dependence of the deterioration rate constant can be described through the Arrhenius equation:

$$\ln k = \ln k_0 - \frac{E_a}{RT} \tag{11.7}$$

where k_0 is a pre-exponential factor; E_a is the activation energy in J mol^{-1}; R is the universal gas constant (8.3144 J mol^{-1} K); T is the absolute temperature. Activation energy can be calculated from the slope of the linear plot of ln$k$ as a function of reciprocal absolute temperature ($1/T$). A higher value of E_a means that the reaction is more temperature sensitive, and a small change in T produces a large change in the rate constant. The ratios between the rate constants k and k_1 corresponding to the absolute temperatures T and T_1 and between storage life values θ and θ_1, can be calculated as follows:

$$\ln\left(\frac{k}{k_1}\right) = -\frac{E_a}{R}\left(\frac{1}{T} - \frac{1}{T_1}\right) = \ln\left(\frac{\theta_1}{\theta}\right) \tag{11.8}$$

where the following relationship was considered: $k \cdot \theta = k_1 \cdot \theta_1$. Equation (11.8) shows that activation energy values can be estimated from shelf-life data evaluated at least at three different temperatures around those normally used for frozen storage of the product under test. Rodriguez and Zaritzky (1983) plotting $1/\theta$ values (PSL data of frozen beef, obtained from Jul, 1969) as a function of $1/T$, estimated that the activation energy for frozen beef deterioration is 39.5 kJ mol^{-1}. Other authors (Giannakourou *et al.*, 2006) reported activation energy values of $E_a = 136.8$ kJ mol^{-1} for ascorbic loss and $E_a = 79.2$ kJ mol^{-1} for colour loss, both in frozen green peas.

The Q_{10} approach is often used for estimation of the temperature acceleration of shelf-life. Q_{10} of a food product indicates the magnitude of the increase in the rate of deterioration (k) when the temperature is increased by 10°C and can be calculated from shelf-life values:

$$Q_{10} = \frac{k \text{ at temp } T + 10^\circ\text{C}}{k \text{ at temp } T} = \frac{\theta \text{ at } T}{\theta \text{ at } T + 10^\circ\text{C}} \tag{11.9}$$

Besides, Q_{10} can be related to the Arrhenius equation and the shelf-life model through the following expression:

$$Q_{10} = \exp\left(\frac{10E_a}{RT(T+10)}\right) = \frac{\theta_T}{\theta_{T+10}} \tag{11.10}$$

Q_{10} values ranging between 2 and up to 20 have been reported for frozen foods: 2 for frozen beef, 4 for pork sausage, 3.2 for ground hamburger, 3.5 for raw poultry, 9 for fatty fish, 3.2 for fried poultry, 8 for pork (Labuza, 1982; Jul, 1984; Labuza and Schmidt, 1985).

Frozen foods are often exposed to a variable temperature environment. In such cases, the change in the quality property P depends on the kinetics of the reaction. The quality function $f(P)$ (zero or first order) of the food stored for the total time t_{tot} under changing environmental conditions can be estimated from the following integral:

$$f(P) = \int_o^{t_{tot}} k(T(t))dt \tag{11.11}$$

where the kinetic constant k is a function of temperature that changes with time.

If the temperature function $T(t)$ is discretised in small time increments (t_i) of constant temperature T_i, then the change in the quality property is given by (Giannakourou *et al.*, 2006):

$$f(P) = k_{ref}\sum_i \exp\left[\frac{-E_a}{R}\left(\frac{1}{T_i} - \frac{1}{T_{ref}}\right)\right]t_i. \tag{11.12}$$

Using equation (11.12), the extent of quality loss (change in the property P given by $f(P)$) during frozen storage can be calculated when both the kinetic equation and the time–temperature history are known.

There is a continuous quality loss in frozen storage that is in general cumulative, thus the total quality loss through freezing, storage, transport and distribution can be calculated by adding the losses at each step of the process.

When the loss of quality follows a zero order reaction, the fraction of shelf-life consumed at each stage in which temperature remained constant, can be estimated as follows:

$$\frac{P_i - P_0}{P_\theta - P_0} = \frac{t}{\theta} \tag{11.13}$$

where P_0 is the initial quality value, P_θ the quality value at the end of shelf-life θ and P_i the quality value after the stage at constant temperature T_i. Only when deterioration reaction is linear with time (zero order reaction), shelf-life loss can be considered equivalent to the loss of quality. In this case the fraction of quality lost is equal to the ratio between the interval of time at a given temperature and the shelf-life at this particular temperature. Then the percentage of PSL that the frozen food retains is given by:

$$\%\text{ quality retained} = \left(1 - \frac{t_T}{\theta_T}\right)100 \tag{11.14}$$

When the temperature history of the product is known, the fraction of shelf-life consumed can be calculated as the sum of the exposure times at each temperature interval t_{Ti} divided by the shelf-life at that temperature θ_{Ti}.

Thus the final quality retained by the frozen food submitted to different time–temperature intervals can be calculated assuming that the rule of additivity is valid for frozen foods (Jul, 1984), which means that the loss of remaining storage life or quality can be calculated from the knowledge of the prior time–temperature sequences the product has been exposed to

$$\%\text{ quality retained} = \left(1 - \sum_{1}^{m} \frac{t_{T_i}}{\theta_{T_i}}\right)100 = 1 - \frac{t_{\text{total}}}{\theta_{\text{T}_{\text{eff}}}} \tag{11.15}$$

An effective storage temperature T_{eff} can be defined as that constant temperature exposure which causes during the total exposure time (t_{total}) the same quality change as the variable temperature condition does.

Knowing the percentage of quality retained by the food submitted to variable temperatures and knowing the total exposure time, it is possible to calculate the storage life at the effective temperature (θ_{Teff}) and by interpolation the effective temperature (T_{eff}).

In conclusion, to quantify the percentage of quality retained by the food after frozen storage at different temperatures, it is not only necessary to know the activation energy values of the reactions but also the kinetic equations that represent the deterioration rate. When this kinetics is of zero order, a simplified expression given by equation (11.15) can be applied. As an indicative case study, the quality loss of frozen peas was analysed by Giannakourou *et al.* (2006) using kinetic equations that described green colour change and L-ascorbic oxidation; besides, activation energy values of these reactions allowed us to predict shelf-life of frozen products for different time–temperature scenarios.

11.3.3 Time–temperature integrators

Temperature of both the food and the environment should be monitored, to obtain an efficient record of the product history. Measurements can be carried out either by mechanical or by electronic equipment with or without the potential of recording and storing an electronic file of data. The three principal kinds of sensors are thermocouples, platinum resistances and thermistors (Giannakourou *et al.*, 2006).

An alternatively way to individually monitor the temperature conditions through the cold chain is the use of a time–temperature indicator/integrator (TTI). The operation of a TTI is based on mechanical, chemical or enzymatic systems that change irreversibly from the time of its activation. The rate of the change has to increase at higher temperatures in a manner similar to most deterioration food reactions. The change is usually expressed as a visible response (colour development and colour movement, etc.). The TTI can keep track of an accumulated time–temperature distribution function to which a perishable product is subjected from the point of manufacture to the display shelf of the retail outlet, or even to the consumer (Fu and Labuza, 1997). These devices are attached on the food itself or outside the packaging. The time–temperature integrators show a physical or a colour change and an essential condition is that the energy of activation of this device shall coincide with the energy of activation of the frozen food deterioration, that means having the same temperature sensitivity as the food they are monitoring (Rodriguez and Zaritzky, 1983). The ultimate purpose of their application is the translation of their reading to the quality status of the food through an appropriate algorithm.

The commercial devices available in the international market and the extensive work that has been published on TTI development for frozen food, have been reviewed by Giannakourou *et al.* (2006).

11.4 PACKAGING OF FROZEN FOOD

Packaging plays a key role in protecting the frozen product from contamination by external sources and from damage during its passage from the food producer to the consumer. The key functions of food package are: protection, preservation and presentation. Protection is necessary to prevent the following: external biological contamination; dehydration caused by moisture loss and freeze burn; oxidation of lipids, flavours, colours and vitamins; loss or gain of aromas; and physical damage (abrasion, fracture and compression) during storage and transport (Krochta, 2006). Frozen food packaging materials must be able to withstand low temperatures of −25°C to −30°C without embrittlement and sometimes high temperatures, such as microwave and boil-in-bag products.

The materials used, besides being specifically for food, must be chemically inert, non-toxic and prevent the transfer of foreign odours or flavours. They must be stable, elastic, tear-resistant, and proof against water vapour and volatile substances. The alterations and losses in aroma and flavour, enzymatic browning and oxidation of ascorbic acid that take place in the presence of oxygen suggest that such packaging materials should be used that are impervious to oxygen or permit removal of the oxygen from inside the package either by creating a partial vacuum or by injecting inert gases. They must also offer protection against light, particularly UV rays. They should be adaptable to different automatic packaging systems, of an appropriate size and shape for easy storing and distribution, and ready for opening. Packages should be effective as thermal insulators to limit possible temperature fluctuations within the product. A variety of primary packaging are currently employed including plain, coated, metallised, laminated plastic films, lidded plastic trays and thermoforms often contained in paperboard sleeves or cartons.

The materials most commonly used for frozen food packaging include (i) paperboard and moulded pulp that have poor barrier properties but good structural resistance and provide protection against physical damage; (ii) plastic materials with excellent moisture, oxygen and aroma barriers; (iii) semi-rigid plastic containers that have good structural properties; and (iv) aluminium that is a total barrier to moisture, oxygen, and aroma.

The most commonly used plastic packaging materials are:

- low-density polyethylene (LDPE), high-density polyethylene (HDPE), and polypropylene (PP): these are excellent moisture barriers but have poor oxygen barrier properties;
- polyvinylidene chloride (PVDC)/polyvinyl chloride (PVC) copolymer that are excellent moisture, oxygen and aroma barriers.
- ethylene vinyl alcohol (EVOH) copolymer and polyamide (Nylon) with poor moisture barrier properties and excellent oxygen and aroma barrier properties;
- polyethylene terephthalate (PET), with good moisture, oxygen and aroma barrier and good heat resistance; the heat resistance is even greater in crystallised polyethylene terephthalate (CPET).

LPDE is one of the most common plastics used, providing good protection against moisture loss as a flexible film pouch or as a coating on paper board.

Heat resistant HDPE and PP film pouches are used for boil-in-bag or microwavable pouches. PP can be used to form trays or coat paperboard trays that are microwavable. Heat resistant PET can be used to form trays or coated paperboard trays that are dual ovenable. Coating the LDPE or HDPE with or PET or PVDC/PVC copolymer increases the oxygen barrier properties. Layers of EVOH or polyamide (that are moisture sensitive) between layers of PE provide a good oxygen barrier.

Irreversible dehydration of frozen meat will occur at the surface unless it is packaged in air-tight, vapour-resistant material; for example, plastic bags or trays in conjunction with stretch wraps. To protect frozen meat from rancid and freezer burn, PE bags and PVC or PVDC films are used with expanded polystyrene (EPS) trays that is a foamed form of PS which has a very low density but is a rigid material.

In the case of poultry, the skin-tight PVDC film package is adequate; poultry pieces are put into bags and transferred to a rotary vacuumising machine which packages the product with clip closures and bag neck trimming. When shrunk, the bags form a second skin around the poultry, which are then either frozen in brine or in blast freezers. Water absorption is negligible and the function of oxygen barrier is sufficient to prevent fats from becoming rancid. Materials used for packaging include PVDC film and a range of laminates with PE. For packaging frozen fish, PVDC is used in vacuum packaging. This system provides a better alternative to glazing process. It eliminates moisture loss on initial freezing, drip loss on thawing, weight loss on glaze, and reduces labour and time needed for traditional glazing. The lightly vacuumised package enables the fish to retain its fresh characteristics throughout the entire distribution. Packaging material, such as PET film, is also used for fish products in pillow pack style. The film is reverse printed on the treated side and laminated with PE. This laminate possesses barrier properties and is puncture-resistant over a wide range of temperatures, giving protection during transportation (Lee, 2006).

The majority of frozen fruits and vegetables are packaged in plastic films, such as deep freeze grades of PE, made into pillow-type packs on vertical form-fill-seal machines.

Vegetables are packaged commonly with LDPE as moisture barriers. Ethylene vinyl acetate (EVA) bags have mechanical resistance for large weights of product with the necessary moisture barrier, have good heat-seal properties, and remain flexible at freezer temperatures. As the contents of these larger packages are not generally consumed all at once, a reclosable feature on the package is often used.

Frozen bread is packaged in LDPE bags; for frozen desserts such as ice cream, frozen sorbets, mousses, and puddings, thermoformed high-impact polystyrene (HIPS) containers are often used (Lee, 2006).

REFERENCES

Anonymous. (1986). Recommendations for the Processing and handling of frozen foods. *International Institute of Refrigeration*, IIR, Paris, France.

Anonymous. (1989). EU Directive 89/108 on the Approximations of the laws of the Members States related to Quick frozen Foodstuffs for Human Consumption, Brussels, Official Journal, L40 pp. 34–37.

Anonymous. (1992). EU Directive 92/1. *Monitoring of Temperatures in the Means of Transport, Warehousing and Storage of Quick Frozen Foods for Human Consumption*, Brussels, Official Journal, L34, pp. 28–29.

Bertola, N.C., Califano, A.N., Bevilacqua, A.E. and Zaritzky, N.E (1996). Textural changes and proteolysis of low moisture Mozzarella cheese frozen under various conditions. *Lebensmittel Wissenschaft und Technologie* 29, 470–474.

Bevilacqua, A.E. and Zaritzky, N.E. (1982). Ice recrystallization in frozen beef. *Journal of Food Science* 47, 1410–1414.

Creed, P. (2006). Quality and Safety of frozen ready meals. In: D.W. Sun, ed. *Handbook of Frozen Food Processing and Packaging*. Boca Raton, FL: CRC/Taylor & Francis Group, pp. 459–476.

Donhowe, D.P. and Hartel, R. (1996). Recrystallization of ice in ice cream during controlled accelerated storage. *International Dairy Journal* 6, 1191–1208.

Erickson, M.C. (1997). Lipid oxidation: flavor and nutritional quality deterioration in frozen food. In: M.C. Erickson and Y.C. Hung, eds. *Quality in Frozen Food*. New York: Chapman & Hall, pp. 141–173.

Fennema, O.R., Powrie, W.D. and Marth, E.H. (1973). *Low Temperature Preservation of Foods and Living Matter*. New York: Marcel Dekker, pp. 3–207.

Fennema, O.R. (1996). Water and ice. In: O. Fennema, ed. *Food Chemistry*, 3rd edn. New York: Marcel Dekker, pp. 17–94.

Ferrero, C., Martino, M. and Zaritzky, N. (1993a). Stability in frozen starch pastes. Effect of freezing, storage and xanthan gum addition. *Journal of Food Processing and Preservation* 17(3), 191–211.

Ferrero, C., Martino, M. and Zaritzky, N. (1993b). Effect of freezing rate and xanthan gum on the properties of corn starch and wheat flour pastes. *International Journal Food Science and Technology* 28, 481–498.

Ferrero, C., Martino, M. and Zaritzky, N. (1994). Corn starch, xanthan gum interaction and its effect on the stability during storage of frozen gelatinized suspensions. *Starch/Stärke* 46, 300–308.

Ferrero, C., Martino, M. and Zaritzky, N. (1996). Effect of hydrocolloids on starch thermal transitions, as measured by DSC. *Journal of Thermal Analysis* 47, 1247–1266.

Ferrero, C. and Zaritzky, N. (2000). Effect of freezing rate and frozen storage on starch-sucrose-hydrocolloid systems. *Journal of the Science of Food and Agriculture* 80, 2149–2158.

Franks, F. (1985). *Biophysics and Biochemistry at Low Temperatures*. Cambridge: Cambridge University Press, pp. 39–52.

Fu, B. and Labuza, T. (1997). Shelf life testing: procedures and prediction methods. In: M.C. Erickson and Y.C. Hung, eds. *Quality in Frozen Food*. New York: Chapman & Hall, pp. 377–415.

Giannakourou, M.C., Taoukis, P., and Nychas, G.J. (2006). Monitoring and control of cold chain freezing. In: D.W. Sun, ed. *Handbook of Frozen Food Processing and Packaging*. Boca Raton, FL: CRC/Taylor & Francis Group, pp. 279–307.

Goff, H.D. (1997). Measurement and interpretation of the glass transition in frozen foods. In: M.C. Erickson and Y.C. Hung, eds. *Quality in Frozen Food*. New York: Chapman & Hall, pp. 29–50.

Goff, H.D. (2006). Quality and Safety of frozen dairy products. In: D.W. Sun, ed. *Handbook of Frozen Food Processing and Packaging*. Boca Raton, FL: CRC/Taylor & Francis Group, pp. 441–457.

Golden, D.A. and Arroyo-Gallyoun, L. (1997). Relationship of frozen food quality to microbial survival. In: M.C. Erickson and Y.C. Hung, eds. *Quality in Frozen Food*. New York: Chapman & Hall, pp. 174–193.

Graiver, N., Zaritzky N. Califano, A. (2004). Viscoelastic behavior of refrigerated and frozen low moisture Mozzarella cheese. *Journal of Food Science* 69(3), 123–128.
Haard, N.F. (1997). Product composition and the quality of frozen foods. In: M.C. Erickson and Y.C. Hung, eds. *Quality in Frozen Food*. New York: Chapman & Hall, pp. 275–295.
Harper, E.K. and Shoemaker, C.F. (1983). Effect of locust beam gum and selected sweetening agents on ice recrystallization rates. *Jounal of Food Science* 48, 1801–1803.
Hartel, R. (2001). *Crystallization in Foods*. Gaithersburg, MD: Aspen Publishers Inc., pp. 192–231.
Jul, M. (1984). *The Quality of Frozen Foods*. London: Academic Press.
Kasapis, S. (2006). Glass Transitions in frozen foods and Biomaterials. In: D.W. Sun, ed. *Handbook of Frozen Food Processing and Packaging*. Boca Raton, FL: CRC/Taylor & Francis Group, pp. 33–51.
Kennedy, C.J. (2000). Freezing processed foods. In: C.J. Kennedy, ed. *Managing Frozen Foods*. Cambridge, England: Woodhead Publishing Limited, pp. 137–158.
Krochta, J. (2006). Introduction to Frozen packaging. In: D.W. Sun, ed. *Handbook of Frozen Food Processing and Packaging*. Boca Raton, FL: CRC/Taylor & Francis Group, pp. 615–640.
Labuza, T.P. (1979). A theoretical comparison of losses in food under fluctuating temperature sequences. *Jounal of Food Science* 44, 1162.
Labuza, T.P. (1982). *Open shelf Life Dating of Foods*. Westport, CT: Food and Nutrition Press.
Labuza, T.P. and Schmidt, M.K. (1985). Accelerated shelf-life testing of foods. *Food Technology* 36(19), 66–74.
Lai, L.S. (2006). Quality and Safety of frozen eggs and egg products. In: D.W. Sun, ed. *Handbook of Frozen Food Processing and Packaging*. Boca Raton, FL: CRC/Taylor & Francis Group, pp. 503–516.
Lanari, M.C., Bevilacqua, A.E. and Zaritzky, N.E. (1990). Pigments modifications during freezing and frozen storage of packaged beef. *Journal of Food Process Engineering* 12, 49–66.
Lanari, M.C. and Zaritzky, N.E. (1991). Effect of packaging and frozen storage temperatures on beef pigments. *International Journal of Food Science and Technology* 26, 629–640.
Lee, K.H. (2006). Plastic packaging of frozen food. In: D.W. Sun, ed. *Handbook of Frozen Food Processing and Packaging*. Boca Raton, FL: CRC/Taylor & Francis Group, pp. 641–651.
Le Meste, M., Champion, D., Roudaut, G., Blond, G. and Simatos, D. (2002). Glass transition and food technology. A critical appraisal. *Journal of Food Science* 67(7), 2444–2458.
Levine, H. and Slade, L. (1986). A polymer physicochemical approach to the study of commercial starch hydrolysis products (SHPs). *Carbohydrate Polymers* 6, 213–244.
Levine, H. and Slade, L. (1988). Principles of cryostabilization technology from structure/property relationships of carbohydrate/water systems –a review. *Cryo-Letters* 9, 21–60.
Levine, H. and Slade, L. (1991). *Water Relationship in Foods*. New York: Plenum Press, pp. 251–273.
Mac Kenzie, A.P. and Luyet, B.J. (1967). Electron microscope study of recrystallization in rapidly frozen gelatin gels. *Biodynamica* 10, 95–122.
Mac Donald, G.A. and Lanier, T.C. (1997). Cryoprotectants for improving frozen food quality. In: M.C. Erickson and Y.C. Hung, eds. *Quality in Frozen Food*. New York: Chapman & Hall, pp. 197–232.
Marsdon, J.L. and Hendrickson, R.L. (1993). Meat and meat products. In: C P Mallet, ed. *Frozen Food Technology*. New York: Chapman & Hall.
Martino, M.N. and Zaritzky, N.E. (1987). Effects of temperature on recrystallization of polycrystalline ice. *Sciences des Aliments* 7, 147–166.
Martino, M.N. and Zaritzky, N.E. (1988). Ice crystal size modifications during frozen beef storage. *Journal of Food Science* 53, 1631–1637, 1649.
Martino, M.N. and Zaritzky, N.E. (1989). Ice recrystallization in a model system and in frozen muscle tissue. *Cryobiology* 26, 138–148.
McKenna, B. (2006). Shelf-life prediction of frozen foods. In: D.W. Sun, ed. *Handbook of Frozen Food Processing and Packaging*. Boca Raton, FL: CRC/Taylor & Francis Group, pp. 603–612.
Mazur, P. (1970). Cryobiology: the freezing of biological systems. *Science* 168, 939–949.
Miles, M.J., Morris, V.J., Orford, P.D. and Ring, S.G. (1985). The roles of amylose and amylopectin in the gelation and retrogradation of starch. *Carbohydrate Research* 135, 271–281.
Morris, V.J. (1990). Starch gelation and retrogradation. *Trends in Food Science and Technology* 7, 1–6.
Pham, Q.T. and Mawson, R.F. (1997). Moisture migration and ice recrystallization in frozen food. In: M.C. Erickson and Y.C. Hung, eds. *Quality in Frozen Food*. New York: Chapman & Hall, pp. 67–100.
Rapatz, G. and Luyet, B.J. (1959). Recrystallization at high subzero temperatures in gelatin gels subjected to various subcooling treatments. *Biodynamica* 8, 85–105.

Reid, D.S., Kotte, K., Kilmartin, P. and Young, M. (2003). A new method for accelerated shelf-life prediction for frozen foods. *Journal of the Science of Food and Agriculture* 83(10), 1018–1021.

Regand, A. and Goff, H.D. (2003). Structure and ice recrystallization in frozen stabilized ice cream model solutions. *Food Hydrocolloids* 17, 95–102.

Reid, D.S. (1983). Fundamental physicochemical aspects of food freezing. *Food Technology* 37, 110–115.

Reid, D.S. (1994). Basic physical phenomena in the freezing and thawing of plant and animal tissue. In: C.P. Mallet, ed. *Frozen Food Technology*. Glasgow: Blackie Academic and Professional, pp. 1–19.

Reid, D.S. (1997). Overview of physical/chemical aspects of freezing. In: M.C. Erickson and Y.C. Hung, eds. *Quality in Frozen Food*. New York: Chapman & Hall, pp. 10–28.

Reid, D.S., Alviar, M.S. and Lim M.H. (1987). The rates of change of ice crystal size in model systems stored at different temperatures relevant to the storage of frozen food. In: *Proceedings of the XVI International Congress of Refrigeration*, Austria. Vol. C. pp. 397–401.

Rodríguez, N. and Zaritzky, N.E. (1983). Development of a time-temperature integrator-indicator for frozen beef. *Journal of Food Science* 48, 1526–1531.

Roos, Y. (1995). *Phase Transitions in Foods*. Food Science and Technology International Series. Academic Press.

Roos, Y. and Karel, M. (1991a). Non-equilibrium ice formation in carbohydrate solutions. *Cryo-Letters* 12, 367–376.

Roos, Y. and Karel, M. (1991b). Applying state diagrams to food processing and development. *Food Tecnology* 45, 6668–71107.

Roos, Y. and Karel, M. (1991c). Amorphous and delayed ice formation in sucrose solutions. *International Journal of Food Science and Technology* 26, 553–566.

Sista, R.V., Erickson, M.C. and Shewfelt, R.L. (1997). Quality deterioration in frozen foods associated with hydrolytic enzyme activities. In: M.C. Erickson and Y.C. Hung, eds. *Quality in Frozen Food*. New York: Chapman & Hall, pp. 101–110.

Slade, L. and Levine, H. (1991). Beyond water activity: recent advances based on an alternative approach to the assessment of food quality and safety. *Critical Reviews in Food Science and Nutrition* 30, 115–360.

Slade, L. and Levine, H. (1995). Glass transitions and water-food structure interactions. In: J.E. Kinsella, and S.L. Taylor, eds. *Advances in Food Nutrition Research*, Vol. 38. San Diego: Academic Press, pp. 103–269.

Sutton, R.L., Lips, A., Piccirillo, G. and Sztehlo, A. (1996). Kinetics of ice recrystallization in aqueous fructose solutions. *Journal of Food Science* 61(4), 741–745.

Sutton, R.L., Lips, A. and Piccirillo, G. (1996). Recrystallization in aqueous fructose solutions as affected by locust bean gum. *Journal of Food Science* 61(4), 746–748.

Van Arsdel, W.B., Copley, M.J. and Olson, R.L. (1969). *Quality and Stability of Frozen Foods Time-Temperature-Tolerance and its Significance*. New York: Wiley Interscience.

Velasco, P.J., Lim, M.H., Pangborn, R.M. and Whitaker J.R. (1989). Enzymes responsible for off-flavor and off-aroma in blanched and frozen-stored vegetables. *Biotechnology and Applied Biochemistry* 11, 118 127.

Wagner, J.R. and Añón, M.C. (1986). Effect of frozen storage on protein denaturation in bovine muscle I: Myofibrillar ATPase activity and differential scanning calorimetric studies. *Journal of Food Technology* 21, 9–18.

Wiliams, D.C., Lim, M.H., Chen, A.O., Pangborn, R.M. and Whitaker J.R. (1986). Blanching of vegetables for freezing- which indicator enzyme to choose. *Food Technology* 40, 130–140.

Xiong, Y.L. (1997). Protein denaturation and functionality losses. In: M.C. Erickson and Y.C. Hung, eds. *Quality in Frozen Food*. New York: Chapman & Hall, pp. 111–140.

Yamamoto, K. (1990). Electron microscopy of thermal aggregation of myosin. *Journal of Biochemistry* 108(6), 896–898.

Zaritzky, N.E. (1982). Mathematical simulation of the thermal behavior of frozen meat during its storage and distribution. *Journal of Food Process Engineering* 6, 15–36.

Zaritzky, N. (2000). Factors affecting the stability of frozen foods. In: C.J. Kennedy, ed. *Managing Frozen Foods*. Cambridge, England: Woodhead Publishing Limited, pp. 111–133.

Zaritzky, N. (2006). Physical–chemical principles in freezing. In: D.W. Sun, ed. *Handbook of Frozen Food Processing and Packaging*. Boca Raton, FL: CRC/Taylor & Francis Group, pp. 3–33.

Zaritzky, N.E. and Ferrero, C. (2006). Glass transition and rheological behavior in forzen starch-sucrose-hydrocolloids systems. In: P. Buera, J. Welti Chanes, P. Lillford and H. Corti eds. *Water Properties of Food, Pharmaceutical and Biological Material*. Boca Raton, FL: CRC Press, Taylor and Francis Group, pp. 309–324.

12 Freeze Drying

Andy Stapley

12.1 INTRODUCTION

Drying, like freezing, is primarily a method of food preservation, and is also one of the oldest methods of food preservation. An advantage of drying over freezing is that the dried product is shelf stable and thus generally requires only a sealed container to keep it for months or even years at ambient temperature. Refrigeration, on the other hand, requires a power supply, which if interrupted can jeopardise the safety and quality of the product. A further advantage of drying is that the removal of water reduces the weight and volume of the product, which aids storage and transport. The main disadvantage of drying, however, is that the product quality (resemblance to the original undried product) is generally not as high upon rehydration as frozen foods are upon thawing. Particular problems are changes in texture (for solid foods) and the loss of aromas.

Most drying processes involve the evaporation of liquid water to water vapour, but freeze drying instead occurs as a result of sublimation of ice directly to water vapour. The basic steps in a freeze drying process are briefly described as follows, and these will be examined in more detail in the rest of the chapter.

(i) The food is first frozen (usually on trays) which results in the formation of ice crystals within the food and the non-ice phase becoming freeze concentrated.

(ii) The frozen material is then exposed to a headspace with a partial pressure of water vapour well below the equilibrium vapour pressure of ice at the temperature of the material (see Fig. 12.1). This causes the sublimation of the ice crystals (primary sublimation) and also desorption of non-crystalline water present within the food matrix (secondary drying). The very low partial pressure of water vapour in the headspace is generally achieved by placing a condenser unit in close proximity to the food. The temperature of the condenser surface should be significantly lower than that of the drying material (typically −60°C)

(iii) In conventional freeze drying a vacuum is maintained in the freeze drying chamber, typically via the use of a vacuum pump. A common misconception is that the primary purpose for the vacuum pump is to remove water vapour. However, water removal is achieved far more effectively by the condenser. The vacuum pump is instead mainly used to remove air (and other non-condensable gases desorbed from the food) from the chamber. The reason is that the gases slow down the rate of transfer of water vapour from the food to the condenser.

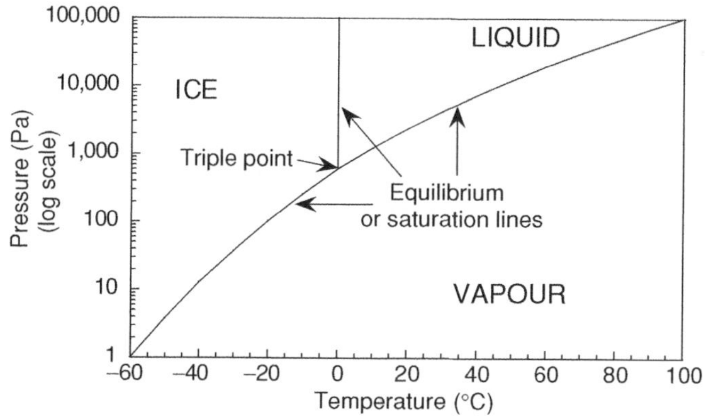

Fig. 12.1 Phase diagram of water showing the variation of equilibrium vapour pressure of water with temperature. At temperatures below the triple point pure water exists as solid ice.

(iv) As sublimation is an endothermic process, heat must be applied to the frozen food to sustain the sublimation process. However, the temperature of the food must always be maintained below its 'collapse' temperature (which is related to the glass transition temperature). Not only must the ice crystals not melt, but also the frozen food matrix must maintain rigidity if a properly freeze dried food is to be obtained. The collapse temperature is not constant, but rises as the moisture content in the food matrix is reduced.

If the process is performed correctly then the resulting freeze dried product will have the following attributes:

(i) For a sufficiently dried product the collapse temperature will have risen far above ambient temperature. The product will therefore have a rigid structure at ambient temperature and be shelf stable.
(ii) If the sample has been consistently maintained below the collapse temperature throughout drying then the ice crystals will leave behind voids (which the crystals once occupied) as they sublime. This results in a highly porous structure with excellent rehydration properties as the water is sucked into the pores by capillary action on wetting. This enables relatively large objects such as meat steaks to be rehydrated more successfully than those dried using other methods. The presence of pores also speeds the sublimation process (see Section 12.2.3) by providing easy routes for water vapour to escape the solid.
(iii) Freeze dried products are maintained at much lower temperatures than other drying processes. Reactions associated with the much higher temperatures that occur in other drying processes (such as the Maillard browning reaction) therefore do not occur to such a large extent in freeze drying.
(iv) Freeze drying is also good at retaining volatile flavour and aroma compounds in comparison to high-temperature drying processes.
(v) The formation of ice crystals during the freezing step can be problematic in materials which already possess a microstructure. This is a particular problem for cellular material such as plants and meat, but can be minimised by fast freezing, which produces small crystals.

Fig. 12.2 Examples of freeze dried foods. (a) coffee, (b) prawns, (c) chicken, (d) peach, (e) strawberry, (f) cranberry, (g) raspberries, (h) chives, (i) tarragon. Pictures (d), (e), (f) and (g) reproduced by kind permission of Molda AG.

Notwithstanding, the overall quality of freeze dried produce is usually superior to that of other dried foods for reasons (ii), (iii) and (iv). Almost any moisture-containing food can be freeze dried, including meat, fruits, vegetables and beverages. However, freeze dried products are usually more expensive than those obtained from other techniques for the following reasons:

(i) The drying chamber must be designed for vacuum operation. The main issues are structural strength (to resist buckling) and the avoidance of air leaks into the system (which brings undesired humidity as well as gas into the chamber).
(ii) The condenser presents a heavy refrigeration burden as it typically operates in the range between −40°C and −60°C, and must extract the same amount of heat in condensation from the condenser coils as that which is provided by heating to sublime the ice from the food. The food must also first be frozen which also incurs an energy cost.
(iii) Freeze drying is a slow process, taking many hours. This requires very large-scale equipment in order to produce acceptable throughput for bulk food manufacture.

Consequently, both the capital and operating costs of freeze dryers are high. So although almost any moisture-containing food can be freeze dried, it is generally only niche foods that are freeze dried on a commercial basis. The most widespread examples include (see Fig. 12.2):

(i) Freeze dried tea and coffee. Coffee is by far the most widely sold freeze dried product.
(ii) Freeze dried meat and vegetables are popular in mountaineering, camping, military and astronaut rations and other 'instant' meals which are readily prepared by the simple addition of boiling water.
(iii) Freeze dried fruits such as strawberries, cranberries and raspberries are an increasingly popular additive to breakfast cereals where milk is added.
(iv) Freeze dried herbs.
(v) Freeze dried pet food for dogs (meat), for cats (fish) and for fish (krill, worms).

In all these examples it can be noted that the freeze dried produce is more convenient to store or carry than the equivalent fresh or frozen produce, and this in many instances compensates for any reduction in quality.

12.2 THE FOOD MATERIAL SCIENCE OF FREEZE DRYING

In order to appreciate the requirements of a freeze drying process, it is important to understand how the freezing and drying environments affect the food material. Perhaps the most important concept, which is key to sample collapse, is that of the glass transition.

12.2.1 The glass transition

The glass transition is a highly useful concept in understanding the behaviour of a wide variety of food materials. It was first used by (synthetic) polymer scientists to describe the relatively sudden change in mechanical properties of amorphous (non-crystalline) polymers from a hard, brittle, 'glassy' state to a soft 'rubbery' material as they are heated over a matter of only a few degrees Celsius. The change is fundamentally different to melting in that there is no enthalpy or volume change (as occurs when crystals melt) and that it only occurs in amorphous materials. The change nevertheless corresponds to a substantial increase in the mobility of molecules, which are almost immobile in the glassy state. The mid-point around which the change in properties occurs is commonly referred to as the glass transition temperature (T_g), although it is important to remember that the change in properties spans a small temperature range of approximately 10–20 K (Roos and Karel, 1991). The concept of a glass transition was not generally used in food science or food processing circles until the 1980s and 1990s, upon when it was extensively disseminated.

The glass transition has now been observed in a wide range of amorphous food materials including sugars as well as food polymers such as proteins and starch. Many food materials are also able to crystallise but this does not tend to occur in freeze drying (for reasons see Section 12.2.2.1).

The glass transition temperature depends strongly on molecular weight, so, for example, dry starch has a T_g of 220°C whereas sucrose has a T_g of 62°C (Roos, 1997). Amorphous water (which is extremely difficult to produce without forming ice crystals instead) has a T_g of around −135°C. When food materials are combined such that they are in contact on a molecular level (such as in a solution) the glass transition of the resulting mixture can be correlated by the Gordon–Taylor equation (Gordon and Taylor, 1952) which calculates a weighted average T_g from the individual components according to their mole fractions and an empirically determined constant (here denoted by κ):

$$T_g = \frac{w_1 T_{g_1} + \kappa w_2 T_{g_2}}{w_1 + \kappa w_2} \tag{12.1}$$

Thus, when small molecules are mixed with larger molecules they depress the apparent glass transition temperature of the larger molecule. The small molecules are said to act as 'plasticisers'. The most effective plasticiser by far in food processing is water which, due to its low T_g value, is able to cause very significant reductions in T_g when mixed with other materials. Water is also, of course, virtually ubiquitous in food formulations. The lowering of T_g by the addition of water to a substance explains why moist foods are generally softer in texture than dry foods (e.g. cooked rice versus uncooked rice grains, fresh bread versus dry bread, chewy sweets versus hard sweets).

In freeze drying the main requirement is to maintain the product in the glassy state (i.e. below the glass transition temperature) during the whole of the drying period. However, in order to fully understand what occurs in freeze drying it is necessary to also consider the role of ice formation. Both processes are best understood with reference to a state diagram.

12.2.2 The use of state diagrams

The state diagram for an example system of sucrose and water system is shown in Fig. 12.3 (Roos and Karel, 1991; Roos, 1997). It shows how the following vary with composition:

(i) The glass transition temperature.
(ii) The equilibrium melting temperature of ice below which ice is stable. Note that this falls below 0°C as the sucrose concentration increases due to freezing point depression.

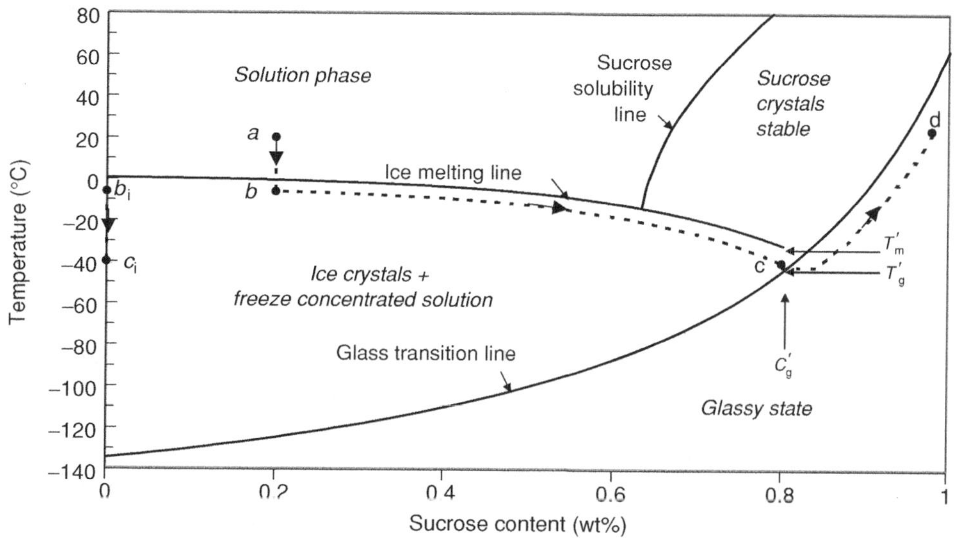

Fig. 12.3 State diagram of sucrose solution (after Roos, 1997). A typical path of a 20% sucrose solution undergoing freezing ((a) to (c)) followed by freeze drying ((c) to (d)) is shown by the dotted line.

(iii) There is also an equilibrium saturation line for sucrose crystals, to the right of which sucrose crystals are stable.

Similar diagrams can also be drawn for most food materials, including multi-component food systems, if all the non-aqueous components are treated as a single 'lumped' component for the purpose of using the Gordon–Taylor equation. This has been applied to fruits, which contain mixtures of cellular pulp, ascorbic and citric acids and various sugars (da Silva *et al.*, 2006), and tuna (Rahman *et al.*, 2003). Although the glass transition line may differ significantly for different materials (due to the different T_g of the dry material), the equilibrium melting line will not vary substantially.

The entire freeze drying process can be followed on this diagram. As an example take as a starting material a 20% sucrose solution at room temperature (at position marked '*a*' on the chart).

12.2.2.1 *Freezing*

Removing heat from the solution initially causes it to cool to '*b*', at a few degrees below the equilibrium melting line. A few degrees of sub-cooling is required to initiate nucleation and crystallisation of ice crystals ('b_i'). The formation of an ice phase depletes water from the solution phase, which consequently becomes more concentrated with respect to sucrose. The solution phase thus now moves in the direction of '*c*' on the chart. Due to the freezing point depression effect the temperature required to maintain ice formation declines steadily, and the solution becomes increasingly viscous as it becomes more concentrated. As the solution approaches the glass transition line at '*c*' the solution becomes so viscous that ice crystallisation ceases and the composition of the viscous solution is now temporarily fixed at C'_g. The ice phase will be at the same temperature at c_i. This is the end of the freezing stage.

Point '*c*' is known as the maximally freeze concentrated solution, as it is not possible to freeze concentrate further. There are a number of points to be made here:

(i) To ensure the maximum degree of freeze concentration, the solution must be held between T'_m and T'_g. At temperatures higher than T'_m freeze concentration will cease at a lower concentration due to insufficient sub-cooling. At temperatures lower than T'_g there is a risk that freeze concentration will cease when the glass transition is met at a lower concentration. However, if the temperature is again raised to the optimal temperature zone then the solution can be 'annealed' to rectify this.

(ii) Assuming the temperature is held between T'_m and T'_g, the maximum concentration of the freeze concentrated phase (C'_g) is independent of the starting composition. However, the starting composition will determine the proportion of ice that can be formed, which can be calculated by a simple mass balance assuming a value of C'_g.

(iii) The state diagram suggests that sucrose crystals can form at the eutectic point. However, as the crystallisation kinetics of sucrose are much slower than that of water, there is generally not sufficient time during the freezing process for sucrose crystallisation to occur. As the rest of the freeze drying process and subsequent storage take place in the glassy state, there is no further opportunity for sucrose to crystallise, and for this reason freeze drying is a popular method of producing food stuffs in the amorphous state (Roos, 1997).

12.2.2.2 *Freeze drying*

Once the solution has been maximally frozen, the freeze drying process can begin by exposing the sample to a low vapour pressure headspace. This results in the sublimation of the pure ice crystals (known as primary drying) and desorption of water from the glassy phase (secondary drying) via exposed surfaces of the matrix. If the freeze concentrated solution is kept below the glass transition temperature then the sublimation of ice crystals leaves behind pores within the rigid structure, which increases the surface area for secondary drying. As the freeze concentrated phase undergoes secondary drying it is able to rise to a higher temperature without collapse occurring as the glass transition temperature rises. Therefore a trajectory such as from '*c*' to '*d*' can be followed.

The sublimation process requires heat for the sample to maintain its temperature. Heat transfer to the sample is thus an important aspect of freeze dryer design.

12.2.3 Importance of the freezing step

The crystallisation of water to form pure ice crystals is a two-step process, requiring crystals to first nucleate and then grow. Both nucleation and growth processes require the solution to be supercooled below the equilibrium melting temperature of the aqueous solution by typically at least a couple of degrees Celsius. The rate of nucleation (the rate of formation of new crystals per unit volume) generally increases exponentially with the degree of supercooling, whereas the growth of crystals (the rate of increase in size) tends to increase in a less dramatic, linear fashion. Thus, as the temperature falls and the degree of supercooling increases, the nucleation rate increases more rapidly than the growth rate. Therefore slow cooling rates (which produces low supercooling) tend to favour growth and results in a small number of large crystals. On the other hand high cooling rates (equals high supercooling) tend to favour nucleation, which produces a larger number of smaller crystals (Oetjen, 1999; Rothmayr, 1975).

When cooling any object, however, the fastest rates of cooling invariably occur at the surface and the slowest rates in the centre. Thus, smaller crystals are generally found at the surface and larger crystals in the interior.

The freezing stage, being the first stage, is critical to the overall success of the freeze drying process, for the following reasons:

(i) The sizes, shapes and distribution of crystals within the food as it freezes, translate directly into the sizes, shapes and distribution of pores in the final product. This not only affects the rehydration behaviour and appearance of the freeze dried product, but also the density.
(ii) Achievement of a connected network of crystals greatly assists the freeze drying process, as this results in a relatively easy path for water molecules to escape from the inside of the food through the pore network that takes its place. Achieving of a connected network of crystals generally requires a high water content in the initial feed (of around 80%). Thus, paradoxically, it can be faster to freeze-dry a product where the liquid feed has a higher initial moisture content (and is able to form a pore network) than one with a lower moisture content (where a network is not formed) and water is forced, in part, to diffuse through the food matrix. For a given material it is usually also found that slower cooling leading to larger crystals is more conducive to a connected network and faster freeze drying (Kochs *et al.*, 1993).
(iii) As mentioned earlier, small crystals must be produced if cellular material is to be freeze dried, to avoid rupture of cell walls. This overrides drying rate considerations for such foods.

12.3 EQUIPMENT AND OPERATION

12.3.1 Introduction

12.3.1.1 *Freezing*

The general principles of freezing are outlined in other chapters in this book and these also apply here. In most instances the food is frozen on trays, either by contacting the tray with a cooled support surface, contacting the food with a cryogenic liquid (such as liquid nitrogen), or by air-blast freezing. There is generally a conductive resistance within the food, which limits the transport of heat from the food. In some freeze drying processes the trays are ribbed or finned to increase heat transfer rates in both freezing and freeze drying stages. The rate of freezing is important in determining ice crystal size, shape and distribution, which has important ramifications for damage to cells and the pore structure of the freeze dried product.

Freeze drying of beverages is generally performed from a concentrated solution which is previously prepared either by vacuum evaporation or by freeze concentration. In the commercial freeze drying of coffee a concentrated solution of typically 40% solids is used for freeze drying. This results in a relatively small fraction of ice forming but porosity is instead achieved by incorporating carbon dioxide under pressure into the coffee solution which forms a foam when the pressure is released. The liquid is quickly frozen to produce a solid foam, which is mechanically broken up into small nuggets (as seen in a typical jar of freeze dried coffee). The freeze drying time is much reduced as a result of carbon dioxide incorporation. The precise method of foaming has a large impact on the porous structure, which also influences the lightness or darkness of colour of the freeze dried product. The porous structure is just discernible in Fig. 12.2a.

12.3.1.2 *Freeze drying*

As mentioned in the introduction the main elements of a freeze dryer are:

(i) A vacuum tight chamber.
(ii) A means of removing air and other non-condensable gases, usually a vacuum pump.
(iii) A condenser to remove water and maintain a low partial pressure of water vapour in the freeze drying chamber.
(iv) Some means of supplying heat.

Freeze dryers are either designed for small-scale laboratory use or for large-scale food production. These are now briefly described. More detailed information can be found in Mellor (1978), Dalgleish (1990) and Oetjen (1999).

12.3.2 Laboratory-scale equipment

Laboratory-scale freeze drier are either flask based or tray based.

In the flask type (see Fig. 12.4) the material is placed in a glass flask, which is connected via a manifold to a condenser and vacuum pump. The use of a manifold allows more than one flask to be freeze dried at the same time. Heat is supplied by convection from the surrounding ambient air through the wall of the flask. This is essentially an uncontrolled method of heating, and the freeze dried material will establish its own temperature. This method is thus not suitable for materials with very low collapse temperatures. When loading the material should be applied in a thin layer around the flask wall (see Section 12.4.3). For solutions this is most easily achieved by freezing the solution *in situ* by rotating the flask containing the solution in a cold bath to freeze the solution in a thin layer on the flask wall. If drying more than one flask at a time it is particularly important to freeze dry nominally identical samples in each of the flasks. If samples show differential rates of drying then problems can arise as the fast drying material provides a rapid source of water vapour, which raises the overall vapour pressure of water in the chamber. This slows down the rate of sublimation of the slower drying materials, which is cooled less and so will rise to a higher temperature than if dried on its own. The possibility of collapse is thus increased.

Tray-type freeze driers come in various levels of sophistication. Some have controlled shelf temperatures (see Fig. 12.5), while others are free-standing shelves within a glass-domed chamber. Where there is more than one free-standing shelf it is also important to freeze dry nominally identical samples on each of the trays for the same reasons as given above for flask drying.

Laboratory freeze driers generally have condensers in the form of a cooled wall. This design allows a considerable quantity of water to be removed in a small space. Much of the

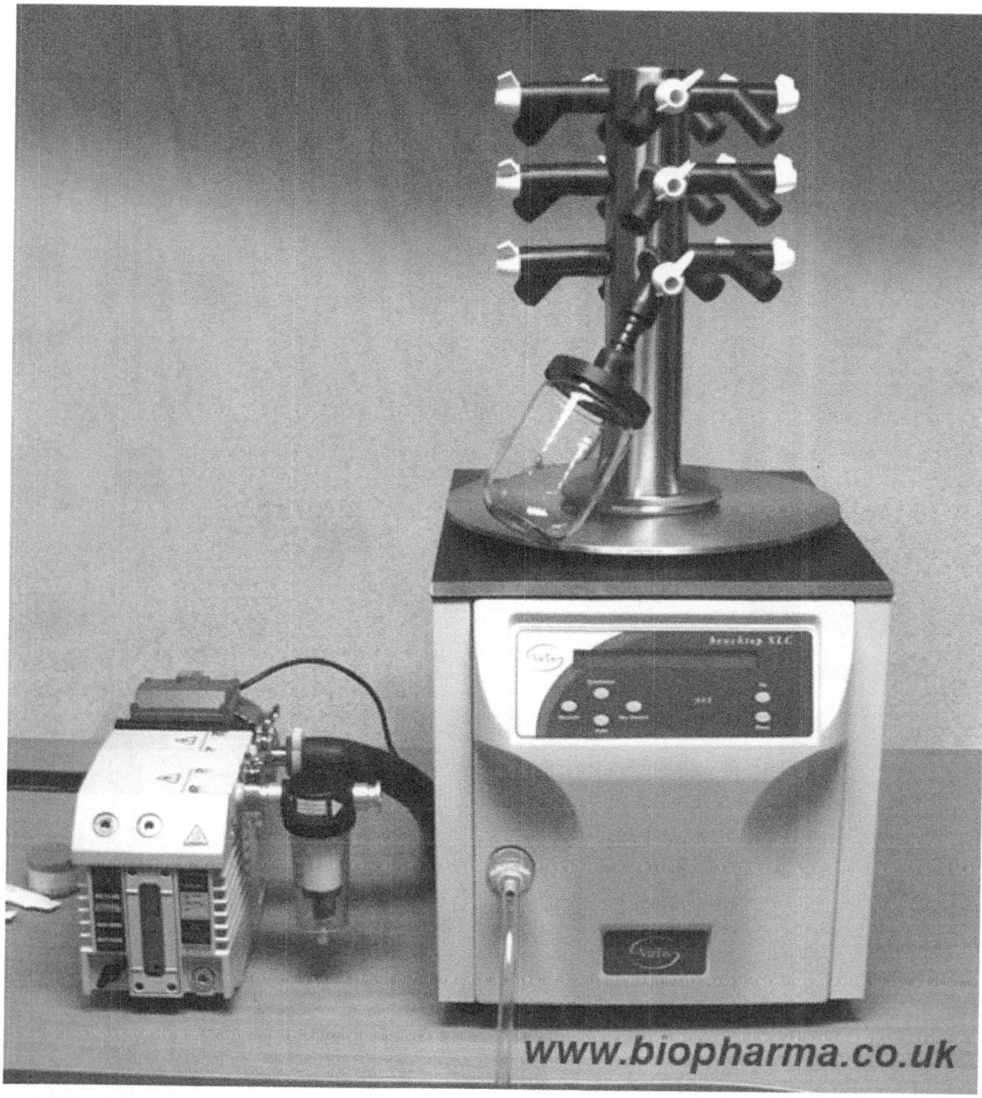

Fig. 12.4 Flask style laboratory freeze dryer, with stainless steel manifold for multiple flask drying (although only one is attached in the picture). The condenser lies in the main (white coloured) body of the apparatus and is connected to the flask through the manifold. To the left is the vacuum pump with oil mist filter attached. Reproduced with kind permission from Biopharma Process Systems Ltd. (www.biopharma.co.uk).

cost of laboratory freeze dryers is due to the refrigeration apparatus that is required. A small vacuum pump (5 m^3 h^{-1}) is all that is required in these installations to evacuate air.

12.3.3 Industrial-scale equipment

Industrial-scale freeze driers are almost exclusively of the tray type (see Figs. 12.6 and 12.7). The trays can be flat, with a raised edge to contain liquid feeds if necessary, or can also be ribbed with metal slats protruding from the floor of the tray. Both batch and continuous

Fig. 12.5 Larger laboratory shelf-style freeze dryer. The shelves are temperature controlled and are located in the upper part of the equipment behind the glass window. The condenser lies in the lower half (behind the black circle). Reproduced with kind permission from Biopharma Process Systems Ltd. (www.biopharma.co.uk).

Fig. 12.6 Production-scale batch freeze dryer showing temperature-controlled product shelves (upper section) and condenser coils (lower section). Reproduced with kind permission from Cuddon Freeze Dry (www.cuddonfreezedry.com).

dryers are used. In the case of continuous dryers, air locks need to be provided at each end for trays to enter and leave without compromising the vacuum within the main chamber. Although many freeze driers share similar features there are many variations in how they are laid out. This is in contrast to freezing processes where equipment is much more standardised.

Heating is provided either by conduction through the supporting shelf (with the shelf heated electrically or by circulating fluid) or by radiant heating from above each shelf. In some installations a dry inert gas is also periodically fed into the chamber to provide heat for sublimation and to aid heat transfer from the heating surfaces to the material.

Condensers can either be within the main chamber itself (see lower section of chamber in Fig. 12.6) or positioned in a separate chamber, which is isolated by a valve – this allows 'pressure rise' measurements to be made (see Section 12.5). The distance between the drying chamber and the condenser should be as short as possible for the condenser to be as effective as possible in reducing the partial pressure in the chamber. The water molecules can typically travel at speeds of over 100 m s^{-1} between the drying shelves and the condenser coils. The condenser coils are generally tubular in nature, and build up an ice deposit of typically 1 cm thickness. A refrigerant passes through the coils, which is cooled in turn by a refrigeration unit. It is crucial that the temperature of the coils is uniform (less than 1 K variation). If parts of the condenser are warmer than others then sublimation will occur from these warmer parts to the cooler parts and the effective area of the condenser is much reduced. In large installations the condenser tubes are divided into sections, each having their own inlet refrigerant supplied in parallel.

(a)

(b)

Fig. 12.7 An example of production-scale freeze equipment. (a) and (b) inside of chamber showing shelves which are heated by circulating water, (c) trays loaded with frozen strawberries awaiting freeze drying, (d) the trays are mounted on carriers which are hung from overhead rails, which enables them to be easily moved around the factory and glide into the dryer between the heated shelves (in this case the trays are radiantly heated from both above and below). Pictures by kind permission of Molda AG.

(c)

(d)

Fig. 12.7 *(continued)*.

Vacuum pumps are usually of the rotary vane type. They remove a fixed volume of air from the chamber per rotation, and so the volumetric rate of removal is constant. The mass of gas per unit volume depends on the pressure by the ideal gas law and so the highest rate of ejection on a mass basis occurs when the chamber is at atmospheric pressure at the commencement of pumping down. The pump is connected to the drying chamber by a valve, which is used to control the chamber pressure. At steady state the vacuum pump needs to overcome the rate of ingress of air into the chamber via leaks or gases which have been desorbed from the food and water vapour.

An alternative means of providing a vacuum is by the use of *steam ejectors*. These use a jet of high-temperature steam passing through a section of the chamber to entrain gas molecules from the chamber. The steam subsequently passes through a Venturi to a condenser and is able to be simply pumped away as a liquid. Typically a two-stage (cascade) ejector arrangement is used whereby the second ejector entrains the steam and entrained gases from the first ejector. This method of pressure reduction is extremely rapid and allows operational pressures to be achieved in 6 minutes or so.

12.4 MATHEMATICAL MODELS OF FREEZE DRYING

Although freeze drying is complex, a surprising degree of insight into the process can be obtained from relatively simple heat and mass transfer models (Mellor, 1978; Karel, 1975). The approach is similar to the Plank equation for freezing, in that it assumes a receding front mechanism, i.e. sublimation occurs at a well-defined front which recedes into the material with time. Ahead of the front is icy material and behind the front is dry material. This is a decent description of the primary sublimation of ice, which has been observed as a receding front on images taken by freeze drying microscopes, but it describes less well the secondary sublimation of water from the concentrated glassy phase. It is also assumed that sensible heat (heat associated with changes in the temperature of the material) can be neglected, i.e. the heat supplied is purely used to sublime ice.

Two main scenarios can be envisaged depending on whether heat applied through the same surface from which moisture escapes the material or via opposite faces.

12.4.1 Heat and mass transfer through the same surface

This occurs when heat is applied to the exposed surface by radiant heating.

Figure 12.8 is a diagram of a slab of food of thickness L undergoing freeze drying. The upper surface temperature of the food is maintained at a temperature T_s and the partial pressure of water vapour in the chamber is p_s. The temperature of the sublimation interface is denoted by T_i which gives rise to a vapour pressure p_i at the interface according to the equilibrium relationship (see Fig. 12.1). The distance of the interface from the upper surface is x.

The rate at which heat is conducted in through the outer dry layer is given by Fourier's Law of thermal conduction:

$$Q = k_{\text{dry}} A \frac{(T_s - T_i)}{x} \tag{12.2}$$

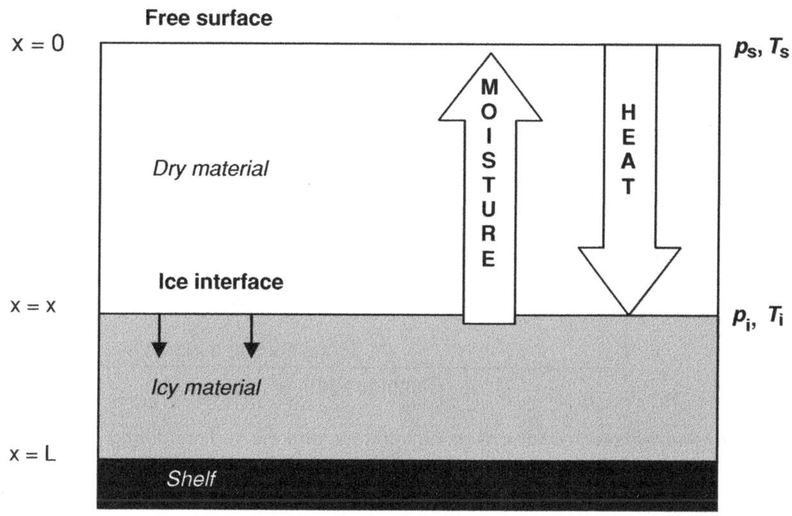

Fig. 12.8 Retreating front model for freeze drying by radiant heating.

This is equated to the rate of energy used in sublimation, i.e.

$$Q = \lambda \dot{M} \tag{12.3}$$

The sublimation rate can also be expressed in two further equations. First, it is equal to the rate of flow of vapour through the dry layer (assuming a pseudo-steady state), which is driven by the gradient in water vapour pressure:

$$\dot{M} = b_{\text{dry}} A \frac{(p_{\text{i}} - p_{\text{s}})}{x} \tag{12.4}$$

We can also relate the rate of sublimation ($\dot{M}$) to the velocity of the sublimation front (dx/dt) with the following steps:

$$\frac{\text{Volume swept out}}{\text{Time}} = A \frac{dx}{dt} \tag{12.5}$$

where A is the area of the front (equivalent to the top surface area of the slab). Multiplying by the dry solids density (mass of dry solids per unit volume) then gives:

$$\frac{\text{Mass of dry solid swept out}}{\text{Time}} = \rho A \frac{dx}{dt} \tag{12.6}$$

Finally, the sublimation rate is obtained by multiplying by the difference in dry-basis moisture content of the food (mass of water per unit mass of dry solids) on either side of the front:

$$\dot{M} = \frac{\text{Mass of water sublimed}}{\text{Time}} = \rho A (m_{\text{ice}} - m_{\text{dry}}) \frac{dx}{dt} \tag{12.7}$$

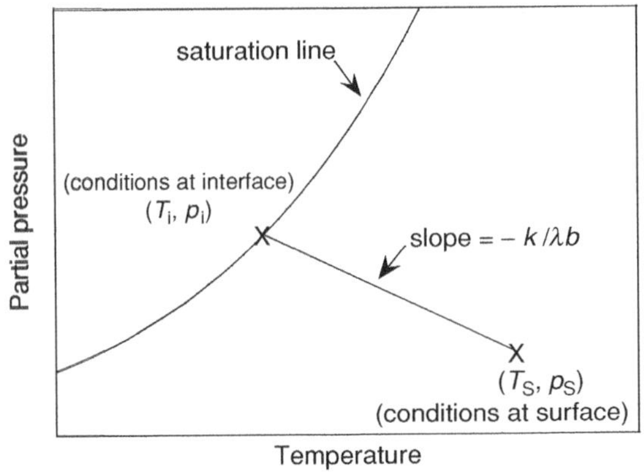

Fig. 12.9 Relationship between conditions at ice interface and the surface during radiant heating. The ice interface must be kept below the collapse temperature.

Combining the heat and mass transfer equations (12.2–12.4) gives:

$$k_{\text{dry}} A \frac{(T_s - T_i)}{x} = \lambda \dot{M} = b_{\text{dry}} A \frac{(p_i - p_s)}{x} \tag{12.8}$$

i.e.

$$(p_i - p_s) = \frac{k_{\text{dry}}}{\lambda b_{\text{dry}}} (T_s - T_i) \tag{12.9}$$

This is a useful equation as it relates the temperature and partial pressure at the ice interface to that at the surface. Furthermore, the relationship is independent of x. It is useful to plot this graphically (see Fig. 12.9) along with the saturation vapour pressure curve of ice, which relates p_i to T_i. Those readers who are familiar with convective drying process may see an analogy with the 'wet-bulb' line relationship between a drying solid and the surrounding air (where the slope of the line is derived from convective heat and mass transfer coefficients). In the freeze drying case the slope of equation (12.9) is made up of 'constants', which relates the ability of the dry layer to conduct heat to the ability to transmit water vapour[1]. Setting values for T_s (controlled via heater power) and p_s (controlled via the condenser temperature and any intervening valves), will thus provide defined values of p_i and T_i. Therefore maintaining constant chamber conditions results in a constant interface temperature.

The thermal conductivity of the dry layer is very low due to its porosity, and is equivalent to that of loft insulation. This can result in large temperature differences between the interface and the surface. Surface temperatures much higher than zero Celsius can be employed without causing sample collapse as the surface should be dry material (assuming sufficient secondary drying has occurred) with a high T_g and collapse temperature. The critical point for sample collapse is the interface temperature, which should be held below the collapse temperature of the maximally freeze concentrated solution. A safety margin should also be employed to

[1] These "constants", however, do depend on system pressure. Raising the pressure increases k_{dry}, but reduces b_{dry} (Karel, 1975).

allow for the fact that in practice secondary drying needs to be given sufficient time to occur behind the primary sublimation front.

Once the interface conditions are established, the drying kinetics (position of the sublimation front with respect to time can be derived by considering either equations (12.2) and (12.3) (heat transfer), or equation (12.4) (mass transfer)). As both are linked (via equation (12.9)) both approaches will yield the same numerical result:

$$\lambda\dot{M} = k_{\text{dry}}A\frac{(T_s - T_i)}{x} = \lambda\rho A(m_{\text{ice}} - m_{\text{dry}})\frac{dx}{dt} \tag{12.10}$$

$$\Rightarrow \int_0^L x\,dx = \int_0^L \frac{k_{\text{dry}}(T_s - T_i)}{\lambda\rho(m_{\text{ice}} - m_{\text{dry}})}dt \tag{12.11}$$

$$\Rightarrow \left[\frac{x^2}{2}\right]_0^L = \frac{k_{\text{dry}}(T_s - T_i)}{\lambda\rho(m_{\text{ice}} - m_{\text{dry}})}t = \frac{L^2}{2} \tag{12.12}$$

$$t = \frac{\lambda\rho(m_{\text{ice}} - m_{\text{dry}})}{2k_{\text{ice}}(T_s - T_i)}L^2 \tag{12.13}$$

Referring back to equation (12.9), this is equivalent to writing

$$t = \frac{\rho(m_{\text{ice}} - m_{\text{dry}})}{2b_{\text{dry}}(p_i - p_s)}L^2 \tag{12.14}$$

Thus, the fastest times are obtained with as large as possible ΔT and Δp, which requires operating at the highest interface temperature possible without causing collapse. It can also be seen from this approximate model that the freeze drying time is proportional to the square of the thickness of the sample. Thus, halving the thickness will reduce drying times by approximately a factor of four. The freeze drying process is therefore best suited to foods which can be prepared in small particles or slices (of the order of a centimetre or less).

12.4.2 Heat and mass transfer through opposite surfaces

This scenario is relevant to heated shelf arrangements. We will thus consider a convective heat transfer resistance between the heating medium and the bottom surface of the solid, and a conductive resistance within the solid (Fig. 12.10).

The heat conduction relation becomes:

$$Q = k_{\text{ice}}A\frac{(T_s - T_i)}{L - x} \tag{12.15}$$

but all the other equations (12.3–12.7) are unchanged.

Thus, if the heat and mass transfer equations are now combined, we have:

$$k_{\text{ice}}A\frac{(T_s - T_i)}{(L - x)} = \lambda\dot{M} = \lambda b_{\text{dry}}A\frac{(p_i - p_s)}{x} \tag{12.16}$$

Note that, in contrast to the previous case the interface conditions now vary with the depth of the front (x) as this does not cancel as before. However, the thermal conductivity used in

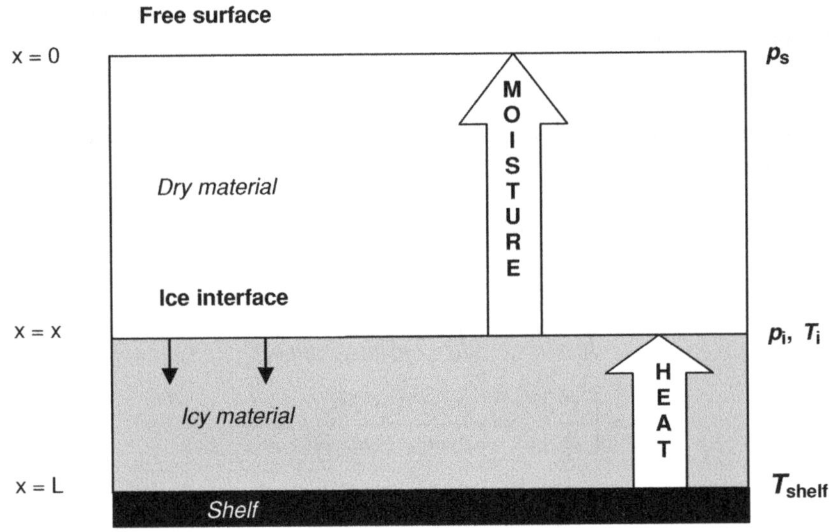

Fig. 12.10 Retreating front model for freeze drying by contact (shelf) heating.

equation (12.16) now refers to that of icy material, which is a very good conductor of heat (k_{ice} is typically of the order of 2 W m^{-1} K^{-1}). Thus, T_i will only be slightly higher than T_s, and one can assume that the shelf temperature controls the interface temperature.

If T_s, and thus T_i and p_i, is kept constant then integration of the mass transfer relation can be performed in an identical manner to produce equation (12.14). This yields the same equation as for the radiant heating case; however, it should be recognised that the two cases differ in how the interface conditions p_i and T_i are established.

12.4.3 Freeze drying of laboratory flasks

The shelf-heating scenario can be further extended to the laboratory flask system. Here the resistance to heat transfer can be modelled by a heat transfer coefficient which is dominated by the natural convection component between the outside of the glass flask and the ambient air temperature in the room where the flask is situated (Fig. 12.11). As before, we neglect the thermal resistance of the icy layer, and the heat capacity of the sample. The heat transfer relationship is again the only equation that changes, and can be written as:

$$Q = UA(T_{air} - T_i) \tag{12.17}$$

where U is the overall heat transfer coefficient between the ambient air (at T_{air}) and the ice front.

Combining with equations (12.3–12.7) yields:

$$UA(T_{air} - T_i) = \lambda \dot{M} = \lambda b_{dry} A \frac{(p_i - p_s)}{x} \tag{12.18}$$

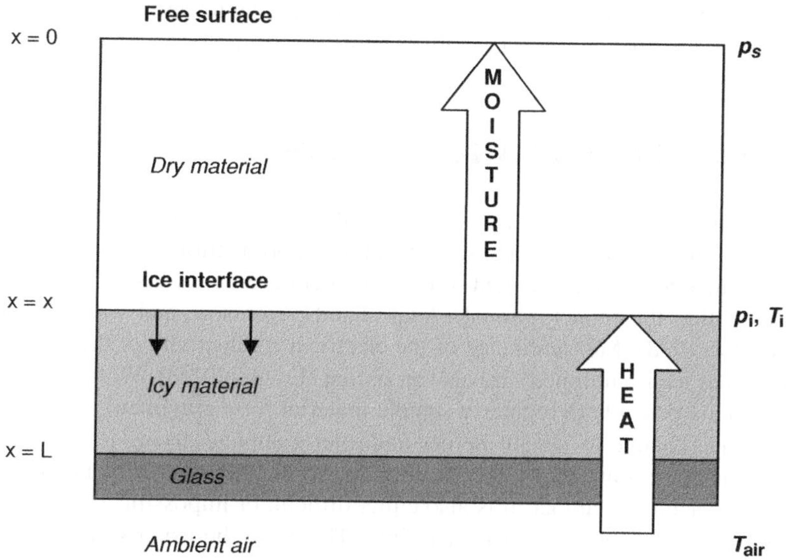

Fig. 12.11 Retreating front model for freeze drying in flasks.

i.e.

$$(p_i - p_s) = \frac{Ux}{\lambda b_{dry}}(T_{air} - T_i) \tag{12.19}$$

By inspection of this equation one can see that a 'wet bulb' line can again be plotted, the slope of which increases with the depth of the front (Fig. 12.12). As the front progresses the rate of sublimation falls. This reduces ΔT and raises T_i. There is a partial feedback effect in that a higher T_i will act to produce a higher Δp, which will stimulate sublimation but this has only limited scope. Collapse of the sample is a danger if the value of x is too high. Collapse

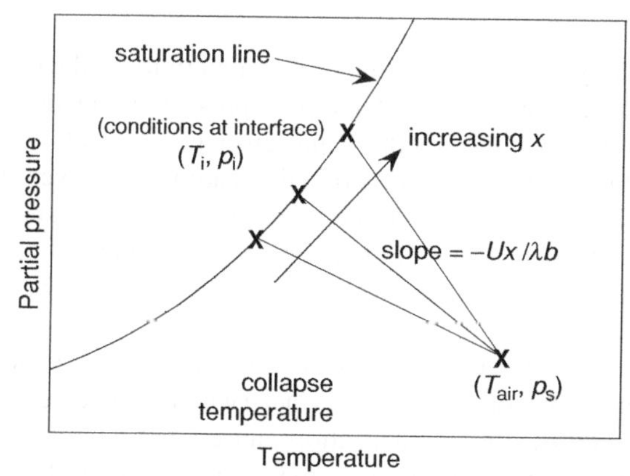

Fig. 12.12 Relationship between interface temperature and front depth during freeze drying in flasks. Above a critical front depth collapse will occur.

will thus occur if the thickness of the material is too large and towards the end of the drying process rather than the beginning. This can be most frustrating for an operator.

12.5 MEASUREMENT AND CONTROL

As in all industrial processes, measurement and the subsequent control of the process is critical to ensure quality standards and safe and efficient production. Typical instrumentation consists of thermocouple measurement of plate temperatures, condenser fluid temperatures, temperatures within the food (optional), total chamber pressure and water vapour partial pressure (Mellor, 1978). Measurements of the electrical conductivity of the drying material (which reduces as ice is sublimed) are also an option (Franks, 1998). With most methods of drying it is usually possible to regularly sample material for a gravimetric moisture content assay (percentage change in weight before and after complete drying in a vacuum oven). However, this is not common in freeze drying due to the inherent difficulties in sampling material from a vacuum chamber. It is also either difficult or impossible to directly monitor the moisture content of the product in real time. Therefore the usual strategy is to use the available instrumentation to perform reproducible runs, and use accumulated experience to refine operating strategies.

Control of temperature is important to maintain conditions in the food safely below the collapse temperature, but not excessively so as drying rates obviously decrease with lowering temperature. Temperature control is most challenging in radiant heating systems, whereby the heater power (rather than temperature) is the controller output. It is shown in Section 12.4.1 that to maintain a constant interface temperature the exposed surface temperature remains approximately constant, but the heater power required steadily decreases as the depth of the dried layer increases and the freeze drying rate decreases.

The monitoring of water vapour pressure is perhaps the best available guide to the progress of freeze drying as the water vapour pressure gradually reduces as the rate of flow of water vapour from the food to the headspace drops. This can be used in conjunction with freeze drying models to deduce, for example, the endpoint of primary sublimation (Genin *et al.*, 1996).

There are various ways of measuring the water vapour pressure:

- *Pirani gauges* work by measuring the thermal conductivity of the gas/vapour in the chamber, which can be correlated to the water vapour pressure in the chamber. This is a cheap, non-invasive and relatively easy method, and gives good results for detecting the end of primary drying, which results in a drop off in the reading. However, the level of sensitivity is not generally sufficient for the accurate monitoring of secondary drying.
- *Dew point hygrometers* measure the change in capacitance of a thin ceramic film in response to the absorption of water, which is a function of the vapour pressure of water in contact with it.
- *Cooled mirror hygrometers* directly measure the dew point in the headspace by cooling a mirror until condensation appears on its surface (which is detected optically). The mirror is then warmed to remove the condensation and cooled when it disappears again. The heating and cooling cycle is continuously maintained to track the dew point. This is the most accurate method of dew-point determination, but also very expensive and thus seldom used for this reason.

- *Direct pressure measurement* by piezo transducers or Bourdon gauges allows measurement of the combined water vapour and air partial pressures within the chamber. This measure can be used to infer the water vapour pressure by assuming negligible air pressure in the chamber (which may or may not be a valid assumption).

This last method is used in the so-called *pressure rise test*, which can be performed usefully on systems where the condenser can be isolated from the drying chamber by a valve (Dalgleish, 1990). The valve to the condenser is closed temporarily and the pressure in the chamber rises as water vapour sublimes from the food into the chamber but is no longer removed by the condenser. After a matter of half a minute the pressure steadies out as equilibrium is obtained and the pressure in the chamber can then be regarded as representative of the equilibrium partial pressure of the subliming ice. This can then be converted to an ice temperature via the p–T relationship for ice. However, this measurement must be quick as the lack of sublimative cooling to counter shelf or radiant heating will cause the sample to warm. The condenser must therefore be quickly reattached to avoid sample collapse. As it is a quick measurement it is best suited to direct pressure measurements as these are rapid, and also the increase in pressure can be solely attributed to water vapour pressure.

12.6 QUALITY ASPECTS

Freeze drying is perhaps the premier drying technique for preserving product quality, which is primarily the reason for its use. Every processing technique, however, incurs quality losses and freeze drying is no exception. The three main quality issues are: (i) damage as a result of the formation of ice crystals during freezing, which may pierce cell walls, (ii) quality losses during the drying process itself, mainly from losses of volatile flavour and aroma compounds, and (iii) the ability to be rehydrated back to the original form. It should not be forgotten, however, that the raw material quality also feeds through to the freeze dried product and is an important aspect. Quality losses can also occur during processing steps prior to freezing such as in the concentration of beverages. For vegetables it is highly necessary that they be blanched (briefly contacted with hot water) prior to freezing to inactivate enzymes.

12.6.1 Quality losses on freezing

The freezing process itself often confers a loss in quality. In particular, the rupture of cell walls due to the formation of ice crystals has already been mentioned in Section 12.2.3. The freeze concentration of solutes also causes changes in ionic strength and can lead to osmotic dehydration of plant cells, especially at low freezing rates, and biochemical changes including protein denaturation. The reader is directed to other chapters in this book and also to Jeremiah (1996) for detailed descriptions of freezing damage for individual food types.

12.6.2 Quality losses on drying

Loss of volatiles

The presence of volatiles is extremely important to the sensory perception of most foods. Not only do they provide a tempting smell, but the detection of aroma compounds in the nasal cavities can account for most of the flavour sensation during consumption (perhaps surprisingly more than from the tongue itself). The loss of volatiles represents the largest

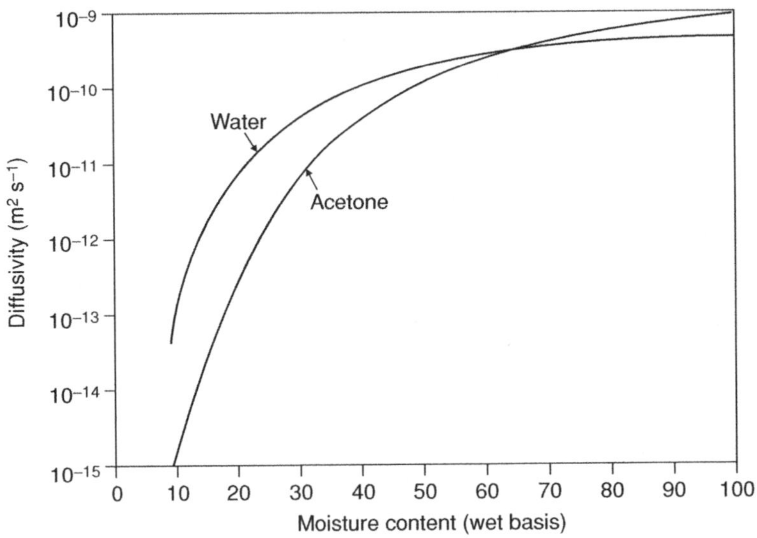

Fig. 12.13 Effect of moisture content on the diffusion coefficients of water and acetone in maltodextrin at 25°C (Adapted from King, 1990).

quality problem with freeze dried foods compared to frozen foods. Thus although rehydration characteristics are good, freeze dried products can sometimes be described as tasteless (Lin *et al.*, 1998).

Retaining volatiles is a problem in any drying process as good mass transfer conditions are desired to remove the water from the food. The question arising is therefore how can the water be removed without also removing the volatile flavour compounds, some of which may be many times more volatile than water. Thijssen and Rulkens (1968) studied this problem and concluded that the relative rates of diffusion of the volatile through the food matrix relative to water was the key factor. This is known as the principle of 'selective diffusion'. Fortunately, as water is a small molecule, its diffusivity is relatively large compared to most volatiles, and this works in favour of retaining the volatile. An additional point is that the diffusivities of both water and volatiles are strong and increasing functions of moisture content. This is because the presence of water increases the mobility of molecules. An example is shown in Fig. 12.13, which shows plots of diffusivity versus moisture content data for water and acetone (a representative volatile) in maltodextrin. Note that as the moisture content reduces, the diffusivity of the larger acetone molecule drops more sharply than that of water. Thus low moisture conditions are more favourable towards retaining volatiles during drying processes.

Freeze drying has an advantage over other drying methods in that the majority of water is removed from the product from the primary sublimation of ice, which, of course, does not contain the volatile components. These volatiles are only present in the freeze concentrated matrix, and so only escape during secondary drying (Coumans *et al.*, 1994). In addition, the starting moisture content for secondary drying is considerably lower than the initial feed water content so less drying is required, and the low starting moisture content of the freeze concentrated matrix means that the selective diffusion ratio is much increased compared to that in the original feed material. Coumans *et al.* (1994) reviewed this topic and suggested that a greater thickness of material between crystals (or pores) produced by slower initial freezing of the product aided volatile retention.

12.6.3 Loss of quality on storage

Freeze dried foods should be stored in a sealed container to prevent further loss of volatiles. Many deteriorative reactions progress very slowly, if at all, if the food is maintained in the glassy state, and especially if stored under an inert atmosphere. Thus, even with long periods of storage, reactions such as the Maillard reaction take place to an extent that is small though finite (Kawai *et al.*, 2005). If the food contains fats, the fats can oxidise over time and become rancid. For this reason freeze drying is not often used for fatty produce.

12.6.4 Rehydration ability

Freeze dried foods are almost universally accepted as superior in their ability to rehydrate in comparison to those dried by other methods. This is as a result of their porous structure (Lin *et al.*, 1998). If cellular damage is kept to a minimum then 90% rehydration of plant material can be routinely achieved.

12.6.5 Avoidance of collapse

Avoidance of collapse – either during freeze drying or in subsequent storage – is by far the most important factor determining the quality of freeze dried produce (Levi and Karel, 1995). Collapse causes a dramatic increase in the loss of volatiles, in shrinkage, in loss of rehydration capability and in allowing deteriorative reactions to occur. For this reason collapse must always be avoided in freeze drying and a safety margin included in freeze drying operations. Various workers report collapse occurring at temperatures 2–3 K above T_g (Pikal and Shah, 1990) or midway between T_g' and T_m' values (Knopp *et al.*, 1998).

12.7 ALTERNATIVE METHODS OF FREEZE DRYING

The primary aim of this chapter has been to set out the main techniques of what may be termed conventional freeze drying. The process, however, remains expensive compared to other methods of dehydration. To counter this a number of variations have been researched.

12.7.1 Microwave heating

Microwave freeze drying was first experimented upon in the 1950s as a way of providing the heat of sublimation more uniformly to a sample. The combination of low thermal conductivities of the gas phase (due to low pressures) and the dry solid phase (due to porosity) means that heat transfer in conventional freeze drying is poor. Microwaves, on the other hand, by-pass both the gas and dry layer, and are readily absorbed by the ice phase. Various researchers (e.g. Cohen and Yang, 1995) have reported increased rates of freeze drying with microwaves compared to conventional methods, but the method still has not been widely commercialised. The main problems are as follows (Karel, 1975):

(i) Microwaves are an expensive form of energy (10–20 times that of steam).
(ii) Microwaves can 'glow discharge' that can lead to ionisation of gases, which can cause a loss of food quality.

(iii) Microwaving is difficult to control as liquid water absorbs microwave radiation more strongly than ice. Inhomogeneities in microwave heating are commonplace, and thus if any ice were to melt accidentally as a result then a thermal 'runaway' effect would ensue as the water would preferentially absorb the microwaves.

12.7.2 Atmospheric freeze drying

Although sublimation in conventional freeze drying occurs under a vacuum it is perfectly possible for sublimation to occur in air at atmospheric pressure so long as the partial pressure of water vapour is below p_{sat}. Indeed, this is by far the most common form of freeze drying on planet earth – it occurs naturally on a massive scale from the snow and ice which cover a large part of the earth's surface. Experiments in atmospheric freeze drying of foods date back to Meryman (1959) who freeze dried small samples of liver and recirculated air back through a packed desiccant bed. An industrial process using atmospheric freeze drying is currently in operation (Dtech A/S, Norway), in which food particles are freeze dried in a fluidised bed by cold air with a relative humidity of 30%. The gas picks up humidity in the fluidised bed but is then recirculated through a condenser to remove water before being reheated. The condenser and heater are integrated via a heat pump and this allows substantial energy savings to be made which is not possible with the conventional freeze dryer design.

Other workers have continued Meryman's idea of using an absorbent material, but included it in close proximity to the drying material in a fluidised bed (e.g. Wolff and Gilbert, 1990; Lombrana and Villaran, 1997; Di Matteo *et al.*, 2003). This means that there are no condenser costs and water vapour molecules do not have to travel far once they have left the food, which increases the drying rate. The process is essentially internally mass transfer controlled for 1 cm size solids. A disadvantage is that the desiccant has to be separated completely from the food afterwards and that the risk of contamination of the food by the desiccant remains.

12.7.3 Spray-freeze-drying

Spray-freeze-drying, as its name suggests, is a two-stage process in which a liquid feed containing a dissolved solid is first sprayed into a cold gas or cryogenic liquid to form frozen particles, which are then freeze dried in a fluidised bed. The small particle size provides the potential for substantially shorter freeze drying times than usually found as freeze drying times approximately scale with the square of the linear dimensions of the object (see equations (12.12) and (12.13)). The process requires efficient heat transfer to the individual particles, which can be achieved by fluidisation. This was applied by Malecki *et al.* (1970) who first spray froze solutions in liquid nitrogen and then freeze dried, using atmospheric air, egg albumin at −20°C and apple juice at −34°C (which proved difficult). A process along these lines has been patented by Glatt AG (Mumenthaler and Leuenberger, 1991). A problem with this process, however, is that the small particles elutriate 1 cm the bed and dry instead in the air filters at the exit of the freeze drying chamber.

12.7.4 Sub-atmospheric fluidised bed freeze drying

A further possible means of improving fluidised bed processes is to apply a partial vacuum. As an example, if the process is operated at 0.1 bar then the mass of gas required would be approximately one-tenth of the amount required at 1 bar, significantly reducing the mass of gas that is required to be recirculated (Anandharamakrishnan *et al.*, 2006). This results in a reduced energy cost for recirculating and drying the gas, and reduces the likelihood of particles being swept away from the fluidised bed. Thus sub-atmospheric operation is particularly suited to small particles. However, a vacuum capability does increase the capital cost and operational difficulties of the plant.

12.8 CONCLUSIONS

Freeze drying still remains a niche area in food processing. The major difficulty will always be the low collapse temperatures that are common in food systems (especially those containing significant quantities of sugars). The low temperatures and consequently low vapour pressures of water limit mass transfer rates. It is possible that new methods of freeze drying improve on drying time, which would bring a greater range of foods into the domain of freeze drying in the future.

Nomenclature

A	area of slab/ice front (m^2)
b	permeability (s)
k	thermal conductivity ($W\ m^{-1}\ K^{-1}$)
L	slab thickness (m)
m	dry basis moisture content ($kg\ kg^{-1}$)
$\dot{M}$	sublimation rate ($kg\ s^{-1}$)
p	vapour pressure of water (Pa)
Q	heat flow (W)
t	time (s)
T	temperature (K)
U	overall heat transfer coefficient ($W\ m^{-2}\ K^{-1}$)
w	mole fraction
x	distance of interface from surface (m)

Greek characters

κ	fittable constant in the Gordon–Taylor equation
ρ	dry solids density (mass of dry solids per unit volume) ($kg\ m^{-3}$)
λ	latent heat of sublimation ($J\ kg^{-1}$)

Subscripts

air	of air (outside flask)
dry	of dry region
g	glass transition
i	at ice interface
ice	of icy region
s	at surface
1,2	component 1,2

REFERENCES

Anandharamakrishnan, C., Khwanpruk, K., Rielly, C.D. and Stapley, A.G.F. (2006). Spray-freeze-drying at sub-atmospheric pressures. In: I. Farkas, ed. *Drying 2006*. Hungary: Szent Istvan University, pp. 636–642.

Cohen, J.S. and Yang, T.C.S. (1995). Progress in Food Dehydration. *Trends in Food Science & Technology* 6(1), 20–25.

Coumans, W.J., Kerkhof, P.J.A.M. and Bruin, S. (1994). Theoretical and practical aspects of aroma retention in spray-drying and freeze-drying. *Drying Technology* 12(1–2), 99–149.

da Silva, M.A., Sobral, P.J.A. and Kieckbusch, T.G. (2006). Water sorption and glass transition of freeze-dried Camu-Camu (*Myrciaria dubia* (Hbk) McVaugh) pulp. *Journal of Thermal Analysis and Calorimetry* 84(2), 435–439.

Dalgleish, J.M. (1990). *Freeze-Drying for the Food Industries*. London: Elsevier.

Di Matteo, P., Donsi, G. and Ferrari, G. (2003). The role of heat and mass transfer phenomena in atmospheric freeze-drying of foods in a fluidised bed. *Journal of Food Engineering* 59(2–3), 267–275.

Franks, F. (1998). Freeze-drying of bioproducts: putting principles into practice. *European Journal of Pharmaceutics and Biopharmaceutics* 45(3), 221–229. 98.

Genin, N., Rene, F. and Corrieu, G. (1996). A method for on-line determination of residual water content and sublimation end-point during freeze-drying. *Chemical Engineering and Processing* 35(4), 255–263.

Gordon, M. and Taylor, J.S. (1952). Ideal copolymers and the 2nd-order transitions of synthetic rubbers .1. Non-crystalline copolymers. *Journal of Applied Chemistry* 2(9), 493–500.

Heldman, D.R. and Hohner, G.A. (1974). Analysis of atmospheric freeze-drying. *Journal of Food Science* 39(1), 147–155.

Jeremiah, L.E. (ed.) (1996). *Freezing effects on food quality*. New York: Marcel Dekker.

Karathanos, V.T., Anglea, S.A. and Karel, M. (1996). Structural collapse of plant materials during freeze-drying. *Journal of Thermal Analysis* 47(5), 1451–1461.

Karel, M. (1975). Heat and mass transfer in freeze drying. In: S.A. Goldblith, L. Rey and W.W. Rothmayr. *International Course on Freeze Drying and Advanced Food Technology*. London: Academic Press, pp. 177–202.

Kawai, K., Hagiwara, T., Takai, R. and Suzuki, T. (2005). The rate of non-enzymatic browning reaction in model freeze-dried food system in the glassy state. *Innovative Food Science & Emerging Technologies* 6(3), 346–350.

King, C.J. (1990) Spray drying food liquids and the retention of volatiles. *Chemical Engineering Progress* 86(6), 33–39.

Knopp, S.A., Chongprasert, S. and Nail, S. L. (1998). The relationship between type TMDSC curve of frozen sucrose solutions and collapse during freeze-drying. *Journal of Thermal Analysis and Calorimetry* 54(2), 659–672.

Kochs, M., Korber, C., Heschel, I. and Nunner, B. (1993). The influence of the freezing process on vapor transport during sublimation in vacuum-drying freeze-drying of macroscopic samples. *International Journal of Heat and Mass Transfer* 36(7), 1727–1738.

Levi, G. and Karel, M. (1995). Volumetric shrinkage (collapse) in freeze-dried carbohydrates above their glass-transition temperature. *Food Research International* 28(2), 145–151.

Lin, T.M., Durance, T.D. and Scaman, C.H. (1998). Characterization of vacuum microwave, air and freeze dried carrot slices. *Food Research International* 31(2), 111–117.

Lombrana, J.I. and Villaran, M.C. (1997). The influence of pressure and temperature on freeze-drying in an adsorbent medium and establishment of drying strategies. *Food Research International* 30(3–4), 213–222.

Malecki, G.J., Shinde, P., Morgan, A.I. and Farkas, D.F. (1970). Atmospheric fluidized bed freeze drying.*Food Technology* 24(5), 601–603.

Mellor, J.D. (1978). *Fundamentals of Freeze Drying*. London: Academic Press.

Meryman, H.T. (1959). Sublimation freeze-drying without vacuum. *Science* 130(3376), 628–629.

Mumenthaler, M. and Leuenberger, H. (1991). Atmospheric spray-freeze drying – a suitable alternative in freeze-drying technology. *International Journal of Pharmaceutics* 72(2), 97–110.

Oetjen, G.W. (1999). *Freeze Drying*. Weinheim: Wiley.

Pikal, M.J. and Shah, S. (1990). The collapse temperature in freeze-drying – dependence on measurement methodology and rate of water removal from the glassy phase. *International Journal of Pharmaceutics* 62, 165–186.

Rahman, M.S., Kasapis, S., Guizani, N. and Al-Amri, O.S. (2003). State diagram of tuna meat: freezing curve and glass transition. *Journal of Food Engineering* 57(4), 321–326.

Roos, Y.H. (1997). Frozen state transitions in relation to freeze drying. *Journal of Thermal Analysis* 48(3), 535–544.

Roos, Y.H. and Karel, M. (1991). Phase-transitions of amorphous sucrose and frozen sucrose solutions. *Journal of Food Science* 56(1), 266–267.

Rothmayr, W.W. (1975). Basic knowledge of freeze drying. In: S.A. Goldblith, L. Rey and W.W. Rothmayr eds. *International Course on Freeze Drying and Advanced Food Technology*. London: Academic Press.

Thijssen, H.A.C. and Rulkens, W.H. (1968). Retention of aromas in drying liquid foods. *de Ingenieur* 80(47), 45–56.

Wolff, E. and Gibert, H. (1990). Atmospheric freeze-drying .1. Design, experimental investigation and energy-saving advantages. *Drying Technology* 8(2), 385–404.

13 Frozen Food Transport

Girolamo Panozzo

13.1 TRANSPORT AND STORAGE

Storage conditions should be maintained during transport. Transport vehicles are essentially cold stores on wheels but are more difficult to maintain at the correct storage temperature than large cold stores due to several factors:

(a) supplementary heat is introduced during loading and unloading the vehicle;
(b) defrosting has a more severe effect on foods compared to refrigerated stores (more restricted space for the coils, more humid air inlet);
(c) possible close contact between the foods and the lateral walls, through loading and displacements of the cargo, connected to the forces that the road circulation generates in the cargo (centrifugal forces).

Regarding (a), two aspects are important: the first is the time that is required for the loading/unloading operation (the time must be the lowest possible) and the second is the level of protection of the cargo during these operations (Fig. 13.1) (Meffert, 1975).

Regarding (c), the problem can be severe when the vehicle is not operating and the sun is irradiating the roof and a lateral wall. In this case the external surface temperature can reach values close to 90°C and the internal surface temperature can be increased as a function of the proximity of the cargo to the wall (Fig. 13.2) (Panozzo and Cortella, 2006).

13.2 MODES OF TRANSPORT

The essential component of any system to transport frozen food is an insulated container. International transport vehicles have to comply with the ATP and ISO rules, where an overall coefficient of heat transfer (K coefficient) is defined and must be lower than 0.4 W m^{-2} K^{-1} (ISO, 1996) (United Nations Economic Commission for Europe, Inland Transport Committee, 1970). The overall coefficient of heat transfer, which represents the insulating capacity of the equipment, is defined by the following formula:

$$K = WS^{-1}\Delta T^{-1}$$

where W is the thermal capacity required in a body of mean surface area S to maintain the absolute difference ΔT between the mean inside temperature T_i and the mean outside temperature T_e, during continuous operation, when the mean outside temperature T_e is constant.

The second main component is a refrigerating unit or a system to store energy at a temperature lower than that prescribed for the transport.

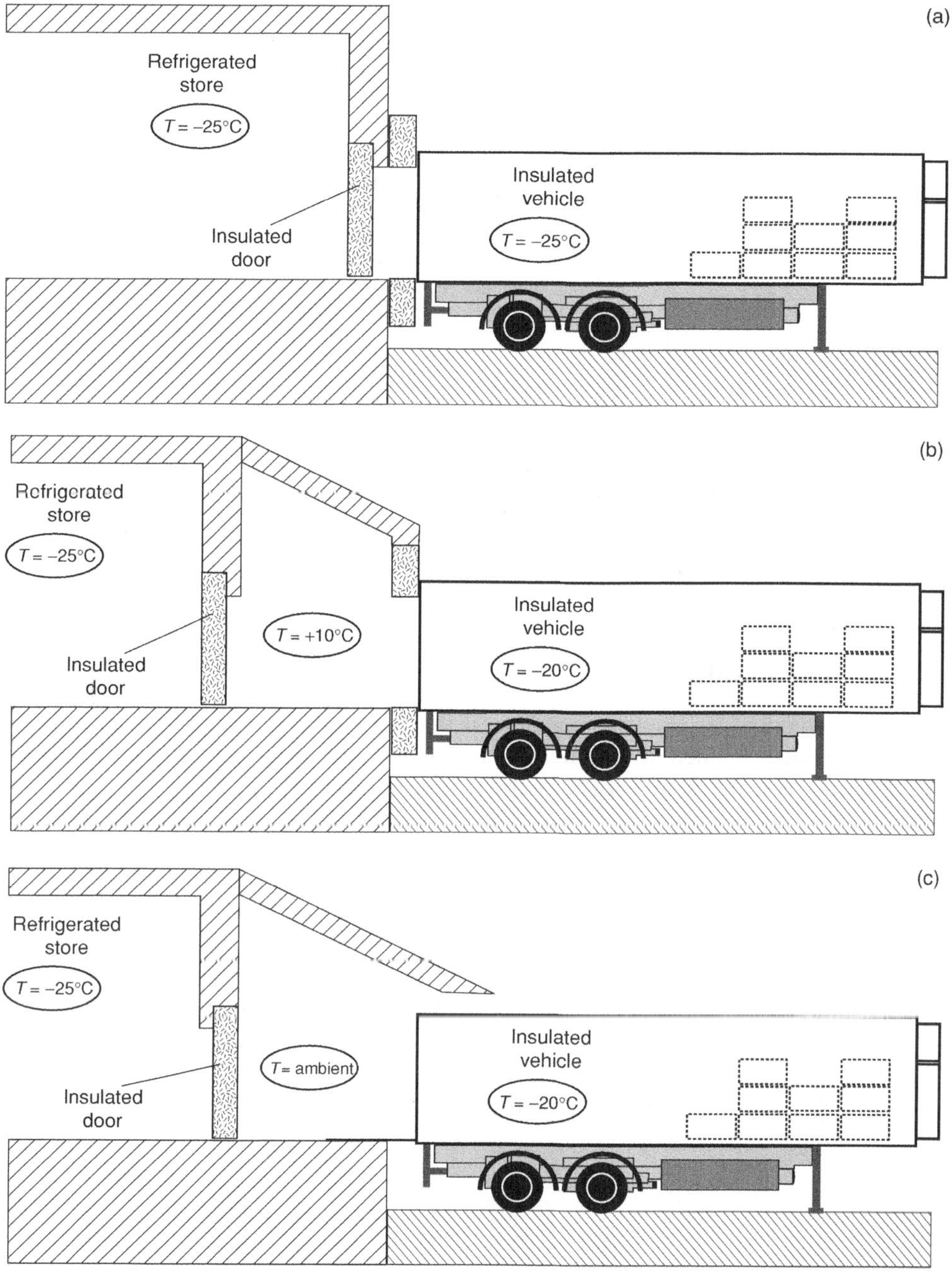

Fig. 13.1 Different layout situations for the loading of insulated vehicles from the refrigerated store: (a) The cargo is brought directly from the store to the vehicle, without any change in the external temperature. The frozen food is highly protected and the loading time can be long; (b) The cargo is crossing a room at a temperature that is intermediate between those of the store and of the external ambient. The protection of the frozen food is weak and the time to load is limited; (c) Situation similar to the case (b), but the degree of protection is lower: (d) The cargo has to cross the yard to be loaded. The frozen food has no protection and the loading time must be as short as possible.

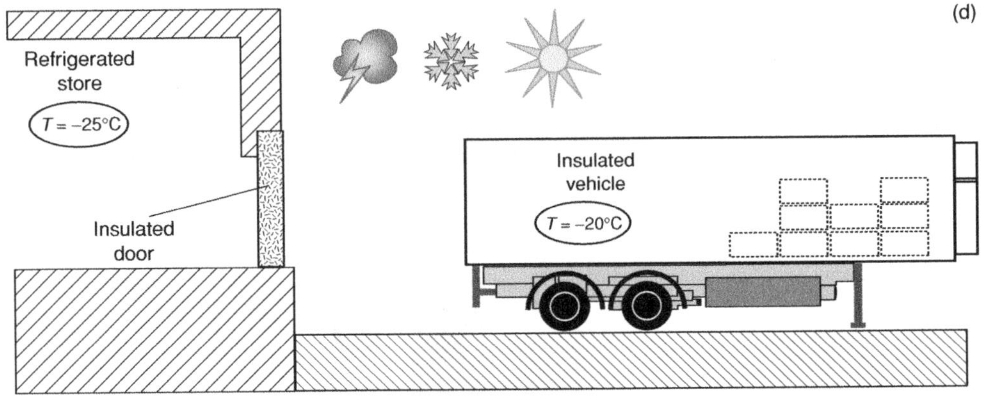

Fig. 13.1 *(continued).*

In some European countries (France, Germany, Italy, Portugal, etc.) national standards are very similar to ATP or ATP is adopted as national law. Also Australian standards have strong connections with ATP. In USA, a lot of voluntary standards are present, but no single standard is universally accepted.

13.2.1 The insulation

The main problem of the insulated container is that the insulating materials that can now be used have thermal conductivity coefficients that are too high and consequently the measured K values in the containers are very close to the limit values.

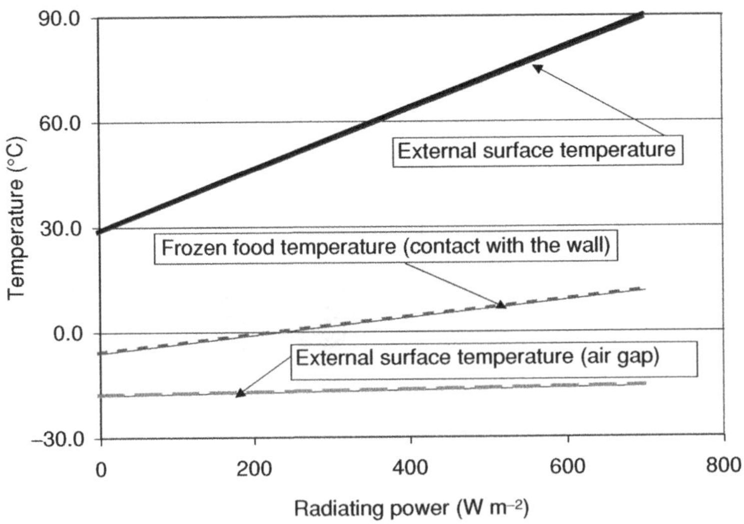

Fig. 13.2 Simulation of the trend of the external wall temperature in an insulated and refrigerated vehicle, versus the solar radiation (the vehicle is not running). The surface frozen food temperatures are also reported, in two different cases: the lower curve is referring to a cargo with an air gap from the wall; the upper one refers to a cargo that is continuously in contact with the wall.

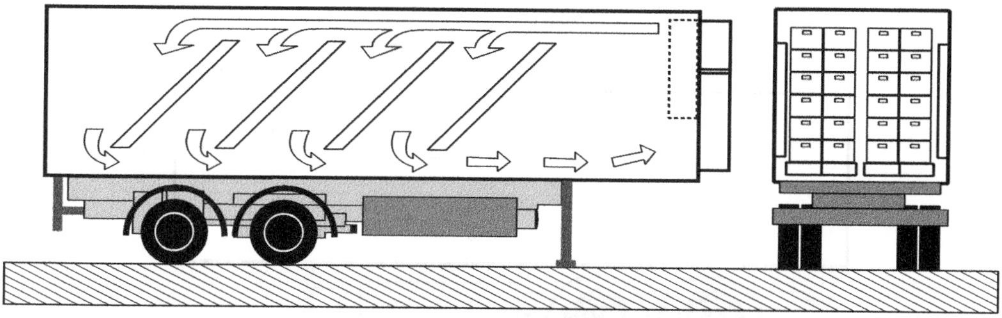

Fig. 13.3 Schematic drawing of the stowing bars between the internal walls surface and the frozen food cargo, in order to create some channels for the refrigerated air circulation. The return channels are made by the free spaces in the pallets.

The *K* values can be improved by enlarging the lateral wall thicknesses, but this then reduces the internal loading area available for storage. In Europe most cargo is palletised. A double pallet is 2.4 m wide whereas the width of European ATP vehicles are 2.6 m. Consequently, the wall thickness cannot be greater than 0.1 m and this is insufficient to provide enough space for pallets and the insulation thickness that is necessary for a low *K* value. Some manufacturers try to overcome this difficulty by increasing the thickness of the other walls (mainly the roof and the floor, as it is difficult to increase the front and rear wall thickness without affecting the ease of movement of the rear door and the correct positioning of the refrigerating unit on the front wall). This solution can reduce the *K* value; it also has an influence on the insulation ageing (the ageing is a function of the average thermal thickness), but is expansive and can increase the weight of the insulated container (Giuri, 2006).

13.2.2 Air circulation

For a frozen cargo, the air flow must circulate around the volume of the cargo, in order to remove the heat that enters from the walls. For this purpose, a sufficient air gap must be maintained between the cargo and the walls (this can be created using a corrugated sheet in the internal surface of the vehicle or positioning some thick stowing bars between the cargo and the wall; Fig. 13.3) (Amodio and Cioffi, 1985). The use of roof ducts is fundamental: these ducts are normally manufactured from textile material and have different lengths: the shorter one should be open at one-third of the length of the roof, and the longer one at three-fourths of the length of the vehicle (Fig. 13.4). The air flow rate is therefore sub-divided into a direct flow from the refrigerating unit immediately after the fans of the evaporator, and into a secondary flow that is directed from the ducts. The direct flow should be about the 20% of the total flow, the shorter duct will provide approximately 50% of the flow and the longer duct the residual 30% (IIR, 1995) (IIR, 1999).

13.2.3 Refrigerating power

The most widely used system to refrigerate the inlet of the vehicle is a vapour compression mechanical system. Several configurations are possible, as shown in Table 13.1 (Stimat and National Research Council Italy, 1982).

For many years R12 was used as the refrigerant in these systems. R502 was an alternative choice, mainly for frozen containers. These refrigerants were banned by the Montreal Protocol

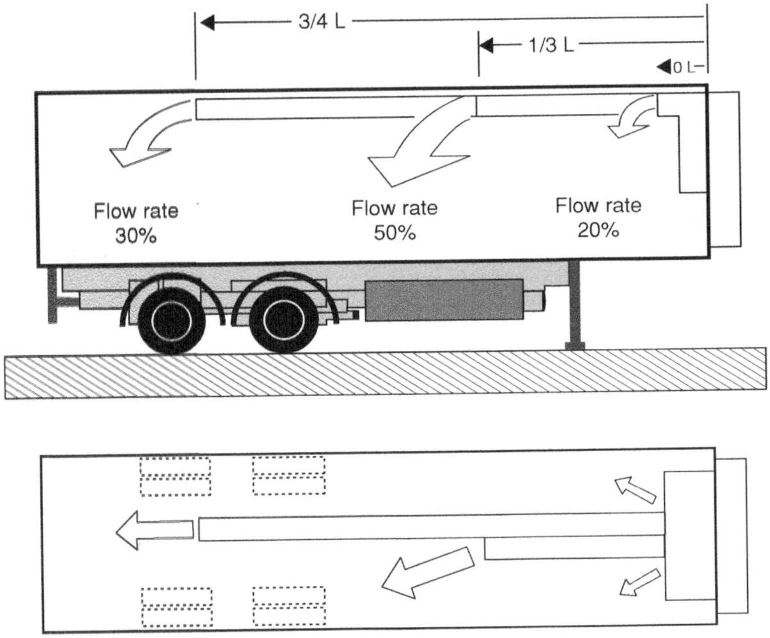

Fig. 13.4 Schematic drawing of the air ducts in the roof of a refrigerated vehicle.

and subsequently were replaced by R404A for higher capacities (large containers) and R134a and for lower capacities (small containers) and marine containers. Occasionally other gases are used as refrigerants but these are rare.

Several alternative refrigerants are currently undergoing development and may have potential in the future. These include:

(a) CO_2 in a transcritical cycle (Micheletto and Rosso, 2005);
(b) air cycle (Spence *et al.*, 2005) (Heap, 2000) (Engelking and Kruse, 1996);
(c) propane or other flammable refrigerants (Kauffeld, 1996) (Tiedemann and Kruse, 1994) (Boldrin *et al.*, 1991).

CO_2, in a very limited number of examples, is also used in a direct form where it is directly expanded into an evaporator in the inlet of the vehicle from a tank of liquid CO_2.

13.2.4 Energy storage

Energy storage systems that maintain a low temperature below that required by the cargo are often used in transport vehicles. Normally these consist of a eutectic solution that changes

Table 13.1

Power sources	Power to the compressor	Compressors	Fans
Diesel engine Vehicle's engine	Diesel engine Electrical motor (stand by) Electrical motor with alternator Hydraulic motor with oil pump (rare)	Open Hermetic Oscillating plate	Axial Centrifugal

phase at a low temperature (Ba, 1991). The solution is contained in a series of plates placed in the roof of the vehicle. Alternatively, the solution can be introduced in plastic tubes (Fig. 13.5) (Grosskopf, 1982). These systems have the following three advantages compared to direct expansion compression refrigeration cycles:

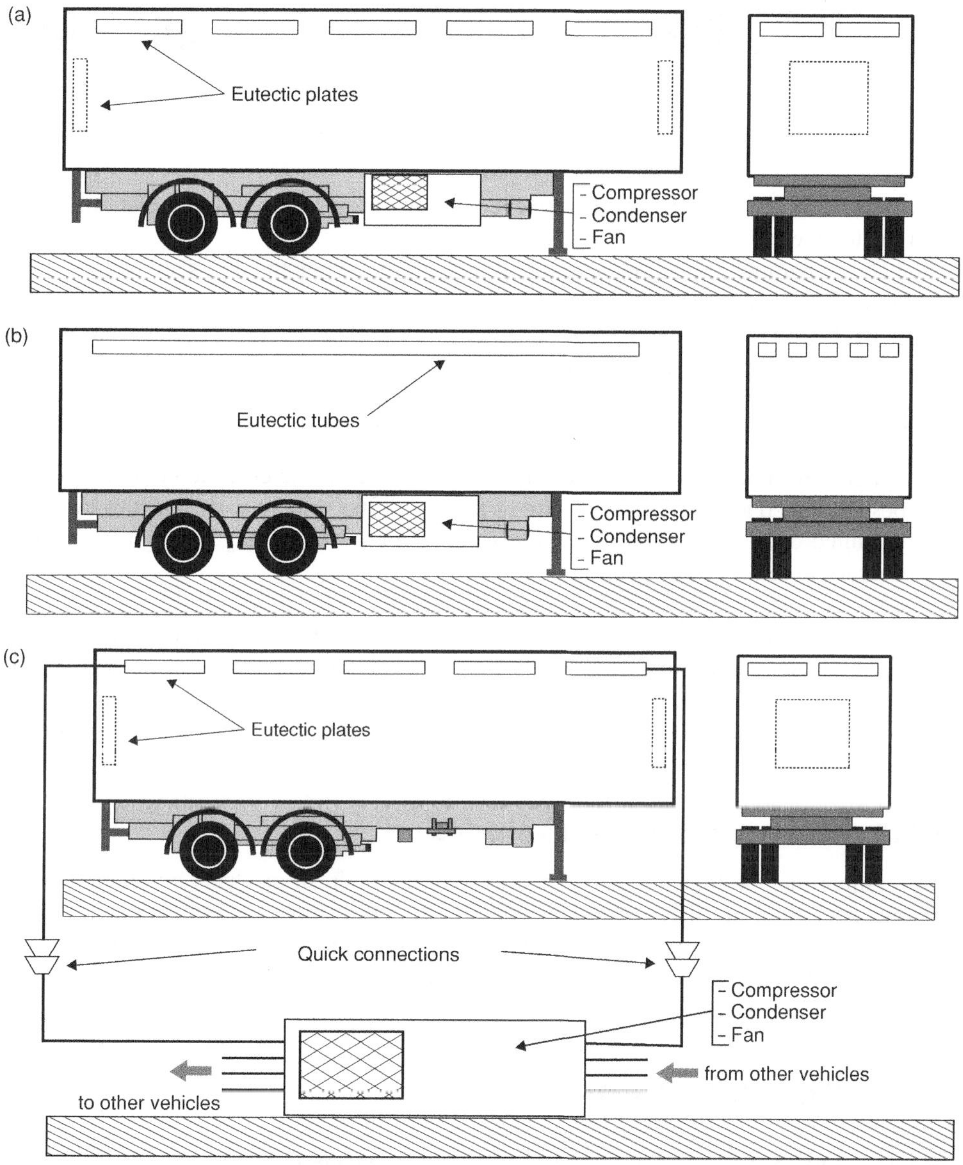

Fig. 13.5 Eutectic systems. (a) Layout of the plates. Normally the plates are placed under the roof. If a further plate is necessary, it can be placed in the rear or front wall, that are normally without doors (dotted plates). (b) Like (a), but with tubes in the place of the plates; (c) Eutectic plates whose compressor and condenser are not on board (like (a) and (b)) but separate. The separate refrigerating unit can feed more vehicles in the same time.

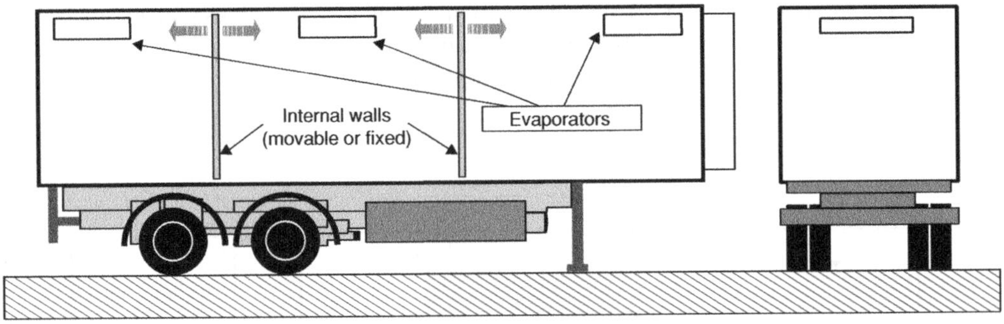

Fig. 13.6 Schema of a multi-temperature or multi-compartment vehicle.

(a) no noise during food distribution (noise reduction is required by some regulations, mainly in the town centres);
(b) the ability to quickly restore the ambient temperature when the doors are opened for the charge or discharge of cargo;
(c) a high reliability: as the energy is stored in a passive system where mechanical breakdowns cannot prevent energy release.

In addition, these are also two peculiar problems:

(a) the higher weight and a lower stability of the vehicle, as the eutectic systems (plates or tubes are joined at the roof: the centre of gravity of the vehicle is shifted upwards);
(b) the limited operational time (when the whole solution has changed its phase, a stop is necessary to refreeze the solution).

The refreezing can be achieved by a vapour compression unit on board, that can work only when fed by mains electricity or can be a separate refrigeration system that can refreeze one or more vehicle in the same time (Wenk, 1976).

The vapour compression system can work with the standard refrigerants, or with newer refrigerants such as CO_2.

13.2.5 Multi-compartment vehicles

Multi-compartment vehicles can transport frozen and/or fresh products by operating more than one evaporator at different (or if needed the same) evaporating temperatures. The advantage of these vehicles is that the different compartments can be charged or discharged separately in such a way that the opened compartment does not influence the other compartments that remain closed (Fig. 13.6).

13.3 THERMAL LOADS

The thermal loads in a frozen food transport vehicle are caused by:

(a) the temperature difference between the outside and inside of the vehicle. To compensate for this temperature difference, a low K value is required, that will in turn help the refrigerating unit maintain a stable energy consumption;

(b) the actual thermal load on a vehicle is very different from the thermal load used in the standard tests: not only the external temperature can be higher than +30°C used in the tests, but also solar radiation is not part of a standard test. The solar radiation can be direct or indirect (from surrounding walls and the ground). The influence of radiation is especially high when the vehicle is stopped and the external air velocity is low: when the vehicle is moving, the high surface heat transfer coefficient reduces the radiative effect on the vehicle (Panozzo and Cortella, 2006);
(c) the air speed outside and inside the container can change the actual K value compared to the tested value, often increasing the value. The air speed during the standard tests is 1–2 m s^{-1}, while the external air speed on the road may be $20 \div 30$ m s^{-1};
(d) external warm air can enter the vehicle during operation due to differences in pressure between the inside and outside of the vehicle (this difference is due to the reduced pressure of the air at low temperature but more to the aerodynamic effects of the air all around the vehicle). An additional effect is the so-called 'pumping' of the lateral walls that are not absolutely rigid and inflect and deflect owing to the alternate eddies that are formed along the lateral walls during operation. This effect 'pumps' the cold air out and the warm air in (this effect is obviously greater in long vehicles) (Bodenheimer; Bachmaier and Bornschlegl, 1965; Schausberger *et al.*, 1969);
(e) when the doors are opened, an exchange from external and internal air occurs, generating a supplementary infiltration heat load for the vehicle. This effect is more important for smaller vehicles that are dedicated to local delivery and is quite negligible for the long-distance transport vehicles that are normally closed at the beginning of the journey and only opened at arrival.

13.4 AGEING

13.4.1 Insulated containers

The importance of the K value has been highlighted previously. The change in the K value with time and 'ageing' of the insulation are important factors that cannot be underestimated. It is impossible to forecast the real ageing of a vehicle as the experimental data vary dramatically and only an average curve can be obtained (Figs. 13.7a,b) to explain the basic physical phenomena of ageing (Panozzo *et al.*; Boldrin *et al.*, 1990; Panozzo, 1999; Giuri, 2006).

An important factor affecting ageing is the average temperature of service of the vehicle and therefore, the aging in frozen food transport is the most severe. Another factor is the distance travelled per year by the vehicle. The physical phenomena of the ageing process are:

(a) migration of the air outside into the cells of the foam of the insulator material;
(b) migration of the blowing gas from the cells:
(c) condensation of water in the cells;
(d) damage to the cells that transform the closed cells in open cells (Klempner and Frisch, 1991).

In the case of (a) and (b) probably only the first phenomenon is important in vehicles that are built by the sandwich technique: the second one is too slow to have any significant effect (Fig. 13.7a).

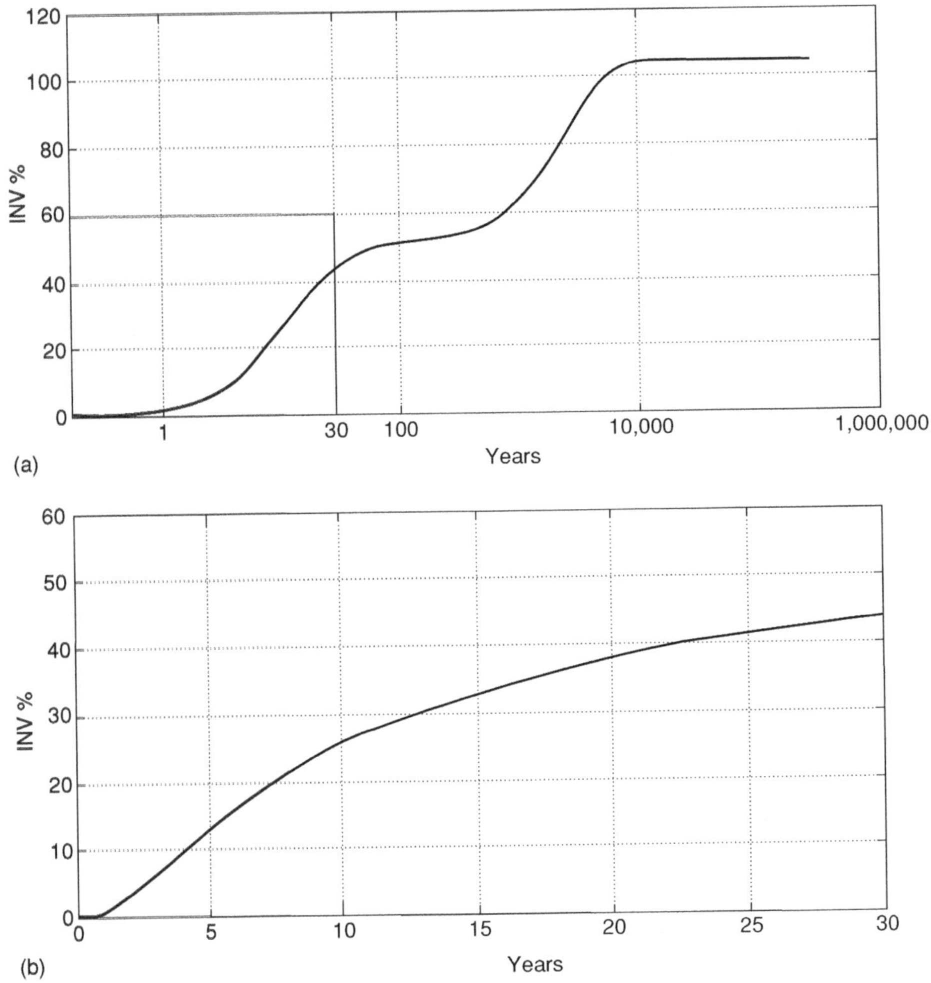

Fig. 13.7 Average ageing curve for insulated vehicles. (a) The first figure is the theoretical ageing curve for a vehicle (logarithmic scale). The first inflection point represents the air inlet in the polyurethane cells; the second one represents the loss of blowing fluid from the cells: the theory shows that air is the main factor responsible for ageing. INV% is the percent increase of the K value with time. In other words, if k_0 is the new vehicle value and k_n is the value after n years, we can write $k_0 = k_n[1 + (\text{INV}\%/100)]$. Consequently, $\text{INV}\% = 100[(k_n - k_0)/k_0]$. (b) The curve (in linear time scale) shows the useful period of the curve (corresponding to the standard life of a refrigerated vehicle). (c) The curve is plotted together with the experimental points: a large dispersion of data is evident. This dispersion makes it impossible to forecast the expected ageing rate for a single sample. The points signed by x-shaped characters are the average values of the experimental values per each year: these points are quite well fitted by the average curve.

Sandwich panels consist of strong skins (fibreglass-reinforced epoxidic resin or metal sheet) bonded to a core of insulating material (polyurethane foam, polystyrene foam).

In the case of water condensation in the cells, Fig. 13.8 demonstrates that not only the thermal conductivity of the foam is affected by the condensation, but also the slope of the curve of the thermal conductivity versus the temperature (the phenomenon is correlated with the characteristic slope of the curve for the solid materials, in this specific case, the ice).

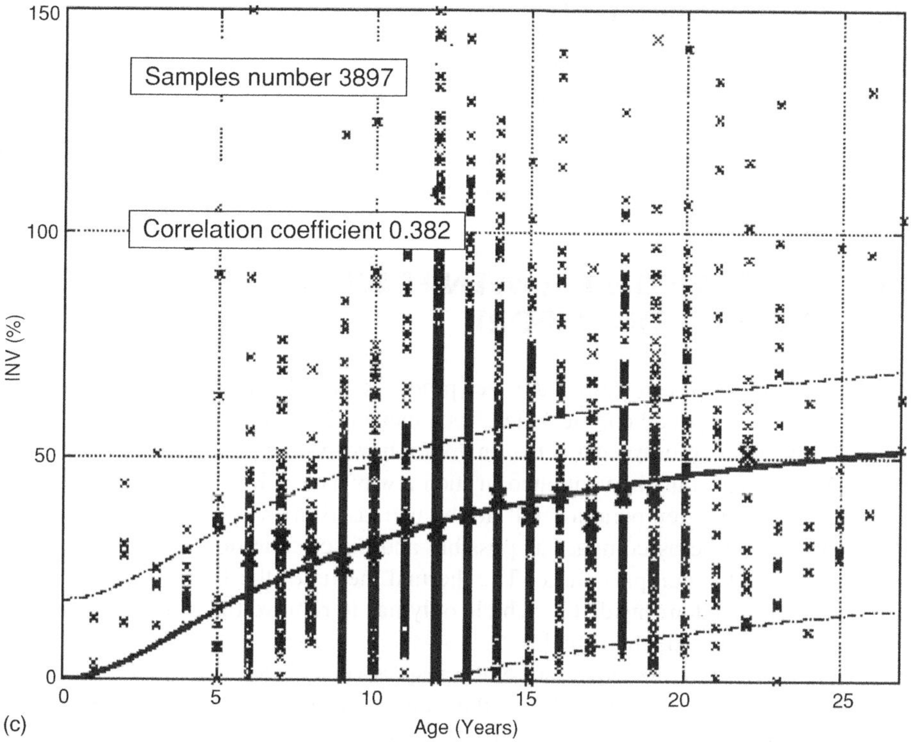

Fig. 13.7 *(continued).*

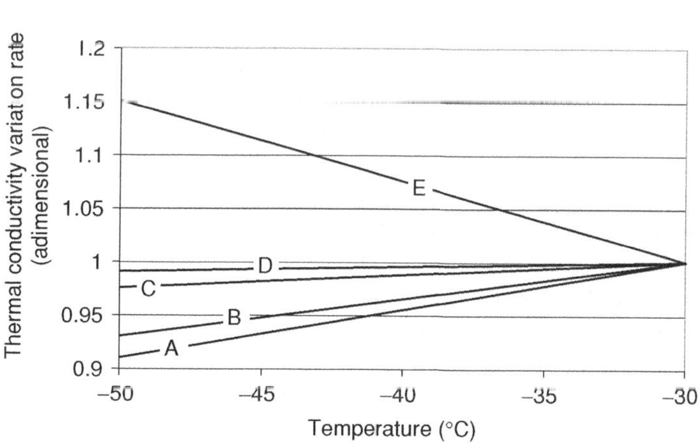

Fig. 13.8 The curves A and B refer to new samples (a bare panel and a new roll container): the slope is positive (the thermal conductivities of the gases in the cells predominate). The curves C and D refer to low aged samples: the slope is positive but is decreasing. The curve E refers to an aged roll container: the slope is negative (the thermal conductivity of ice in the cells is predominant, and the slope for the ice is negative).

13.4.2 Refrigerating units

In practice, the performance of refrigerating units deteriorates over time and can be attributed primarily to a lack of maintenance of the units, a situation that is reversible. Only a small percentage of the deterioration is due to the real irreversible ageing of the units (Fig. 13.9). In all cases, if a good maintenance program is carried out, ageing of the refrigeration units would be just a fraction of the ageing of the insulation (Fig. 13.10) (Panozzo *et al.*, 2003).

13.5 COMPARISON BETWEEN FRESH AND FROZEN PRODUCT TRANSPORT

The main difference between fresh and frozen product is that the fresh product (fruits, vegetables, not completely ripened cheese) often has a metabolic charge (this charge is increasing with increasing temperature) while frozen foods have no supplementary charge. Consequently, a fresh cargo must be ventilated in such a way as to allow ventilation to every part of the cargo so that the temperature will not be increased in any position. On the contrary, a frozen cargo must be as compact as possible and air circulation is only required to the external surface of the compact cargo. The thermal inertia of the frozen cargo is large and therefore the ventilated air inside the vehicle only has to remove the heat transferred across the insulation from outside the vehicle.

Test standards require the temperature of transport vehicles to be not higher than −20°C. If the transport vehicle were to maintain the same conditions as many refrigerated stores, the transport temperature would be −25°C to −30°C. This situation implies that the K value should be lower than prescribed by the standards, and that, the loss of performance of the refrigerating unit at lower temperature must be considered (Fig. 13.11 shows a typical curve of the power and COP of a unit).

13.6 STANDARDS AND CLASSIFICATION

The most used standard is ISO 1492 'Thermal containers' that provides, together with data about the mechanical properties of the containers, thermal characteristics such as the K value of the container and its air leakage rate. The level of the air leakage rate is used to correct the K value. This standard is the basis for many certification and testing registers.

The ATP (Agreement on the international carriage of perishable foodstuff and on the special equipment to be used for such carriage) is an international agreement. The treaty has currently been signed by the following European countries: Albania, Austria, Belarus, Belgium, Bosnia and Herzegovina, Bulgaria, Croatia, Czech Republic, Denmark, Estonia, Finland, France, Georgia, Germany, Greece, Hungary, Ireland, Italy, Latvia, Lithuania, Luxembourg, Monaco, Montenegro, the Netherlands, Norway, Poland, Portugal, Romania, Russian Federation, Serbia, Slovakia, Slovenia, Spain, Sweden, the former Yugoslav Republic of Macedonia and the United Kingdom. The following non-European countries have also signed the ATP: Azerbaijan, Kazakhstan, Morocco, Tunisia, United States, Moldova and Uzbekistan. Switzerland signed the Agreement but has not ratified it. In Europe, Iceland, Malta, Cyprus and Ukraine have not signed the Agreement so far.

The text is deposited at the United Nations in New York and changes are discussed annually at an international conference in Geneva. Although ATP is mandatory only for the countries

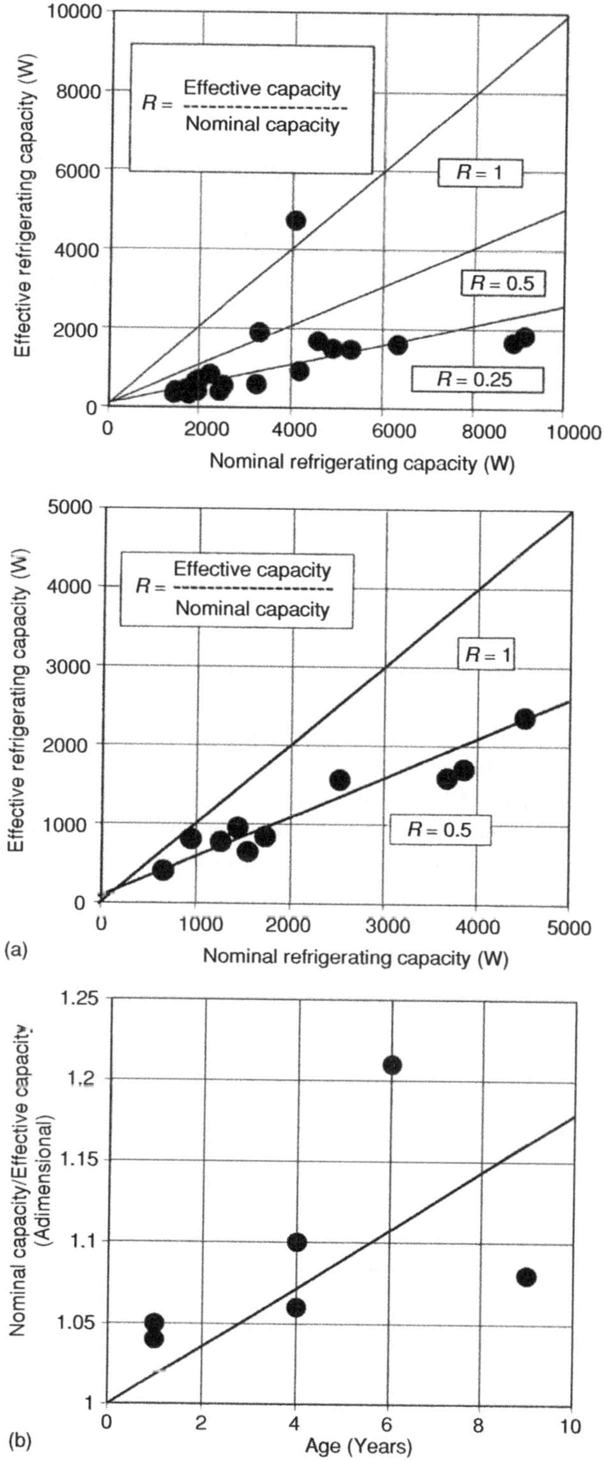

Fig. 13.9 Ageing rate for the refrigerating units. (a) The ratio between the nominal capacity and the measured one is very low (about 0.25 for FNA vehicles and 0.5 for FRC vehicles). The experiments showed that the lack of maintenance was the most important reason of ageing. When the maintenance is carried out correctly, the figures are lower. (b) Experimental points and extrapolation line for the ageing of the refrigerating units after maintenance.

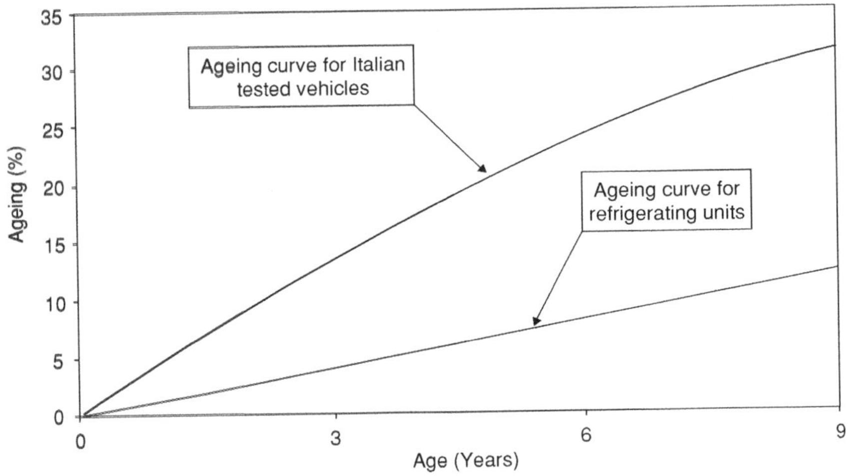

Fig. 13.10 Comparison between the ageing rate for the insulated vehicles and for the refrigerating units.

that have signed the Agreement, it is becoming a standard *de facto* and many national rules are derived from the basic text. The ATP prescribes steady state measurements of the K value and of the refrigerating power of the refrigerating units, and also particular procedures to maintain in service older vehicles.

After the approval in the EU in 1992, of two new directives, the main part of these directives was introduced into the ATP text. However, the philosophies of the ATP and of the EU directives are different; the ATP concentrates on physical characteristics of transport

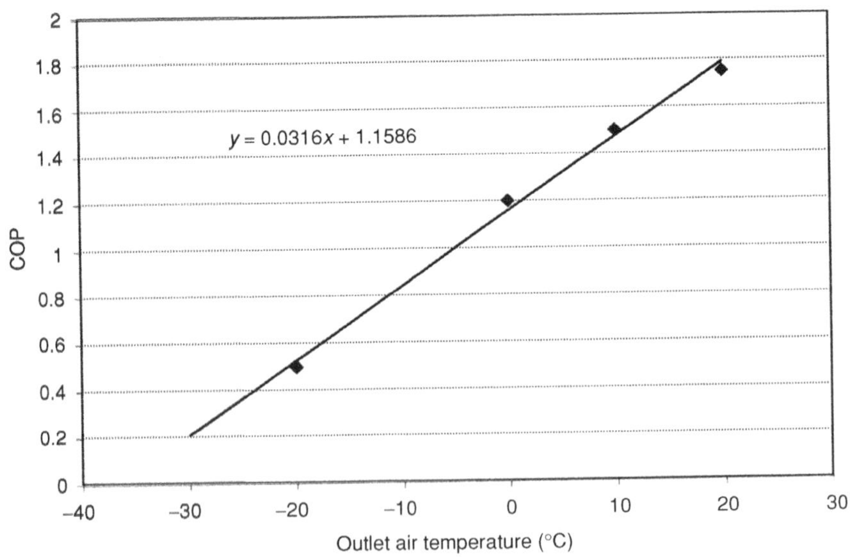

Fig. 13.11 Typical curve of the coefficient of performance (COP) of a refrigerating unit versus the outlet air temperature.

vehicles whereas the directives takes into consideration only the temperature of the food during transport and the systems used to measure them.

The complete titles of the two directives are:

- *Commission Directive 92/1/EEC* (OJ L34, p28, 11/2/1992) of *13 January 1992* on the monitoring of temperatures in the means of transport, warehousing and storage of quick-frozen foodstuffs intended for human consumption
- *Commission Directive 92/2/EEC* (OJ L34, p30, 11/2/1992) of *13 January 1992* laying down the sampling procedure and the Community method of analysis for the official control of the temperatures of quick-frozen foods intended for human consumption

The ATP classifications for frozen foods are:

Table 13.2

IR	Insulated reinforced vehicle	K equal or lower than 0.4 W m^{-2} K^{-1}	
FRC	Mechanically refrigerated insulated reinforced vehicle	K equal or lower than 0.4 W m^{-2} K^{-1}	Refrigerating power higher than: 1.35 KS ΔT; 1.75 KS ΔT
RRC	Refrigerated insulated reinforced vehicle	K equal or lower than 0.4 W m^{-2} K^{-1}	The energy store must be sufficient to maintain the inlet temperature under −20°C for at least 12 h, with an external temperature = +30°C and an internal heat source = 1.35 ΔT

Note: K is the overall heat transfer coefficient; S is the average surface of the container. ΔT is the difference between the average internal and external temperatures during the test.

13.7 SYSTEMS OF TRANSPORT

A frozen cargo can be transported by road, rail, water or air. There can be specific systems or intermodal systems.

In the first case we can have trucks, wagons, ships etc.; in the second case, we have different types of containers.

13.7.1 Road transport

The different types of trucks can be divided into (DIN 8959, 2000):

(a) short-distance delivery transport,
(b) medium-distance transport
(c) long-distance transport.

In case (a), this transport is characterised by short vehicles that are devoted to the transport from the last refrigerated store to the display cabinets in the supermarkets and food shops or directly to home. For many foods that are stored at low temperature, the eutectic plates system is often used.

In case (b), medium-length vehicles are normally chosen (e.g. 6–8 m) and these vehicles transport foods from producer to store or from store to store. Mechanical compression refrigeration systems are mainly used in this type of transport.

In case (c), the semi-trailer is the preferred vehicle and can carry up to 33 pallets. The possibility to overcome this limit does not seem realistic unless the maximum dimensions of the trucks are changed dramatically (see Section 12.2.1).

The semi-trailer is a trailer without a front axle. A large proportion of its weight is supported by a road tractor. A semi-trailer is equipped with legs that can be lowered to support it when it is unhooked from the tractor. The maximum standard length for the insulated container is 13.60 m (together with the refrigeration unit).

This type of transport can act over long or medium distances and can often partially charge or discharge its cargo at stages during the journey. These different logistic systems are illustrated in Figs. 13.12a–d. Often the journey is optimised from a logistic point of view by means of a communication system such as GSM or similar or a satellite communication system (Fig. 13.13).

13.7.2 Rail transport

The main differences between road and rail transport are shown in Table 13.2.

For rail transport the wagons are refrigerated by either mechanical or stored energy. Mechanical refrigeration for each single wagon is the predominant technology. In the past, systems that fed power to single refrigerating units in each wagon from a central wagon, used diesel as fuel (Berecnchtein and Sapozhnikov, 1975). Another old system was based on iced water (stored in the two compartments at the extremities of the wagon).

It had several advantages:

(a) greater liability;
(b) lower maintenance;
(c) absolute interchangeability;
(d) the possibility to be fed by solid CO_2, becoming, without any change, a vehicle for frozen foods.

The main differences with the road transport are in Table 13.3.

13.7.3 Water transport

A distinction is made between transport on internal waterways (rivers, channels, lakes) and the sea.

The most usual form of water transport is sea transport. In this case ships can be insulated and refrigerated (and was the case for the first refrigerated transport in the world) or can be container ships. A curious system is in use in Venice (Fig. 13.14), where all the transport is carried out by water. The boats that carry frozen foods have an insulated (in some case also refrigerated) box that is built with the same technology as road vehicles. The box is not the usual parallelepiped shape (the roof is formed by two inclined planes) and the doors are in the roof.

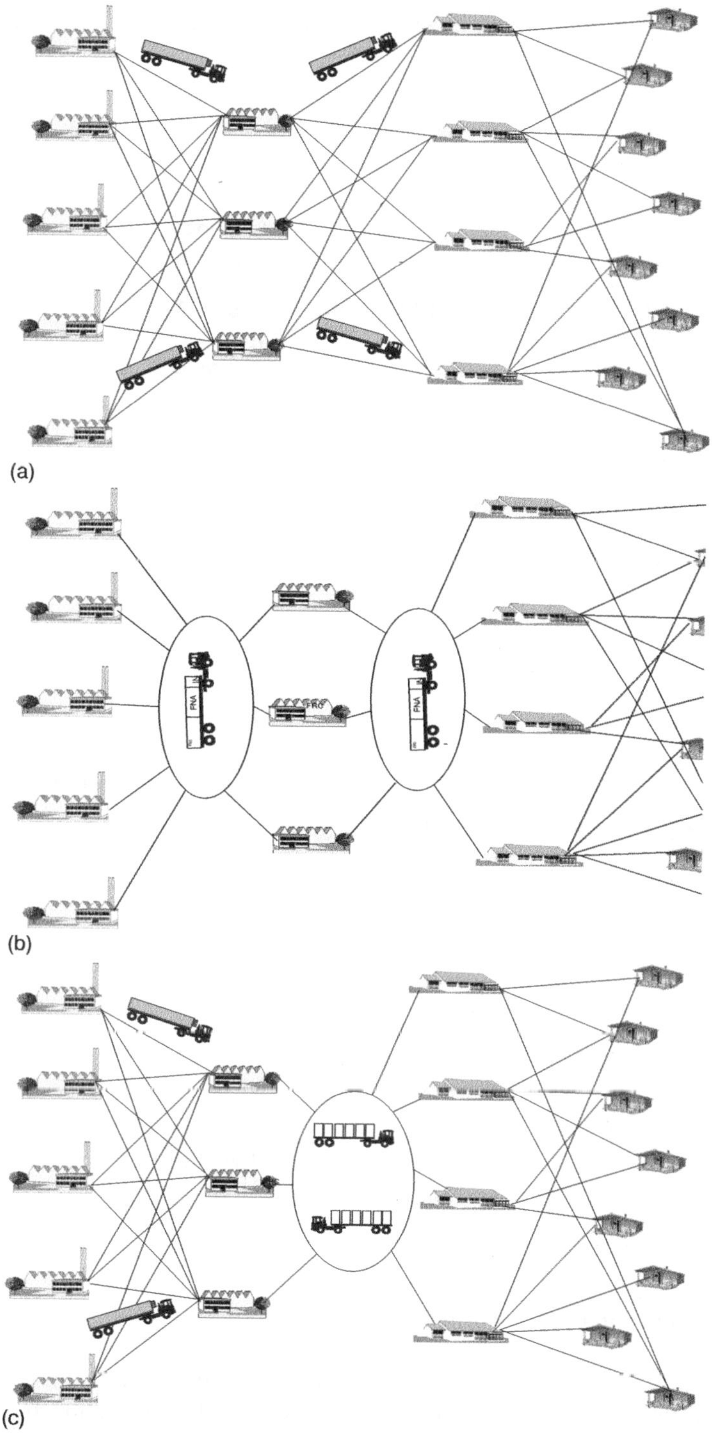

Fig. 13.12 Different schemes of transport of frozen foods. (a) Standard circulation of semi-trailers of different owners for different producers, stores and shops. (b) Typical circulation for multi-compartment vehicles: the logistic simplification is evident. (c) The roll containers circulation: only the phase of the delivery to the shops is changed with reference to the situation (a). (d) Typical circulation for home delivery vehicles. The owner of the vehicles is also the producer of the frozen foods or the vehicles are in exclusive service for the producer. (e) Legend for the previous schemes.

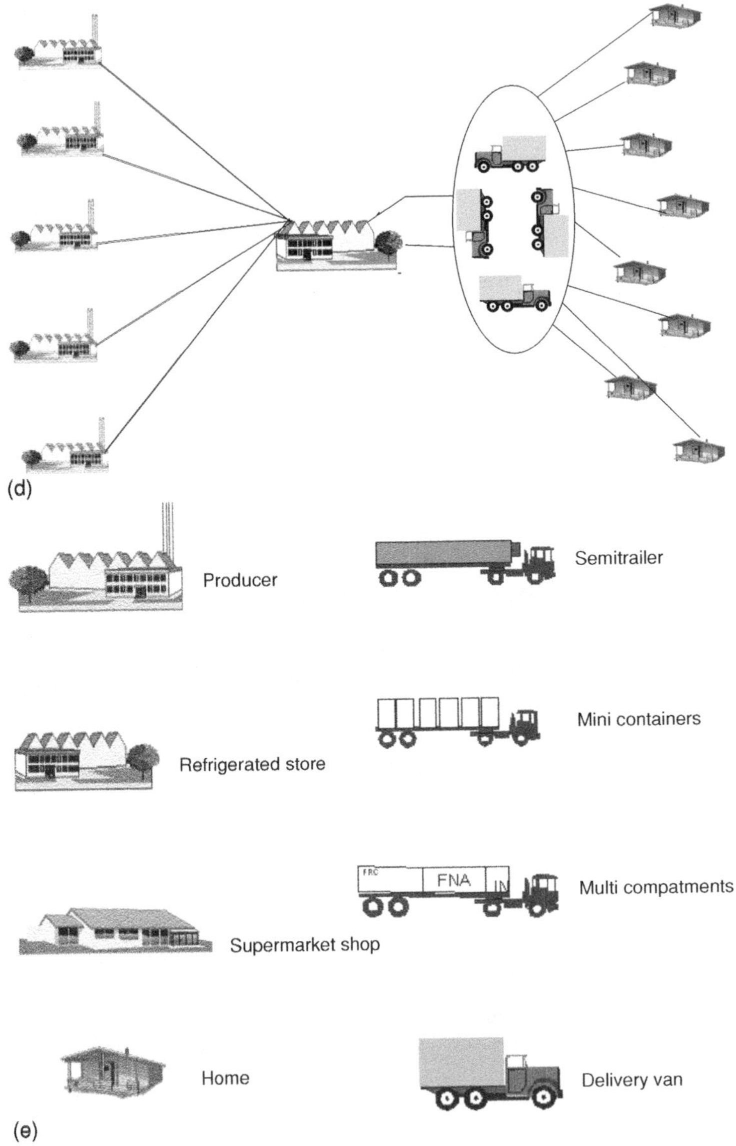

Fig. 13.12 *(continued).*

Container ships can have two different layouts:

(a) self-refrigerated containers, each with its own refrigerating unit. In this case, the greatest problem is the space between each unit that must be sufficient to allow fresh air to be fed to each condenser, and in the lower decks the ventilation must be sufficient to remove the thermal energy released by the condensers;

(b) insulated containers, specifically designed for ship transport. Each container has two circular holes in the upper and lower side of the front wall. These circular holes are connected with two flexible tubes that can supply cold air to the container and evacuate

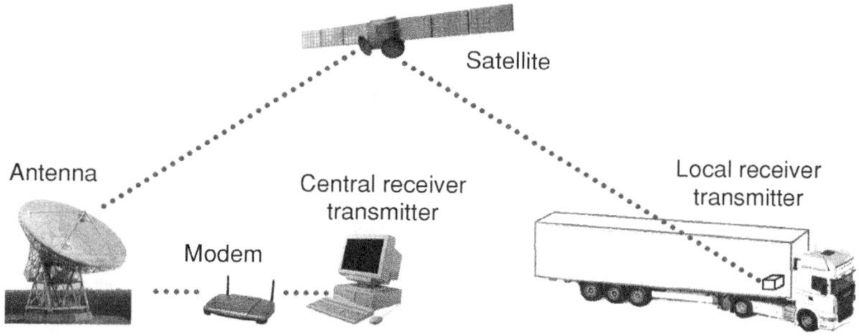

Fig. 13.13 Scheme of continuous contact between the vehicles and its headquarters by satellite. The system was devised as anti-theft device and was used for the information exchange in order to let the vehicle load and discharge the cargo partly, without coming back to the headquarter. In the future, the system could be used also for recording the data of the cargo (temperature, humidity, . . .).

Table 13.3 Advantages and disadvantages of refrigerated rail transport.

Disadvantages of rail transport	Advantages of rail transport
Lower flexibility	Greater thickness of the walls (lower K value)
Loading only by lateral doors	Greater dimensions (double that of road vehicle)
	Greater rigidity (possible lower ageing?)

the exhaust air. Whilst the containers are on the docks, the containers can be attached to special equipment to achieve the same function.

13.7.4 Intermodal transport

Intermodal transport can be carried out using standard vehicles that can be carried on rail wagons (e.g. the piggyback system) or ships; however, these systems are heavy and not easy to use. However, most intermodal transport uses insulated boxes that can be transferred easily from ships to the rail wagons to road vehicles.

Fig. 13.14 An old picture of frozen food transport in Venice. Reproduced with permission of Cold Car (www.coldcar.it).

Table 13.4 Classification of the vehicles for the frozen foods transport, as prescribed by ATP.

Type of container	Use	Standard (dimensions)	Standard (Thermal characteristics)
(a) ISO thermal containers	Road, rail, water	Yes (ISO 1942)	Yes (ISO 1492)
(b) IATA containers	Air, road	Yes (but adapted to each plane)	No
(c) Swap bodies	Rail, road	Yes (UIC)	Yes (ATP)
(d) Roll containers	Road to road	No	Yes (ATP)
(e) Mini containers	Road to road	No	Yes (EN 12571)

These containers can be sub-divided as shown in Table 13.4.

Case (a): the dimensions are standard for width and height (8′) and the length is a multiple of 10′ (to a maximum of 40′). The thermal characteristic required by ISO 1942 is perfectly compatible with ATP. They can be attached to the frame of the vehicles by four corner fittings and can be stacked in a row of six containers.

Case (b): they are normally not parallelepiped as are adapted to the airplane shape. Insulation is poor (or absent) and the refrigeration can in some cases be solid CO_2. The main thermal problems concern the airport as during the flight the external temperature is very low.

Case (c): they are movable boxes that can be transferred from railways frames to road frames by mechanical forks or lifts, but they cannot be stacked.

Case (d): they are containers whose maximal dimension does not exceed 2 m, without any system for moving and blocking. They can be insulated but also refrigerated (mechanical systems, Peltier systems, and CO_2 systems). An example of a roll container is shown in Fig. 13.15. Roll containers are not standardised and there is some doubt that they can be considered ATP vehicles; a proposal to introduce the roll container into the ATP is under discussion. Their structure can be similar to that of road vehicles (assembling from sandwich panels) or can be made by a 'rotomoulding system'. The typical characteristic of this system is the construction of an external shell for the insulating foam by moulding polyethylene in big boxes that by rotating, can generate a cave shell in a monolithic system. The shell can be filled by polyurethane, expanded *in situ*, while the walls are contained in a rigid system to avoid deformations during the expansion (Panozzo *et al.*, 2006).

Case (e): they are similar to the roll containers but cannot be defined as vehicles. The issue of where to put the border between a real transport systems and simple packaging systems is subject of an ongoing debate. This problem impacts on the standard tests for these boxes. ATP standards can be used or specific standards such as EN 12571. The philosophies of the two standards are completely different. While ATP requires steady state measurements of the global heat transfer coefficient, EN 12571 requires only transient tests concerning the time that a certain quantity of a specific material can be maintained at a temperature higher (or lower) than a specific value. A typical EN 12571 test is reported in Fig. 13.16.

The main advantage of the use of mini containers is that the container can be filled directly in the cold store and opened directly at the end destination, without any break in the cold

(a)

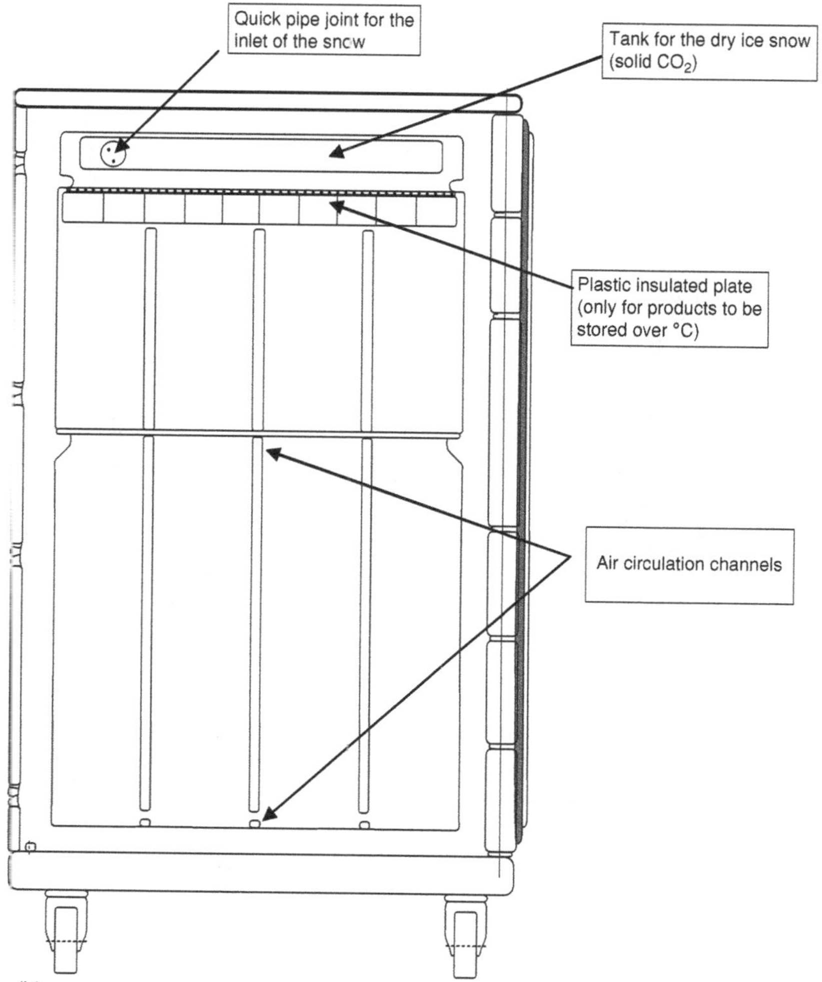

(b)

Fig. 13.15 Roll container picture, and schematic of its principal components.

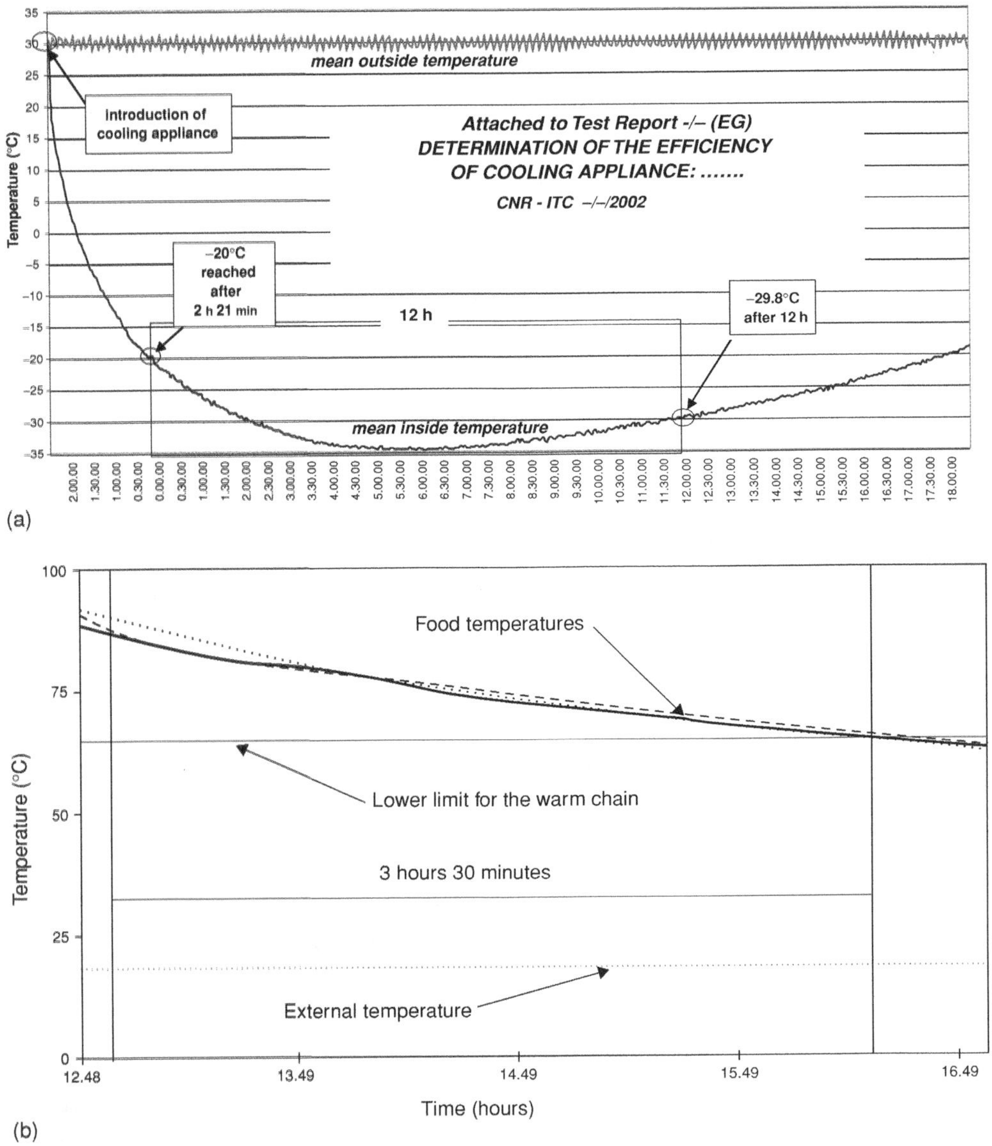

Fig. 13.16 Results of the standard tests of a mini container. (a) Typical ATP efficiency tests for a CO_2 system in a roll container. This test must be associated with a K value measurement. (b) Typical test for small containers for food transport. In this case the measured curve is related to warm transport (the temperature must not be lower than 60°C).

chain. The disadvantage is the high tare of the system and the necessity to return the empty containers to source.

The possibility to use containers with a K value greater than 0.4 W m^{-2} K^{-1} in a vehicle with a similar K value (the combined K value would be lower than 0.4 W m^{-2} K^{-1}) for the transport of the frozen foods is currently under discussion.

13.8 ENERGY LABELLING

The energy labelling system is widely used for domestic appliances and is mandatory in several countries, including the EU.

Using as a basis an energy labelling scheme for refrigerated display cabinets, a labelling scheme for transport vehicles was studied and adopted. The basic value for the labelling was the *K* value with the results from the air-tightness test added (van Gerwen *et al.*, 1999). The basic philosophy was that no expensive measurement should be required. Only a very simple measure of the absorption coefficient of the external surface was proposed. All the other parameters were derived from geometrical considerations and from the presence or absence of some installations (e.g. bulk head, air ducts). The transport vehicle was classified using two different and separate labels: one connected to the insulation and the other to the refrigerating unit. . The largest problem was the effect of ageing which could not be predicted: the energy consumption of a refrigerated vehicle is mostly dependent on the aged *K* coefficient, but this value cannot be forecast for a new vehicle, at least at the actual stage of the technology. This issue is currently being studied by a number of organisations and countries (German Government, Germanischer Lloyds, Transfrigoroute International, International Institute of Refrigeration).

13.9 LIMIT VALUES

13.9.1 Global heat transfer coefficients

The recommended *K* values for frozen and fresh foods are 0.4 and 0.7 W m^{-2} K^{-1}, respectively. Much discussion is on (especially regarding the *K* value for frozen foods) the difficulties in obtaining these *K* values using normal lateral wall thicknesses and the almost practical impossibility of maintaining these values after 6 years of service (the ATP prescribes the first check of in-service vehicles after 6 yr).

Theoretically, these *K* values should be justified by some physical reason connected with the physical and biological changes during the transport of foodstuffs. However, the values appear to be based on the fact that historically pallets were not widely used and 100 mm of cork was the main form of insulation. Plotting the *K* values that are possible in these conditions (Fig. 13.17) we find the values are very close to the standard limits.

Although *K* values are based on what was physically achievable with early vehicles rather than on any physiological changes in the food there are good reasons to maintain low *K* values. These include:

(a) the possibility of contact between frozen foods and walls (see paragraph 12.1, point c);
(b) being able to maintain food temperature during any breakdowns or increased heat loads on the vehicle;
(c) reduced heat gain from ambient and consequently reduced refrigerating power and energy consumption.

13.9.2 Safety coefficient

The refrigeration unit safety coefficient is currently 1.75 (the refrigerating power must be higher than 1.75 $KS\Delta T$), and 1.35 for an efficiency test. The value 1.35 was an empirical

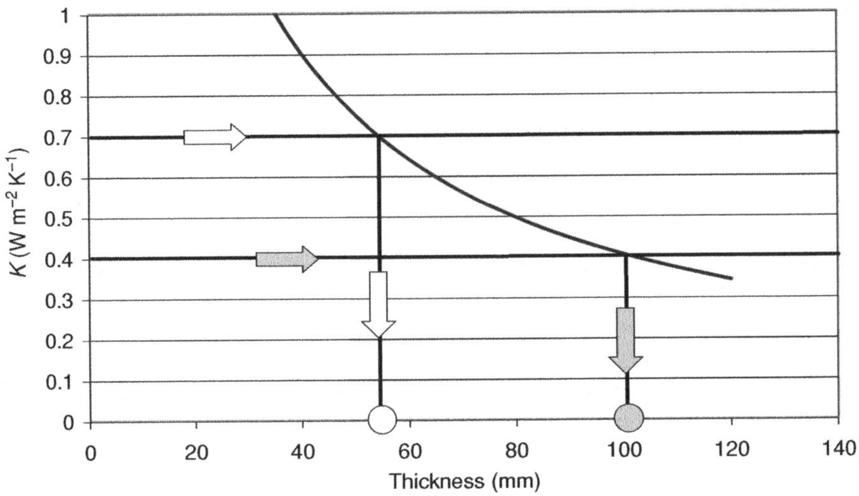

Fig. 13.17 *K* value for a vehicle insulated by cork versus the walls thickness. 100 mm thickness gives directly 0.4 W m^{-2} K^{-1} *K* value.

evaluation of the contribution to the thermal load on a vehicle due to solar radiation and by ageing of the insulation. No reason is reported for the value 1.75 (Rudnik, 1958) .

A larger value is sometimes requested in order to increase safety but can result in greater energy consumption.

13.10 THE FUTURE OF MATERIALS USED IN REFRIGERATED TRANSPORT

The Montreal Protocol and its successive modifications banned many of the insulation blowing agents used in the past. Many replacement blowing agents produce insulation with higher thermal conductivities and consequently *K* values have increased (Fig. 13.18). In the near future substances that have high global warming potential may be banned. Consequently, it is possible that only hydrocarbons, air, water, and CO_2 will be available as blowing agents for insulating panels in the future. This situation will probably lead to a further increase in the global heat transfer coefficient with all the problems illustrated in Section 13.9.1.

For a better understanding of this situation it is important to point out that the low thermal conductivity of expanded foam is caused mainly by the thermal conductivity of the blowing fluid in the gas phase. An empirical law states that the thermal conductivity of a gas is a function of its molecular weight (Figs. 13.19a,b) (Panozzo and Cortella, 2005). In these figures it is evident that there is no possibility of finding gases better than R11, at reasonable cost. Consequently, different insulation technologies should be explored. These include:

(a) vacuum insulation (some experiments have been carried out in the transport without appreciable success);
(b) aerogels (if the mechanical properties of these materials could be highly increased);

(c) reduction of the insulation cell dimensions (if the cell diameter in the insulating foam could be reduced by a factor of 100, the thermal conductivity of this hypothetical foam should be the same of a vacuum foam, and the blowing agent should have no importance, also ageing should be reduced for these materials) (Smoluchowski, 1898) (De Ponte, 1987).

Refrigeration system refrigerants

The refrigerants used in the refrigerating units of vehicles have also been affected by the Montreal Protocol but the replacement fluids have had little practical effect on the performances

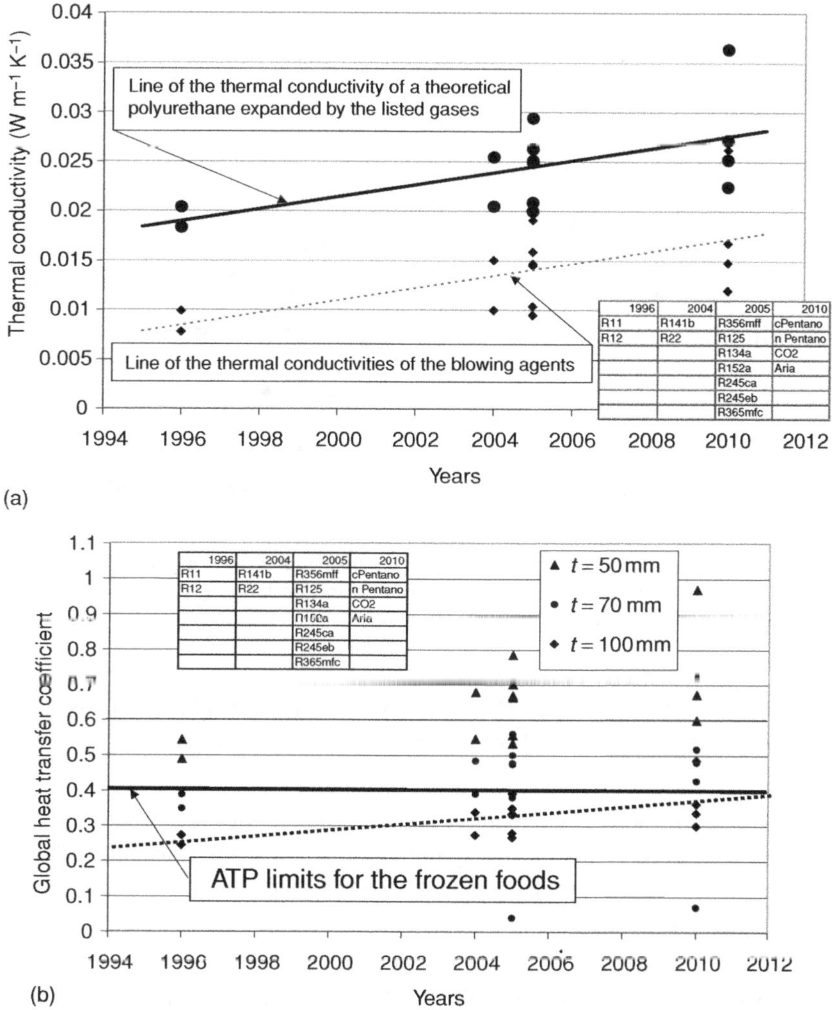

Fig. 13.18 Average values respectively for the thermal conductivity of the blowing fluid (gas), of the expanded polyurethane and of the global heat transfer coefficient for insulated vehicles, when the blowing fluid is changed. It can be seen that it is difficult to be lower than the limit value for frozen especially in the future, when some fluids now in use will be probably banned (and this will be impossible with a thickness of 50–60 mm).

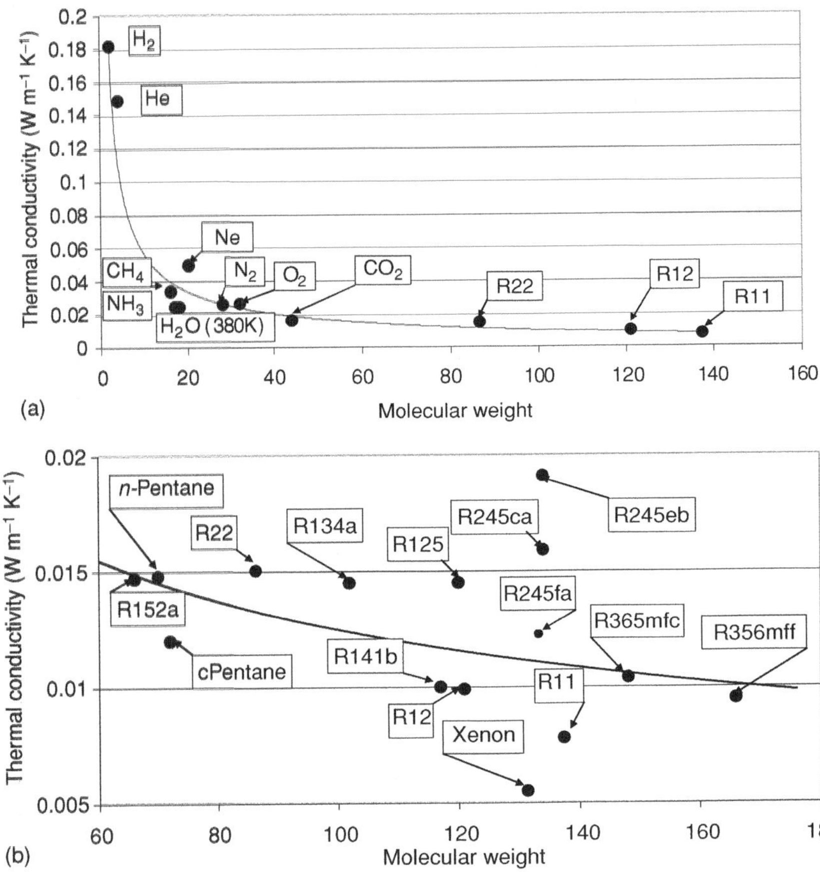

Fig. 13.19 Thermal conductivities of the gases versus their molecular weights. Panel (b) refers to the higher molecular weights, in the region that is interesting for the blowing process: only the rare gases could work better than R11.

of the vehicles. CO_2 (Lorentzen, 1994) is already used in some vehicles and is forecast to be used more widely in the future. CO_2 has some important advantages:

(a) the dimensions of the heat exchangers and all the parts of the refrigerating unit can be reduced (the reduced volume is a very important factor in the transport technology);
(b) CO_2 is important in transport as it is a fire extinguisher;
(c) CO_2 is very cheap.

In addition, in the future particular attention will be paid to the performances of refrigeration units in warm climates where the COP of the units will change due to the hot ambient conditions (Fig. 13.20) (Panozzo and Minotto, 2003). The global energy balance of a refrigerated vehicle should be rearranged in these more severe conditions, either from the point of view of the choice of the refrigerant or from that of the ratio between the refrigerating capacity and the heat gain through the insulation.

Mr Gabriele Minotto of ITC CNR provided the figures presented in this chapter.

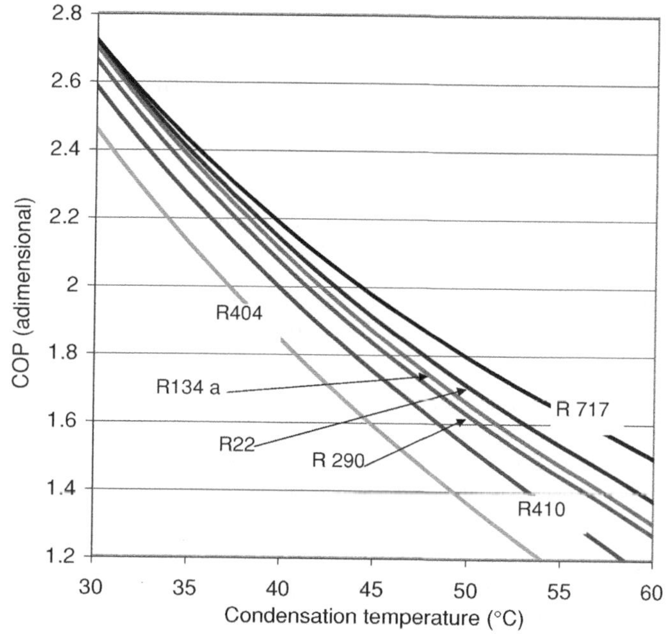

Fig. 13.20 COP referred to the most used refrigerants versus the condensation temperature. Clearly, the condensation temperature is a function of the external air temperature. When the external air temperature is increasing, the COP is decreasing with a different slope. If we do not take into account R717 (Ammonia) and R22 (possible future ban), the performances of the other fluids are diverging more and more with increasing temperature. The best performances of R134a at high temperatures could explain its wider use in marine containers.

REFERENCES

Amodio, U. and Cioffi, E. (1985). T.W.V. experiments – decisive conclusions reached. In: *Proceedings of the Meetings of Commissions D2, D3 IIR and International Center for Transportation Studies*, Roma. March 13–15, pp. 31–48.

Ba, A. (1991). Studio di un sistema di accumulo termico per un frigorifero domestico fotovoltaico per utilizzo in climi caldi. (Study of a thermal storage system for a domestic photovoltaic refrigerator to be used in warm climate). Il Freddo. November–December 1991, 45(6), 588–595.

Bachmaier, W. and Bornschlegl, A. (1965). The influence of air leakage on the K-factor of special equipment for the road transport of perishable foodstuffs. Bull IIF-IIR Annexe 65-3, Cracow, 135–147.

Berecnchtein, M.G. and Sapozhnikov, S.A. (1975). Nouvelle section frigorifique a cinq wagons pour le transport ferroviaire del l'URSS. In: *Proceedings of the XIV International Congress of Refrigeration.* Moscow. pp. 335–344.

Bodenheimer, B.A. Coping with air exchange. Refrigerated Transporter. June 77, 32–34.

Boldrin, B., Minotto, G., Panozzo, G., Toniolo, B. and Varotto, G. (1990). Ageing of insulated vehicles. statistical elaboration of tests results. Boldrin, B. *et al.* (1990). Ageing of insulated vehicles. statistical elaboration of tests results. IIF-IIR Commission B2,C2, D1, D2/3, Dresden 1990–9.

Boldrin, B., Minotto, G., Panozzo, G., Toniolo, B. and Varotto, G. (1991). Propane as an alternative refrigerant in foodstuffs transport. In: *Proceedings of the XVIII International Congress of Refrigeration*, Vol. 4. Montreal. pp. 2062–2067.

De Ponte, F. (1987). Ricerca sui materiali isolanti: stato dell'arte e prospettive. (Today research on insulating materials and an insight into the future). La Termotecnica, May pp. 25–30.

DIN 8959. Insulated food carriers, 2000-9.

Engelking, S. and Kruse, H. (1996). Development of air cycle technology for transport refrigeration. *Proceedings of the 1996 International Refrigeration Conference, Purdue University.*; pp. 349–356.

Giuri, S. (2006). Analisi statistica sulla velocità di degrado delle caratteristiche termiche dei veicoli isolate; (Statistical analysis about the ageing velocity the thermal characteristics of insulated vehicles). Degree Thesis. University of Padova DFT. 10 March.

Grosskopf, P. (1982). Zur Whal der richtigen Transportkaelteanlage. DKV Tagungsbericht. 9. Jahrgang . Essen. 407–427.

Heap, R.D. (2000). Transport refrigeration: maintaining and enhancing living standards. *Temperature* 22(7), 19–20 (2000.07).

IIR (1995). Guide to refrigerated transport.

IIR (1999). Control of the cold chain for quick-frozen foods handbook.

ISO. (1996). 1496-2 E. Series 1, Freight Containers —Specification and testing— Part 2: Thermal containers. p. 51.

Kauffeld, M. (1996). The use of hydrocarbons in transport cooling. *KI Luft Kältetech* 32(3), 107–110.

Klempner, D. and Frisch, K.C. (1991). *Polimeric Foams*. Hanser Publishers, p. 41.

Lorentzen, G. (1994). Revival of carbon dioxide as a refrigerant. *International Journal of Refrigeration* 17(5), 292–301.

Meffert, H.F.Th. (1975). Recent developments in the transport of perishable under refrigeration. In: *Proceedings of the XIV International Congress of Refrigeration*, Moscow pp. 301–320.

Micheletto, A. and Rosso, G. (2005). State-of-the-art of low-temperature delivery bodies using a two-stage transcritical carbon dioxide cycle: development and applications + erratum. Bull. IIF-IIR, March, 85 (2005-3), pp. 4–15.

Panozzo, G. (1999). Parameters affecting the ageing of the insulated vehicles. In: *XX International Congress of Refrigeration*. Sydney. 1999.

Panozzo, G. and Cortella, G. (2005). Refrigerated transport of deep frozen foods: problems with the materials for thermal insulation after the international ban of the HCFC'S. *EUROFREEZE 2005: Individual Quick Freezing of Foods*. Sofia Bulgaria. Jan 13–15.

Panozzo, G. and Cortella, G. (2006). Connections between standards and technologies in perishable foodstuffs transport. In: *Proceedings of the 3rd Central European Congress on Food*, Sofia (Bulgaria). pp. 1–8, 22–24 maggio.

Panozzo, G. and Minotto, G. (2003). The refrigerated transport in warm countries. Short course about 'Refrigerated Transport'. In: *21st Intenational Congress of Refrigeration*. USA: Washington D.C. August 17–22.

Panozzo, G., Amantia, M., Barizza, A. and Rossi, S. (2003). Ageing of mechanical refrigerating units for refrigerated transport. In: *21st Intenational Congress of Refrigeration*. Washington D.C. August 17–22.

Panozzo, G., Minotto, G., Martini, R. and Giordana, E. (2006). Dry ice snow in mini containers. *Innovative Equipment and Systems for Comfort and Food Preservation*. New Zealand: Auckland. February 16–18.

Panozzo, G., Boldrin, B., Minotto, G., Toniolo, B. and Biancardi, N. Ageing of insulated vehicles. theoretical model and experimental analysis. In: *XIX International Congress of Refrigeration*, Vol. II. The Hague 95. pp. 583–589.

Rudnik, H.E. (1958). Die entwicklung von kaeltesaetzen fuer schienen- und strassenfahrzeuge. (Development of refrigerating plants for railway and trucks). Kaeltetechnik. 10 Jarhgang Heft 8/1958, pp. 244–248.

Schausberger, H., Kriha, J. and Zellhofer, J. (1969). Investigations into the problem of leakage of refrigerator vans at the international vehicle testing station of ORE/UIC in Vienna. Bull IIF-IIR Annexe 69-3, Vienna, 133–142.

Smoluchowski, M. (1898). On conduction of heat by rarefied gases. Philosophical Magazine, Vol. 46 pp. 192–206. Available from http://en.wikipedia.org/wiki/Marian_Smoluchowski.

Spence, S.W.T., Doran, W.J. and Artt, D.W. (2005). Performance analysis of a feasible air-cycle refrigeration system for road transport. *International Journal of Refrigeration* 28(3), 381–388.

Stimat and National Research Council Italy. (1982). Agrotrans.

Tiedemann, T. and Kruse, H. (1994). Long-term solutions with hydrocarbons. *Purdue Refrig. Short Course* 266–308.

van Gerwen, R.J.M., van der Sluis, S.M., Schiphouwer, H., Bennahmias, R., David, R., Guilpart, J., Slama, T. and Panozzo, G. (1999). Energy labelling of refrigerated transport equipment. Final report. TNO Report. TNO-MEP-R 99/065. February.

United Nations Economic Commission for Europe, Inland Transport Committee. (1970). Agreement on the International Carriage of Perishable Foodstuffs and on the Special Equipment to be used for such Carriage. Geneva: United Nations. Including amendments thereto up to Sept 2003. p. 81.

Wenk, C. (1976). Befoerderung im Werksverher. Besondere befoerderungsmittel fuer leicht verderbliche Lebensmittel. 11 Teil II, 1976-TUEV. 102–114.

14 Frozen Retail Display

Giovanni Cortella

14.1 INTRODUCTION

Frozen food requires strict temperature control during the whole cold chain. Specially designed refrigerated cabinets are employed for display during retail sale, with the twofold purpose of effective display and correct storage of food. Unfortunately, the two functions are in some ways contrasting. For effective display a cabinet with high product visibility and wide front opening, possibly in the absence of any kind of door or lid, and equipped with a bright lighting system, is required (Rigot, 1990). All these features cause a significant heat load on the cabinet, often giving rise to unwanted temperature gains, in excess of the recommended values for the storage of the goods. Furthermore, removing the additional heat income requires extra refrigerating power with consequent higher energy consumption. Retail cabinets are intensive energy users, and efforts should be made to reduce energy consumption while maintaining effective display and storage abilities.

Shop managers are much more interested in the display function of cabinets, because they need to encourage product sale. This is particularly true for frozen food, which often competes with fresh food. On the other hand, customers should be assured that correct storage conditions have been maintained throughout the whole cold chain, the display cabinet included. It is common experience that storage temperature of frozen food in retail cabinets is sometimes out of control, particularly in open cabinets, which are critical from this point of view. Temperature distribution is frequently strongly uneven, with the most exposed packages at a temperature well above the prescribed value. The responsibility for this situation rests not only with the manufacturer, but also with the shop manager and with the person in charge of maintenance.

This chapter aims to illustrate how the cabinet operates, highlighting the concepts which make the basis for its design, and which should be part of the general knowledge of installers and users. Where possible, suggestions on how to improve the storage conditions and reduce the energy consumption will be provided. Finally, some information will be given about the recent environmental issues, which claim a reduction in the environmental impact of these units, with repercussions on their design and operation.

14.2 CLASSIFICATION

Manufacturers offer a great variety of display cabinets, suitable to all the possible applications in retail premises. Numerous criteria can thus be adopted for the classification of this equipment, e.g. the load temperature, the cabinet geometry, the presence of doors or lids, the

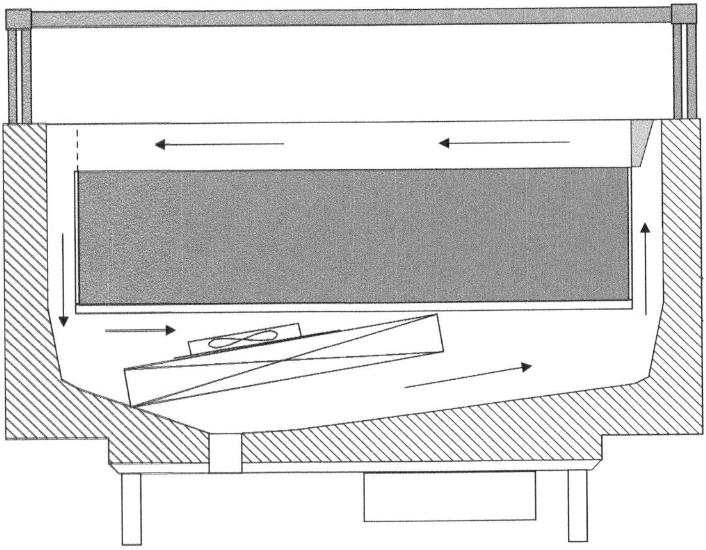

Fig. 14.1 Horizontal open top cabinet.

type of air distribution, the type of refrigerating equipment (Rigot, 1990; Gac and Gautherin, 1987; ASHRAE, 2006; IIR, 1986).

With reference to the geometry, display cabinets can be sub-divided into four classes (ISO, 2005a):

- horizontal, with opening at the top and accessible from above (Fig. 14.1);
- vertical, either semi-vertical multi-deck or roll-in (Fig. 14.2);
- semi-vertical, whose overall height does not exceed 1.5 m, with either a vertical or inclined display opening;
- combined, e.g. consisting of a horizontal and a vertical cabinet (Fig. 14.3).

All of them can be open, when the frontal opening does not have any physical barrier, or closed, when access to the load volume is gained by opening a door or a lid.

The refrigerating power is supplied to the load volume by means of air movement around the goods. A cooling coil (evaporator) is placed out of the load volume, and air is cooled and circulated either by natural or forced convection. In the first case, the cooling coil is placed above the load volume, usually at the back side, and air is driven by negative buoyancy to descend over the goods. This kind of air circulation is not the best choice, because it allows for a lower refrigerating effect. It is usually intended for displaying sensitive goods such as pastry, meat and ice cream, where an excessive air velocity could lead to undesired surface dehydration. Forced air circulation is employed in all the other cases. Air is cooled by exchanging heat in one or more cooling coils, and then is supplied to a discharge grille placed on the upper front of upright cabinets and the rear top of well cabinets. Therefore one or more air curtains are obtained, which refrigerate the load compartment, and create a barrier against warm air entrainment from the ambient in the case of open-type cabinets.

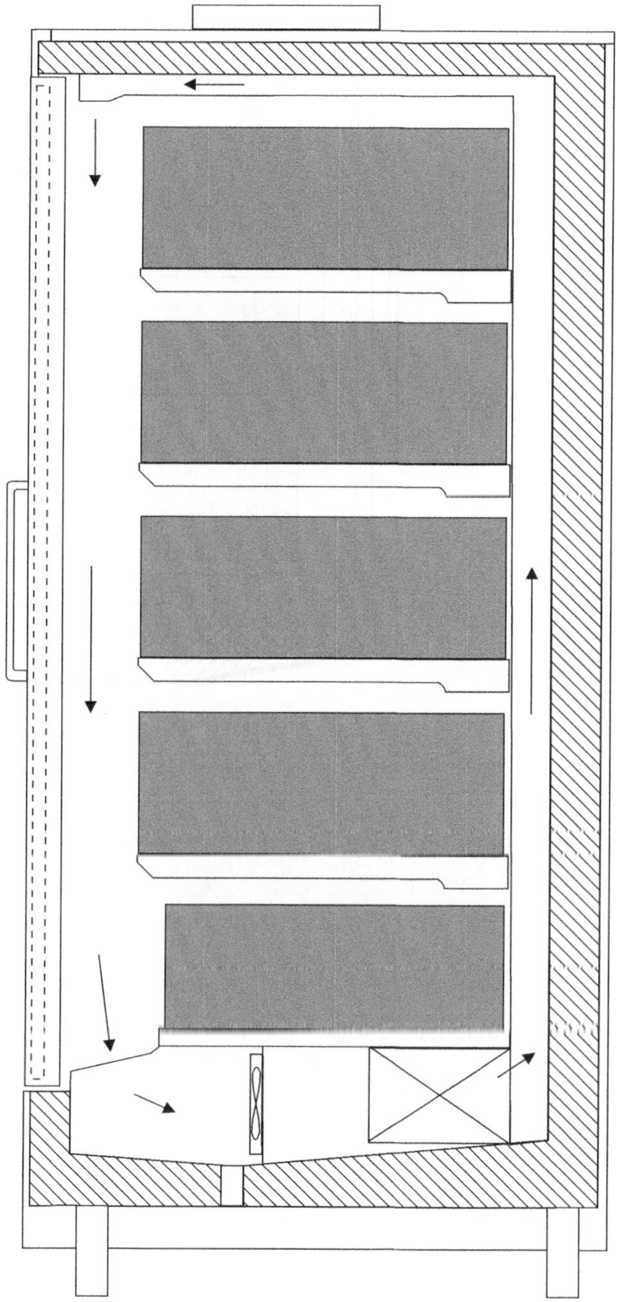

Fig. 14.2 Vertical multi-deck glass door cabinet.

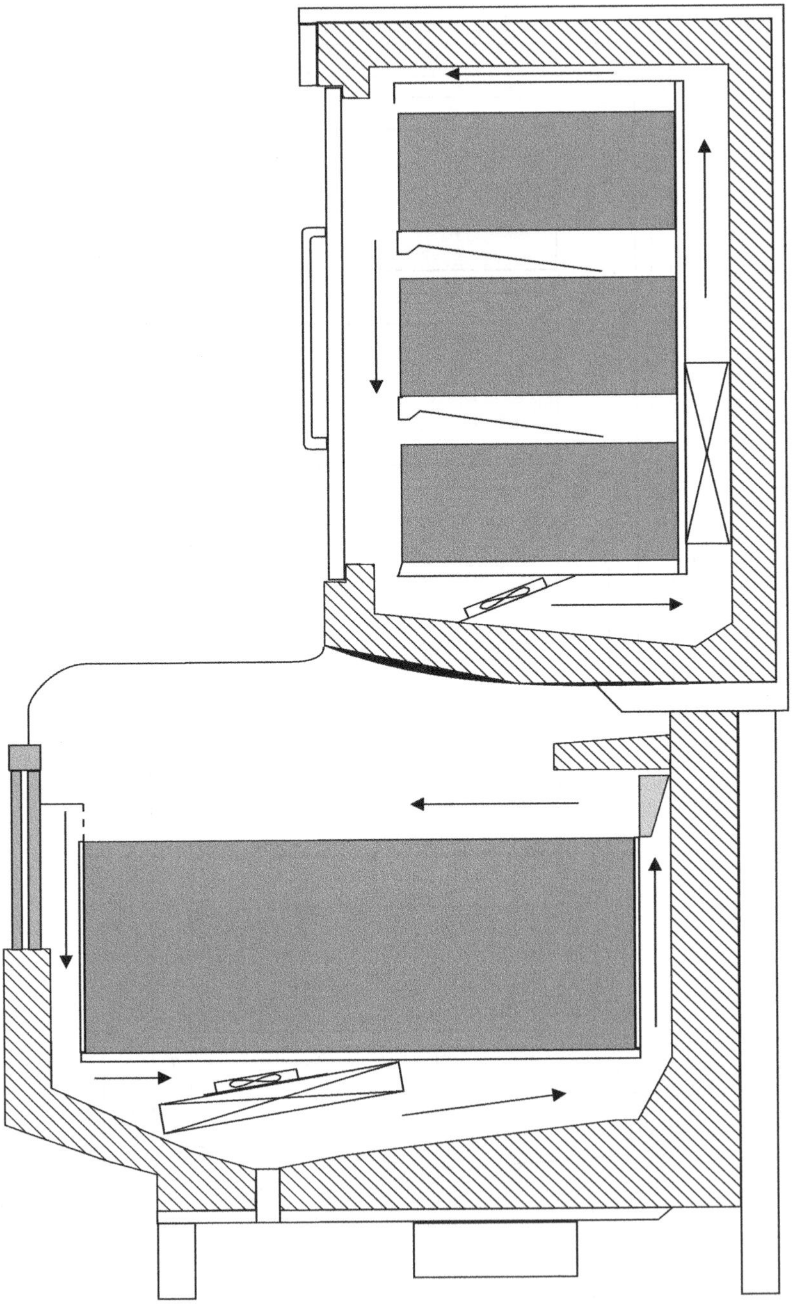

Fig. 14.3 Combined horizontal open top and vertical glass door cabinet.

When frozen food has to be displayed, only a small number of classes of cabinets can be effectively used. First of all, frozen food sale is essentially a self-service sale, therefore serve-over counters are not considered. Second, when temperature in the load compartment is around −20°C, heat is exchanged to a great extent with the external ambient, by conduction, air entrainment and radiation. A huge amount of refrigerating power is necessary, which cannot usually be exchanged by natural convection, thus requiring forced air circulation. In order to establish optimal storage conditions for the goods and reduce the energy consumption as much as possible, horizontal cabinets are preferred. In this case, due to air stratification, infiltration of ambient air is low and as for open units, and major part of the heating load is due to radiant heat. Unfortunately, the display function is less effective, because only the upper layer of products is visible by the customer, and the ratio load volume/floor space is not too good. Horizontal cabinets are preferred for high-turnover products, and for soft-packaged products, which are best stacked horizontally. The alternative option for a best display is represented by vertical multi-deck cabinets. Some vertical cabinets for frozen food are not fitted with any door or lid and in this case the warm air entrainment from the ambient becomes critical, the energy consumption is probably the highest among all the cabinets, and the correct storage conditions are difficult to guarantee. To prevent uninterrupted air entrainment from the ambient, vertical cabinets for frozen food are commonly fitted with glass doors, which are sometimes treated with a reflective external layer to reduce radiant heat exchange.

Temperature requirements are almost all common for frozen foods, with the exception of ice creams which require a lower storage temperature (−20 to −24°C) (ASHRAE, 2006; IIR, 1986) and specifically designed cabinets for serving soft ice cream at −15± 1°C (often referred to as 'dipping cabinets'). These units are open at the rear, to allow staff to scoop ice cream portions. Cooling of the upper side of the ice cream containers is crucial in these units, because melting due to radiant heat must be avoided, while reducing temperature fluctuations to prevent ice crystals growth.

In the following sections horizontal open or closed cabinets, and vertical closed cabinets will be considered, which are the most widely used units for frozen food display.

14.3 DESIGN

According to the vocabulary of the EN 23953 Standard (ISO, 2005a), a refrigerated display cabinet is a *cabinet cooled by a refrigerating system which enables chilled and frozen foodstuffs placed therein for display to be maintained within prescribed temperature limits.* This definition states that display cabinets are intended only to *maintain* food at the correct storage temperature, not to reduce its temperature from a higher value. Therefore goods must be loaded at the correct temperature. However, even if the cabinet is not designed for cooling down goods, it is equipped with a refrigerating unit of significant power. The reason for this requirement resides in a considerable heat gain from the ambient, which takes place in a variety of ways.

14.3.1 Heat balance

For a correct evaluation of the refrigerating power requirement, it is necessary to assess the overall heat balance of the display cabinet, i.e. identify and quantify all the possible heat exchanges between the refrigerated load volume and its surroundings (Billiard and Gautherin, 1993; Clodic and Pan, 2002).

Since the ambient temperature is always higher than that of the load volume, heat exchange can take place by conduction through the solid walls of the cabinet. However, such walls are sandwich panels or triple glazed glass doors with low overall thermal conductivity, thus leading to a minor heat gain when compared to the other sources. This heat gain acts on the refrigerating power requirement only, and scarcely influences the load temperature, because air ducts and air curtains completely surround the load volume.

Air infiltration takes place through the front opening in the case of open cabinets, or each time the door/lid is open in the case of closed cabinets. This is one of the major heat sources especially for open cabinets. In such a case, one or more air curtains are established, with the purpose of creating a barrier against warm air entrainment from the external environment (Navaz *et al.*, 2005). However, air curtains are effective only in horizontal cabinets, where cold air stratification helps preventing warm air infiltration, while their efficacy is a crucial problem for vertical cabinets for frozen food. Their correct exploitation is one of the key-points for a successful design of the display cabinet (Cortella, 2002). Since air curtains are effective only to some extent, vertical cabinets for frozen food are usually of the closed type only. In some countries, also the horizontal cabinets for frozen food are commonly equipped with glass lids. In this case, cooling of the load in some small units can be obtained by conduction, with the evaporator placed on the side walls ('chest freezers').

During the time that the shop is closed, it is helpful to reduce air infiltration in open cabinets. For this reason, open cabinets are often equipped with 'night covers' or 'night curtains' to cover the open area in horizontal and vertical cabinets.

Radiative heat transfer takes place through openings and transparent glass. It is due to the temperature difference between the load surface and the room ceiling, walls or any other object situated in front of the cabinets, here included lighting appliances. It can assume significant values, because it is governed by the difference in the fourth power of the absolute temperature of the two surfaces, and by their emissivity. The emissivity of a surface is an attribute to determine the ability of the surface to emit and absorb thermal radiation. It ranges from 0 (perfect reflector) to 1 (the black body, a perfect absorber and emitter). In the case of frozen food, the temperature difference between the load and the warm surfaces can be significant and, furthermore, the emissivity of the packages and walls is usually quite high (about 0.9). This is one of the most important reasons why an uneven temperature distribution of food is encountered in frozen food display cabinets. Packages situated in the uppermost layer (in horizontal cabinets) or in the front (on the shelves of a vertical cabinet) are subject to radiative heat transfer, and show a temperature few degrees higher when compared to the other packages in the cabinet (Nesvadba, 1985).

Further heat gains take place from internal sources, such as lighting, defrosting and demisting devices.

High-intensity lighting is usually preferred by retailers, who ask, as an example, for lights placed under each shelf in vertical multi-shelf cabinets. This choice causes a huge heating effect on the packages by radiation, and should be avoided, together with the use of spotlights. High-efficiency lights, such as fluorescent lights or LEDs, should be preferred (Narendran and Raghavan, 2002).

Defrosting devices are always installed in the cabinets, in order to keep the cooling coils and the air ducts clear of any deposition of ice.

Demisting devices are used in the presence of glass door, in order to facilitate the evaporation of water mist which deposits on the internal side of the door when it is open. Further, electric heaters often heat up the doorstop to prevent ice formation, or other surfaces, which might get in contact with the customers and feel unpleasantly cold.

The heat produced by all the sources listed above need to be removed by the refrigerating apparatus of the display cabinet, so as to maintain the goods at a constant temperature.

14.3.2 Air distribution

Heat removal from the load volume is performed by means of air circulation. It has been already mentioned that air circulation can take place by natural convection or can be forced by fans. In the case of display cabinets for frozen food, air is usually forced by fans, in order to establish an adequate heat exchange at the cooling coil and obtain an even air distribution. In the case of open cabinets, forced air circulation is necessary to create the air curtains.

All the display cabinets shown in Figs. 14.1–14.3 are refrigerated by forced air circulation.

Air is moved by fans placed underneath the load volume, and is cooled in a cooling coil where a refrigerating medium provides heat removal. It is then supplied to a plenum with the purpose of generating a uniform distribution and finally it is forced through a honeycomb to a symmetrical discharge grille. Here one or more air curtains are obtained, and air is returned to fans through a return grille placed at the bottom of the air curtain. Air curtains have different purpose depending on the type of cabinet.

14.3.2.1 *Open cabinets*

In the case of open cabinets, air curtains provide a barrier against warm air entrainment from the ambient. Cold air moves downwards because of its velocity at the discharge grille and because of buoyancy, but along its path the curtain width increases, and its velocity decreases. The curtain also impacts onto shelves and food, thus increasing the possibility that only a fraction of the discharged air will be returned at the return grille, while the residual will overspill the bottom of the cabinet. In order to respect a mass balance, an equal mass of air from the ambient will be entrained in the air curtain and conveyed to the return grill. This occurrence will cause an unpleasant feeling to the customers (the so-called 'cold feet effect') and a noticeable increase in the refrigeration power requirement, due to the necessity of cooling and dehumidifying the ambient air which has been entrained.

It has been shown that air infiltration accounts for 60–75% of the total refrigeration load (Rigot, 1990; Axell and Fahlen, 2002), thus emphasising the necessity for the best design of this component of the cabinet. The ratio of the mass flow rate of ambient air entrained to the total mass flow rate at the return grill is defined the 'induction factor', and it is a helpful indicator of the effectiveness of an air curtain (Rigot, 1990). For the sake of an experimental evaluation, it can be written in terms of air temperature:

$$\frac{m_{\text{ambient}}}{m_{\text{return}}} = \frac{t_{\text{return}} - t_{\text{discharge}}}{t_{\text{ambient}} - t_{\text{discharge}}}$$

In horizontal open cabinets for frozen food usually only one air curtain is established, taking advantage of air stratification which prevents warm air infiltration. In this case, particular care should be taken in choosing the air velocity, in order to avoid strong turbulence, which could promote heat and mass exchange with the ambient.

In vertical open cabinets for frozen food usually up to three air curtains are established, at different air temperatures. The internal curtain is maintained at a low temperature (i.e. about −30°C), and performs heat removal from the goods, contrasting heat gains from radiation, lighting and defrosting. The central curtain is maintained at approximately 0°C, and acts as the barrier to prevent air infiltration from the ambient. The external curtain usually consists of ambient air drawn from the top of the cabinet. Its purpose is to damp cold air curtain fluctuations, which could amplify air entrainment, and to protect the air inner curtains from ambient air movement.

14.3.2.2 *Closed cabinets*

In the case of closed cabinets, forced air distribution follows the same path as described for open cabinets. Only one air curtain is designed in this case, whose function is mainly to distribute cooled air throughout the load volume while contrasting heat gain from radiation and lighting on the frontal surface of goods. Nevertheless, during door or lid opening the air curtain acts as a barrier against air infiltration, even if the strong turbulence induced by the door movement prevents an effective operation in vertical cabinets.

In vertical cabinets, when the door is open, glass fogging occurs, because condensation of atmospheric humid air takes place on the internal cold surface of the door. This is a major concern for the display of products. As soon as the door is closed, rapid defogging must be achieved to re-establish proper visibility of the product. Defogging is achieved by a combination of air movement supplied by the air curtain, and glass heating by means of electric heaters embedded in the door.

Glass fogging is a crucial problem also on the external side of the doors, in hot and humid climatic conditions. For this reason, glass doors and lids are often triple glazed, in order to reduce heat transmission through the structure and keep the external surface temperature above the dew point temperature of the ambient, thus avoiding vapour condensation.

14.3.3 Refrigerating equipment

Display cabinets need refrigeration equipment to adequately cool air which, in turn, contrasts the numerous heat gains so as to keep food at the correct temperature. Depending on the refrigerating equipment, display cabinets can be classified as integral units incorporating their own condensing units or as remote cabinets with remote condensing units (ISO, 2005a, 2005b). In the cabinets of the first type, an entire refrigerator is contained in the unit, which only needs a power supply connection and for this reason is also named a 'stand alone' unit. If this unit is of the open type, a drainage piping is sometimes required, to evacuate water after the defrosting operation (in closed cabinets a smaller amount of water is produced, which can be removed by evaporation). In remote cabinets, the unit is connected to a central refrigerating system which serves the whole low-temperature cabinets within a supermarket; only the evaporator and the expansion device are contained in the unit, while compressor and condenser are placed outside the selling area.

In order to reduce the amount of refrigerant required for the system, it is possible to make use of a secondary fluid, which is cooled in a centralised refrigerating system and circulated, to distribute the refrigerating power to all the cabinets (indirect refrigerating system). This kind of system is less energy efficient; however, it permits an important reduction in the total amount of refrigerant used, and makes possible the use of toxic or flammable refrigerants, since the refrigerating system can be confined (Nyvad and Lund, 1996). This is an important issue because the refrigerants adopted in remotely operated refrigerating systems are almost exclusively halogenated hydrocarbons where strong pressure is being put towards the use of natural refrigerants such as ammonia, carbon dioxide and hydrocarbons. This topic will be dealt with in a later section.

Whatever refrigerating means is adopted, one or more cooling coils are used in the cabinets to refrigerate air. Because their surface temperature is well below 0°C (usually below −35°C), the cooling coils are subject to frost deposition due to air dehumidification. This is particularly true in the case of open cabinets, where air infiltration from the ambient takes place, and is emphasized in warm and humid climates. A defrosting device is therefore needed, whose operation should be cyclical and able to keep the coil and any air ducts clear of ice. Apart from the energy consumption of this device, an important consequence of its use is the unavoidable temperature rise of the products, due to the lack of any cooling effect during the defrosting cycles, and to the undesirable heat supply.

14.4 OPERATION

Retail cabinets are intended for the display of perishable food at the correct storage temperature. Various standards are in force in order to compare the performance of different units, and to classify their ability to display goods at different temperatures. Even after the cabinet is designed and its compliance with the standards is verified, its performance is not guaranteed at every operating condition. The influence of ambient conditions on the performance is crucial, because unexpected radiative load or air movement can cause excessive temperature fluctuations. The performance of a display cabinet strongly depends also on its installation, operation and maintenance. Furthermore, other important factors that affect the performance of a cabinet are the way it is loaded, the use of night covers and the operation of the defrosting device.

14.4.1 Standards in force

Various standards are in force, e.g. the EN Standard 23953 (ISO, 2005a, 2005b) and the ASHRAE Standard 72-2005 (ASHRAE, 2005). In the following section the EN Standard will be discussed, which is in force in Europe. The ASHRAE Standard is similar in scope, although there are some differences between the exact test conditions for the two standards.

The EN Standard aims to give requirements for the construction and the performance of cabinets, to specify test conditions for checking that such requirements are satisfied and to list the characteristics to be declared by the manufacturer (ISO, 2005b). The performance of the cabinets is measured in terms of storage temperature of goods and of energy consumption. Strict prescriptions are given to perform the 'temperature test', which has to be carried out in a test room whose climate conditions are specified and controlled, and with the cabinet fully loaded with packages (that have the same thermal properties as lean meat) made of water, cellulose and some additives, whose composition, dimensions and thermal characteristics are detailed. Doors or lids, if any, shall be opened cyclically with prescribed time period and angle.

Different climate conditions are considered in the standard, and the climate classes are listed in Table 14.1. Depending on the test result, the cabinet is assigned to a temperature class among those listed in Table 14.2. The temperature limits refer to the highest and lowest temperature of the warmest and coldest package in a 24-hour test.

It is important to note that test conditions may be considerably different from normal operating conditions in a retail store. Compliance with the requirements of the standard does not imply a correct operation when the cabinet is installed in the store, because of different climate conditions, loading arrangement, thermal properties of the load and

Table 14.1 Climate classes according to the EN 23953 Standard (ISO, 2005b).

Climate class	Dry bulb temperature (°C)	Relative humidity (%)	Dew point (°C)
0	20	50	9.3
1	16	80	12.6
2	22	65	15.2
3	25	60	16.7
4	30	55	20.0
5	40	40	23.9
6	27	70	21.1
7	35	75	30.0
8	23.9	55	14.3

settings of the unit. It is thus clear that the user has a responsibility to ensure the correct storage conditions and should operate the unit with full awareness of the consequences of any choices. In the following, some suggestions are given about proper installation and operation, with the purpose of obtaining the best storage conditions with the lowest possible energy consumption.

14.4.2 Installation

Display cabinets should be certified for compliance with the Standard in force, and should be assigned a temperature class and a climate class. In other words, compliance with the standard means that the cabinet is able to keep food at the defined storage temperature (temperature class), when it is operated in an environment with the ambient conditions defined in the climate class.

Thus, the first step for a retailer is the choice of the cabinet belonging to the most appropriate temperature class, bearing in mind that cabinets are not designed to lower food temperature, but only to maintain it.

The second step is the choice of the climate class which better fits the actual ambient conditions in the shop. This choice should be made with full awareness, especially in the case of retail premises situated in warmer countries or not supplied with an air conditioning system.

The third step is correct installation, to reproduce to the highest degree the test conditions at which the cabinet has been designed. In particular, air conditions and radiative load should be controlled, especially for open cabinets.

Table 14.2 Temperature classes according to the EN 23953 Standard (ISO, 2005b).

Class	Highest temperature of the warmest package (°C)	Lowest temperature of the coldest package (°C)	Lowest temperature of the warmest package (°C)
L1	≤−15	na	≤−18
L2	≤−12	na	≤−18
L3	≤−12	na	≤−15
M1	≤+5	≥−1	na
M2	≤+7	≥−1	na
H1	≤+10	≥+1	na
H2	≤+10	≥−1	na
S	Special classification		

Air conditions to control are: temperature, humidity and velocity (Axell and Fahlen, 2002; Gautherin and Srour, 1995). High ambient temperature is often the cause of an increased risk of load temperature fluctuation, and it always leads to a higher energy consumption. The effect of high ambient humidity is also significant (Howell *et al.*, 1999). Because of air entrainment in the load volume, water vapour condensation occurs on the air ducts and on the cooling coil, thus leading to frost build-up. The higher the air humidity is, the faster ice builds up, with consequent premature reduction of the air flow rate due to blockage, thus requiring more frequent defrosting operations. Finally, ambient air velocity is crucial especially in the case of open cabinets. It has been demonstrated that even slow air movements (<0.2 m s^{-1}) parallel to the opening interfere with the air curtains. Air entrainment can be promoted, and the induction factor significantly increased. 3D effects can also take place in the air curtain, leading to an uneven load temperature distribution (D'Agaro *et al.*, 2006a). If the ambient air movement is directed against the cabinet opening, disruption of the air curtain can take place, with severe consequence on the storage conditions of goods.

In general, the cabinets should be placed as far as possible from doors, windows, air diffusers, and customers' habits should be investigated so as to install cabinets at the end of their shopping paths (Larson *et al.*, 2005). In this way, only customers interested in buying frozen food will walk in front of the cabinets, thus reducing ambient air movement. Furthermore, frozen goods will be the last items to be picked up, and the time of their exposure to ambient temperature will be reduced.

An air conditioning system in the retail shop could be helpful with regard to the optimal operating conditions, provided that air diffusers do not promote excessive air movement close to the cabinet opening. In many cases, especially for large stores, it is preferable to adopt a separate air conditioning system for the selling area where display cabinets are placed. Depending on their refrigerating unit, display cabinets can be equivalent to a positive or negative heat load to the air conditioning system. Self-contained ('stand-alone') units represent a source of heat, because the electric power consumption and the refrigeration load are discharged to the ambient via the condensing coil. On the contrary, remote condensation units represent a source of cold, because of cold air dispersion from the air curtains or due to door openings. In this case, it is usual to feel the so-called 'cold feet effect' when walking close to the cabinets, because of cold air stratification close to the floor. The air conditioning system should attempt to rectify this particular situation and provide comfortable conditions for customers, while maintaining low relative humidity and adequate ambient temperature (Tassone, 1997). Innovative air distribution system to counteract the 'cold feet effect' have been investigated, with return grills on the floor to avoid air stratification.

Radiative load originates from high-temperature sources. Apart from the use of low-emissivity packaging for the products, it can be reduced only by avoiding direct exposure of the load surface, that is placing the cabinet out of direct sunlight, as far as possible from windows, and avoiding the use of incandescence or halogen lights directed towards the goods. Also, radiation shields can be effective (Faramarzi and Woodworth-Szieper, 1999). In the past, some authors suggested the use of canopies with low-emissivity surfaces over horizontal open-top cabinets, to reduce radiative heat exchange between the load surface and the supermarket ceiling. This solution allowed for a reduction of the surface temperature of products up to 4 K (Gac and Gautherin, 1978), and can be particularly effective when the supermarket roof is exposed to direct sun radiation. Use of closed cabinets with low-emissivity glass doors or lids is indeed much more effective, even if visibility and access are reduced because a physical barrier is created between the customer and the product.

Fig. 14.4 Load limit line.

14.4.3 Loading

Loading food in a display cabinet is a crucial phase in the cold chain, and particular care should be taken to ensure the preservation of food quality.

First of all, as already mentioned, a display cabinet is intended only to *maintain* the temperature of goods within the prescribed limits. This means that food shall be loaded at the correct storage temperature, and that freezing or cooling of goods from a higher temperature (e.g. on delivery) shall be performed in other refrigerated facilities. When frozen food is transferred from the refrigerated rooms to display cabinets, loading should be performed in a very short time, because the rise in temperature of frozen food left in ambient temperature is faster than it might apparently seem to be; small-sized products could even thaw. Never leave frozen foods outdoors under direct sunlight!

Every cabinet should be marked with one or several 'load limit line(s)' on the inside face of the load volume, to indicate the load limit (Fig. 14.4) (ISO, 2005b). Complying with this requirement is essential for correct operation of the cabinet. Food stored beyond this limit will not be maintained at the correct temperature and will probably disrupt the air curtain, affecting the performance of the cabinet and the storage temperature of the whole load.

Food in display cabinets is subject to storage conditions worse than those in a refrigerated room, mostly because of temperature fluctuations. Moreover, goods in the front of the cabinet are subject to radiation, and their temperature is a few degrees higher than that of other packages. Stock rotation thus becomes important, using the FIFO (First In First Out) principle.

Radiative heat is one of the most important factors which affect food display negatively. If low-emissivity packages are used, food packaging could contribute effectively to the reduction of radiative heating. It is more and more familiar to see frozen food wrapped in aluminium bags, whose emissivity (about 0.2–0.3) is much lower than that of paper (about 0.9–0.95). In open cabinets, radiative heat gain is approximately proportional to the emissivity of food packages. The use of low-emissivity material allows a 4–5°C reduction in food temperature at the front or top of the cabinet.

14.4.4 Defrosting setting

During normal operation, humid air entrained from the external environment is subject to water vapour condensation on the cold surfaces inside the cabinet, especially on the return air ducts and on the cooling coil, which are the first components air encounters. Because these surfaces are below 0°C, condensation results in frost formation, with consequent reduction of the circulation of refrigerated air, instability of the air curtain and a loss of performance of the cabinet. For this reason, anti-sweat heaters are to be run continuously, adding a permanent heating load to the display case. Furthermore, regular defrosting cycles are needed, during which the refrigerating equipment is switched off and the cooling coil is heated. Heating can be performed by means of electrical heaters placed around the coil and in its vicinity, or by reversing the refrigerating cycle, i.e. circulating warm liquid or condensing refrigerant inside the coil (Baxter and Mei, 2002; Gage and Kazachki, 2002). The second system is more

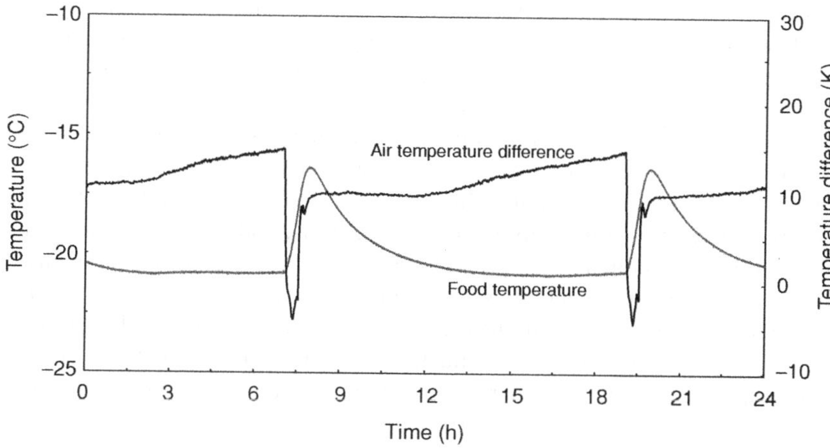

Fig. 14.5 Food temperature and air temperature difference between return and discharge during correct operation.

effective, because heating is provided 'from inside' and the coil surface is cleared of ice more rapidly.

Defrosting cycles affect the performance of the cabinet and the temperature of food. Figure 14.5 shows the temperature of one package of frozen food in the upper layer of a horizontal display cabinet during normal operation. The temperature rises during the defrosting cycles and it takes a few hours to assume the right value. Figure 14.5 also shows that the temperature difference between the return air and the discharge air, which is a valid indicator of the correct operation of the cabinet. It can be seen that the air curtain temperature increases about 10 K over time, and that this increase is boosted as frost builds up, due to the reduction in the flow rate, till defrosting is operated again.

Figure 14.6 shows the same temperature values in the case of excessive frosting, due to uncontrolled air infiltration because of incorrect loading or inadequate ambient conditions.

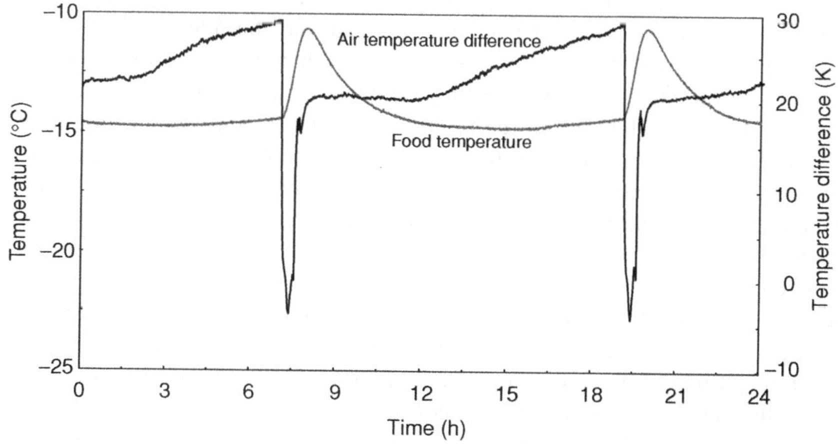

Fig. 14.6 Food temperature and air temperature difference between return and discharge during improper operation.

In this case the air curtain temperature difference is about three times higher, which is the evidence of insufficient flow rate, and the load temperature is too high, with a rising trend (Camporese *et al.*, 1995). The account of comparison between the two conditions underlines the importance of a correct operation of the defrosting device.

A new concept using two evaporators in parallel, where one evaporator is cooling while the other is defrosting, has been proposed and showed successfully to perform energy saving and reduction of food temperature fluctuations (Clodic *et al.*, 2005a). As regards the defrosting frequency, it is commonly set on a time basis, and more sophisticated systems are employed only occasionally. The growing use of electronic controls in display cabinets and in the refrigerating equipment allows for the use of smart systems operating on a different basis (e.g. on the air temperature difference). Such defrost strategies lead to better food temperature control, in particular to the reduction of temperature fluctuations, and to a lower energy consumption owing to the synergy with the refrigerating equipment capacity control (Camporese *et al.*, 1995; Tassou *et al.*, 2001).

14.4.5 Monitoring and maintenance

The comparison between the temperature measurements reported in Figs. 14.5 and 14.6 emphasise the need for a continuous and effective monitoring of the operation of display cabinets.

The EN Standard 23953 (ISO, 2005b) deals with this topic and states that the cabinet shall incorporate a temperature display instrument showing the air temperature in refrigerated display cabinets, to provide an indication of the operation and functioning of refrigerating equipment and information on its operating state. It also warns that '*as a rule, measured air temperature is not identical with foodstuff temperature in refrigerated display cabinets*'. The temperature sensor is usually located so as to measure return air temperature, which is expected to be the warmest position in the air flow. However, it has been demonstrated above that the temperature difference between return and discharge air could represent an advanced indicator of the operating conditions of the unit.

Food temperature control should be the true objective of monitoring; however, it is quite difficult and expensive to directly measure food temperature. Only air temperature is actually measured, and a correlation is established between this variable and load conditions. Cheap data loggers are available, and test packages with similar thermal properties as food can be easily prepared. In this way, the time–temperature history of the load could be easily monitored. However, food loaded in display cabinets is subject to a significantly uneven temperature distribution, thus requiring measurements at several locations. For this reason, it should be preferable to provide every frozen food package with a time–temperature sensor. Such devices are commercially available (time–temperature indicators), and usually they change their colour depending on both the temperature reached by the package and the time this temperature has been maintained, thus giving witness of a significant interruption of the whole cold chain.

In the case of open cabinets, the use of devices aiming to seal the cabinet opening during shop closing time is very effective for the reduction of both warm air infiltration and radiative heat gain, thus promoting the reduction of the number of defrosting operations, of the food temperature fluctuations and of the energy consumption.

In the case of horizontal open cabinets, the unit is usually equipped with 'night covers', i.e. well-insulated opaque lids, that must be put in place by an operator. In the case of vertical open cabinets, usually opaque curtains are supplied instead of the lids, which can be equipped with electric devices for an automatic movement. Operation with the night covers is very effective for the improvement of open cabinets' performance, provided that they are well fitted. Their use is strongly recommended, costs for their purchase and daily positioning are paid back by lower energy consumption and improved food quality.

As regards maintenance, a regular inspection on a daily basis is necessary to check the load temperature and the correct operation of the unit. Less frequently, a thorough cleaning is required. The refrigerating unit has to be stopped, the cabinet emptied, and all the accessible surfaces washed and sanitised. In the case of incorporated condensing units, the condenser must be cleaned to remove dust.

In the event of a short-term breakdown, removal of food can be avoided provided that night covers are immediately applied, customer access prevented and fans switched off. In this case, the thermal capacity of the goods will delay the temperature rise.

14.5 ENVIRONMENTAL ISSUES

Refrigerated display cabinets exploit a well-established technology; however, there is still room for improvement. Significant upgrading is required with respect to their environmental impact, in particular with reference to the use of halogenated hydrocarbons as refrigerants, and to the energy consumption. Nonetheless, the recently raised issue of life cycle impact has opened a new avenue for further enhancement.

14.5.1 Refrigerants

Commercial refrigeration accounts for about 17% of the worldwide consumption of refrigerants for refrigeration and air conditioning.

Centralised direct expansion systems have a wide range of refrigerating capacities, the refrigerant charge varying from 100 kg to about 1500 kg. Stand-alone equipment covers many different types including vending machines, ice machines, etc. In this case the refrigerant charges vary from 200 g up to 1 kg, but 10–12 million such units are in use globally (Kruse, 2005).

From another point of view, retail stores can be classified as hypermarkets, supermarkets and convenience stores. In 2003 in Europe, the distribution of display cabinets in such stores was 14%, 32% and 54%, respectively. In hypermarkets, the average nominal charge of refrigerant was 0.27 kg m^{-2} selling area, and the accidental emission rate was 23–37% of the total charge per year. The average refrigerant charge in supermarkets was 0.29 kg m^{-2}, the emission rates having decreased from the year 1994 to 2002, currently between 18 and 35% per year. In the case of convenience stores, it is difficult to estimate an average charge of refrigerant, because it depends strongly on the type of cabinets used. In the case of stand-alone units, a maximum 1% emission rate is estimated, while in the case of remote condensing units a 15% emission rate can be guessed (Clodic *et al.*, 2005b). It must be recalled that some countries have put high focus on leakage reduction, also through the introduction of taxation on refrigerants, and that this policy is giving appreciable results.

The environmental issues of ozone depletion, covered for the first time by the Montreal Protocol signed in 1987, and of greenhouse gas emissions, covered by the Kyoto Protocol signed in 1997, caused the need for substantial changes in all refrigeration systems. The widely used halogenated hydrocarbon refrigerants were found to contribute greatly to both ozone depletion and global warming.

Chlorofluorocarbons (CFC) and hydrochlorofluorocarbons (HCFC) have been progressively banned in the developed countries, and only hydrofluorocarbons (HFC) are currently accepted as refrigerants because of their zero ozone depletion potential (ODP). However, it is commonly realized that HFCs are not a sustainable technology in the long run, because of their not negligible global warming potential (GWP). The attention is now focusing on natural refrigerants, i.e. carbon dioxide, hydrocarbons and ammonia.

Carbon dioxide is the most promising fluid. Because of its low critical temperature (about 32°C) and high critical pressure (about 7.4 MPa), carbon dioxide requires particular devices to perform a refrigerating cycle. However, it has been used as a refrigerant since 1850, and its peculiarity of being safe, i.e. non-toxic and non-flammable, made it soon become the ideal fluid for marine installations. Low energy efficiency was the main drawback, and the introduction of the most promising synthetic refrigerants after World War II caused its decline. Nowadays, it has regained favour because of its low environmental impact (Lorentzen, 1994; Pearson, 2005), and particular emphasis is being put on the improvement of energy efficiency.

CO_2 can be employed either as a secondary fluid or as refrigerant. This allows for a much wider choice of possible refrigerating plants for commercial refrigeration.

One of the main limitations of indirect cooling systems is the low energy efficiency, due to the poor heat exchange of the secondary fluids and to their viscosity, which increases the pumping power requirement. CO_2 has a much better heat transfer behaviour than the conventional single-phase fluids, because it is possible to take advantage of its phase change. A secondary loop with carbon dioxide can be set up, where the refrigerant evaporates at low temperature in the cooling coil inside the cabinets, and condenses in heat exchangers situated in a machine room, where heat is removed by means of a primary refrigerating plant. However, it must be realised that if the plant were to stop operating that a large pressure increase could occur from evaporation of the liquid contained in the system. In this case, release of carbon dioxide to atmosphere is usually planned.

Carbon dioxide can be used effectively in the primary refrigerating circuit, as well as in direct expansion systems. When rejecting heat directly to the atmosphere, a sub-critical cycle is possible whenever the outdoor temperature is below 15°C (Girotto *et al.*, 2004). This condition occurs for a significant number of hours throughout the year both in Europe and in North America. On the contrary, when the condensing temperature approaches the critical value, a trans-critical refrigeration cycle is established. Condensation is replaced by an almost isobaric heat rejection at high temperature, followed by an isenthalpic expansion due to throttling, where liquid formation occurs. Special high-pressure components and compressors must be used, the maximum pressure in the circuit being around 8–10 MPa. Comparison between the use of CO_2 and of the conventional HFC R404A leads to an estimation of about 8% higher energy consumption and 10% higher costs. Non-ideal heat transfer takes place especially in the gas cooler and in the compressor (Fartaj *et al.*, 2004). Improvement can be achieved, by modifications to the basic cycle using various devices (Zha *et al.*, 2005) or through two-stage systems, with a sub-critical low-temperature stage and trans-critical high-temperature stage. In this configuration, the same refrigerating plant can supply to both the low-temperature and the high-temperature cabinets, for frozen and chilled food, respectively. An intermediate pressure vessel can be adopted, and liquid CO_2 pumped into a secondary loop serving the chilled food units (Schiesaro and Kruse, 2002).

Use of two-stage cycles involves the need for efforts to optimise the operating conditions of the refrigerating system, with particular reference to the gas cooler pressure (high pressure) and to the intermediate pressure (Sarkar *et al.*, 2004; Cavallini *et al.*, 2005).

Problems related to high pressure or to the trans-critical cycle can be by-passed by making use of cascade cycles, where CO_2 is used as a refrigerant in the low-temperature system, and another refrigerant on the high-temperature side. With such a system, the refrigerating power can again be supplied to both the low-temperature and the high-temperature cabinets. The CO_2 cycle serves the low-temperature cabinets, and rejects heat as a load to the high-temperature refrigerating plant. If natural refrigerants are to be taken into consideration for both cycles, ammonia or propane can be used on the high-temperature side (Sawalha *et al.*, 2005). However, in this case, a secondary fluid is necessary to feed the display cabinets, due to safety requirements.

14.5.2 Energy consumption

Storage of frozen and chilled food accounts for approximately 40–50% of the electricity use in supermarkets. The total electrical energy required by a supermarket varies from 360 to 520 kW h^{-1} m^{-2} depending on the selling area (from >2000 to <600 m^2, respectively) (Axell and Lindberg, 2005). One vertical display cabinet for frozen food with a glass door (on average 2.4 m^2 total display area) requires from 54 to 68 kW h^{-1} per day for operation (European Committee of Air Handling and Refrigeration Equipment Manufacturers). About 45% of this value is direct energy consumption, i.e. energy required for lighting, defrosting and demisting devices, fans and so on. This means that most of the refrigerating load in a closed cabinet is produced by the cabinet itself, and does not originate from air infiltration. There is thus room for great improvement, which can involve:

- the design of the cabinet for the reduction of the refrigeration load;
- the design of the refrigerating equipment for the increase of its efficiency;
- the synergy among all the energy conversion systems in the supermarket, for the reduction of the global energy requirement.

The best design of cabinets originates from a thorough knowledge of the flow patterns both in the air curtains and in the air ducts. Low air velocity, even air distribution, full exploitation of the evaporator surface are all factors which affect the total energy consumption significantly. Air curtains are a very complex system, and designing them requires particular skills. Many advanced tools are available for the prediction and measurement of air flow. Computational fluid dynamics (CFD) is a useful tool for the prediction of flow patterns and heat and mass transfer, and many authors have shown CFD modelling to be a valuable means for the improvement of air curtains (Cortella, 2002; Foster, 2005; D'Agaro *et al.*, 2006b). Particle image velocimetry is a very accurate method for flow velocity measurements (Field and Loth, 2006), and a valuable tool for the adoption of the best CFD boundary conditions, especially at the discharge grill.

To push towards a better design of display cabinets in terms of energy consumption, standards have included some definitions which are the basis for energy labelling of display cabinets (ISO, 2005b). As an example, the EN ISO 23953 Standard (ISO, 2005b) defines the way to measure the total energy consumption (TEC, kW h^{-1} per day) of a cabinet, split into direct energy consumption (DEC) and refrigeration electrical energy consumption (REC). The energy consumption can be compared to the total display area (TDA, m^2) and the visibility of products (VPA), thus considering also the display function in the evaluation of the effectiveness of the cabinet. The ratio TEC/TDA has been adopted by Eurovent as a parameter for the comparison of different cabinets in the 'Refrigerated display cabinets' programme. Various manufacturers agree on a voluntary basis to participate in this programme, and have their products rated on the basis of the above-mentioned parameter (European Committee of Air Handling and Refrigeration Equipment Manufacturers).

The design of the refrigerating equipment has been discussed above. Various system configurations have been consolidated; and many improvements have been made recently to enhance their efficiency. One more option is available when taking into consideration direct expansion systems, where each cabinet is supplied with an expansion device. In this case, it has been demonstrated that the use of electronic expansion valves in place of thermostatic valves reduces energy consumption significantly and permits a complete control of the refrigerating plant (Tahir and Bansal, 2005). In this way, different strategies for regulation, pull down and defrosting can be adopted, and the control system can be adapted to the operating conditions (outdoor and indoor climate, presence of customers, goods loading...) obtaining even better global performance and further savings.

It has been mentioned above that the refrigerating plants for storage and display of food in a supermarket account for 40–50% of its total electrical energy consumption. The remaining fraction of electrical energy consumption is apportioned between space air conditioning and heating (5–10%), lighting (30–35%) and services. A noticeable reduction in the global energy consumption can be achieved, and a lot of efforts are being made to identify effective and reliable solutions. Many studies have been carried out on this topic (Baxter, 2003), and one of the most promising solutions is the recovery of heat rejected at high temperature by the refrigerating system, which offers an attractive resource for use in store space heating. Heat pumps can be installed, which use the rejected heat to supply the heating system with water at high temperature. Energy savings in the refrigeration system can also be achieved with this configuration, because of low condensing pressures, along with the energy benefits already seen through heat recovery.

Another point of practical, economic viability of an integrated combined heating and cooling system in a supermarket is to drive the compressor of the refrigerating system by a gas engine whose waste heat is used to satisfy the hot water and heating requirements (Maidment *et al.*, 2001). A payback period of around 4 years has been estimated for this system, which could be reduced with further optimisation.

Combined heat and power (CHP) systems can be set up in a similar way, when a gas engine drives a generator and cogeneration of heat and electric power is achieved. In this way electricity becomes available also for lighting and services, but a great amount of heat is produced, which will be rejected to the environment when the heating system is not operating. Integration with an absorption chiller gives place to a combined cooling heat and power (CCHP) system (Maidment and Tozer, 2002; Maidment *et al.*, 1999). This system allows for heat recovery during the warm season when the heating system is not operating, by means of an absorption chiller, which serves the chilled food cabinets. Such systems give primary energy savings of over 15%, while also achieving attractive payback times.

In the medium to longer term, emerging technologies such as mini gas turbines and fuel cells will further benefit the application of CCHP to commercial refrigeration.

14.5.3 Life cycle issues

The use of environmentally compatible refrigerants and the reduction of their emission, together with the reduction of energy consumption, are the two most important factors in the requirement for low environmental impact of the operation of display cabinets. However, in the framework of a *sustainable development*, a comprehensive evaluation of the impact caused by a display cabinet should take into account the whole life of the unit. A control strategy of pollution and use of raw materials must be integrated at the design stage, to achieve cleaner technologies and control the use of resources. Life cycle assessment (LCA) allows the evaluation of the environmental impact of equipment comprising the manufacturing and emissions of refrigerants, and the construction, installation, operation, dismantling and disposal of the equipment (Frischknecht, 2000). An inventory has to be made of all the materials and the energy sources which are used during the whole life cycle of the unit. Further on, the evaluation of the environmental impact is performed considering some impact categories, i.e. the use of energy resources, the damages to the ecosystem (global warming potential, ozone depletion potential, toxicity, photochemical smog production, acidification, eutrophication) and the impacts on human health. This impact can be summarised in a single index, calculated by weighing the various above-mentioned impacts. Different weight compositions are used, depending on the relative importance given to the various impact categories.

LCA has already been applied to the evaluation of the environmental impact both of supermarket refrigerating systems (Frischknecht, 2000; Diehlmann *et al.*, 2005) and of display cabinets (Watkins *et al.*, 2005; Watkins *et al.*, 2004; Paternicò, 2005). When display cabinets are considered, both of the open and closed type, with internal or remote condensing unit, it has been assessed that most (>97%) of the total environmental impact derives from the energy required for operation. For this reason, the most effective intervention for the reduction of the environmental impact is the reduction of the electric energy consumption. The boundaries of the system can be restricted neglecting the operation phase, in order to magnify the impact of use of raw materials, from assembly to disposal. For display cabinets, the most important impact derives from the use of copper, which is a basic material in piping and coils. Other materials could be considered; of course the influence of this possible modification on the energy consumption should be first evaluated.

LCA is used also as a tool for obtaining other certifications. As an example, the Environmental Product Declaration (EPD) is defined as 'quantified environmental data for a product with pre-set categories of parameters based on the ISO 14040 series of standards, but not excluding additional environmental information' (The Swedish Environmental Management Council). It is a declaration that the manufacturer prepares following harmonised calculation rules (Product Category Rules, PCR) and which is certified by a third party review. Usually, product category rules require the use of LCA for the assessment of the environmental impact. Currently, EPDs for domestic refrigerators have been certified, while the PCR rules for commercial refrigeration are still not available.

14.6 CONCLUSIONS

Retail cabinets are required to perform effective product display, while maintaining the correct storage temperature. This is not an easy task, because the two functions are in a certain way contrasting. The most effective display is achievable using open cabinets, with large display area, great visibility and easily reachable products. On the other hand, such features lead to high food temperature fluctuations, uneven temperature distribution, excessive sensitivity to ambient conditions and high energy consumption. The difficult task for the designer is to reach a compromise solution for each type of cabinet. Display cabinets are designed so that the amount of refrigerating power they exploit is exactly that required for the correct operation at standard conditions. This is especially due to the need for reduction both in costs and in energy consumption. For this reason, the responsibility for the best performance rests also with the persons in charge of installation, operation and maintenance.

The choice of the refrigerating equipment depends strongly on the dimensions of the shop and on its layout, and can have significant repercussion on the energy consumption of the system. Installation requires following some easy guidelines to reduce heat infiltration due to air movement and radiation.

A correct operation needs awareness of the effects of incorrect loading, of insufficient defrosting, and of the missed application of night covers.

Proper maintenance means continuous monitoring of the cabinet operation, periodical cleaning and ability to adjust the setting of the unit.

Therefore the correct food storage conditions during display depend on many actions performed by various people, and do not rely only on the refrigerator.

A lot of work has been carried out to develop sophisticated tools to assist the designer in the evaluation of the efficacy of the unit and in the assessment of its environmental impact, towards a sustainable development of this technology. However, there is still potential for improvement, and the necessity to use new refrigerants and control the environmental impact is a motivation for important advances.

REFERENCES

ASHRAE. (2005). ASHRAE Standard 72. – Method of Testing Commercial Refrigerators and Freezers.

ASHRAE. (2006). American Society of Heating, Refrigerating and Air Conditioning Engineers, *Handbook 2006 Refrigeration*, Ch. 46, Retail Food Store Refrigeration and Equipment, ASHRAE, Atlanta.

Axell, M. and Fahlen, P. (2002). Climatic influence on display cabinet performance. In: *Proceedings of the IIF-IIR Meeting 'New Technologies in Commercial Refrigeration'*, IIR Paris, pp. 181–190.

Axell, M. and Lindberg, U. (2005). Field measurements in supermarkets. In: *Proceedings of the IIF-IIR Meeting 'Commercial Refrigeration'*, IIR Paris, Paper p. 48.

Baxter, V.D. (2003). IEA Annex 26: *advanced supermarket refrigeration/heat recovery systems – Final report volume 1 – executive summary*, IEA Annex 26.

Baxter, V.D. and Mei, V.C. (2002). Warm liquid defrosting technology for supermarket display cases. In: *Proceedings of the IIF-IIR Meeting 'New Technologies in Commercial Refrigeration'*, IIR Paris, pp. 147–151.

Billiard, F. and Gautherin, W. (1993). Heat balance of an open type frozen food display cabinet. In: *Proceedings of the IIF-IIR Meeting 'Cold chain refrigeration equipment by design'*, IIR Paris, pp. 322–332.

Camporese, R., Cortella, G., Bigolaro, G. and Scattolini, M. (1995). Controllo del regolare funzionamento di banchi frigoriferi. Il Freddo, Linea, 4, 341–348.

Cavallini, A., Cecchinato, L., Corradi, M., Fornasieri, E. and Zilio, C. (2005). Two stage transcritical carbon dioxide cycle optimisation: a theoretical and experimental analysis. *International Journal of Refrigeration* 28, 1274–1283.

Clodic, D. and Pan, X. (2002). Energy balance, temperature dispersion in an innovative medium temperature open type display case. In: *Proceedings of the IIF-IIR Meeting 'New Technologies in Commercial Refrigeration'*, IIR Paris, pp. 191–199.

Clodic, D., Borlein, C. and Zoughaib, A. (2005a). Defrosting of vertical low temperature open display case. What are the needed improvements?. In: *Proceedings of the IIF-IIR Meeting 'Commercial Refrigeration'*, IIR Paris, Paper p. 58.

Clodic, D., Barrault, S., Palandre, L. and Kuijpers, L. (2005b). Quantification of refrigerant banks and emissions in the commercial refrigeration sector in Europe – emission predictions for the period from 2005 to 2020. In: *Proceedings of the IIF-IIR Meeting 'Commercial Refrigeration'*, IIR Paris, Paper p. 76.

Cortella, G. (2002). CFD-aided retail cabinets design. *Computers and Electronics in Agriculture* 34, 43–66.

D'Agaro, P., Cortella, G. and Croce, G. (2006a). Two- and three-dimensional CFD applied to vertical display cabinets simulation. *International Journal of Refrigeration* 29(2), 178–190.

D'Agaro, P., Cortella, G. and Croce, G. (2006b). Numerical simulation of glass door fogging and defogging in refrigerated display cabinets. *Applied Thermal Engineering* 26, 1927–1934.

Diehlmann, A., Stadtländer, C. and Schwiegel, M. (2005). Eco – efficiency considerations for european supermarket refrigeration systems. In: *Proceedings of the IIF-IIR Meeting 'Commercial Refrigeration'*, IIR Paris, Paper p. 17.

European Committee of Air Handling and Refrigeration Equipment Manufacturers, www.eurovent-cecomaf.org. *Product Certification Programmes, Retail Display Cabinets*, www.eurovent-certification.com.

Faramarzi, R.T. and Woodworth-Szieper, M.L. (1999). Effects of low-emissivity shields on the performance and power use of refrigerated display case. *ASHRAE Transactions* 105(1), 533–540.

Fartaj, A., Ting, D.S.-K. and Yang, W.W. (2004). Second law analysis of the transcritical CO_2 refrigeration cycle. *Energy Conversion and Management* 45, 2269–2281.

Field, B.S. and Loth, E. (2006). Entrainment of refrigerated air curtains down a wall. *Experimental Thermal and Fluid Science* 30, 175–184.

Foster, A.M., Madge, M. and Evans, J.A. (2005). The use of CFD to improve the performance of a chilled multi-deck retail display cabinet. *International Journal of Refrigeration* 28, 698–705.

Frischknecht, R. (2000). Life Cycle Assessment for commercial refrigeration systems operated in Switzerland. In: *Symposium on Industrial Ecology and Material Flows*, Helsinki, 30 Aug–3 Sept.

Gac, A. and Gautherin, W. (1978). Amélioration du fonctionnement des meubles ouverts de vente au détail des produits surgelés. In: *Proceedings of the IIF-IIR Meeting Commissions C2, D1, D2*, Budapest (H), IIR Paris, pp. 237–248.

Gac, A. and Gautherin, W. (1987). *Le froid dans les magasins de vente de denreés périssables*, Pyc edn. Paris.

Gage, C. and Kazachki, G. (2002). Warm liquid defrost for commercial food display cases: experimental investigation at 32.2°C. In: *Proceedings of the IIF-IIR Meeting 'New Technologies in Commercial Refrigeration'*, IIR Paris, pp. 169–178.

Gautherin, W. and Srour, S. (1995). Effect of climatic conditions on the operation of refrigerating equipment in a hypermarket. In: *Proceedings of the 19th International Congress of Refrigeration*, IIR Paris, Paper pp. 705–712.

Girotto, S., Minetto, S. and Neksa, P. (2004). Commercial refrigeration system using CO_2 as the refrigerant. *International Journal of Refrigeration* 27, 717–723.

Howell, R.H., Rosario, L., Riiska, D. and Bondoc, M. (1999). Potential savings in display case energy with reduced supermarket relative humidity. In: *Proceedings of the 20th International Congress of Refrigeration*, IIR Paris, Paper p. 113.

IIR-International Institute of Refrigeration. (1986). *Recommendations for the Processing and Handling of Frozen Foods*. IIR/IIF Paris.

ISO. (2005a). European Standard EN ISO 23953-1. Refrigerated display cabinets; Part 1: Vocabulary.

ISO. (2005b). European Standard EN ISO 23953-2. Refrigerated display cabinets, Part 2. Classification, requirements and test conditions.

Kruse, H. (2005). Commercial refrigeration – on the way to sustainability. In: *Proceedings of the IIF-IIR Meeting 'Commercial Refrigeration'*, IIR Paris, Keynote Lecture.

Larson, J.S., Bradlow, E.T. and Fader, P.S. (2005). An exploratory look at supermarket shopping paths. *International Journal of Research in Marketing* 22, 395–414.

Lorentzen, G. (1994). Revival of carbon dioxide as a refrigerant. *International Journal of Refrigeration* 7(5), 292–301.

Maidment, G.G. and Tozer, R.M. (2002). Combined cooling heat and power in supermarkets. *Applied Thermal Engineering* 22, 653–665.

Maidment, G.G., Zhao, X., Riffat, S.B. and Prosser, G. (1999). Application of combined heat and power and absorption cooling in a supermarket. *Applied Energy* 63, 169–190.

Maidment, G.G., Zhao, X. and Riffat, S.B. (2001). Combined cooling and heating using a gas engine in a supermarket. *Applied Energy* 68, 321–335.

Narendran, N. and Raghavan, R. (2002). Solid state lighting for refrigerated display cases. In: *Proceedings of the IIF-IIR Meeting 'New Technologies in Commercial Refrigeration'*, IIR Paris, pp. 66–71.

Navaz, H.K., Henderson, B.S., Faramarzi, R., Pourmovahed, A. and Taugwalder, F. (2005). Jet entrainment rate in air curtain of open refrigerated display cases. *International Journal of Refrigeration* 28, 267–275.

Nesvadba, P. (1985). Radiation heat transfer to products in refrigerated display cabinets. In: *Proceedings of the IIF-IIR Meeting 'Storage Lives of Chilled and Frozen Fish and Fish Products'*, IIR Paris, pp. 323–329.

Nyvad, J. and Lund, S. (1996). Indirect cooling with ammonia in supermarkets. In: *Proceedings of the IIF-IIR Meeting 'Applications for Natural Refrigerants'*, IIR Paris, pp. 207–217.

Paternicò, P. (2005). *Environmental Impact Assessment of Display Cabinets during their Life Cycle: LCA Analysis*, Mechanical Engineering Degree Thesis, Tutor prof. G. Cortella, University of Udine (in Italian).

Pearson, A. (2005). Carbon dioxide—new uses for an old refrigerant. *International Journal of Refrigeration* 28, 1140–1148.

Rigot, G. (1990). *Meubles et Vitrines Frigorifiques*, Pyc edn. Paris.

Sarkar, J., Bhattacharyya, S. and Gopal, M.R. (2004). Optimization of a transcritical CO_2 heat pump cycle for simultaneous cooling and heating applications. *International Journal of Refrigeration* 27, 830–838.

Sawalha, S., Rogstam, J. and Nilsson, P-O. (2005). Laboratory tests of NH_3/CO_2 cascade system for supermarket refrigeration. In: *Proceedings of the IIF-IIR Meeting 'Commercial Refrigeration'*, IIR Paris, Paper p. 20.

Schiesaro, P. and Kruse, H. (2002). Development of a two stage CO_2 supermarket system. In: *Proceedings of the IIF-IIR Meeting 'New Technologies in Commercial Refrigeration'*, IIR Paris, pp. 11–21.

Tahir, A. and Bansal, P.K. (2005). MEPR versus EEPR valves in open supermarket refrigerated display cabinets. *Applied Thermal Engineering* 25, 191–203.

Tassone, A. (1997). Hypermarchés: le casse-tête de la climatisation. *Revue Pratique du Froid* 845, 35–38.

Tassou, S.A., Datta, D. and Marriott, D. (2001). Frost formation and defrost control parameters for open multideck refrigerated food display cabinets. In: *Proceedings of the Institution of Mechanical Engineering*, Vol. 215, Part A, pp. 213–222.

The Swedish Environmental Management Council. *Environmental Product Declaration*. Available from www.environdec.com.

Watkins, R., Tassou, S.A. and Datta, D. (2004). Life cycle analysis of a commercial refrigerated display cabinet. In: *HPC 2004 – 3rd International Conference on Heat Powered Cycles*, Cyprus, October.

Watkins, R., Tassou, S.A. and Pathak, S. (2005). Environmental impacts and life cycle assessment of refrigerated display cabinets. In: *Proceedings of the IIF-IIR Meeting 'Commercial Refrigeration'*, IIR Paris, Paper p. 61.

Zha, S., Harfner, A. and Nekså, P. (2005). Study on transcritical R-744 (CO2) voorhees heat pump cycle. In: *Proceedings of the IIF-IIR Meeting 'Commercial Refrigeration'*, IIR Paris, Paper p. 33.

15 Consumer Handling of Frozen Foods

Onrawee Laguerre

15.1 INTRODUCTION

There is a substantial amount of published information on the storage conditions of food during its shelf-life. Most of this information concerns the products under the responsibility of the producer, the processor, the distributor, the wholesaler or the retailer. There is comparatively little information on what happens following purchase by the consumer from a supermarket until storage in a domestic refrigerator and the effect of the storage conditions on food quality.

Domestic refrigerators are widely used in industrialised countries. There are approximately 1 billion domestic refrigerators worldwide (IIR, 2002) and the demand in 2004 was 71.44 million units (11.2 million in China, 10.7 million in USA, 4.43 million in Japan, 3.36 million in India, 3.14 million in Brazil . . . ; JARN, 2005). In developing countries, the production is rising steadily: total production rose 30% in 2000 (Billiard, 2005). In France, there are 1.7 refrigerators per household (AFF, 2001). The domestic freezer is a piece of equipment that facilitates life and makes it possible to deal with unforeseen situations.

Nowadays, two important phenomena concerning the domestic refrigerator are observed:

- the volume of the appliance has tended to increase. The volume can vary from 230 litres to more than 400 litres, while the height can be up to 2 m (Lucas, 1996). This can be explained by a change in consumer lifestyles: consumers tend to shop less often and purchase larger quantities of products. The distance between shops and homes is increasing because of the growth of hypermarkets located in suburbs compared with that of supermarkets located in the centres of cities;
- sales of frozen products are increasing every year. This requires higher storage capacity. The top four European countries in terms of consumption of frozen foods par capita, excluding poultry, are Sweden (46.9 kg), Ireland (46.6 g), the UK (45.5 kg) and Denmark (45.2 kg) (*Newsletter of the IIR*, No. 25, January, 2006). The leading segments were pizza followed by frozen potatoes.

Epidemiological data from Europe, North America, Australia and New Zealand indicate that a substantial proportion of food born disease is attributable to improper food preparation practices in consumers' homes (Redmond and Griffith, 2003). Data also illustrate that a large proportion of consumers lack knowledge of adequate refrigeration temperatures. Surveys carried out in various countries on the temperature and the microbial contamination in the refrigerating compartment under real use conditions show an alarming situation (Flynn *et al.*,

1992; James and Evans, 1992; Lezenne Coulander, 1994; O'Brien, 1997; Sergelidis *et al.*, 1997; Laguerre *et al.*, 2002; Jackson *et al.*, 2007; Azevedo *et al.*, 2005).

Compared to surveys on the refrigerating compartment, few surveys have been carried out on temperatures in domestic freezers (Olsson and Bengstsson, 1972). The microbial risk during food preservation in the domestic freezer is negligible since the growth is inhibited when the temperature is lower than −10°C (Geiges and Schuler, 1988). However, food quality deterioration is found to happen often during storage: such deterioration includes frost formation, rancidity, colour modification etc. It is therefore necessary to understand the physical and biological phenomena that take place during storage in domestic freezers.

The objective of this chapter is to present the state of the art of knowledge on frozen food preservation at home. Several subjects are dealt with: consumer behaviour, the cold production system, temperatures in domestic freezers, frost formation in frozen food during storage in domestic freezers, the influence of storage conditions on food quality, and home freezing and thawing.

15.2 CONSUMER BEHAVIOUR

A consumer survey published in the *Grand Froid* magazine (Anonymous, 1989) showed that the freezer has several uses within the family: to preserve purchased frozen products (74.5% of surveyed people), to produce frozen food from purchased fresh food (51.8%), to produce frozen food from personal production (18%) and to preserve left-overs (14.7%). The products which the consumer freezes at home are primarily meats, chicken (72% of surveyed people) and vegetables (60.5%).

The INCA national survey (Volatier, 2000) was carried out on the food consumption of people living in France. Concerning the frozen food handling by the consumer during transport to home, this survey showed that:

- certain consumers left frozen food products at ambient temperature for too long a time. 35% of the surveyed people left food products for <15 minutes, 55% between 15 and 45 minutes and 5% >45 minutes. The distance between the store and home governed the time interval between the time of purchase and preservation in a freezer:
- consumers totally trusted their refrigerator, without ensuring adequate maintenance;
- consumers did not take any particular precautions when freezing products.

This survey also showed that young people (<45-yr old), workers and people in rural areas were less conscious of the importance of product temperature. The opposite is true of people living alone and older consumers (>45 yr) who placed their frozen food in a freezer as soon as possible (about 50% of them leave the frozen food less than 15 min at ambient temperature). 81% of the people surveyed bought frozen products and the ice cream at the last moment before leaving the supermarket. 67% of the people in the survey used an insulated bag for transportation of frozen foods. Only 12% of those surveyed checked the temperature in their freezers. Women were found to pay more attention to compliance with product temperature than men (Brousseau and Volatier, 1999).

15.3 COLD PRODUCTION SYSTEM IN DOMESTIC REFRIGERATORS/FREEZERS

There are several types of refrigerator/freezer available in the market: single-door chiller with an ice box freezer, single-door freezer, double-door fridge-freezers and chest freezers (Fig. 15.1). In addition, there is a growing trend for 'American style' larger side-by-side chillers and freezers. The most common refrigerators and freezers have four major parts to

Single-door fridge with icebox

Single-door freezer

Fig. 15.1 Different types of refrigerator/freezer available in market.

Double-door fridge-freezer (freezer can be located top or bottom)

Chest freezer

Fig. 15.1 *(continued).*

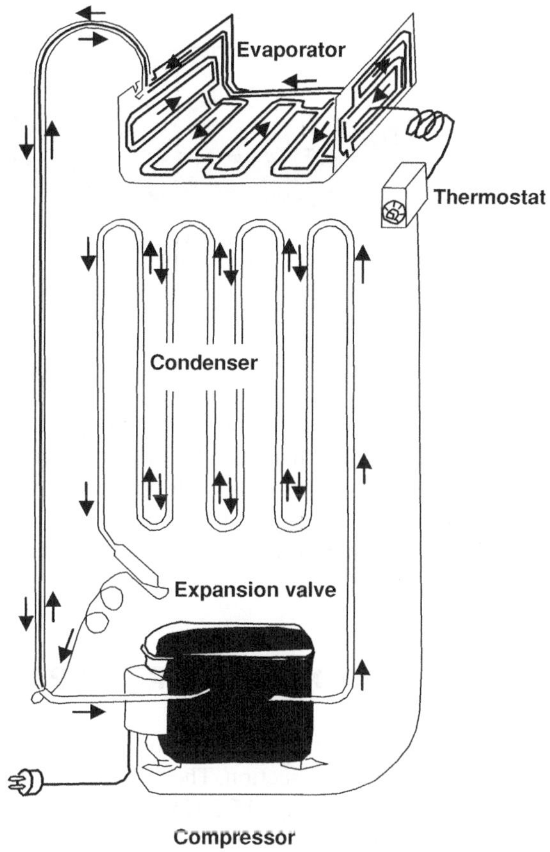

Fig. 15.2 Cold production system in a domestic refrigerator/freezer.

their refrigeration system – a compressor, a condenser, an expansion valve and an evaporator (Fig. 15.2). In the evaporator section, a refrigerant (commonly R600a or R134a in newer refrigerators, R12 in older models) is vapourised to absorb heat added to the refrigerator due to heat transfer across the refrigerator's walls and infiltration through the door and seals. The refrigerant boils at −18°C to −20°C when pressurised at 0.9 to –1 bar, so the evaporator temperature is maintained at or near that temperature if the appliance is working correctly. At the next stage, an electric motor runs a small piston compressor and the refrigerant is pressurised. This raises the temperature of the refrigerant and the resulting superheated, high-pressure gas (it is still a gas at this point) is then condensed to a liquid in an air-cooled condenser. In most refrigerators and freezers, the compressor is in the base and the condenser coils are at the rear of the appliance. From the condenser, the liquid refrigerant flows through an expansion valve (almost always a capillary tube), in which its pressure and temperature are reduced and these conditions are maintained in the evaporator. The whole process operates continuously, by transferring heat from the evaporator section (inside the refrigerator to the condenser section (outside the refrigerator), by pumping refrigerant continuously through the system described above. When the desired temperature is reached, the pump stops and so does heat transfer.

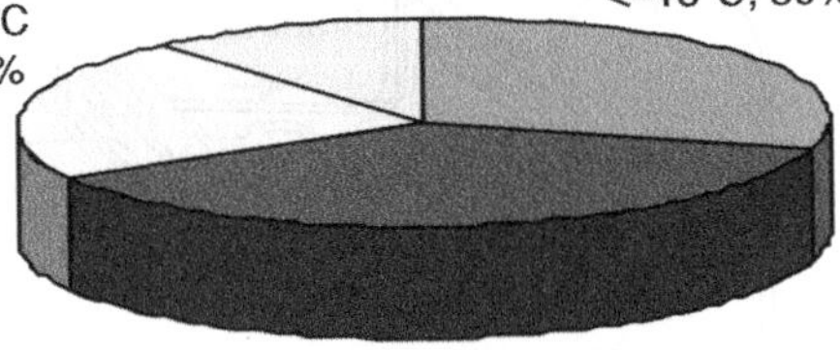

Fig. 15.3 Survey results of air temperature distribution in the home freezer in the USA. (Data from Olsson and Bengstsson, 1972).

Refrigerator/freezers may be equipped with one or two compressors. In the case of one-compressor, the operating cycle is controlled by the air temperature in the refrigerating compartment whereas in two-compressor models each operating cycle is controlled independently by the air temperature in the refrigerating compartment and in the freezer. The temperature in each compartment is, therefore, better regulated and the price is higher than the one-compressor appliance.

15.4 TEMPERATURE IN DOMESTIC FREEZERS

A small amount of data is available on surveys of the temperatures in the freezing compartments compared to those in the refrigerating section. The distribution of temperature observed in home freezers in the USA is shown in Fig. 15.3 (Olsson and Bengstsson, 1972). It can be seen that only 30% of surveyed freezers run below −18°C, which is the recommended temperature for frozen product preservation.

It should be remembered that refrigerator design in the USA and in Europe is quite different. The size of appliances in the USA is generally larger than in Europe and refrigerators in the USA are usually equipped with a fan to circulate air within the refrigerator. Therefore, the temperatures shown above are only an indication. Several internal temperature levels can be obtained in domestic freezers available in the market. These levels are indicated by the stars marked on the freezing compartment (Table 15.1). Of course, the air temperature in the freezer determines the storage period for frozen foods: the higher the air temperature, the shorter the storage period.

The air temperature fluctuations in a domestic freezer equipped with one compressor are generally more significant than those equipped with two compressors. An example of these fluctuations is presented in Fig. 15.4. Both refrigerators are two-door models, with a refrigerating compartment on the top and a freezing compartment on the bottom. Air-temperature stratification can be observed in the compartment in both cases. The air on the top shelf was slightly higher than that on the middle one (cold air is heavier). The characteristics of these two refrigerators are presented in Table 15.2. The wall of the freezing compartment of these refrigerators was composed of an inner liner (1 mm of polystyrene, $\lambda = 0.15$ W m^{-1} K^{-1}), foam (polyurethane, $\lambda = 0.02$ W m^{-1} K^{-1}, 5.8 cm for the one-compressor refrigerator and 6.3 cm for the two-compressor refrigerator) and a metal outer sheet (0.7 mm, $\lambda = 50$ W m^{-1} K^{-1}). The air temperature inside the appliance was regulated by a thermostat. When this

Table 15.1 Temperature levels in the freezing compartment available in the market, nature and storage period of preserved foods.

Star marked on the freezing compartment	Internal temperature	Preserved products	Storage period (IIR, 1986)
*	<6°C	Ice production. Preservation of frozen foods	3 days
**	≤12°C	Ice production. Preservation of frozen foods	1 week
***	≤18°C	Ice production. Preservation of frozen products until their expiry date. Does not allow freezing of fresh products	1 month
****	≤18°C	Ice production. Preservation of frozen products until their expiry date. Allow freezing of fresh products, at least 4.5 kg per 100 L and within 24 h. Ability to reach −24°C	3 months

temperature was higher than the maximum setting, the compressor was 'on' until the minimum setting value was reached, when it switched 'off'. These maximum and minimum differential settings were fixed by the manufacturer. In the case of the one-compressor model where one refrigeration system operated both the refrigerator and freezer, the operating cycle was regulated entirely by the air temperature in the refrigerating compartment. In the case of the two-compressor model where the refrigerator and freezer operated on separate refrigerating systems, the operation of the refrigerator and freezer were independently regulated by the air temperature in each compartment.

The analysis of recorded air temperature over 2 days is presented in Table 15.3. Although the average temperature of the freezing compartment of the one-compressor refrigerator was approximately 3°C lower than that of the two-compressor refrigerator, the amplitude of variation was higher while the frequency was lower. It was clearly shown, in this example, that the temperature was less stable in the freezing compartment of the one-compressor refrigerator.

Table 15.2 Characteristics of refrigerators.

	One-compressor refrigerator	Two-compressor refrigerator
External dimensions (height × width × depth)	185 cm × 60 cm × 60 cm	195 cm × 60 cm × 60 cm
Internal dimensions of the freezing compartment	62 cm × 48 cm × 38 cm	64 cm × 47 cm × 42 cm
Thermostat setting	+4°C (impossible to set the temperature of the freezing compartment)	+4°C (refrigerating compartment) and −18°C (freezing compartment)
Power of the compressor	120 W	160 W

Table 15.3 Air temperature in the freezer and duration of the compressor cycle.

	T_{air} average[a]	T_{air} min[a]	T_{air} max[a]	'on' period/ 'off' period	% 'on' period
One-compressor refrigerator	−25.0°C	−32.3°C	−15.0°C	55 min/70 min	44%
Two-compressor refrigerator	−21.8°C	−28.5°C	−17.0°C	15 min/30 min	33%

[a] Two measurements – one on the top shelf and one on the middle shelf.

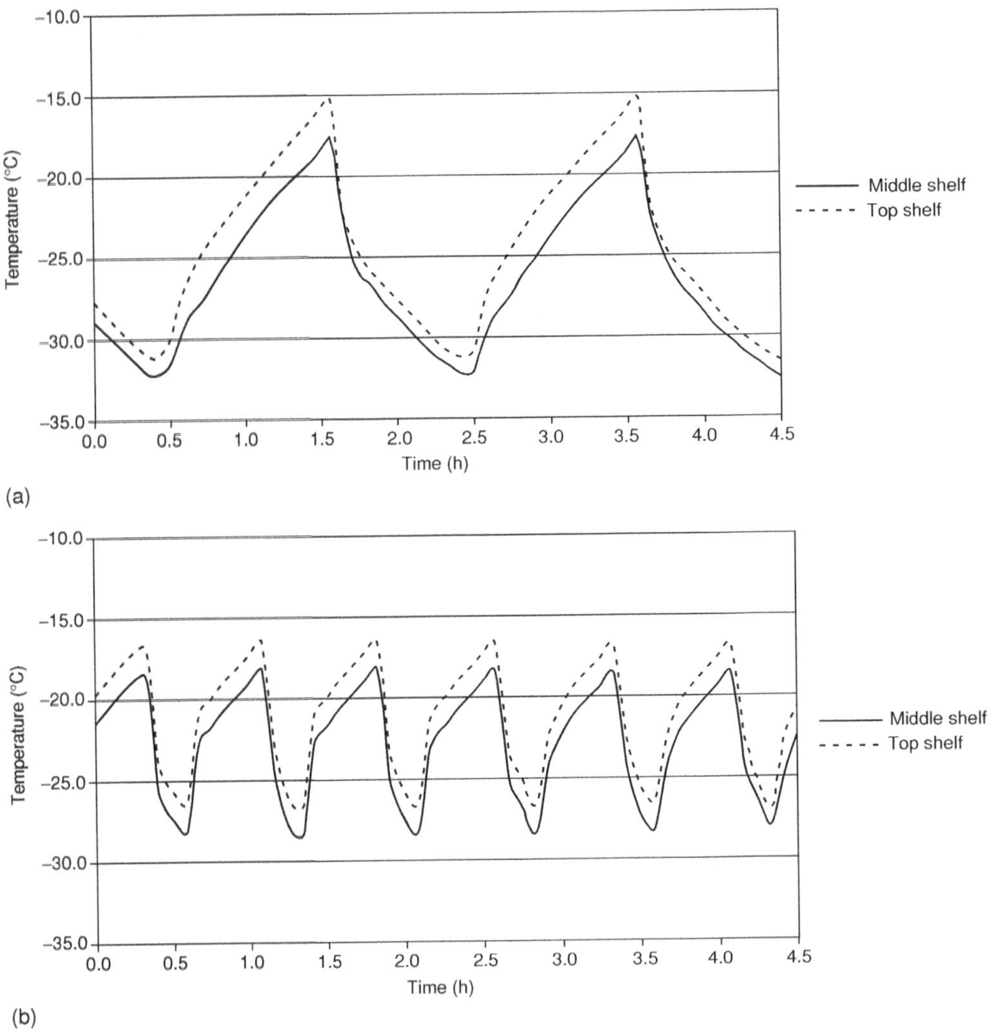

Fig. 15.4 Air-temperature variations in the freezing compartment: (a) one-compressor refrigerator; and (b) two-compressor refrigerator.

15.5 FROST FORMATION IN PRODUCT PACKAGE

When foodstuffs are preserved in a freezer, ice crystals develop gradually in the package. These phenomena, visible to the consumer, take place when the product surface temperature is lower than the dew temperature of the surrounding air. In a domestic freezer, the compressor 'on' and 'off' cycles contribute to temperature fluctuations of air, packages and products. Furthermore, there is a difference between the amplitude of air-temperature variation and that of a product due to thermal inertia. This leads to frost formation, which is often accompanied by surface dehydration for certain products and product weight loss.

Several studies were carried out to investigate the influence of average temperature, amplitude and frequency of temperature fluctuations on the quality of several packed foods: Poovarodom *et al.* (1990) on chopped steak, Bak *et al.* (1999) on shrimps, Bustabad (1999) on meat and Martins *et al.* (2005) on green beans. These authors reported that the higher the average temperature, the greater the amplitude of the fluctuations and the lower the frequency of the fluctuations lead to more rapid quality deterioration. They also observed that the weight loss increased rapidly with storage time (6% after a 9-month storage period at –20°C, Poovarodom *et al.*, 1990). The weight then becomes stable due to crust development on the product surface, which is a barrier to water sublimation. These authors also observed a small quantity of frost when the storage temperature was low; the amplitude of fluctuation were small and the frequency of fluctuations were high. The product dehydration resulting from the weight loss may induce colour modification, rancidity and destruction of ascorbic acid (Martins *et al.*, 2005). This colour modification (discoloured and parched area on the product surface) is called 'freezer burn' and occurs when food is not tightly wrapped, and stored too long in the freezer with large temperature fluctuations (Rahman and Shafiur, 1999).

Frost formation in frozen food stored in domestic freezers is influenced by several parameters: product characteristics, insulation of packages, positions of products in packages, the weight of product, mean air temperature, frequency and amplitude of temperature fluctuations in the freezer. The influence of some of these parameters was studied by Laguerre and Flick (2006) who reported on the variation in product weight loss over 3 months. The results shown hereafter were obtained from this study.

To study the influence of product characteristics, two types of frozen products bought at a supermarket were used: melon and potato balls. For the melons (spheres, 20–25 mm diameter, approximately 8 g), the fresh products were cleaned, sorted, cut into balls and frozen quickly after harvest. For the potato ball (spheres, 23–26 mm diameter, approximately 7 g), the product was prepared from mashed potatoes and crumbed on the surface.

To study the influence of package insulation, two types of boxes were used:

- Box without insulation: made of high density polystyrene, internal dimensions 132 mm × 132 mm × 78 mm, 2.5-mm wall thickness.
- Box with insulation: the same box cited previously was insulated on the six faces using an expanded polystyrene sheet, 10-mm thickness.

To study the influence of temperature fluctuations in the freezing compartment, the two refrigerators mentioned previously (Fig. 15.4) were used.

During the experiment, the boxes containing the product were placed on the top and the middle shelves of the freezing compartment, while the refrigerating compartment was empty. On average, the product layout was five balls long and wide and three balls high. The product weight was approximately 400 g.

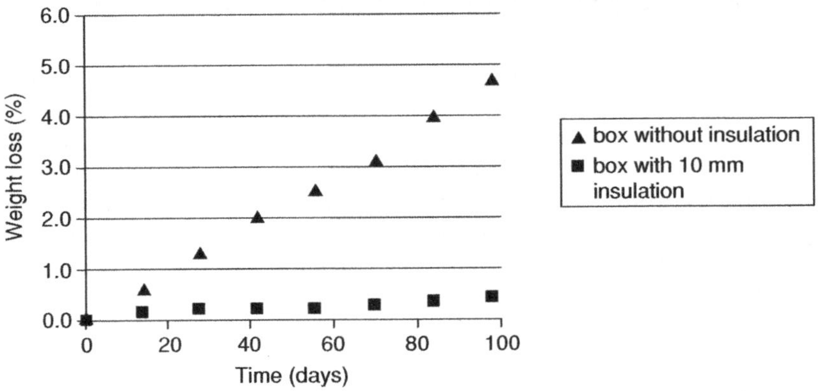

Fig. 15.5 Influence of package insulation on product (melon balls) weight loss during storage in a freezing compartment (two-compressor refrigerator).

The product weight, after frost elimination, was measured once every week or 2 weeks during the 3-month storage period. The results of this study are described in the following section.

15.5.1 Influence of package insulation

Figure 15.5 presents the weight loss of melon balls in the box with and without insulation. For the box with insulation, the weight loss was negligible even after a 3-month storage period (<0.5% after a 3-month storage period). There was a noticeable increase in the weight loss of the box without insulation, which reached 5% after a 3-month storage period. In this box, it was observed that the frost deposited primarily on the bottom horizontal wall. Much less frost formed on the top horizontal wall and no frost was observed on the vertical walls. Moreover, more frost deposits occurred on the product located at the centre of the box than that on the sides.

15.5.2 Influence of the product characteristics

The weight loss of frozen melon and potato balls in the box without insulation is presented in Fig. 15.6. It can be seen that the weight loss of potato balls is lower than that of melon. This can be explained by the fact that the crumbed surface of the frozen potato balls was a barrier to water flux, which slowed down water sublimation.

Only a small quantity of frost was observed in the product with low surface water activity (potato balls) since the partial vapour pressure (driving force of water migration) was low in this case.

15.5.3 Influence of refrigerator characteristics

The weight loss of melon in boxes without insulation stored in the freezing compartment of the one- and two-compressor refrigerators is presented in Fig. 15.7. It can be seen that the weight loss was slightly higher when the product was stored in the one-compressor refrigerator. This was because there were more temperature variations in the one-compressor refrigerator (greater amplitude of air temperature outside the box and lower frequency).

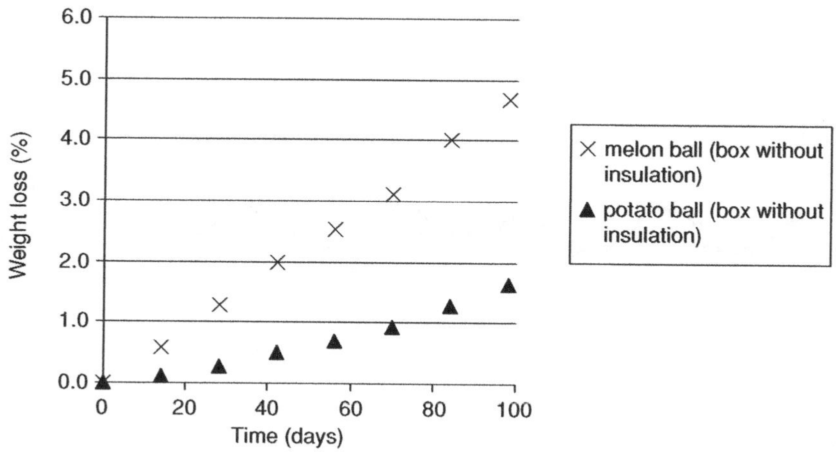

Fig. 15.6 Influence of product characteristics on weight loss during storage in the freezing compartment of a two-compressor refrigerator.

15.5.4 Air and product temperature variations

The variations in the air and product core temperatures (melon balls) inside the box without insulation are presented in Fig. 15.8. It can be seen that the air temperature (T_a) fluctuations are more significant than those of the product (T_p). Figure 15.8(a) shows that the fluctuation was more noticeable near the side walls (T_{p1} and T_{a1}) compared to that in the middle of the box (T_{p2} and T_{a2}). This was primarily due to thermal inertia. Figure 15.8(b) shows that the product and air temperatures were lower at the bottom of the box. Moreover, the temperature fluctuations at the bottom (T_{p3}, T_{a3}) were more significant than those at the top (T_{p4}, T_{a4}). This appeared to be due to the air layer (approximately 10-mm thickness) at the top of the box, which played an insulating role.

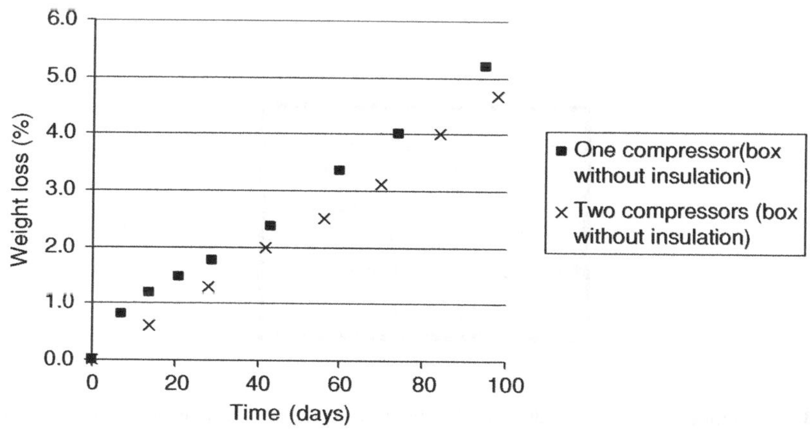

Fig. 15.7 Influence of refrigerator characteristics (one- and two-compressor models) on product weight loss (melon balls) during storage in a freezing compartment.

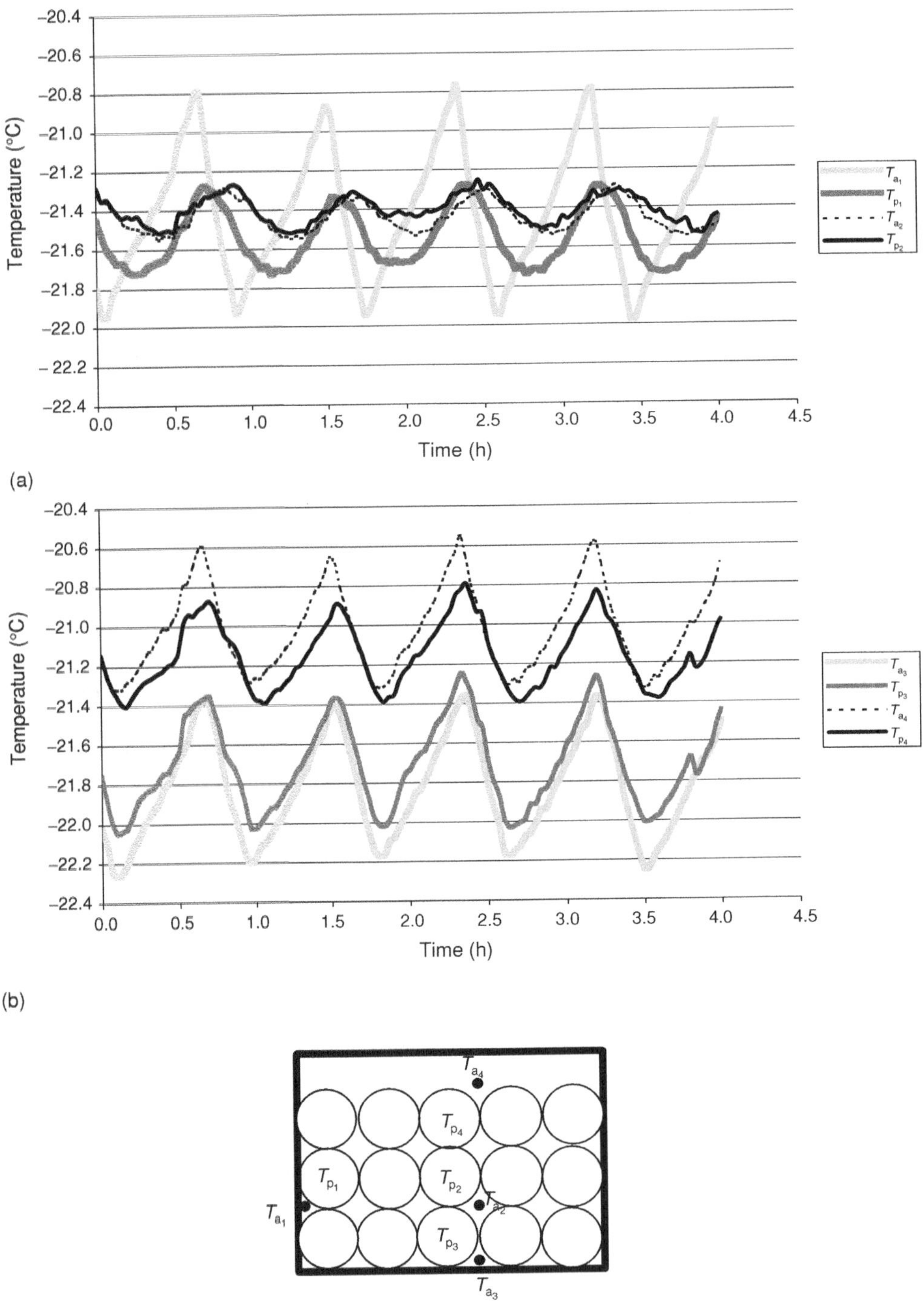

Fig. 15.8 Variations in air and product (melon balls) core temperature at different positions in a box without insulation.

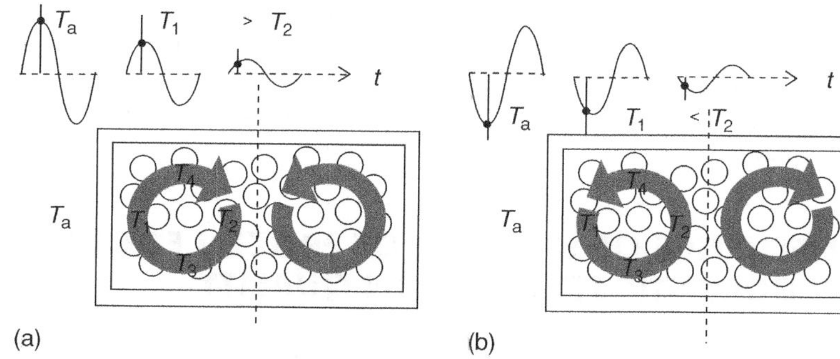

Fig. 15.9 Airflow by natural convection inside a product box preserved in a freezer (a) at the end of a period during which the compressor was turned off; and (b) at the end of a period during which the compressor was turned on.

15.5.5 Interpretation of results

It can be observed in Fig. 15.8 that the air and product temperature changes were rather periodic. The fluctuation frequency was the same for all the points and corresponded to the frequency of the on/off cycle of the compressor. The minimum temperature was not obtained at the same time for all positions. However, there was a delay corresponding to the effects of inertia: when the compressor was turned on, after the air near the regulating sensor reached the maximum differential value resulting in the air in the freezing compartment being first cooled down. In turn, the air cools down the outer part of the product in the box, and finally the core region of the product was cooled down. Thus, the temperature of the core region reached its minimum value when the compressor had already turned off and the air temperature was already increasing. Figure 15.9 shows the supposed airflow in a package of particle products.

At least four regions can be distinguished in the product box:

- core region, where the temperature variation was low;
- base region, where the mean temperature was lower than the middle area. In this position, the temperature variation can be significant if the box is not well insulated;
- top region, where the mean temperature was higher than at the base. In this position, the temperature variation was relatively low because of the air layer (between the product and the top package wall), which provided additional insulation;
- side region, where the mean temperature was close to that in the middle and the temperature variation can be significant.

For the box without insulation, frost deposited particularly on the product located in the central zone of the box and on the bottom horizontal surface of the box. Frost developed in the zone where the temperature was low in the box and/or where the temperature variation was low. This was the case at the bottom of the box (thermal stratification) and at the centre (thermal inertia). For the insulated box, only a small quantity of frost was observed, since the air and product temperature were relatively homogeneous and constant.

A use condition such as door opening contributes not only to the air temperature fluctuations in the freezer; it also contributes to the frost formation on the evaporator which reduces the efficiency of the exchange between evaporator and air. The frost formation leads to increased energy consumption of the appliance equipped with electrical defrost. However, this may not be the case for manual defrost appliances such as chest freezers or models with an ice box. The influence of the door opening on the temperature variations and on the energy consumption was studied by Liu *et al.* (2004) and Saidur *et al.* (2002). They recommended that the door be kept closed for as long as possible to minimise the frost formation and the energy consumption.

15.6 INFLUENCE OF STORAGE OF FROZEN FOODS IN DOMESTIC FREEZERS ON QUALITY

At home, we deal with combinations of an almost infinite number of variables. A wide range of foods of varying quality at the purchase point come into the home. These foods are then stored for different periods of time in various ways, according to the habits, beliefs and knowledge of the consumer. Collecting and sorting the scientific and sociological information, to assess the effect of domestic storage on food quality is a daunting prospect (Cook, 1978).

During frozen storage, there is a gradual and irreversible loss of quality with time. The loss of quality arises from the individual or combined effects of physical, physico-chemical, chemical or biochemical changes. However, there is usually one limiting factor that influences the storage life of frozen food (e.g. desiccation, enzyme-induced changes). Proper application of temperatures and proper packaging can retard the quality degradation.

The mechanisms of quality degradation are already presented in another chapter. Only studies concerning the influence of storage in a domestic freezer on the quality of frozen foods are presented here.

The freezing rate, storage temperature and duration of storage in domestic freezing were shown to affect certain quality aspects of the final product such as colour, texture, tenderness, juiciness and flavour. For meat, some authors (Verbeke *et al.*, 1984) observed that the freezing and thawing rates used in their study had no demonstrable effect on sensory quality (drip loss, colour and tenderness). This indicates that domestic freezing and thawing procedures were of minor importance for the eating quality of the meat. Some studies reported different observations, for example, Kandeepan and Biswas (2005) found that the odour and flavour decreased with storage period (negative effect) while the texture, tenderness and juiciness increased with storage period (positive effect). These authors also found that microbial contamination decreased during storage in a freezing compartment while it increased in the refrigerating compartment. These authors recommended a storage period of up to 30 days in a freezer in order to satisfactorily maintain the quality of meat. The effect of fluctuating and constant temperature during storage in chest freezer (−30°C and −10°C) and in ultra-low temperature freezer (−60°C) on food quality was also investigated by Gormley *et al.* (2002). Several products were studied: frozen raw salmon, smoked mackerel, stewed pork pieces, ice cream, pizza, hollandaise sauce, strawberries and blanched broccoli. These authors reported that the temperature regimes had a significant effect on peroxide and free fatty acid values. However, they had a minimal effect on texture, colour, water-holding capacity and drip loss on thawing on most of the products.

The microbiological risk associated with storage of frozen food is negligible. Several studies showed that at −8°C, the reproduction of bacteria in food is limited. The investigation of Geiges and Schuler (1988), Schmidt-Lorenz (1963) and Schmidt-Lorenz and Gutschmidt (1968, 1969) showed that at −5°C, bacterial reproduction of the order of 3–5 log units per gram is possible. It can therefore be presumed that the lowest temperature at which bacteria can reproduce is around −8°C. Yeast can better tolerate cold than bacteria. The same authors (Schmidt-Lorenz, 1963; Schmidt-Lorenz and Gutschmidt, 1968, 1969) found that the reproduction of yeast was limited at approximately −10°C.

The storage temperature of −18°C that is specified for deep-frozen food in most countries implies that micro-organisms will not reproduce under these conditions. The allowance in many legal specifications for basic temperatures of up to −15°C during short periods when the frozen products are being moved shows that constant temperatures cannot be achieved in practice. Temperature fluctuations can occur during transport by the consumer from the store to the domestic freezer and also during freezer defrosting operations.

Experience of earlier phases of deep freezing technology has shown that it is not possible to store food for long periods at a temperature that is only slightly below the growth limit of micro-organisms. Food thus stored shows noticeable signs of deterioration that is independent of the growth of micro-organisms and can be traced to the activity of enzymes. Hall and Alcock (1987), who investigated the influence of microbial enzymes on the sensory quality of food after 6 months' storage at −15°C, noticed that deterioration occurred at a slightly reduced rate even during storage at −29°C.

Temperature abuse does have a negative influence and will contribute to quality deterioration. If temperature abuse is not severe or the duration is short, the quality degradation may not be significantly noticeable. To confirm this, a study (Sorensen, 2002) involving 14 products was carried out (orange and melon juices, prepared dish, tuna steaks, breaded whiting, cheesecake, peas, wrinkled, scampi, ice cream, cod, broccoli, gateaux, broad beans). These products were exposed to the same temperature abuse model, simulating a poor freezer chain, including simulation of home transport (2 h at room temperature). In most cases, the taste panel found no difference between samples subjected to temperature abuse and control samples (kept at a constant temperature around −20°C). However, differences in the appearances of the frozen samples were often recorded.

15.7 HOME FREEZING

The number of consumers using home freezers is increasing. Home freezing of food products is becoming widespread nowadays since it concerns 83% of families (Volatier, 2000). Many consumers purchase fresh retail cuts of meat, place them in a freezer and later thaw them before cooking. This survey also showed that home freezing is in relation to the number of people in a family (more people, more home freezing), age (young people, <25-yr old, less freezing), the professional status of the person (workers practice less freezing). 11% of surveyed people said that they do not use any particular precautions when freezing.

An example of temperature monitoring during the shelf-life of a package of chilled meat product from factory to consumer is shown in Fig. 15.10 (Derens *et al.*, 2003). In this study, a data logger was discreetly inserted inside the product at a factory, then, it followed the cold chain until the consumption point. In fact, the consumer discovered the presence of the data logger only when he or she opened the package. An explanation displayed inside the package allowed him or her to send the data logger back to the authors for analysis of the product temperature. The expertise of these authors enabled the following time–temperature history of the product to be developed:

- 0–10 hours, factory.
- 10–20 hours, transport. The product temperature was stable at 2°C.

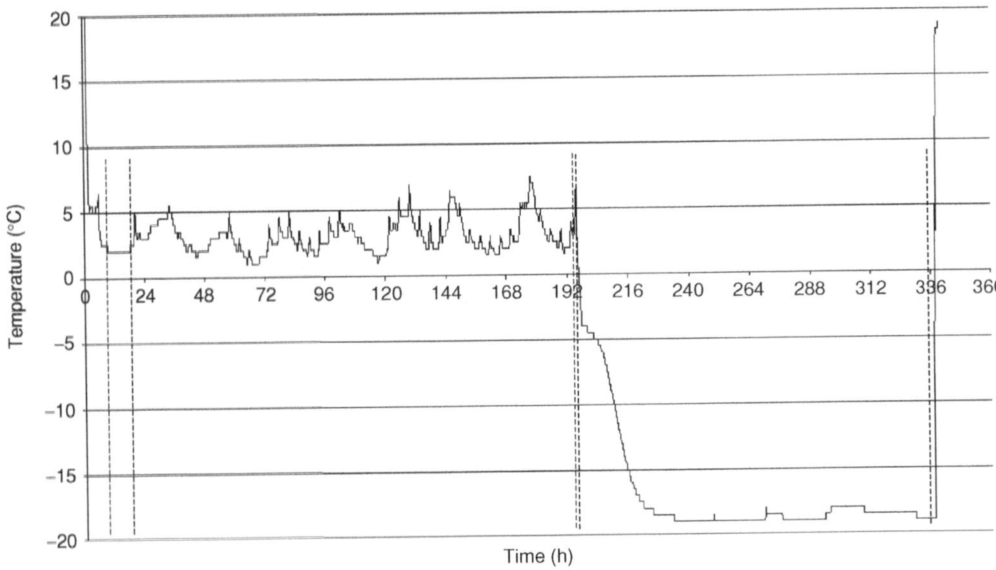

Fig. 15.10 Temperature monitoring of a meat product from factory until consumer (source: Derens *et al.*, 2003).

- 20–195 hours (7 days), display cabinet. The temperature cycles observed showed that there was one defrosting operation per day.
- 195–196 hours transport in the consumer's car.
- 196–333 hours (6 days), domestic freezer. It took 40 hours to reduce the product temperature from +7°C to −19°C.
- > 335 hours, thawing. The product temperature increased from −19°C to 20°C

Home freezing in the freezer compartment can also be undesirably slow. Placing warm food in a freezer can also warm the other products stored in the freezer until the freezing process has taken the heat from the additional load. Slow freezing results in a loss of nutrients, especially for fruits and vegetables, as well as structural damage due to the formation of large ice crystals, loss of juices (drip) and loss of sensory quality.

Recently, several manufacturers have introduced models with 'quick freeze' and 'quick thaw' capabilities. To verify the efficiency of this technology, Anderson *et al.* (2004) studied the time required for freezing and thawing of different meat products for five different models of household refrigerators (all are side-by-side refrigerator/freezer designs). Two refrigerators had 'quick thaw' and three had 'quick freeze' capabilities. It was found that some refrigerator models froze and thawed foods significantly faster than others. Heat transfer coefficients ranged from 8 to 15 W m^{-2} K^{-1} during freezing and from 5 to 8 W m^{-2} K^{-1} during thawing. Mathematical predictions for freezing and thawing time in the refrigerators gave results similar to those obtained in experiments. These authors also observed that the heat transfer coefficient had greater impact on freezing time than the environmental temperature, while the thawing time was influenced more by the environmental temperature. The highest drip loss from meat during thawing was observed when the rate of thawing was highest. Therefore, one must be careful when increasing air movement within a thawing chamber, because this may result in increased drip loss and evaporation, especially if the product is not fully sealed. However, the temperature should be sufficiently low to prevent microbial growth.

Power failures can occur, and in this case the temperature of products and potential for microbial growth in domestic freezers rises. In order to acquire information on whether or not the frozen products can still be consumed after a power failure over certain period, Bedinghaus and Ockerman (1991) carried out a study on frozen meat products stored in home freezers. They investigated this concern by using an upright home freezer. Fourteen trials were conducted with packages of either beef or pork with varying trial load weights (43–128 kg). Meat samples were frozen, the power was turned off and the door remained closed. The door was then opened (daily for 9 days) only when packages within each lot were removed for analysis of temperature, microbial count and pH. These authors concluded that, after about 36 hours without electricity, bacterial populations reached log 6 and the meat product's wholesomeness for consumption or possible refreeze became questionable.

15.8 HOME THAWING

During thawing, the exudation from all food is a result of physical phenomena which take place during freezing and thawing. The quantity of exudation is influenced by several factors (Durosset, 1997):

- freezing process (slow freezing rate, more exudation);
- temperature fluctuations during storage period (more fluctuations, more exudation);
- freezing storage period (longer storage period, more exudation);
- thawing process (quick thawing rate, more exudation);
- dimensions of the product (smaller size, more exudation);
- nature of the product (more water content, more exudation);
- storage period used for the thawed product (longer period, more exudation).

As mentioned previously, the microbial risk is negligible during frozen storage since microbial growth is inhibited when the temperature is lower than $-10°C$. However, this risk is considerable during thawing. Two factors are favourable for microbial growth in thawed products:

- cellular membrane damage, particularly where a slow freezing rate is applied. This damage allows the penetration of micro-organisms inside the tissue. Then, microbial growth accelerates with higher temperatures;
- exudates of some products are rich in nutritive substances promoting bacterial multiplication.

To limit microbial growth during thawing, particularly growth of psychrotrophic bacteria, thawing should be performed under the following conditions:

- short time period;
- product temperature (surface and centre) less than $2°C$.

Exudation causes not only product weight loss and microbial risk; it also leads to sensorial and nutritional deterioration. One implication of EU legislation is that the surface temperature of meat should not rise above 7°C and offal above 3°C during thawing. As some consumers may use inappropriate thawing methods, store the thawed foods too long at a temperature that is too high, and have no access to fast freezing, a message specifying 'do not refreeze foods after defrosting' on the package must be clear (EU Directive 89/108). The Canadian Food Inspection Agency advises consumers to discard any thawed food that has remained at room temperature for more than 2 hours. Refrigerators typically operate at 0–5°C; however, thawing in a refrigerator can be undesirably slow. In addition, these foods occupy refrigerator space and may cross-contaminate foods stored in the refrigerator.

When thawing is recommended, it is preferable to thaw products in the refrigerating compartment to avoid a product temperature that is too high, particularly at the surface. If thawing at ambient temperatures must be done, the duration should not exceed 3–4 hours (Sorensen, 2002). This may be good advice for consumers, as it ensures that the temperature of the food does not become so high as to result in excessive microbial growth, especially of pathogenic bacteria. After thawing, the product should be cooked or consumed directly.

15.9 CONSUMER RECOMMENDATIONS

Poor temperature control of foods in the home is a major cause of food poisoning. A consumer guide for food handling was proposed by Brady (1995). Some recommendations for shopping, storage temperature, preparation, thawing, cooking, serving, reheating were presented by the author. After purchase of a frozen product from a supermarket, the recommended product temperature must be maintained by the consumer until storage in the domestic freezer. It is necessary to protect the food during the carry-home period by keeping this period short (i.e. by buying frozen foods at the end of a shopping trip). It is also necessary to protect the product by using insulated cool bags or boxes during transport to the home. Frozen foods should be transferred into a suitable cold storage appliance immediately on arriving home. If no precautions are taken when the consumer transports the product to the home, product temperatures may rise rapidly. For example, the temperature of spinach (a 500-g package) can increase from −18°C to −15°C within 10 minutes (Gac, 1994).

There are a few surveys on home freezer management in relation to the quality of food coming from the freezer. An analysis of numerous surveys and comments from a considerable number of consumers indicate that there are consumer practices that lead to poorer quality food from the freezer. The causes are listed in Table 15.4.

Table 15.4 Causes of poor quality food from the home freezer (Source: Cook, 1978).

(1) Purchase of lower quality frozen foods:
- low-quality raw material;
- poor cold chain, e.g. retail display at temperatures that are too high.

(2) Home freezing of poor quality raw materials:
- over-mature garden produce;
- poor-quality meat.

(3) Freezing at temperatures that are too high.
(4) Keeping food too long in the freezer.
(5) Freezing too much food too often.

As for industrial products, food to be frozen should be in perfect hygienic condition and of prime quality. The temperature of many home freezers can be regulated. According to the weight of product to be frozen, the freezer should be set at its lowest temperature up to 1 day before adding a new batch of food for freezing and returned to the usual setting, about −18°C, the following day.

Joints of meat and thick fish fillets, or whole fish cook more satisfactorily in the thawed state. It is especially important to thaw all poultry completely; otherwise cooking is unlikely to be complete at the centre. This may lead to a health hazard. Small chops, steaks or fish may be cooked from frozen, or partially thawed, provided lower cooking temperatures and longer times are allowed.

The door or lid of the freezer must be kept closed. It should only be opened when required to load or remove something and closed immediately afterwards. The door should not be left open for longer than necessary. During door openings temperature fluctuations in chest freezers are less significant than those of upright models due to the cold air contained within the well of the chest freezer whereas the cold air is not so well contained in upright freezers. Also unfrozen food should not be placed adjacent to frozen food in the freezer as the frozen food will warm up and its quality will deteriorate. Food for freezing should be placed in the fast freeze compartment until frozen.

In the case of breakdown or power failure, the freezer should not be opened. A fully loaded unopened freezer will maintain an adequate temperature for many hours.

Consumers should defrost their freezer from time to time (unless it is a frost-free type). Defrosting should be done when more than 5 mm of ice has accumulated on the walls. The timing of this operation should be chosen to coincide with low stocks. Any remaining stocks should be placed in the refrigerator or in insulated 'cool bags' while defrosting is carried out. Ice should be scraped off the side walls with a wooden or plastic spatula before washing out with warm water or a solution of 30 g bicarbonate of soda for 2 litres of water and then dry with a soft cloth. A knife should not be used to scrape the frost as it may result in damage to the refrigeration system.

When buying a freezer, consumers should make sure it will work in the position desired, possibly in a hot kitchen, in a built-in unit, under a work surface or in an outbuilding that is cold in winter. There are a range of optimal ambient temperatures for refrigerator/freezer operation. This is indicated by the manufacturers using the following symbol: N (16–32°C), NS (10–32°C), ST (18–38°C) and R (18–43°C). The last one (R) is suitable for the hot and humid conditions (up to 80% relative humidity). The storage temperature of the frozen food is more important than the size, shape, convenience, 'add-on gimmick' or price of the freezer (Gigiel, 1998). The temperature in the freezer and the energy consumption of the appliance are greatly influenced by the environmental conditions. The consumer must choose a dry, well-ventilated and temperate site, allowing at least 5 cm space for free air circulation around the condenser. This then allows a high rate of heat extraction from the appliance.

Increased consumer education will result in demand for refrigerators that maintain food at correct low temperatures. It is technically possible for manufacturers to supply these types of domestic refrigerators, but they are more expensive. For the purchase decision-making by the consumer, the degree of food temperature control is less important than the price. It would be preferable to address this issue in the future. Finally, consumer good practice in the operation of freezers can contribute to reducing microbiological risk and quality loss from foods stored in the home.

15.10 ENERGY LABELLING

Energy consumption of domestic refrigerators and freezers has attracted considerable attention worldwide due to environmental awareness. Consumers are encouraged to use more energy-efficient appliances which also ensure overall product quality. The manufacturers perform energy-consumption tests on a random sample of their appliances, which they report on the energy-consumption labels attached to every unit on sale. This is a common practice in many countries nowadays. It is difficult to compare the energy efficiency of different models of refrigerators/freezers precisely because each of them varies in its features and utility. The 'star rating' is an indicator derived from an algorithm that relates the energy consumption with the internal volume. This indication is used to complement information on the energy consumption per year (kW h per year). Some studies (Meier and Heinemeier, 1988; Meier and Jansky, 1993) showed that the energy consumption in field use correspond to that measured in the laboratory (for some models, the energy consumption is lower in field use).

The test procedure to prepare the energy-consumption labels depends on the test standard relevant to the country where the refrigerators/freezers are being used. There are a number of energy-consumption test standards around the world. This leads to different energy consumption determined from one standard to the other when the same cabinet is used. This is due to different factors included in the test conditions such as compartment internal temperatures, number and locations of thermocouples and door openings. It is therefore interesting to establish correlation between the energy consumption obtained from various standards.

REFERENCES

AFF (Association Française du Froid). (2001). Conseil National du Froid. *Livre blanc sur les fluides frigorigènes*. Paris, p. 51.

Anderson, B.A. *et al.* (2004). Thawing and freezing of selected meat products in household refrigerators. *International Journal of Refrigeration* 27(1), 63–72.

Azevedo, I. Regalo, M., Mena, C., Almeida, G., Carneiro, L., Teixeira, P., Hogg, T. and Gibbs, P.A. (2005). Incidence of *Listeria* spp. in domestic refrigerators in Portugal. *Food Control* 16, 121–124.

Bak, L.S., Anderson, A.B., Anderson, E.M. and Bertelsen G. (1999). Effect of modified atmosphere packaging on oxidative changes in frozen stored cold water shrimp (*Pandalus borealis*). *Food Chemistry* 64, 169–175.

Bedinghaus, A.J. and Ockerman, H.W. (1991). Temperature, pH and bacterial populations of meat as influenced by home freezer failure. *Journal of Food Science* 56(6), 1508–1510.

Billiard, F. (2005). Refrigerating equipment, energy efficiency and refrigerants. *Bulletin International Institute of Refrigeration* 85(1), 12–25.

Brady, P.L. (1995). A quick consumer guide to safe food handing, *USDA Home and Garden Bulletin*, No. 248, October.

Bustabad, O.M. (1999). Weight loss during freezing and storage of frozen meat. *Journal of Food Engineering* 41, 1–11.

Brousseau, A.D. and Volatier, J.L. (1999). Le consommateur français en 1998- une typologie de préférence. *Cahier de recherche CREDOC*, No. 130, June.

Cook, D.J. (1978). The effect of domestic storage on the nutritional value, palatability and acceptability of food. *Journal of Consumer Studies and Home Economics* 2, 287–296.

Derens, E. Guilpart, J., Proson, E. and Palagos, B. (2003). La chaine du froid, du fabricant au consommateur: les résultats de l'audit ANIA-Cemagref. Final Report.

Durosset, P. (1997). La décongélation, Journée technique du 15 October – Nantes.

Flynn, O.M., Blair, I. and McDowell, D. (1992). The efficiency and consumer operation of domestic refrigerators. *International Journal of Refrigeration* 15(5), 307–312.

Gac, A. (1994). Emballage des produits congelés ou surgelés. *Revue Générale du Froid* Juin, 20–22.

Geiges, O. and Schuler, U. (1988). Behaviour of microorganisms during long-term storage of food products at subzero temperatures. *Microbiologie, Aliments, Nutrition* 6, 249–257.

Gigiel, A. (1998). A practical guide to the cold chain from factory to consumer, chapter Guidance for the home freezer. In: R.L. Fuller, ed. *Concerted Action CT96*. p. 1180.

Gormley, R., Walshe, T., Hussey, K. and Butler, F. (2002). The effect of fluctuating and constant frozen storage temperature regimes on some quality parameters of selected food products. *Lebensmittel Wissenschaft und Technologie* 35, 190–200.

Anonymous. (1989). The consumer and their appliances. *Grand Froid* 5(41), 57–59, in French.

Hall, L.P. and Alcock, S.J. (1987). The effect of microbial enzymes on the quality of frozen foods. *Food Microbiology* 4, 209.

IIR (International Institute of Refrigeration). (2002). *Report on Refrigeration Sector Achievements and Challenges*. p. 77.

IIR (International Institute of Refrigeration). (2006). Markets: frozen foods. *Newsletter of the IIR (International Institute of Refrigeration)* No. 25, January, 2006.

IIR, International Institute of Refrigeration. (1986). Recommendations for the Processing and Handling of Frozen Foods, 3rd ed.

JARN (Japan Air Conditioning, Heating & Refrigeration) (2005), no 157, p. 41.

Jackson, V., Blair, I.S., McDowell, D.A., Kennedy, J. and Bolton, D.J. (2007). The incidence of significant food borne pathogens in domestic refrigerators. *Food Control* 18(4), 346–351.

James, S.J. and Evans, J. (1992). Consumer handling of chilled foods: temperature performances. *International Journal of Refrigeration* 15, 299–306.

Kandeepan, G. and Biswas, S. (2005). Effect of low temperature preservation on microbial and sensory quality of buffalo meat, Livestock *Research for Rural Development* 17(11), 7.

Laguerre, O., Derens, E. and Palagos, B. (2002). Study of domestic refrigerator temperature and analysis of factors affecting temperature: a French survey. *International Journal of Refrigeration* 25, 653–659.

Laguerre, O. and Flick, D. (2006). Frost formation on frozen products preserved in domestic freezers. *J Food Engineering*, in press.

Lezenne Coulander de, P.A. (1994). *Koelkast Temperature Thuis, Report of the Regional Inspectorate for Health Protection*. Leeuwarden, The Netherlands.

Liu, D.Y., Chang, W.R. and Lin, J.Y. (2004). Performance comparison with the effect of door opening on variable and fixed frequency refrigerators/freezers. *Applied Thermal Engineering* 24, 2281–2292.

Lucas, J. (1996). Evolution des appareils de froid ménager. *Revue Générale du Froid*, May, 25–28.

Martins, R.C., Almeida, M.G. and Silva, C.L.M. (2005). The effect of home storage conditions and packaging materials on the quality of frozen green beans. *International Journal of Refrigeration* 27(8), 850–861.

Meier, A. and Heinemeier, K. (1988). Energy use of residential refrigerators: a comparison of laboratory and field use. *ASHRAE Trans.* 94(2). Also in LBL-25613 from the Lawrence Berkeley Laboratory.

Meier, A. and Jansky, R. (1993). Field performance of residential refrigerator: a comparison with the laboratory test. *ASHRAE Trans.* 99(1). Paper no. CH-93-20-1.

O'Brien, G.D. (1997). Domestic refrigerator air temperatures and the public's awareness of refrigerator use. *International Journal of Environmental Health Research* 7, 141–148.

Olsson, P. and Bengstsson, N. (1972). *Time Temperature Conditions in the Freezer Chain. Report No. 309*. SIK-The Swedish Food Institute, Gothenberg.

Poovarodom, N., Letang, G., Bouvier, J. and Billiard, F. (1990). Influence de l'emballage et des fluctuations de température de conservation sur la qualité des denrées alimentaires surgelées, IIF, Paris, Coll. Science et technique du froid, In: *Progrès dans la science et la technique du froid en industrie alimentaire*, 559–567.

Rahman, M.S. (1999). Food preservation by freezing. In: M. Shafiur Rahman, ed. *Handbook of Food Preservation*. New York: Marcel Dekker, pp. 259–284.

Redmond, E.C. and Griffith, C. (2003). Consumer food handling in the home: a review of food safety studies. *Journal of Food Protection*, 66(1), 130–161.

Saidur, R., Masjuki, H.H. and Choudhry, I.A. (2002). Role of ambient temperature, door opening, thermostat setting position and their combined effect on refrigerator-freezer energy consumption. *Energy Conversion and Management* 43, 845–854.

Schmidt-Lorenz, W. (1963). Microbieller Verderb gefrorener Libensmittel während der Gefrielagerung. *Kältetecnik* 15, 39.

Schmidt-Lorenz, W. and Gutschmidt, J. (1968). Microbieller und sensorische Veränderungen gefrorener Lebensmittel bei Lagerung im Temperaturbereich von $-2.5°C$ bis $-10°C$. *Lebensmittel Wissenschaft und Technologie* 1, 26.

Schmidt-Lorenz, W. and Gutschmidt, J. (1969). Microbieller und sensorische Veränderungen gefrorener Brathähnchen une Poularden bei Lagerung im Temperaturbereich von $-2.5°C$ bis $-10°C$. *Fleischwirtschaft* 49, 1033.

Sergelidis, D., Abrahim, A., Sarimvei, A., Panoulis, C., Karaioannoglou, P and Genigeorgis, C. (1997). Temperature distribution and prevalence of *Listeria spp*. in domestic, retail and industrial refrigerators in Greece. *International Journal of Food Microbiology* 34, 171–177.

Sorensen, L.B. (2002). Frozen food legislation. Review article, *Bulletin IIR* 4, 4–17.

Verbeke, R., Van Hoop, J. and Lauwers, H. (1984). Influence of different freezing and storage conditions on quality and characteristics of beef cuts. *Thermal Process Quality Foods*, Elsevier GB 522–527.

Volatier, J.L. (2000). Enquête Individulle et National sur les Consommations Alimentaires. INCA Survey, Jean-Luc Volatier, coordinateur. Lavoisier, Paris.

Index

Printed and bound in the UK by
CPI Antony Rowe, Eastbourne

Printed and bound by CPI Group (UK) Ltd, Croydon, CR0 4YY

15/09/2026

14962399-0001